U1827244

Post-translational Modifications of Proteins

METHODS IN MOLECULAR BIOLOGY™

John Walker, SERIES EDITOR

457. **Membrane Trafficking,** edited by Ales Vancura, 2008

456. **Adipose Tissue Protocols,** Second Edition, edited by Kaiping Yang, 2008

455. **Osteoporosis,** edited by Jennifer J. Westendorf, 2008

454. **SARS- and Other Coronaviruses:** *Laboratory Protocols,* edited by Dave Cavanagh, 2008

453. **Bioinformatics,** *Volume 2: Structure, Function, and Applications,* edited by Jonathan M. Keith, 2008

452. **Bioinformatics,** *Volume 1: Data, Sequence Analysis, and Evolution,* edited by Jonathan M. Keith, 2008

451. **Plant Virology Protocols:** *From Viral Sequence to Protein Function,* edited by Gary Foster, Elisabeth Johansen, Yiguo Hong, and Peter Nagy, 2008.

450. **Germline Stem Cells,** edited by Steven X. Hou and Shree Ram Singh, 2008.

449. **Mesenchymal Stem Cells:** *Methods and Protocols,* edited by Darwin J. Prockop, Douglas G. Phinney, and Bruce A. Brunnell, 2008.

448. **Pharmacogenomics in Drug Discovery and Development,** edited by Qing Yan, 2008.

447. **Alcohol: Methods and Protocols,** edited by Laura E. Nagy, 2008.

446. **Post-translational Modifications of Proteins:** *Tools for Functional Proteomics,* Second Edition, edited by Christoph Kannicht, 2008.

445. **Autophagosome and Phagosome,** edited by Vojo Deretic, 2008.

444. **Prenatal Diagnosis,** edited by Sinhue Hahn and Laird G. Jackson, 2008.

443. **Molecular Modeling of Proteins,** edited by Andreas Kukol, 2008.

442. **RNAi: Design and Application,** edited by Sailen Barik, 2008.

441. **Tissue Proteomics: Pathways, Biomarkers, and Drug Discovery,** edited by Brian Liu, 2008.

440. **Exocytosis and Endocytosis,** edited by Andrei I. Ivanov, 2008.

439. **Genomics Protocols,** Second Edition, edited by Mike Starkey and Ramnath Elaswarapu, 2008

438. **Neural Stem Cells:** *Methods and Protocols*, Second Edition, edited by Leslie P. Weiner, 2008

437. **Drug Delivery Systems,** edited by Kewal K. Jain, 2008

436. **Avian Influenza Virus,** edited by Erica Spackman, 2008

435. **Chromosomal Mutagenesis,** edited by Greg Davis and Kevin J. Kayser, 2008

434. **Gene Therapy Protocols:** *Volume 2: Design and Characterization of Gene Transfer Vectors*, edited by Joseph M. LeDoux, 2008

433. **Gene Therapy Protocols:** *Volume 1: Production and In Vivo Applications of Gene Transfer Vectors*, edited by Joseph M. LeDoux, 2008

432. **Organelle Proteomics,** edited by Delphine Pflieger and Jean Rossier, 2008

431. **Bacterial Pathogenesis:** *Methods and Protocols*, edited by Frank DeLeo and Michael Otto, 2008

430. **Hematopoietic Stem Cell Protocols,** edited by Kevin D. Bunting, 2008

429. **Molecular Beacons:** *Signalling Nucleic Acid Probes, Methods and Protocols,* edited by Andreas Marx and Oliver Seitz, 2008

428. **Clinical Proteomics:** *Methods and Protocols*, edited by Antonio Vlahou, 2008

427. **Plant Embryogenesis,** edited by Maria Fernanda Suarez and Peter Bozhkov, 2008

Post-translational Modifications of Proteins

Tools for Functional Proteomics

Second Edition

Christoph Kannicht
Molecular Biochemistry Department
Octapharma Research and Development
Berlin, Germany

Editor

 Humana Press

Christoph Kannicht
Molecular Biochemistry Department
Octapharma Research and Development
Berlin, Germany
christoph-kannicht@octapharma.com

Series Editor
John Walker
28 Selwyn Avenue
Hatfield, Hertfordshire AL10 9NP, UK

ISBN: 978-1-58829-719-8   ISBN: 978-1-60327-084-7 (e-book)
ISSN 1064–3745

Library of Congress Control Number: 2007938940

Cover illustration: The influence of protein glycosylation on the spot pattern of glycoproteins in 2D gel
electrophoresis (above). Different colored spots represent the spot pattern of a glycoprotein mixture without
(black) and following sialidase treatment (red) or deglycosylation (purple). The oligosaccharide structure
depicted below represents a typical N-linked glycan and explains the action of the sialidase treatment
(1), i.e. cleavage between galactose and sialic acid, and the complete cleavage of the glycan from the protein
by deglycosylation (2).

Printed on acid-free paper.

9 8 7 6 5 4 3 2 1

springer.com

Preface

*Post-translational Modifications of Proteins: Tools for Functional Proteomics,
Second Edition,* is a compilation of detailed protocols needed to detect and analyze the most important co- and post-translational modifications of proteins. For
reasons of simplicity, although not explicitly mentioned in the title, both kinds of
modifications are covered, regardless of whether they occur during or after biosynthesis of the protein. My intention was to cover the most significant protein
modifications, focusing on the fields of protein function, proteome research, and
characterization of pharmaceutical proteins.

The majority of all proteins undergo co- and/or post-translational modifications.
The protein's polypeptide chain may be altered by proteolytic cleavage, formation
of disulfide bonds, or covalent attachment of phosphate, sulfate, alkyl groups, lipids, carbohydrates, polypeptides, and others. Knowledge of these modifications is
extremely important, because they may alter physical and chemical properties,
folding, conformation distribution, stability, activity, and, consequently, the function of the proteins. Moreover, the modification itself can act as an added functional
group. Examples of the biological effects of protein modifications include: phosphorylation for signal transduction, ubiquitination for proteolysis, attachment of
fatty acids for membrane anchoring or association, glycosylation for protein half-
life, targeting, cell–cell and cell–matrix interaction, and carboxylation in protein–
ligand binding, to name just a few. Full understanding of a specific protein
structure–function relationship requires detailed information not only about its
amino acid sequence, which is determined by the corresponding DNA sequence,
but also on the presence and structure of protein modifications. Consequently,
analysis of post-translational modifications of proteins is essential for proteomic
research, the development of new drugs, and for the production, registration, and
monitoring of therapeutic pharmaceutical proteins.

In general, post-translational modifications of proteins can be classified according to their chemistry or the targeted amino acid. They can be subdivided into
reversible or irreversible reactions, enzymatic or nonenzymatic reactions, according to their subcellular location or functional aspects of the modification. Though
the organization of the chapters considers both the frequency and the chemical
nature of the particular post-translational modification, it still remains arbitrary.
The individual chapters of this book provide detailed step-by-step instructions for

the analysis of the most important protein modifications, e.g., the assignment of disulfide bonds in proteins (Chapter 1). The detection and analysis of protein phosphorylation by selective fluorescent staining in 2D-gels and by advanced mass spectrometry, respectively, is covered by Chapters 2 and 3. Chapters 4 to 7 describe analysis of protein sulfation, α-amidation, γ-glutamate, ß-hydroxyaspartate, and lysine hydroxylation. Protein ubiquitination, sumoylation, and ISGylation are covered by Chapters 8 to 10, analysis of protein methylation and acetylation by Chapter 11. Methods for analysis of lipid modifications to both the carbohydrate and lipid portion as well are given in Chapters 12 and 13 on S-acylation and glycosylphosphatidylinositols respectively. Chapters 14 to 21 describe analysis of protein glycosylation in great detail. Starting with the detection of protein glycosylation (Chapter 14), analysis of carbohydrate composition (Chapter 15) cleavage, labeling, separation, and sequence analysis of *N*-linked glycans are described (Chapters 16 to 18). Analysis of protein *O*-glycosylation in general and specific detection of *O*-linked N-acetyglucosamine residues follow (Chapters 20 and 19, respectively). Analysis of *O*-glycosidically linked *N*-acetylglucosamine (O-GlcNAc) deserves special mention. *O*-GlcNAc is a transient modification, which is involved in several cellular functions as transcription, translation, nuclear transport, and cell signalling. Because of its exceptional position within the glycosylation of proteins it is treated in a separate chapter. Chapter 21 provides a method to analyze the oligosaccharides that are present at specific single glycosylation sites in a protein. Chapters 22 to 24 give practical approaches, i.e., how to analyze and monitor glycosylation of recombinant proteins from different cell lines. Finally, a topic of general interest is treated in the last chapter. Chapter 25 describes the use web-based protein databases for analysis of post-translational modifications of proteins. Web-based databases give information on protein modifications and allow the prediction of post-translational modifications on yet uncharacterized proteins, based on the fact that post-translational modifications occur at specific amino acids, amino acid sequences, or specific 3D-structures of the protein, respectively.

Let me give special mention to two areas of research of high current interest: the fields of (1) proteomics and (2) the characterization of biological pharmaceuticals. (1) With respect to proteomics, research in the field of genomics has lead to knowledge of the complete human DNA sequence. Measurement of the mRNA pool at a specific status of the cell, the "transcriptome," was found to not necessarily reflect the cells' actual protein expression pattern. In proteomic research, the description of expression levels of proteins related to a defined cell or tissue status will be incomplete without knowledge of post-translational modifications of those proteins. The increasing interest in post-translational modifications of proteins in this field is reflected by use of the term "phosphoproteomics." Phosphoproteomics describes the analysis of the sites and amount of protein phosphorylation under different biological conditions. (2) An additional important practical application of post-translational modification analysis is to ensure product quality of therapeutic pharmaceutical proteins. The exact structure of a protein pharmaceutical cannot be defined without knowledge of all post-translational modifications. Recombinant proteins intended for therapeutic use in humans must be accorded particularly thorough investigation. Product quality depends on accurate post-translational modification in the respective expression system during production, e.g., insect, several mammal, or human cell lines. Note that different expression systems may vary in their ability to carry out post-translational modifications and that the applied cell-culture conditions also influence these modifications. Thus, post-translational modifications of recombinant proteins have to be monitored during production and documented for registration. In their guidance Q6A for the pharmaceutical industries, the international conference on harmonization of technical requirements for registration of pharmaceuticals for human use (ICH) states, that "For desired product and product-related substances, details should be provided on primary, secondary and higher-order structure; post-translational forms (e.g., glycoforms); biological activity, purity, and immunochemical properties, when relevant." Consequently, almost each and every post-translational modification of a protein is of concern for the regulatory agencies. Moreover, glycoengineering, the directed modification of protein glycosylation, or the artificial attachment of polymers to therapeutic proteins demand analytical tools for their characterization as well.

Growing knowledge of the biological roles of protein modifications, on the one hand, and the development and availability of sophisticated, sensitive analytical methods on the other hand, are already leading to increased interest in co- and post-translational modifications of proteins. *Post-translational Modifications of Proteins: Tools for Functional Proteomics* intends to serve as practical guide for researchers working in the field of protein structure–function relationships in general, in the rapidly growing field of proteomics, as well as scientists in the pharmaceutical industries.

Christoph Kannicht

Contents

Preface.   v

Contributors.   xiii

1   Disulfide Bond Mapping by Cyanylation-induced Cleavage
    and Mass Spectrometry.   1
    Jiang Wu

2   Detection of Post-translational Modifications by Fluorescent
    Staining of Two-Dimensional Gels.   21
    Archana M. Jacob and Chris W. Turck

3   Identification of Protein Phosphorylation Sites by Advanced
    LC ESI MS/MS Methods.   33
    Christoph Weise and Christof Lenz

4   Analysis of Tyrosine-*O*-Sulfation.   47
    Jens R. Bundgaard, Jette W. Sen, Anders H. Johnsen,
    and Jens F. Rehfeld

5   α-Amidated Peptides: *Approaches for Analysis*.   67
    Gregory P. Mueller and William J. Driscoll

6   γ-Glutamate and β-Hydroxyaspartate in Proteins.   85
    Francis J. Castellino, Victoria A. Ploplis, and Li Zhang

7   Lysine Hydroxylation and Cross-linking of Collagen   95
    Mitsuo Yamauchi and Masashi Shiiba

8   Mass Spectrometric Determination of Protein Ubiquitination   109
    Carol E. Parker, Maria R. E. Warren, Viorel Mocanu,
    Susanna F. Greer, and Christoph H. Borchers

9   Analysis of Sumoylation   131
    Andrea Pichler

10 Detection and Analysis of Protein ISGylation.   139
    Tomoharu Takeuchi and Hideyoshi Yokosawa

11 Analysis of Methylation, Acetylation, and Other Modifications
    in Bacterial Ribosomal Proteins.   151
    Randy J. Arnold, William Running, and James P. Reilly

12 Analysis of S-Acylation of Proteins   163
    Michael Veit, Evgeni Ponimaskin, and Michael F. G. Schmidt

13 Metabolic Labeling and Structural Analysis
    of Glycosylphosphatidylinositols from Parasitic Protozoa   183
    Nahid Azzouz, Peter Gerold, and Ralph T. Schwarz

14 2-Dimensional Electrophoresis: *Detection of Glycosylation
    and Influence on Spot Pattern*   199
    Klemens Löster and Christoph Kannicht

15 Carbohydrate Composition Analysis of Glycoproteins
    by HPLC Using Highly Fluorescent Anthranilic Acid (AA) Tag . . .   215
    George N. Saddic, Shirish T. Dhume, and Kalyan R. Anumula

16 Enzymatical Hydrolysis of *N*-Glycans from Glycoproteins
    and Fluorescent Labeling by 2-Aminobenzamide (2-AB).   231
    Rolf Nuck

17 Separation of N-Glycans by HPLC   239
    Martin Gohlke and Véronique Blanchard

18 Enzymatic Sequence Analysis of *N*-Glycans by Exoglycosidase
    Cleavage and Mass Spectrometry—detection
    of Lewis X Structures   255
    Christoph Kannicht, Detlef Grunow, and Lothar Lucka

19 Immunochemical Methods for the Rapid Screening
    of the *O*-Glycosidically Linked *N*-Acetylglucosamine
    Modification of Proteins   267
    Michael Ahrend, Angela Käberich, Marie-Therese Fergen,
    and Brigitte Schmitz

20 Analysis of *O*-Glycosylation   281
    Juan J. Calvete and Libia Sanz

21 Characterization of Site-specific *N*-Glycosylation 293
    Katalin F. Medzihradszky

22 **Monitoring Glycosylation of Therapeutic Glycoproteins
    for Consistency by HPLC Using Highly Fluorescent
    Anthranilic Acid (AA) Tag** 317
    Shirish T. Dhume, George N. Saddic, and Kalyan R. Anumula

23 **Comparability and Monitoring Immunogenic *N*-linked
    Oligosaccharides from Recombinant Monoclonal Antibodies
    from Two Different Cell Lines using HPLC with Fluorescence
    Detection and Mass Spectrometry** 333
    Bruce R. Kilgore, Adam W. Lucka, Rekha Patel,
    Bruce A. Andrien, Jr., and Shirish T. Dhume

24 **Mass Spectrometry and HPLC with Fluorescent Detection-Based
    Orthogonal Approaches to Characterize *N*-linked Oligosaccharides
    of Recombinant Monoclonal Antibodies** 347
    Adam W. Lucka, Bruce R. Kilgore, Rekha Patel, Bruce A. Andrien, Jr.,
    and Shirish T. Dhume

25 **Web-based Computational Tools for the Prediction and Analysis
    of Post-translational Modifications of Proteins** 363
    Vladimir A. Ivanisenko, Dmitry A. Afonnikov,
    and Nikolay A. Kolchanov

Index 385

Contributors

Dmitry A. Afonnikov
Institute of Cytology and Genetics SB RAS, Novosibirsk State University, Novosibirsk, Russia

Michael Ahrend
Department of Biochemistry, Institute of Animal Sciences, University of Bonn, Bonn, Germany

Bruce A. Andrien, Jr.
Alexion Pharmaceuticals, Cheshire, CT

Kalyan R. Anumula
Inhibitex R&D, Alpharetta, GA

Randy J. Arnold
Department of Chemistry, Indiana University, Bloomington, IN

Nahid Azzouz
Laboratory for Organic Chemistry, Swiss Federal Institute of Technology, Zürich, Switzerland

Véronique Blanchard
Charité-Universitatsmedizin Berlin, Zentralinstitut fur Laboratoriumsmedizin und Pathobiochemie, Berlin, Germany

Christoph H. Borchers
Department of Biochemistry and Biophysics, University of North Carolina at Chapel Hill, Chapel Hill, NC

Jens R. Bundgaard
Department of Clinical Biochemistry, Copenhagen University Hospital, Copenhagen, Denmark

Juan J. Calvete
Instituto de Biomedicina de Valencia, C.S.I.C., Valencia, Spain

Francis J. Castellino
Department of Chemistry and Biochemistry, University of Notre Dame, Notre Dame, IN

Shirish T. Dhume
Alexion Pharmaceuticals, Cheshire, CT

William J. Driscoll
Department of Anatomy, Physiology and Genetics, Uniformed Services University of the Health Sciences, Bethesda, MD

Marie-Therese Fergen
Department of Biochemistry, Institute of Animal Sciences, University of Bonn, Bonn, Germany

Peter Gerold
Chiron Behring GmbH & Co., Marburg, Germany

Martin Gohlke
Dynavax Technologies, Berkeley, CA

Susanna F. Greer
Lineberger Comprehensive Cancer Center, University of North Carolina at Chapel Hill, Chapel Hill, NC

Detlef Grunow
Charite Universitätsmedizin Berlin, Institute for Biochemistry and Molecular Biology, Berlin, Germany

Vladimir A. Ivanisenko
Institute of Cytology and Genetics SB RAS, Novosibirsk State University, Novosibirsk, Russia

Archana M. Jacob
Max Planck Institute of Psychiatry, Munich, Germany

Anders H. Johnsen
Department of Clinical Biochemistry, Copenhagen University Hospital, Copenhagen, Denmark

Angela Käberich
Department of Biochemistry, Institute of Animal Sciences, University of Bonn, Bonn, Germany

Christoph Kannicht
Molecular Biochemistry Department, Octapharma Research and Development, Berlin, Germany

Bruce R. Kilgore
Alexion Pharmaceuticals, Cheshire, CT

Nikolay A. Kolchanov
Institute of Cytology and Genetics SB RAS, Novosibirsk State University, Novosibirsk, Russia

Christof Lenz
Applied Biosystems, Darmstadt, Germany

Klemens Löster
Human GmbH, Magdeburg, Germany

Adam W. Lucka
Alexion Pharmaceuticals, Cheshire, CT

Lothar Lucka
Charite Universitätsmedizin Berlin, Institute for Biochemistry and Molecular Biology, Berlin, Germany

Katalin F. Medzihradszky
Department of Pharmaceutical Chemistry, School of Pharmacy, University of California San Francisco, San Francisco, CA & Proteomics Research Group, Biological Research Center, Szeged, Hungary

Viorel Mocanu
Department of Biochemistry and Biophysics, University of North Carolina at Chapel Hill, Chapel Hill, NC

Gregory P. Mueller
Department of Anatomy, Physiology and Genetics, Uniformed Services University of the Health Sciences, Bethesda, MD

Rolf Nuck
Charité – Universitaetsmedizin Berlin, Institut für Molekularbiologie und Biochemie, Berlin-Dahlem, Germany

Carol E. Parker
Department of Biochemistry and Biophysics, University of North Carolina at Chapel Hill, Chapel Hill, NC

Rekha Patel
Alexion Pharmaceuticals, Cheshire, CT

Andrea Pichler
Max F. Perutz Laboratories, Medical University of Vienna, Department of Medical Biochemistry, Vienna, Austria

Victoria A. Ploplis
Department of Chemistry and Biochemistry, University of Notre Dame, Notre Dame, IN

Evgeni Ponimaskin
Universität Göttingen, Zentrum Physiologie und Pathophysiologie Göttingen, Germany

Jens F. Rehfeld
Department of Clinical Biochemistry, Copenhagen University Hospital,
Copenhagen, Denmark

James P. Reilly
Department of Chemistry, Indiana University, Bloomington, IN

William Running
Department of Chemistry, Indiana University, Bloomington, IN

George N. Saddic
Charles River Laboratories, Malvern, PA

Libia Sanz
Instituto de Biomedicina de Valencia, C.S.I.C., Valencia, Spain

Michael F. G. Schmidt
Freie Universität Berlin, Institut für Immunologie und Molekularbiologie, Berlin,
Germany

Brigitte Schmitz
Department of Biochemistry, Institute of Animal Sciences, University of Bonn,
Bonn, Germany

Ralph T. Schwarz
Med. Zentrum für Hygiene und Med. Mikrobiologie, Philipps-Universität
Marburg, Marburg, Germany

Jette W. Sen
Department of Clinical Biochemistry, Copenhagen University Hospital,
Copenhagen, Denmark

Masashi Shiiba
Dental Research Center, University of North Carolina, Chapel Hill, NC

Tomoharu Takeuchi
Department of Biochemistry, Graduate School of Pharmaceutical Sciences,
Hokkaido University, Sapporo, Japan

Chris W. Turck
Max Planck Institute of Psychiatry, Munich, Germany

Michael Veit
Freie Universität Berlin, Institut für Immunologie und Molekularbiologie, Berlin,
Germany

Maria R. E. Warren
Department of Biochemistry and Biophysics, University of North Carolina at
Chapel Hill, Chapel Hill, NC

Christoph Weise
Institut für Chemie und Biochemie, Freie Universität Berlin, Berlin, Germany

Jiang Wu
Wyeth Research, Cambridge, MA

Mitsuo Yamauchi
Dental Research Center, University of North Carolina, Chapel Hill, NC

Hideyoshi Yokosawa
Department of Biochemistry, Graduate School of Pharmaceutical Sciences,
Hokkaido University, Sapporo, Japan

Li Zhang
Department of Chemistry and Biochemistry, University of Notre Dame, Notre
Dame, IN

1

Disulfide Bond Mapping by Cyanylation-induced Cleavage and Mass Spectrometry

Jiang Wu

Summary Oxidation of sulfhydryl groups to form a disulfide bond is one of the most common post-translational modifications in proteins. Disulfide bonds play important roles in stabilizing three-dimensional structure and modulating bioactivity of the cystinyl proteins. The determination of disulfide bond linkage is therefore an integral part of structural characterization of proteins. A mass spectrometry-based strategy utilizing chemical cleavage at cysteine residues following cyanylation reaction is described for the identification of both sulfhydryl and disulfide bond linkage in proteins. The method has been particularly powerful for the assignment of disulfide bonds in proteins containing adjacent or closely spaced cysteines.

Key Words Disulfide; cyanylation; chemical cleavage; mass mapping.

1 Introduction

Determination of the three-dimensional structure of proteins is essential for elucidating their biological activity and function. Oxidation of sulfhydryl groups to form a disulfide bond is one of the most common post-translational modifications in proteins. Disulfide bonds play important roles in stabilizing three-dimensional structure and modulating bioactivity of the cystinyl proteins. The determination of disulfide bond linkage is therefore an integral part of structural characterization of proteins *(1,2)*.

Traditionally, methodology for disulfide mapping involves either enzymatic or chemical cleavage of the protein backbone between half-cystinyl residues to obtain peptides that contain only one disulfide bond. Labor-intensive efforts are then pursued to isolate and sequence the various cystinyl peptides by Edman degradation or mass spectrometry, or both. The identified peptides are then assembled to specific segments of the protein *(3,4)*. Although this approach has been used with great success, it suffers from serious drawbacks and is often inadequate for many challenging problems. A major constraint is the requirement for proteolytic

From: *Post-translational Modifications of Proteins.*
Methods in Molecular Biology, Vol. 446.
Edited by: C. Kannicht © Humana Press, Totowa, NJ

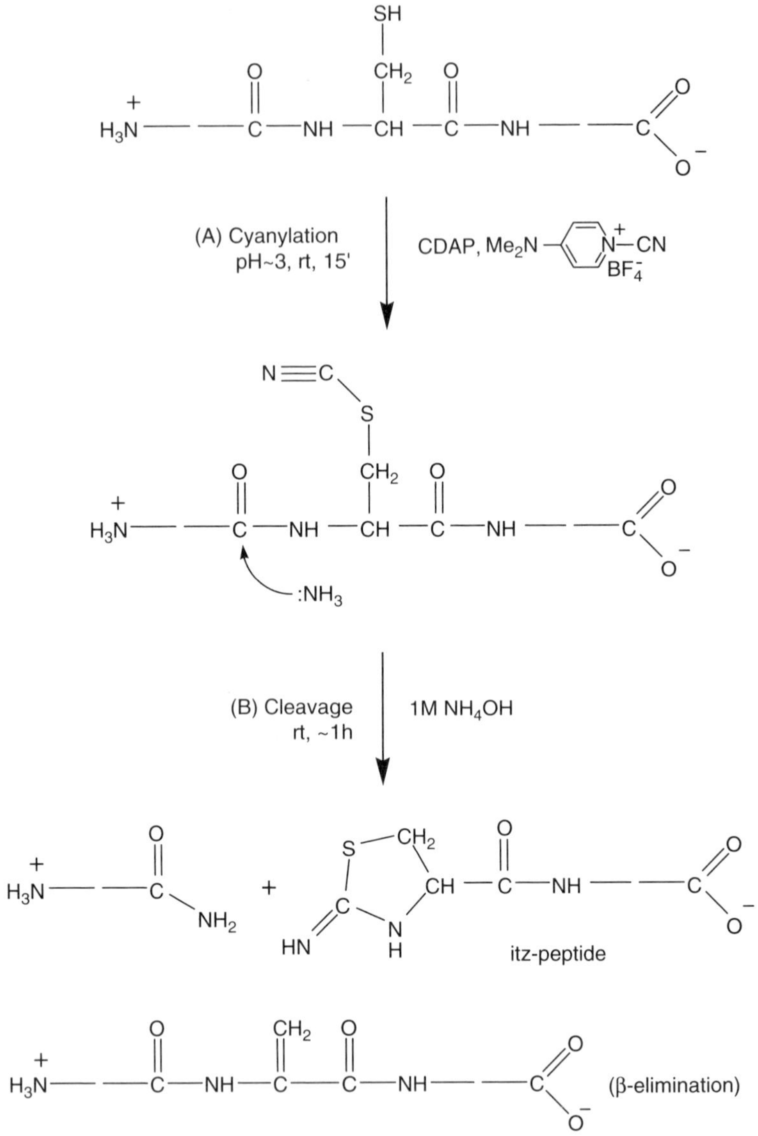

Fig. 1.1 Schematic illustration of (**A**) cyanylation and (**B**) cleavage reaction

cleavage between all cysteine residues, especially in cases where the cysteines reside close to one another in the sequence or disulfide bonds are highly knotted because the possibility of an intervening proteolytic cleavage site becomes less likely *(4)*. Proteins containing adjacent cysteines are usually refractory to this methodology. Secondly, the conditions most frequently used for specific proteolytic digestion (i.e., pH ~8) promote thiol-disulfide exchange and disulfide scrambling *(3,4)*. As a result, the disulfide linkage thus determined may be prone to artifacts.

In spite of the tremendous progress in the field of protein characterization and global protein identification over the past decade, disulfide bond assignment still represents a challenging task in functional proteomics.

Chemical cleavage at cysteine residues has been used to great advantage in numerous studies exploring the structure and function of the proteins. Jacobson *(5)* first showed that 2-nitro-5-thiocyanobenzoic acid (NTCB) specifically cyanylates protein sulfhydryl groups under mildly alkaline conditions, leading to a base-catalyzed cleavage of the protein backbone at the N-terminal side of the S-cyano-cysteine residues to form an amino-terminal peptide and a series of 2-iminothiazolidine-4-carboxyl peptides (itz-peptides). A competitive reaction, β-elimination of HSCN from the S-cyanocysteine group, may also occur under the same conditions, depending primarily upon the structure of amino acids on the N-terminal side of cyanylated cysteine (*see* **Note 1**). As an alternative, 1-cyano-4-dimethylamino-pyridinium tetrafluoroborate (CDAP) was later proposed to be advantageous over the NTCB reagent for the specific cyanylation of sulfhydryl groups under acidic conditions (pH 3–5) to minimize sulfhydryl/disulfide exchange *(6–8)*. Recently, extensive studies were carried out to optimize the cleavage of cyanylated cysteinyl proteins *(9–11)*. The cleavage reaction, which involves nucleophilic attack to the carbonyl carbon of the amide followed by a concerted cyclization, can be greatly accelerated in methylamine, giving rise to an α-methylamidated N-terminal peptide *(11)*. Figure 1.1 shows the general scheme of cyanylation and cleavage reactions that have been used in our laboratory.

The cyanylation and subsequent cleavage are specific to free cysteine residues, that is, the selective chemical cleavage at modified cysteinyl residues can be achieved in the presence of cystinyl residues *(5,12)*, a feature that is particularly useful for identifying free cysteines and those involved in disulfide bonds. Based on the cyanylation-induced cleavage, a mass spectrometry-based strategy was described for the identification of both sulfhydryl and disulfide bond linkage in proteins *(12,13)*. The method has been particularly powerful for the assignment of disulfide bonds in proteins containing adjacent or closely spaced cysteines *(14–16)*.

2 Materials

Sillucin was isolated and purified according to the described methods *(17)*. Other chemicals and proteins were obtained from Sigma and used without further purification.

2.1 Peptides and Proteins

1. Rabbit muscle creatine phosphokinase.
2. Bovine pancreatic ribonuclease A (RNase A), type III-A.
3. Bradykinin.
4. Bovine pancreatic insulin.
5. Horse skeletal myoglobin.

2.2   Solvents and Buffers

1. HPLC solvent A: 0.1% TFA in water.
2. HPLC solvent B: 0.1%TFA in acetonitrile.
3. Aqueous acetonitrile (1/1, v/v) containing 0.1% TFA.
4. 0.1M Citrate buffer (pH 3.0) containing 6M guanidine-HCl.
5. 1M NH$_4$OH aqueous solution.
6. 1M NH$_4$OH solution containing 6M guanidine-HCl.
7. 0.1M 1-Cyano-4-dimethylamino-pyridinium tetrafluoroborate (CDAP) solution in 0.1M citrate buffer (pH 3.0). Prepare fresh before use.
8. Saturated α-Cyano-4-hydroxycinnamic acid (α-CHCA) solution prepared fresh in aqueous CH$_3$CN (1/1, v/v) containing 0.1% TFA prior to use.
9. 0.1M *Tris*(2-carboxyethyl)phosphine hydrochloride (TCEP) aqueous solution.
10. 0.1M TCEP solution in 0.1M citrate buffer (pH 3.0).

2.3   Instrumentation

1. MALDI-TOF MS: MALDI experiments were performed on a Voyager Elite time-of-flight (TOF) mass spectrometer (PerSeptive Biosystems Inc., Framingham, MA). The accelerating voltage in the ion source was set to 25 kV. Data were acquired in the positive linear mode of operation. The spectra were collected just slightly above the laser threshold necessary for analyte ion production. Mass calibration was achieved by using external standards of bradykinin (*m/z* 1,061.2), bovine pancreatic insulin (*m/z* 5,734.5), and horse skeletal myoglobin (*m/z* 16,952).
2. High-Pressure Liquid Chromatography (HPLC): The separation of modified proteins was carried out using a reversed-phase Vydac C18 column (10-μm particle size, 300-Å pore, 4.6 × 250 mm) under a linear gradient elution controlled by a Waters 6000 system. The UV detection was at 215 nm. The mobile phases A and B were 0.1% TFA aqueous solution and CH$_3$CN containing 0.1% TFA, respectively. The HPLC conditions were optimized for individual proteins.

3   Methods

3.1   Identification of Free Cysteine Residues by Mass Mapping of Cyanylation-induced Cleavage Products

For proteins containing both sulfhydryls and disulfide bonds, the first step is to determine the number and location of sulfhydryl groups. The cyanylation and chemical cleavage of the denatured original protein under non-reducing conditions, and subsequent mass mapping of the cleavage products allow the number and locations of sulfhydryl groups to be deduced (12).

3.1.1   Experimental Procedure

1. Dissolve 5 nmol of protein in 10 μL of 0.1M citrate buffer (pH 3.0) containing 6M guanidine-HCl.
2. Add a 10-fold molar excess (over equivalent cysteine content) of 0.1M CDAP in 0.1M citrate buffer (pH 3.0) to the protein solution. Incubate the mixture for 10–15 min at ambient temperature.
3. Remove immediately the excess reagent from the modified protein by HPLC (see Note 2). Collect HPLC fractions corresponding to the modified protein and remove solvent under reduced pressure (speed vac concentrator).
4. Reconstitute the protein in 10 μL of 1M NH$_4$OH containing 6M guanidine-HCl, incubate the mixture for 1 h at ambient temperature, and remove the excess ammonia in the speed vac (see Note 3).
5. Add 10 μL of 0.1M TCEP aqueous solution. Incubate at 37°C for 30 min to promote the reduction of the disulfide-linked peptide fragments.
6. Dilute a 1-μL aliquot of the above solution with 100 μL of CH$_3$CN:H$_2$O (1/1, v/v) containing 0.1% TFA (see Note 4). Mix equal volumes of the diluted protein sample and α-CHCA matrix and allow to air-dry.
7. Acquire MALDI spectra. Calculate the theoretical masses by adding to the M.W. of the parent peptide 25 Da for each site of cyanylation, adding 25 Da for peptide fragments containing the iminothiazolidine moiety, subtracting 34 Da for β-elimination.

3.1.2   Analysis of Creatine Phosphokinase

Figure 1.2 shows the MALDI spectrum of the cleavage products of rabbit muscle creatine phosphokinase, a protein containing 380 amino acids. The cysteines reside at positions 73, 145, 253, and 282, respectively, among which Cys282 is in the active site. The primary structure of this protein was established from cDNA clones, which do not provide structural information regarding the cysteine status (17). The MALDI spectrum of the cleavage products showed peaks at m/z 3412.3, 8159.2, 8231.9, 10721, and 12620, corresponding to the fragments consisting of residues itz-253-281, itz-73-144, 1–72, itz-282-380, and itz-145-252, respectively. In addition, doubly charged fragments are also readily identified in the spectrum. It is clear that the cyanylation and cleavage must have taken place at each of the cysteine residues. Therefore, all 4 cysteine residues are in the reduced form and there is no disulfide-bonded cystine in rabbit muscle creatine phosphokinase.

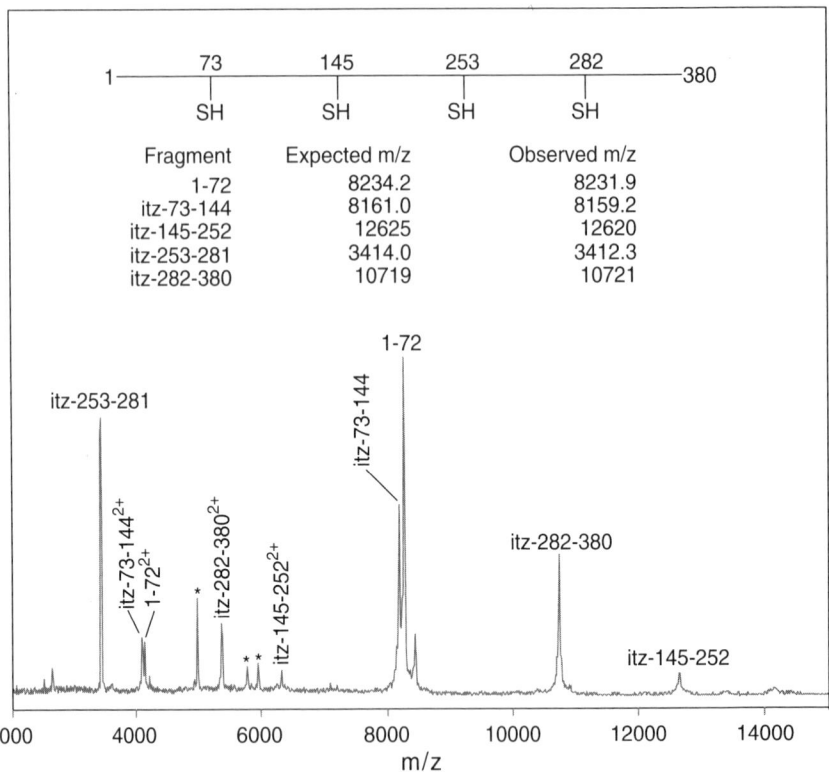

Fig. 1.2 MALDI spectrum of rabbit muscle creatine phosphokinase (MW. 42,977), after cyanylation by CDAP and subsequent cleavage in $1M$ NH$_4$OH solution. The peaks marked with asterisk are due to "carry-over" from an impurity in the sample. itz, iminothiazolidine derivative. (Reproduced with permission from (9))

3.2 Disulfide Bond Mapping in Proteins Containing Multiple Cystines by Partial Reduction and Cyanylation-induced Chemical Cleavage

The analysis of a protein containing multiple disulfide bonds requires some means of reducing a particular disulfide bond followed by applying the cyanylation/cleavage methodology to corresponding pair of nascent cysteines (13). Partial reduction by the water soluble TCEP under acidic conditions provides a convenient means of generating an array of partially reduced protein isoforms (13). The general procedure is shown in Figure 1.3 for the simplest hypothetical peptide containing two cystines. The strategy has been applied to the characterization of disulfide structure in a number of proteins and is particularly useful in proteins with a cysteine-knotted core or adjacent cysteinyl residues and in determining disulfide linkage in protein folding intermediates. (14–16, 18, 19).

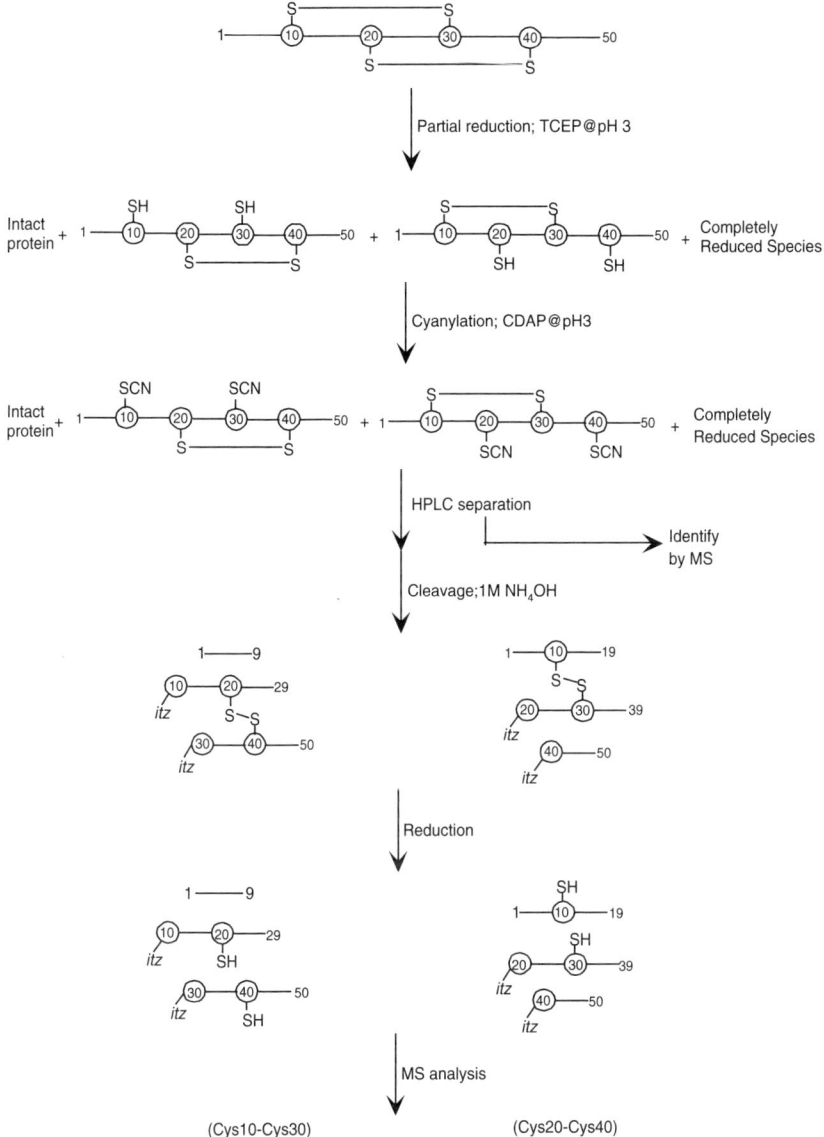

Fig. 1.3 Overview of the methodology for mapping disulfide bonds in proteins

In our experiments, the partial reduction procedure involves establishing a kinetically limited reduction reaction from which only a small portion of each disulfide bond (e.g., 10% each) is reduced whereas the majority of the protein remains intact *(13,21,22)*. Experimental conditions, such as incubation time, temperature, and stoichiometry of reducing agents, can be controlled so that singly reduced isoforms of the protein are predominant products. The singly reduced isoforms of the original protein are most desirable as the cyanylation and subsequent cleavage can be applied in a straightforward manner as described above (*see* **Note 5**).

3.2.1 Experimental Procedure

The experimental conditions are summarized below:

1. Dissolve 10 nmol of protein sample in 10 µL of 0.1*M* citrate buffer (pH 3.0) containing 6*M* guanidine-HCl.
2. Add an equivalent of 0.1*M* TCEP (in 0.1*M* citrate buffer, pH 3.0) for the cystine content in the protein. Incubate at room temperature for 10–15 min (*see* **Note 6**). The TCEP solution can be stored under N_2 at −20°C for weeks with little deterioration.
3. Add a 20-fold molar excess (over total cysteine content) of freshly prepared 0.1*M* CDAP solution in 0.1*M* citrate buffer (pH 3.0) to the protein solution (*see* **Note 7**). Cyanylation of the nascent sulfhydryl groups is achieved by incubation at ambient temperature for 10–15 min.
4. Separate the partially reduced and cyanylated species by reversed-phase HPLC with linear gradient elution (*see* **Section 2.3** and **Note 8**). Collect the HPLC fractions manually.
5. Identify the HPLC fractions by MALDI-MS (+52 Da for a singly reduced/cyanylated species; +104 Da for a doubly reduced/cyanylated species, etc).
6. Reconstitute a solution of the singly reduced/cyanylated protein residues in 2 µL of 1*M* NH_4OH containing 6*M* guanidine-HCl, then add another 5 µL of 1*M* NH4OH solution. Cleavage of peptide chains is carried out by reaction for one hour at ambient temperature. The excess ammonia is removed in a speed vac (*see* **Note 3**).
7. Add 5 µL of 0.1*M* TCEP aqueous solution and incubate at 37°C for 30 min to release the truncated peptide chains still linked by the remaining disulfide bonds (*see* **Note 9**).
8. Dilute a 1-µL aliquot of the above solution to 100 µL with 1:1 (v/v) $CH_3CN:H_2O$ containing 0.1% TFA for analysis by MALDI.

3.2.2 Advantages of the Methodology

This methodology offers several unique advantages compared with other approaches:

1. The chemistry and data interpretation, in majority of the cases, are simple, fast, and straightforward. Cleavage of the peptide chain takes place only at reduced and cyanylated cysteinyl sites, which, in principle, yields 3 fragments (and 2 β-elimination products if partial cleavage occurs) for each singly reduced/cyanylated protein isomer. The mass of each fragment is related to the position of the 2 cyanylated cysteinyl residues, which in turn can be used to deduce a particular

disulfide bond linkage. β-Elimination, an alternative to peptide chain cleavage, provides mass spectral data corresponding to overlapped peptides and serves as a confirmation for the disulfide bond assignment.

2. The experimental conditions minimize problems related to disulfide bond scrambling because both partial reduction of disulfide bonds and cyanylation of nascent sulfhydryls are performed in an acidic medium.

3. The procedure can be applied to proteins containing adjacent or closely spaced cysteines, for which conventional methodology fails (*see* **Section 3.3.**).

3.2.3   Synopsis of the Procedure Applied to RNase A

RNase A (Mr 13,683) contains 124 amino acids, among which eight cysteines are linked by 4 disulfide bonds (Scheme 1). Figure 1.4 shows the separation of residual intact RNase A and its partially reduced/cyanylated isoforms. The peaks marked 1–4, each representing ~10% of the intact protein, show +52 Da shift from the original protein, and corresponding to the reduction of a different disulfide bond and cyanylation of the corresponding two cysteines. Mass mapping of the truncated peptides allows one to deduce the disulfide pair that had undergone reduction, cyanylation, and cleavage. Figure 1.5A–D are MALDI spectra of peptide mixtures resulting from cleavage of each one of the 4 singly reduced/cyanylated isoforms of RNase A corresponding to HPLC peaks 1–4 in Figure 1.4, respectively. Table 1.1 lists the calculated and observed m/z values for fragments resulting from the cleavage of the peptide chains at cysteine sites corresponding to reduction of different disulfide bonds (*see* **Note 10**).

In the Figure 1.5A, 3 peaks at m/z 2,705.3, 6,548.5, and 4,527.4 are assigned to fragments 1–25, itz-26–83, and itz-84–124, respectively (Table 1.1). From these data, one can deduce that reduction, cyanylation, and peptide chain cleavage occur at Cys26 and Cys84. The overlapped peptide fragments resulting from β-elimination support the assignment. For instance, the MALDI peak at m/z 9176.7 corresponds to a partially cleaved peptide, β(1–83), resulting from peptide backbone cleavage at Cys84, but β-elimination at Cys26; thus, this peak corroborates information provided by the peaks representing residues 1–25 and itz-26–83. Likewise, the MALDI peak at m/z 10,998.6 is another partially cleaved peptide, β(itz-26–124), with cleavage at Cys26 and β-elimination at Cys84; similarly, this peak corroborates information provided by the peak representing itz-26–83 and

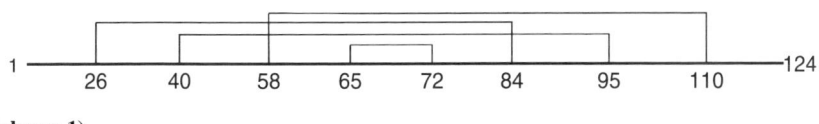

1 ———————————————————————————124
    26   40   58   65   72   84   95   110

(Scheme 1)

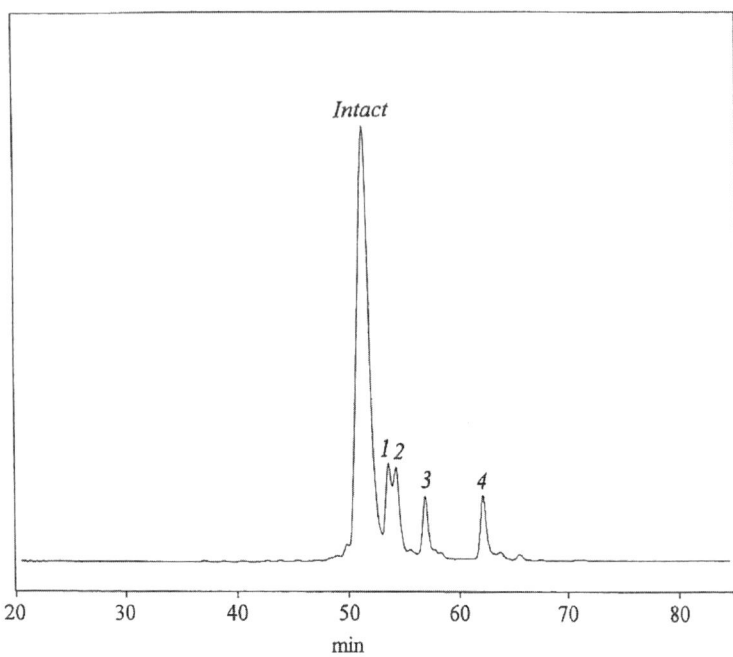

Fig. 1.4 HPLC chromatogram of denatured RNase A and its partially reduced/cyanylated isoforms. The HPLC separation was carried out on a reversed-phase Vydac C18 column at a flow
rate of 1.5 mL/min with a linear gradient 20–40% B in 90 min, where A = 0.1% TFA aqueous
solution and B = 0.1% TFA in CH$_3$CN. The peak marked *Intact* is intact RNase A. Peaks 1–4
represent singly reduced/cyanylated RNase A isomers, as identified by MALDI-TOF analysis.
(Reproduced with permission from *(13)*)

itz-84–124. Overall, a disulfide bond linkage between Cys26-Cys84 can be
unambiguously deduced.

The MALDI spectrum in Figure 1.5B, corresponding to the cleavage products
represented by HPLC peak 2, shows two main peaks at *m/z* 5,907.7 and 7,083.8,
assigned to fragments itz-72–124 and 1–64, respectively. Another expected fragment itz-65–71 is missing. However, it is still possible to deduce the Cys65-Cys72
linkage from these 2 fragments because no other combination affords such masses.
Two partially cleaved products at *m/z* 6,617.9 and 7,790.0, attributable to residues
β(itz-65-124) and β(1–71), respectively, are particularly informative for the confirmation of the assignment in this case.

The HPLC peaks 1 and 2 in Figure 1.4 were not resolved completely. This is
reflected in the MALDI spectra (Figs. 1.5A and 1.5B) of the cleavage products of
these two fractions, each of which carries small fragments corresponding to the
cleavage products of the other fraction.

With similar strategy, two other disulfide bond linkages, Cys40-Cys95 and Cys58-
Cys110, can readily be recognized from Figures 1.5C and 1.5D, respectively.

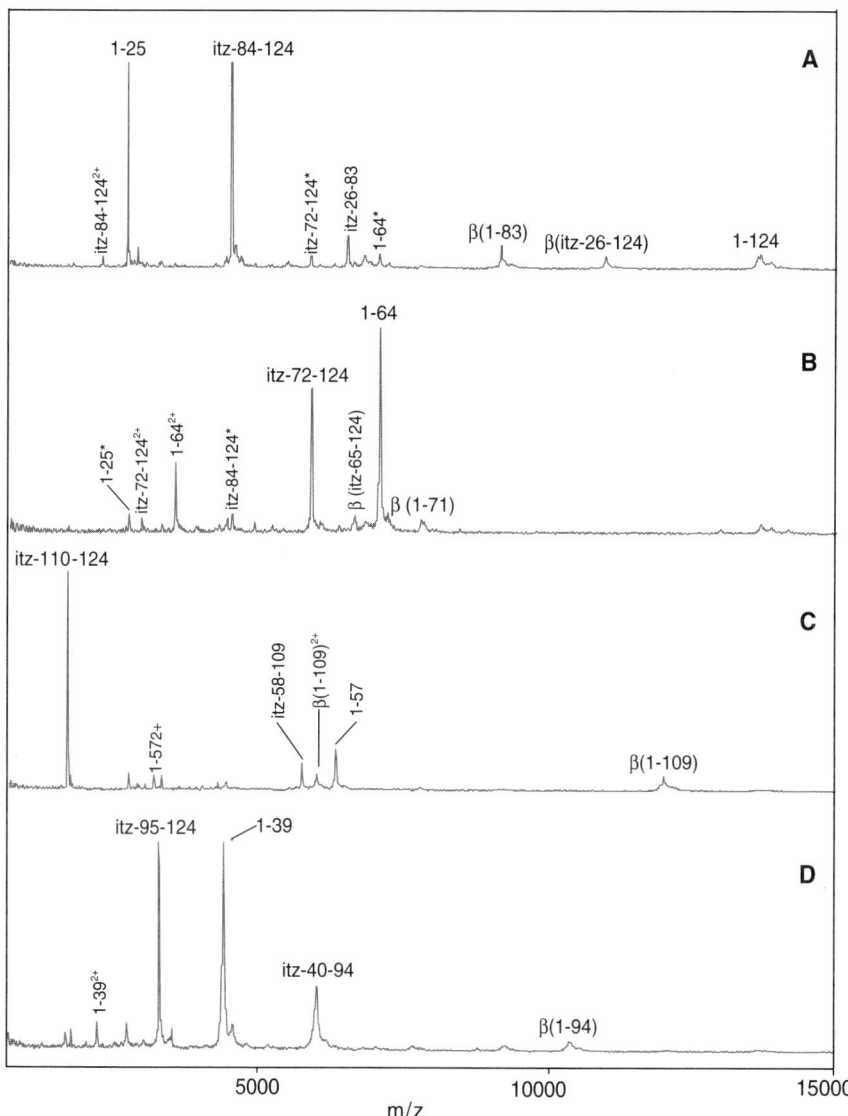

Fig. 1.5 The MALDI mass spectra of peptide mixtures resulting from the cleavage of the 4 singly reduced/cyanylated RNase A isoforms, corresponding to the HPLC peaks 1–4 in Figure 1.4, respectively. The peaks with asterisk represent the "carry-over" from other HPLC fractions. See also Table 1.1 for calculated and observed *m/z* values. (Reproduced with permission from *(13)*)

Table 1.1 Calculated and Observed *m/z* Values for Possible Fragments Resulting from the Cleavage Reaction of RNase A Chains at Sites of Designated Cysteine Pairs

Disulfide pair	Fragment	Calculated m/z	Observed m/z
	1–25	2706.8	2705.3
	itz-26-83	6547.3	6548.5
Cys26–Cys84	itz-84-124	4526.0	4527.4
	β(1–83)	9176.2	9176.7
	β(itz-26-124)	10995.3	10998.6
	1–64	7083.9	7083.8
	itz-65-71	789.8	nd
Cys65–Cys72	itz-72-124	5906.5	5907.7
	β(1–71)	7795.7	7790.0
	β(itz-65-124)	6618.3	6617.9
	1–57	6353.1	6351.1
	itz-58-109	5767.4	5766.8
Cys58–Cys110	itz-110-124	1659.8	1659.8
	β(1–109)	12042.4	12036.7
	β(itz-58-124)	7349.1	nd
	1–39	4413.9	4414.4
	itz-40-94	6063.6	6061.2
Cys40–Cys95	itz-95-124	3302.7	3303.7
	β(1–94)	10399.5	10430.5
	β(itz-40-124)	9288.3	9293.4

3.3 Assignment of Disulfide Bonds in Highly Knotted Proteins Containing Adjacent Cysteines

Proteins containing adjacent cysteine residues are amenable to the partial reduction/cyanylation/ cleavage/mass mapping approach *(14–16)*. The procedures used are the same as those described in **Section 3.2**. However, the stoichiometry of TCEP and CDAP must be optimized based on the properties of the individual proteins.

An example illustrated here is the analysis of the disulfide structure of sillucin, a highly knotted 30-residual protein containing 3 adjacent cysteines and 4 disulfide bonds *(15,20)*. Four major components were observed by HPLC separation after partial reduction and cyanylation of sillucin (Fig. 1.6). Analysis by MALDI-MS indicated that the HPLC fraction IP represents the intact peptide, whereas the peaks 1–3 represent species that are 52, 104, and 208 Da heavier that the intact peptide, corresponding respectively to singly, doubly, and completely reduced and cyanylated sillucin.

(Scheme 2)

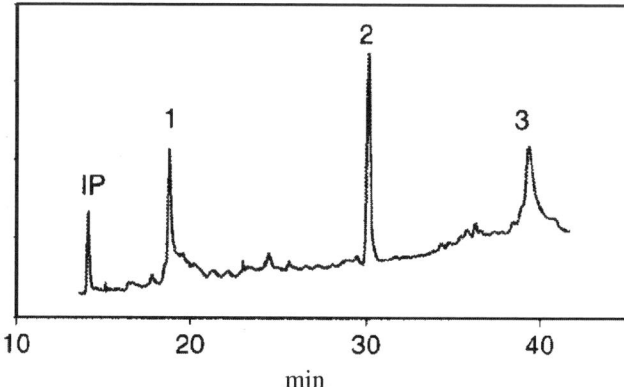

Fig. 1.6 HPLC chromatogram of denatured sillucin (IP) and its partially reduced and cyanylated isoforms. The separation was carried out on a Vydac C18 column at a flow rate of 1.0 mL/min with a linear gradient of 15–40% B in 50 min, where A is 0.1% TFA in water and B is a 90% (v/v) acetonitrile/0.1% TFA mixture. Peaks IP, 1, 2, and 3 represent the intact peptide, a singly reduced/cyanylated species, a doubly reduced/cyanylated species, and the completely reduced/cyanylated species, respectively, as determined from analysis by MALDI-MS. (Reproduced with permission from *(15)*)

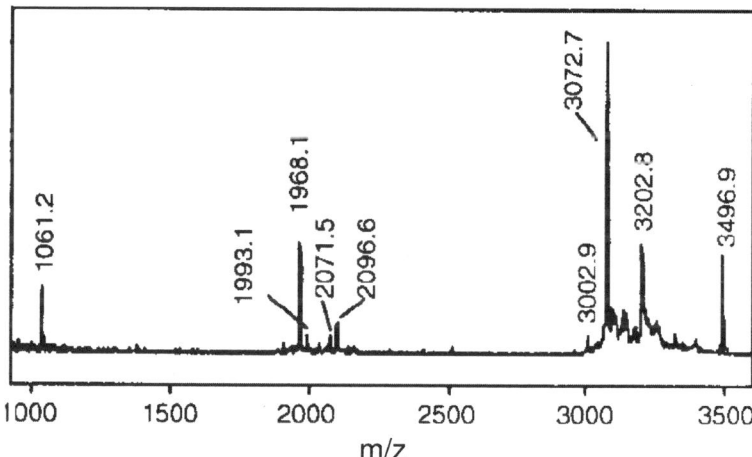

Fig. 1.7 MALDI-MS spectrum of the peptide mixture resulting from the cleavage and complete reduction of the singly reduced and cyanylated sillucin species represented by HPLC peak 1 in Figure 1.6. (Reproduced with permission from *(15)*)

The MALDI-MS spectrum resulting from the cleavage followed by complete reduction of the HPLC fraction 1 showed 2 major peaks at m/z 1,968.1 and 3,072.7 (Fig. 1.7). The peak at m/z 1,968.1 could be assigned to itz-13–29 or itz-14–30, while the peak at m/z 3,072.7 can be attributed to the β-elimination product of 1–29, inferring the cleavage at Cys30. Thus the disulfide bond between Cys13 and Cys30 must have been reduced during partial reduction to form the singly reduced and cyanylated species (HPLC fraction 1). As a matter of fact, another expected fragment of residues 1–12 was detected after the mixture of cleavage products was separated by HPLC (data not shown).

The MALDI-MS spectrum in Figure 1.8A corresponds to the analysis of the cleavage products (before complete reduction) of the doubly reduced and cyanylated isoform (HPLC fraction 2 in Fig. 1.6). The peak at m/z 1,300.5 can be assigned to itz-13–23-ox (calculated MH^+ =1,300.4 Da), where "ox" indicates the fragment contains residual disulfide bonds. As the analysis was carried out without the complete reduction of the sample, the presence of itz-13–23-ox implied the existence of a disulfide bond between Cys14 and Cys 21. The assignment is confirmed by MALDI-MS analysis of the completely reduced cleavage products (Fig. 1.8B). The peak at m/z 1,302.5 (2 mass units higher than the corresponding peak at m/z 1,300.5 in Figure 1.8A) attributes to the reduction of intrachain Cys14-Cys21 disulfide bond. Given the knowledge of Cys14–Cys21, the peak at m/z 1,369.7 can be interpreted as itz-12-(β@13)-23-ox (calculated MH^+ =1,369.5), where β@13 is

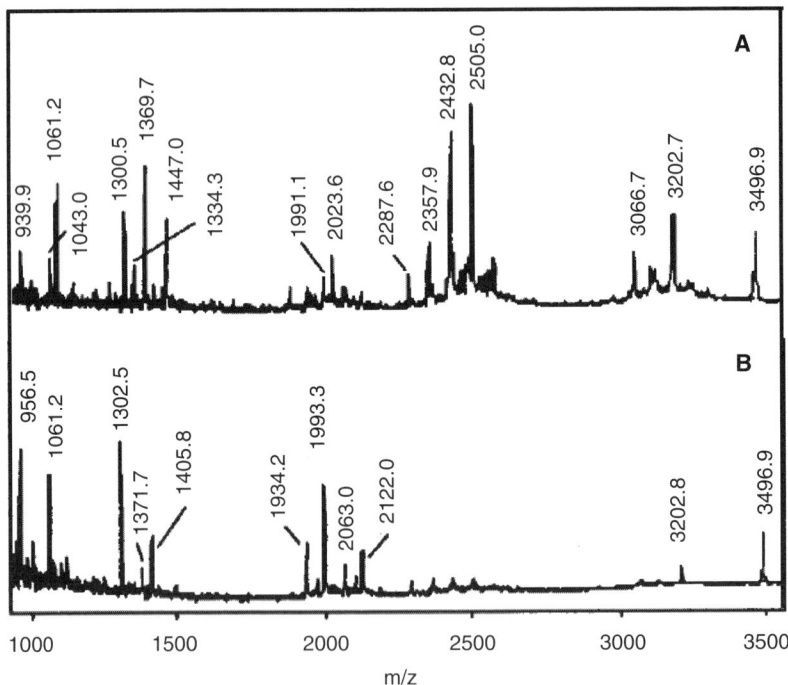

Fig. 1.8 (**A**) MALDI-MS spectrum of the peptide mixture resulting from the cleavage of the doubly reduced and cyanylated sillucin species represented by HPLC peak 2 in Figure 1.6. (**B**) MALDI-MS spectrum of the peptide mixture following complete reduction. (Reproduced with permission from (*15*))

designated to the β-elimination at Cys13. Similarly, the MALDI peak at m/z 1,371.7 in Figure 1.8B results obviously from reduction of the Cys14-Cys21 disulfide bond in itz-12-(β@13)-23-ox. Thus, Cys12, Cys13, Cys24, and CysZ must have formed 2 disulfide bonds. Given the knowledge of Cys13–Cys30 described above, CysZ must be Cys30. Thus the third disulfide bond, Cys12-Cys24, can be deduced. The fourth disulfide bond, Cys2-Cys7, can be deduced by default. Overall, the linkage of 4 disulfide bonds in sillucin is identified as Cys2–Cys7, Cys12–Cys24, Cys13–Cys30, and Cys14–Cys21.

3.4 Algorithm-Assisted Elucidation of Disulfide Structure

The concept of a "negative signature mass algorithm" (NSMA) was introduced to elucidate disulfide structure of cystinyl proteins by processing mass spectral data from partial reduction and cyanylation-induced cleavage products (23). In addition to automated data interpretation, a key advantage of the NSMA is its capacity to interpret mass spectra from mixtures of the cleavage fragments without separating the partially reduced/cyanylated isoforms and without knowledge of the extent of partial reduction. The capability of the NSMA approach to data interpretation is demonstrated here with the data from the analysis of RNase A described in **Section 3.2.**

The NSMA operates by ruling out certain possible linkages based on the recognition of mass spectral peaks at mass values corresponding to cyanylation-induced cleavage fragments containing free cysteine residues. For example, in RNase A, there are 4 and 2 cysteine residues respectively in itz-26–83 and itz-84–124 (Scheme 1). Masses of such fragments are defined as "negative signature masses." In contrast to conventional approach, the NSMA does not directly rule-in linkages, but rather eliminates linkages from a list of all possible theoretical linkages, with the goal of ruling out enough linkages so that only one disulfide structure can be constructed from the remaining list of potential disulfide linkages. If cysteine residues containing a free sulfhydryl group are part of a recognized fragment, the algorithm rules out linkage between the terminal cysteines (those at which cyanylation-induced cleavage occurred) and the internal cysteine residue(s). For example, the fragment itz-26–83 of the RNase A contains internal sulfhydryls at residues 40, 58, 65, and 72 (*see* **Note 11**). Thus detection of fragment itz-26–83 implies that no disulfide bonds in the original protein could have existed between Cys26 and Cys40 or Cys 58 or Cys65 or Cys72, nor between Cys84 and Cys40 or Cys58 or Cys65 or Cys72. Otherwise, quite different cleavage products would have resulted. A program using the algorithm allows calculation of all possible cleavage fragments based on the protein sequence and then compares this theoretical list to experimental mass spectral data. Then the program internally performs mass mapping using the NSMA logic and returns a list of possible disulfide linkages that fit the input data.

To illustrate the process of eliminating invalid combinations of disulfide linkages, 28 possible linkages in RNase A are listed in Table 1.2. After applying the NSMA,

Table 1.2 List of Possible Disulfide Linkages for RNase A and the Associated m/z Value for the Corresponding Negative Signature Mass[a]

S-S No.	Potential disulfide linkages	Identified negative signature masses (Da)									
		4526.0	5906.5	7083.9	6547.3	3302.7	4413.9	6063.6	5767.4	6353.1
		Itz-84-124	Itz-72-124	I-64	Itz-26-83	Itz-95-124	I-39	Itz-40-94	Itz-58-109	1-57
1	Cys26–Cys40				X					
2	Cys26–Cys58				X		X			
3	Cys26–Cys65			X	X					X
4	Cys26–Cys72				X					
5	Cys26–Cys84									
6	Cys26–Cys95									
7	Cys26–Cys110									
8	Cys40–Cys58							X		X
9	Cys40–Cys65			X				X		
10	Cys40–Cys72							X		
11	Cys40–Cys84				X			X		
12	Cys40–Cys95									
13	Cys40–Cys110									
14	Cys58–Cys65			X					X	
15	Cys58–Cys72								X	
16	Cys58–Cys84				X				X	
17	Cys58–Cys95							X	X	
18	Cys58–Cys110									
19	Cys65–Cys72									
20	Cys65–Cys84				X					
21	Cys65–Cys95							X		
22	Cys65–Cys110								X	
23	Cys72–Cys84		X		X					
24	Cys72–Cys95		X					X		
25	Cys72–Cys110		X						X	
26	Cys84–Cys95	X						X		
27	Cys84–Cys110	X							X	
28	Cys95–Cys110					X			X	

seven disulfide linkages remain (highlighted in Table 1.2), leaving 35 possible combinations of four linkages, most of which are invalid (*see* **Note 12**). For instance, disulfide linkages 5, 6, 7, and 12 cannot form a valid disulfide structure because Cys26 would be linked to more than 1 cysteine. Examination of all 35 possible combinations by the exhaustive matching algorithm reveals that only 1 combination (5, 12, 18, 19 in Table 1.2) is valid, which is the correct disulfide structure of RNase A. Thus, these results show that from batch processing the mass spectral data from all reduced and cyanylated isoforms, the disulfide structure of RNase A can be distinguished from the other 104 isomeric forms by detecting mass spectral peaks for only 3 negative signature masses: 6,547.3 Da calculated for (itz-26–83), 5,767.4 Da for (itz-58–109), and 6,063.6 Da for (itz-40–94).

4   Notes

1. For most of the denatured peptides examined, the expected cleavage products are predominant, however, amino acids with rigid or bulky side chains, such as Pro and Tyr on the N-terminal side of cysteine, are more resistant to amide cleavage, giving β-elimination as a main product *(9)*.
2. Although the cyanylation is generally specific to sulfhydryl groups, a large excess of CDAP (>50-fold over protein sulfhydryl groups) and excessive incubation time (>2h) may result in minor modification of other amino acid side chains. It is recommended that the excess reagent be removed immediately after cyanylation to minimize side reactions. Ultrafiltration or use of a desalting cartridge may be superior to a C18 column for reagent removal and purification of hydrophobic proteins.
3. Methyamine was recently described to be superior to ammonia for base-catalyzed cleavage reaction of the cyanylated proteins *(11)*. The cleavage in 1*M* methylamine solution has much higher reaction rate, minimized side reactions, and significantly improved yield.
4. Upon dilution of sample, good quality MALDI signals can be observed, presumably because of the reduced salt concentration. Alternatively, desalting with a ziptip can be performed to enhance MALDI signals.
5. Singly-reduced protein isomers are advantageous over other reduced protein isomers, because only a few fragments are obtained after cyanylation/cleavage, leading to the most simple mass spectrum. In principle, a singly reduced isomer contains two nascent cysteine residues, which, after cyanylation and cleavage, only generate three cleavage products plus two possible β-elimination products. To minimize structural diversity of disulfide bonds, proteins under study must be denatured by dissolution in 6*M* guanidine so that difference in the accessibility of TCEP to the disulfide bonds is minimized.
6. Reduction by water-soluble TCEP can be carried out at pH 3.0 to minimize disulfide bond scrambling *(13)*. Furthermore, at pH 3.0, the reduction of disulfide bonds is kinetically controlled which makes partial reduction possible.

As the cystine content may not be known *a priori*, and the rate of disulfide reduction may vary, the extent of reduction should be monitored and optimized for individual proteins. As a rule of thumb, an equivalent of TCEP for the cystine content is a good initial stoichiometry. The extent of reduction may be readily adjusted by controlling the reaction time and temperature.

7. A 5-fold molar excess of CDAP would drive the cyanylation reaction to completion. However, it should be noted that CDAP reacts instantly with residual TCEP. Therefore, a larger amount of CDAP is applied to ensure complete cyanylation.

8. Opening a disulfide bond disrupts protein structure, exposes the protein's interior hydrophobic amino acids, and changes the protein's hydrophobicity to different extents. Therefore, different isoforms may exhibit different retention times on reversed-phase HPLC columns. Some of the more hydrophobic proteins, especially after reduction and cyanylation, exhibited a long retention time or even irreversible retention on a C18 column in the water/acetonitrile mobile phase. In such a case, a C4 column with water/1-propanol mobile phase may be utilized for faster elution *(24)*. Most of the proteins can be eluted by 40% aqueous 1-propanol within a reasonable time.

9. Although the TCEP is recommended for the partial reduction step, other reducing agents such as DTT may be used at this stage to promote the complete reduction of remaining disulfide bonds.

10. As the number of cysteine residues increases, the potential combinations of disulfide bond linkages increase exponentially. However, a cursory analysis of the MALDI spectrum of the cleavage products can exclude many of the isomeric possibilities. Alternatively, an algorithm-assisted strategy described in **Section 3.4** can be used for the data interpretation.

11. In general, the NSMA is most effective when dealing with singly reduced species because cleavage products are likely to have the greatest number of internal free cysteine residues, allowing the NSMA to eliminate the highest number of possible disulfide linkages.

12. The NSMA doesn't require the detection of all possible negative signature masses to eliminate enough disulfide linkages to deduce the correct disulfide structure. As shown in Table 1.2, most rows have more than two Xs, indicating that the data set is highly redundant. If only 3 (instead of 9) negative signature masses (namely itz-26–83, itz-58–109, and itz-40–94) had been detected due to less successful chemistry or mass spectral analysis, the same number of disulfide linkages would still have been eliminated.

References

1. Creighton, T. E. (1984) Disulfide bond formation in proteins, in *Methods in Enzymol, 107,* (Wold, F. and Moldave, K. eds.), Academic, San Diego, CA, pp.305–329.
2. Gorman, J. J., Wallis, T. P., and Pitt, J. J. (2002) Protein disulfide bond determination by mass spectrometry. *Mass Spectrom. Rev.* 21, 183–216.

3. Smith, D. L. and Zhou, Z. (1990) Strategies for locating disulfide bonds in proteins, in *Methods in Enzymology, vol. 193*, (McCloskey, J. A. ed.), Academic, NY, pp.374–389.
4. Hirayama, K. and Akashi, S. (1994) Assignment of disulfide bonds in proteins, in *Biological Mass Spectrometry: Present and Future* (Matsuo, T., Caprioli, R. M., Gross, M. L., and Seyama, Y., eds), Wiley, NY, pp.299–312.
5. Jacobson, G. R., Schaffer, M. H., Stark, G. R., and Vanaman, T. C. (1973) Specific chemical cleavage in high yield at the amino peptide bonds of cysteine and cystine residues. *J. Biol. Chem.* 248, 6583–6591.
6. Wakselman, M., Guibe-Jampel, E. (1976) 1-Cyano-4-dimethylamino-pyridinium salts: new water-soluble reagents for the cyanylation of protein sulphydryl groups. *J. C. S. Chem. Comm.* 21–22.
7. Nakagawa, S., Tamakashi, Y., Hamana, T., Kawase, M., Taketomi, S., Ishibashi, Y., Nishimura, O., Fukuda, T. (1994) Chemical cleavage of recombinant fusion proteins to yield peptide amides. *J. Am. Chem. Soc.* 116, 5513–5514.
8. Nakagawa, S., Tamakashi, Y., Ishibashi, Y., Kawase, M., Taketomi, S., Nishimura, O., Fukuda, T. (1994) Production of human PTH (1–34) via a recombinant DNA technique. *Biochem. Biophys. Res. Commun.* 200, 1735–1741.
9. Wu, J., Watson, J. T. (1998) Optimization of the cleavage reaction for cyanylated cysteinyl proteins for efficient and simplified mass mapping. *Anal. Biochem.* 258, 268–276.
10. Pipes, G. D., Kosky, A. A., Abel, J., Zhang, Y., Treuheit, M. J., Kleemann, G. R. (2005) Optimization and application of CDAP labeling for the assignment of cysteines. *Pharma. Res.* 22, 1059–1068.
11. Gallegos-Perez, J. L., Rangel-Ordonez, L., Bowman, S. R., Ngowe, C. O., Watson, J. T. (2005) Study of primary amines for nucleophilic cleavage of cyanylated cystinyl proteins in disulfide mass mapping methodology. *Anal. Biochem.* 346, 311–319.
12. Wu, J., Gage, D. A., Watson, J. T. (1996) A strategy to locate cysteine residues in proteins by specific chemical cleavage followed by matrix-assisted laser desorption/ionization time-of-flight mass spectrometry. *Anal Biochem* 235, 161–174.
13. Wu, J., Watson, J. T. (1997) A novel methodology for assignment of disulfide bond pairings in proteins. *Protein Sci* 6, 391–398.
14. Yang, Y., Wu, J. Watson, J. T. (1998) Disulfide mass mapping in proteins containing adjacent cysteines is possible with cyanylation/cleavage methodology. *J. Am. Chem. Soc.* 120, 5834–5835.
15. Qi, J., Wu, J., Somkuti, G. A., Watson, J. T. (2001) Determination of the disulfide structure of sillucin, a highly knotted, cysteine-rich peptide, by cyanylation/cleavage mass mapping. *Biochemistry.* 40, 4531–4538.
16. Borges, C. R., Qi, J., Wu, W., Torng, E., Hinck, A. P., Watson JT. (2004) Algorithm-assisted elucidation of disulfide structure: application of the negative signature mass algorithm to mass-mapping the disulfide structure of the 12-cysteine transforming growth factor beta type II receptor extracellular domain. *Anal Biochem.* 329, 91–103.
17. Putney, S., Herlihy, W., Royal, N., Pang, H., Aposhian, H. V., Pickering, L., Belagaje, R., Biemann, K., Page, D., Kuby, S., Schimmel, P. (1984) Rabbit muscle creatine phosphokinase cDNA clone, primary structure and detection of human homologues. *J. Biol. Chem.,* 259, 14317–14320.
18. Schutte, C. G., Lemm, T., Glombitza, G. J., Sandhoff, K. (1998) Complete localization of disulfide bonds in GM2 activator protein. *Protein Sci.* 7, 1039–1045.
19. Wu, J., Yang, Y., Watson, J. T. (1998) Trapping of intermediates during the refolding of recombinant human epidermal growth factor (hEGF) by cyanylation, and subsequent structural elucidation by mass spectrometry. *Protein Sci.* 7, 1017–1028.
20. Somkuti, G. A., Walter, M. M. (1970) Antimicrobial polypeptide synthesized by Mucor pusillus NRRL 2543. *Proc Soc Exp Biol Med.* 133,780–785.
21. Gray, W. R. (1993). Disulfide structures of highly bridged peptides: A new strategy for analysis. *Protein Sci.* 2, 1732–1748.
22. Gray, W. R. (1993) Echistatin disulfide bridges: selective reduction and linkage assignment. *Protein Sci.* 2, 1749–1755.

23. Qi, J., Wu, W., Borges, C. R., Hang, D., Rupp, M., Torng, E., Watson, J. T. (2003) Automated data interpretation based on the concept of "negative signature mass" for mass-mapping disulfide structures of cystinyl proteins. *J. Am. Soc. Mass Spectrom.* 14, 1032–1038).
24. Rubinstein, M. (1979) Preparative high-performance liquid partition chromatography of proteins. *Anal. Biochem.* 98, 1–7.

2
Detection of Post-translational Modifications by Fluorescent Staining of Two-Dimensional Gels

Archana M. Jacob and Chris W. Turck

Summary Post-translational modifications (PTMs) are key to the regulation of functional activities of proteins. Quantitative and qualitative information about PTM stages of proteins is crucial in the discovery of biomarkers of disease. Recent commercial availability of fluorescent dyes specifically staining PTMs of proteins such as phosphorylation and glycosylation enables the specific detection of protein regulations taking place with respect to these modifications. Activity and molecular and signalling interactions of many proteins are determined by their extent of phosphorylation. In our search for biomarkers of neurodegenerative diseases such as Multiple Sclerosis (MS), using its animal model, Experimental autoimmune encephalomyelitis (EAE), we have applied the phopshorylation specific fluorescent dye, ProQ Diamond, to study changes taking place in the phosphoproteome. Subsequent Colloidal Coomassie staining of the same gels detects the changes at the whole proteome level. We have detected many changes taking place in the CNS tissue of the EAE animals at the whole proteome as well as at the phosphoprotcome level that has given valuable insights into the patho-physiological mechanism of EAE and possibly also MS.

Key Words Phosphoproteome; ProQ Diamond; 2D gel electrophoresis; fluorescent stain; in-gel digestion; peptide extraction.

1 Introduction

The analysis of post-translational modifications is of high significance in proteomic studies aimed at the discovery of protein markers relevant in the pathogenesis of diseases. Although limited in the coverage of the whole proteome, one of the main strengths of 2-dimensional polyacrylamide gel electrophoresis (2D PAGE) is the ability to visualize protein isoforms. However, because of their very low stoichiometry, the post-translationally modified isoforms often remain undetected on 2D gels using classical staining methods. Commercially available fluorescent stains such as, ProQ Emerald, (Molecular Probes) and ProQ Diamond, (Molecular Probes)

From: *Post-translational Modifications of Proteins.*
Methods in Molecular Biology, Vol. 446.
Edited by: C. Kannicht © Humana Press, Totowa, NJ

have facilitated the specific detection and identification of glycosylation and phosphorylation of proteins separated on 2D gels, respectively. Proteins separated on 2D gels for proteomic analysis can be stained first by ProQ and subsequently by Colloidal Coomassie stain or any other protein stain.

ProQ Diamond stain binds specifically to proteins with phosphate groups on serine, threonine, and tyrosine residues. Because of the fluorescent nature ProQ Diamond stain can detect phosphorylated proteins present in as low as 4 ng per spot. The staining intensity correlates with the number of phosphate groups present in the respective proteins (1,2). This high sensitivity is especially very critical in the case of phosphorylated proteins because of their very low abundance (3,4). ProQ Diamond staining allows the comparative expression profiling of the phosphoproteome both in a quantitative and qualitative manner. Moreover, its high sensitivity increases the proteome coverage. As the dye binds noncovalently to the phosphate groups, it is compatible with subsequent mass spectrometric analysis.

Mouse CNS tissue protein extracts are separated on 2D gels and stained for phosphoproteome and whole proteome (Figs. 2.1a and 2.1b). The two images are digitally colored and overlaid (Fig. 2.1c) to determine the relative position of the phosphoproteins on the whole protein stained image. The spots of interest are excised and prepared for mass spectrometry analysis. Thus, this method detects in parallel the expression level changes and altered phosphorylation modifications taking place under different physiological conditions (Figs. 2.2 and 2.3).

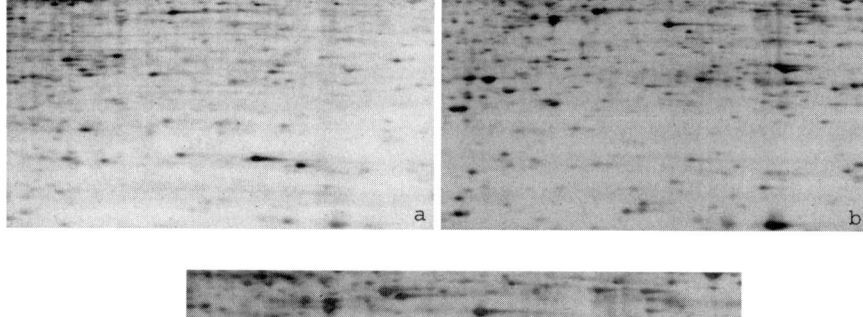

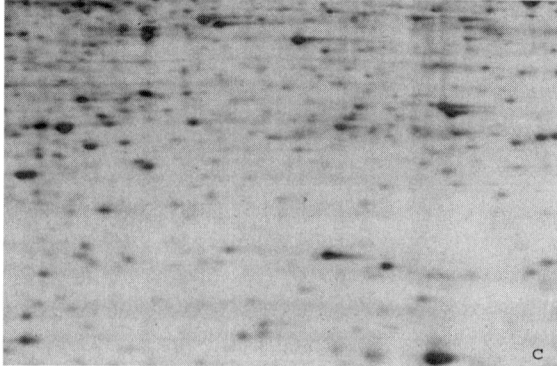

Fig. 2.1 Phosphoproteome and whole proteome. (**a**) ProQ Diamond stained image of mouse spinal cord 2D gel. (**b**) Image of the same gel stained with Coomassie Blue. (**c**) The 2 images overlaid on top of each other. Image C is used to determine the relative position of the ProQ Diamond stained spots on the Coomassie stained image for spot picking

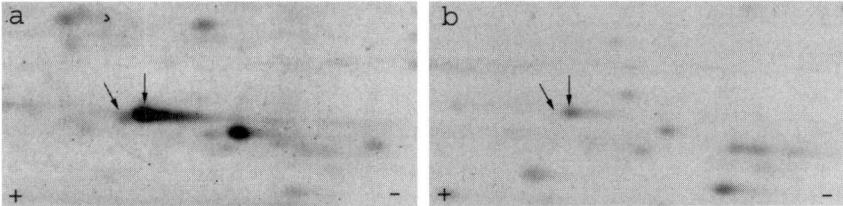

Fig. 2.2 (a) ProQ Diamond stained gel image. (b). Same gel stained subsequently with Coomassie stain. The arrows indicate a protein that migrates at two different positions on a 2D gel. The more acidic spot is visible only on the ProQ Diamond stained image

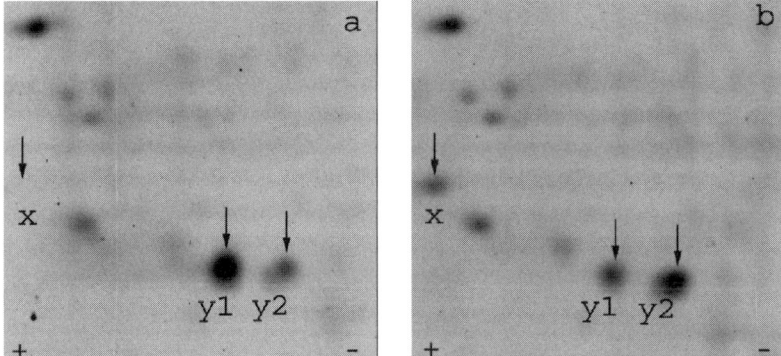

Fig. 2.3 Quantitative and qualitative differences of the phoshoproteome between control and EAE mouse brain. (a) 2D gel image of EAE brain proteins stained with ProQ. (b) 2D gel image of control brain proteins stained with ProQ. Spot **x** is up-regulated in EAE, an example for a quantitative difference in expression. Spots **y1** and **y2** represent the same protein. It differs between diseased and control animals in its extent of phosphorylation, a typical example of a phosphorylaton change, where the protein moves more to the acidic end

2 Materials

2.1 Sample Preparation of Brain and Spinal Cord Sample for 2-Dimensional Electrophoresis

Unless otherwise mentioned all reagents are purchased from Bio-Rad, Hercules, CA

1. Isoelectric focussing (IEF) buffer: $7M$ urea, $2M$ thiourea, $100\,mM$ dithiothreitol (DTT), 4% (w/v) 3-[(3-cholamidopropyl)dimethylammonio]-1-propanesulfonate hydrate (CHAPS), 0.05% biolytes 3–10, 0.001% (v/v) bromophenol blue (for color). Prepared and stored as 1-mL aliquots at −80°C. Once thawed should not be frozen again (*see* **Note 1**).
2. Protease inhibitors: All protease inhibitors are added to the IEF buffer in $1 \times$ concentration just before use.

   a. Pepstatin (Roche): 1000× stock prepared by dissolving 1 tablet in 1 mL etha-
      nol and stored at −20°C up to 3 mo.
   b. Complete (Roche): 25× stock prepared by dissolving 1 tablet in 2 mL Double
      distilled water and stored in aliquots at −20°C up to 3 mo.
   c. Phenylmethylsulfonyl fluoride (PMSF), (Roche):100 mM stock (100×) pre-
      pared in methanol or ethanol and stored at 4°C.

3. Tissue sample grinding kit (GE Healthcare Amersham Biosciences) (*see* **Note 2**).

2.2   2-Dimensional Gel Electrophoresis

2.2.1   Isoelectric Focusing

1. IPG strips of desired pH range (4–7 pH range for brain tissue and 5–8 pH range
   for spinal cord).
2. Filter wicks.
3. Isoelectric focusing apparatus.
4. Equilibration trays.

2.2.2   Equilibration

1. Equilibration buffer base (EQB): 50 mM Tris-HCl, pH 8.8, 6M urea, 2% sodium
   dodecyl sulphate (SDS), 20% (v/v) glycerol. Store as 20- and 40-mL aliquots at
   −20°C. While aliquoting, the solution should be constantly stirred using a magnetic stirrer. Glycerol will otherwise accumulate at the bottom. Thaw before use
   and vortex to get a clear solution.
2. Equilibration buffer 1: EQB containing 2% (w/v) DTT. Dissolve few hours
   before use at room temperature and keep at dark. Working volume is at least
   6 mL per gel strip.
3. Equilibration buffer 2: EQB containing 2.5% (w/v) iodoacetamide (Bio -Rad).
   Dissolve few hours before use at room temperature and keep at dark. Working
   volume is at least 6 mL per gel strip.

2.2.3   SDS PAGE

1. Tris buffers: 1.5M Tris-HCl, pH 8.8, and 0.5M Tris-HCl, pH 6.8. Store at room
   temperature.
2. 10% SDS. A ready made stock of 20% w/v SDS is diluted 1:2 with water and
   stored at room temperature.
3. Thirty percent acrylamide/bis solution (37.5:1 with 2.6% C) (Genaxxon) (*see*
   Note 3).
4. *N,N,N,N′*-Tetramethyl-ethylenediamine (TEMED).

5. Ammonium persulfate (APS): Prepare 10% w/v solution in water before use.
6. Water-saturated isobutanol. Shake equal volumes of water and isobutanol in a glass bottle and allow to separate overnight. Use the top layer. Store at room temperature.
7. Agarose overlay buffer: 0.5% (w/v) agarose is dissolved in Tris glycine SDS (TGS) running buffer by boiling in a microwave. Few drops of bromophenol blue are added to the buffer for color. Store at 4°C. Melt in a microwave before use and maintain at 60°C before use.
8. Running buffer (10×): 250 mM Tris, 1.920M glycine, 1% (w/v) SDS pH 8.3. Store at room temperature.
9. Prestained molecular weight markers.
10. Hinged spacer plates.
11. Gel casting chamber.
12. Gel combs for 2D gels with 1 reference well (Protean plus comb, Bio –Rad).

2.3   ProQ Diamond Staining

1. Fixing buffer: 10% acetic acid, 50% methanol. Stored at room temperature or prepared before use (*see* **Note 4**).
2. ProQ Diamond stain (Molecular Probes), store protected from light at 4°C (*see* **Note 5**).
3. Destain 1: 20% Acetonitrile (ACN) and 50 mM sodium acetate pH 4.0 (*see* **Note 6**). For stock buffer solution, dissolve 1M sodium acetate in double distilled water and adjust the pH to 4.0 using fuming hydrochloric acid (HCl) and store at room temperature.
4. Staining trays compatible with methanol

2.4   Colloidal Coomassie Staining

1. Colloidal solution: 17 mM ammonium sulphate, 2% phosphoric acid and 34% v/v methanol (*see* **Note 7**).
2. R-250 Brilliant Coomassie, (Sigma).

2.5   Spot Processing

1. Destain 2: 1:1 solution of 20 mM NH_4HCO_3, pH 8.00, and 100% ACN. Mix equal volumes and leave and store at room temperature.
2. 1 mM NH_4HCO_3 pH 8.0 and trypsin (Sequencing grade modified trypsin, Promega). Dissolve trypsin in 1 mM NH_4HCO_3 at 1 µg/µL concentration and store in 5-µL aliquots at −20°C.
3. 2% trifluoroacetic acid (TFA), (Merck) and 5% formic acid (HCOOH), (Merck).

3   Methods

3.1   Sample Preparation

1. Thaw required amount of IEF buffer at room temperature (300 μL per sample). Dissolve DTT precipitates by vortexing well (*see* **Note 8**).
2. Add protease inhibitors in 1× concentration.
3. Centrifuge the tissue grinding tubes for a minute and remove the storage liquid (*see* **Note 9**).
4. To each of the grinding tube add 200 μL of IEF buffer and vortex vigorously so that the beads are suspended in the buffer.
5. Drop the tissue sample into the grinding tubes directly from liquid nitrogen (*see* **Note 10**).
6. Grind the tissue for 1 to 2 min into a solubilized homogenate.
7. Centrifuge the samples at 14,000 rpm for 10 min and collect the supernatant into a fresh tube (*see* **Note 11**).
8. Add another 100 μL of IEF buffer into the grinding tubes containing residual tissue and grind for 1–2 min.
9. Centrifuge the samples at 14,000 rpm for 10 min and collect the supernatant to the same tube as in step 7.
10. Centrifuge the combined supernatant at 14,000 rpm for 10 min to remove any bead particle present. The samples can be stored at −80°C at this step if not proceeded immediately.

3.2   2 DE Electrophoresis

3.2.1   Isoelectric Focusing

1. Take the required amount of protein from the protein homogenate and make up the volume of the samples to 300 μL with IEF buffer without protease inhibitors.
2. Incubate the samples at room temperature for 30 min (*see* **Note 12**).
3. Centrifuge at 14,000 rpm for at least 10 min (*see* **Note 13**).
4. Add samples into the respective IEF tray wells and place the IPG strip with gel side down on to the sample. After 1 h of incubation overlay the strips with mineral oil and rehydrate for 12 h actively at 50 volts.
5. Insert filter wicks wet with water at each electrode when rehydration is over.
6. Focus proteins up to 60,000 Vh. (PROTEAN Plus IEF cell, Bio –Rad). After the focussing is complete the strips can be stored at −80°C for several months.

3.2.2   Equilibration

This step should be performed just before proceeding to step 15 in **Section 3.2.3.**

1. Prepare EQB1 and EQB2.
2. Thaw the strips for equilibration.
3. Reduction: The strips are incubated in 6 mL of EQB 1 in dark for 10 min at room temperature with gentle shaking. The buffer is removed by simply pouring out.
4. Carbamidomethylation: The strips are incubated with 6 mL of EQB 2 in dark for 10 min at room temperature with gentle shaking. Pour out the buffer and wet the strips with 1 × TGS buffer.

3.2.3 SDS PAGE

This part follows the use of Bio -Rad PROTEAN plus multi-casting chamber, Gradient former and PROTEAN plus Dodeca Cell.

1. Assemble the hinged spacer plates in the multi-casting chamber. A separation sheet should be placed between the plates and between the first plate and the surface of the chamber to ensure easy detachment of the plates after gel casting. Fill the remaining space with acrylic blocks and separation sheets, in order to flush the stack to the front of the chamber. Seal the chamber using a tubing and close the chamber by screwing down on opposite sides.
2. Connect the gradient former to the multicasting chamber port using a tubing with stopcocks attached to each end. Keep the gradient former at a higher altitude than the casting chamber to allow smooth flow of the gel solution. Turn valves of the stopcocks to off position.
3. Prepare the gel solution: The following composition is for preparing one 12% gel (20 cm × 20.5 cm) of 1 mm thickness. Mix 24 mL of acrylamide/bis solution, 15 mL of 1.5M Tris buffer pH 8.8, 20.4 mL water, 600 μL of 10% SDS, and 15 μL of TEMED solution in an Erlenmeyer flask and mix well.
4. Degas the solution for 15 min by connecting to vacuum.
5. Add 520 μL of APS per gel volume and mix well.
6. Pour the gel mixture gently into the gradient former and open the valves.
7. Stop the flow by closing the stopcocks, leaving enough space for stacking gel.
8. Overlay the top of the gels with water-saturated isobutanol (*see* **Note 14**). The gels polymerize in one hour. The gels can be stored at this point at 4°C overnight.
9. Disassemble the chamber and remove the plates. Pour out isobutanol and wash the top of the gel with excess water. Dry the stacking gel area with a filter paper or vacuum. Clamp the top sides of the gel plates using paper clamps.
10. Prepare 5% staking gel buffer. For one gel, mix 1.25 mL of acrylamide/bis solution, 1.88 mL of 0.5M Tris pH 6.8, 4.3 mL water, 75 μL of SDS, 10 μL of TEMED and mix well.
11. Degas the solution and add 25 μL of APS per gel volume.
12. Apply the stacking gel solution and insert combs without making air bubbles (*see* **Note 15**). Leave the gels to polymerize for at least 3 h.

13. When the gels are polymerized remove the combs and wash and dry the top of the gels.
14. Equilibrate the strips as described in **Section 3.2.2.**
15. Fill the space on top of the gels with agarose. Dip the strips for few seconds in 1×TGS buffer and insert on top of the stacking gel. Press down the strips gently without making bubbles between the strip and the top of the stacking gel. The strips should stay in contact with the top of the gel along the whole length of the strip.
16. Keep the protein standard well dry by inserting a filter paper piece into the well to absorb agarose. Change filter paper when soaked. Apply 10 μL of protein marker and seal the well with agarose. The gels are ready for electrophoresis when agarose is polymerized
17. The gels are placed into the Dodeca cell, running chamber and run at 50V in 1×TGS buffer either overnight or until the dye front has passed the stacking gel. The cell is connected to a buffer re-circulation pump and a thermostat set at 10°C. The voltage is then hiked to 200V and run until the dye-front reaches the bottom of the gels (*see* **Note 16**).

3.3 *ProQ Diamond Staining*

All steps are carried out at room temperature.

1. Add 200 mL of fix solution to the labeled staining trays. This is the volume sufficient to cover the large format gels of size 20 cm × 20.5 cm.
2. Gels are detached from the glass plate by gently scraping the edge of the gels. Flush water between the gel and the glass plate on all sides to enable smooth separation of the gels from the plate without damaging the gels. Transfer the gels to respective staining trays containing fix solution.
3. The gels are fixed overnight by gentle shaking. During this process the gels shrink in size.
4. Discard the fix solution and rinse the gels in excess double distilled water for 30 min by gentle shaking at room temperature. Repeat three times.
5. The following steps should be carried out with minimum exposure of light. Discard water and add 200 mL of ProQ Diamond stain to each gel and keep them immediately in dark for 3 h at room temperature with gentle shaking (*see* **Note 17**).
6. Prepare the destain.
7. Remove the stain and add 250 mL of destain to each gel. Shake the gels in dark for 30 min. Repeat the procedure two more times (*see* **Notes 18** and **6**).
8. Discard the destain and rinse the gels in excess water for 15 min. Change water and keep the gels at dark.
9. Gels are scanned with the help of a fluorescent scanner (FX – Bio -Rad) (*see* **Note 19**).

3.4 Colloidal Coomassie Staining

1. The gels are left overnight in water with gentle shaking at room temperature.
2. Wash the gels again for 30 min in water and discard the water.
3. Add 250 mL of colloidal solution to the gels and shake them for 1 h at room temperature.
4. Weigh 250 mg of Coomassie blue powder for each gel and sprinkle it on top of the gels. The powder dissolves in the solution. Leave the gels shaking for 2–3 d.
5. Discard the stain and transfer the gels to another tray containing excess water and wash for 1 h.
6. Change water and prepare the gels for scanning.
7. Gels are scanned using a densitometer (GS-800, Bio –Rad) (*see* **Notes 19** and **20**).
8. Seal the gels in a plastic bag and store at 4°C.

3.5 Image Analysis

The method described here uses PD Quest (Bio –Rad) as the image analysis software (*see* **Note 21**).

1. ProQ diamond stained images and the Coomassie images are analyzed separately as two different analysis sets. Gels are cropped to the same size using the same crop settings. A match-set is created and the spots of interest are selected.
2. To determine the position of the ProQ stained spots on the visible Coomassie gel the gel images are overlaid on top of each other.
3. The original ProQ image and the Coomassie image of the same gel are opened in PD Quest. The gels are cropped at the exact boundaries of the gels (*see* **Note 22**).
4. The file sizes of the two images are adjusted to the same values.
5. The two images are selected under 2 different color filters in the multicolor channel view. This overlays the 2 images on top of each other. Spots visible by both stains appear as mean of the two primary colors of the parent spots.
6. Zoom through different parts of the gels and mark every spot that overlaps, on print- outs of both ProQ and Coomassie images.
7. For spots that are visible only on ProQ image, determine their relative position with respect to that of the neighboring Coomassie stained spots and mark them on the Coomassie image.
8. Highlight the position of all the spots of interest on the Coomassie image.

3.6 Spot Picking, Digestion and Peptide Extraction for Mass Spectrometry

Because of their very low abundance the ProQ stained spot should be processed with care taken to achieve maximum yield of peptides out of the gel pieces and to minimize the amount of salt and tryptic peptides in the peptide mixture (*see* **Note 23**).

1. Cut the end of a 1 mL pipet tip to a diameter of around 1.5 mm (see **Note 24**).
2. Pick out the spots following their marked position on the Coomassie image.
3. Destain the gel spots 2 times in destain 2 (1:1 solution of ACN and 20 mM NH$_4$HCO$_3$). The gel spots are incubated in the destain 2 for 30 min at room temperature.
4. Dry the gel pieces under a hood for several hours or overnight.
5. Add 5 μL of 1 mM NH$_4$HCO$_3$ containing up to 50 ng trypsin to each gel piece (*see* **Note 25**). Leave the spots for 15 min at 4°C (*see* **Note 26**).
6. Digest the gel spots for 5 h at 37°C (*see* **Note 27**).
7. After digestion, centrifuge the tubes to collect all the liquid at the bottom.
8. Add 1 μL of 2% TFA or 5% HCCOH to each tube.
9. Vortex at 37°C for 30 min and sonicate for 3–4 min (*see* **Note 28**).
10. Repeat step 9.
11. Analyze the samples by mass spectrometry.

4 Notes

1. The amount of DTT used in IEF buffer should be changed according to the desired pl range. For basic region greater than pl 8.0 it is better to use 50 mM DTT instead of 100 mM. This minimizes streaking at the basic region of the gel.
2. Other methods suitable for tissue disruption can be adapted, for example, sonication.
3. Acrylamide is a neurotoxin in the liquid form. Avoid inhaling or contact with skin. Pour under a hood.
4. Methanol is toxic, do not inhale.
5. Do not store the dye for too long or buy in big bulk unless required. Take care to order fresh batch of dye that is good for 6 mo. Old dye leads to unspecific background binding.
6. Acetonitrile is a neurotoxin. Use a mask while using acetonitrile and pour under a hood. While using large volumes of ACN keep the windows open. Discard as harmful liquid waste.
7. Methanol should be added in the end, right into the middle of the solution while stirring with magnetic bars. This leads to precipitation of salt like snow flakes in the middle. Keep stirring until the solution turns clear.
8. Sample preparation with IEF buffer should be carried out optimally at around 22°C. Do not heat the samples above 30°C or cool below 16°C. Urea modifies proteins at temperature above 30°C and upon cooling DTT precipitates out.
9. These grinding tubes are prepared for the use of a maximum tissue weight of 100 mg. If the tissue weight exceeds 100 mg, it is desirable to grind the tissue into a homogenized powder using liquid nitrogen and then to use an aliquot for protein extraction

10. When the samples are in the form of a powder or when using cells, the beads along with the buffer can be transferred to the original tube containing the sample. In this case the tubes should be 1.5-mL tubes, where the pestle fits well at the bottom.

11. In the case of tissue rich with fat and lipids they form a very thin layer on top after centrifugation. Avoid taking this layer if present to increase the quality of 2D gels.

12. This incubation is very important for good separation of the proteins by isoelectric focusing.

13. This is to remove any possible solid particles that might obscure the pores of the IPG strips.

14. Do not apply too much pressure on top of the gel layer while applying isobutanol as it might result in uneven surface. All gels must have the same volume of overlay solution.

15. Because the stacking gel mixture is added individually to each gel this step should be done relatively fast while preparing many gels at a time.

16. We have found many phosphorylated proteins in the very low molecular weight region of the gels. Therefore it is better to stop the run before the dye-front has reached the very bottom of the gel. On the other hand if proteins in the high molecular weight region are the focus it is desirable to run up to the very bottom.

17. Shake well before use. Small particles precipitate out during storage, which leads to speckle formation on the gels. Volume used is enough to cover the gels. We have also used 100 mL of fresh dye together with 100 mL of used dye, which is not older than 6 mo, and it works equally well.

18. Overnight destaining leads to loss of signal especially of very low abundant proteins and increased background staining of unspecific binding.

19. The gels may vary in their size slightly between each other. It is important that the same area on the scanner is scanned for all the gels to be compared irrespective of their original size. Later during gel analysis the gels are cropped at their boundaries.

20. The size of the gels after Coomassie staining is different from their size after ProQ staining.

21. Alternatively the gels can be overlaid with the help of any 2D analysis software that has a Warping tool, (Non linear, Progenesis, Delta 2D etc).

22. This step is very important for the overlay of ProQ and Coomassie stained images of the same gel using software without warping tool.

23. Work clean to avoid keratin contamination.

24. Small gel size is desirable for keeping low reaction volume during in-gel digestion and extraction. The gel pieces should be collected in plastic tubes from Eppendorf, to avoid polymers during mass spectrometry.

25. Keep the amount of trypsin as low as possible for ProQ stained spots that are absent on Coomassie staining. These proteins are extremely low in abundance, therefore excess of tryptic peptides will lead to suppression of signals during mass spectrometry.

26. The auto-cleavage activity of trypsin is inhibited by keeping the samples at 4°C while the enzyme enters the gels.
27. Overnight incubation of trypsin at 37°C enhances its auto-cleavage leading to more tryptic peptides. 5hr incubation is sufficient to complete in-gel digestion of proteins.
28. Sonication of the gel pieces have proved most successful in the recovery of peptides in our experience.

References

1. Steinberg T. H., Agnew B. J., Gee K. R., Leung W. Y., Goodman T., Schulenberg B., Hendrickson J., Beechem J. M., Haugland R. P., and Patton W. F. (2003). Global quantitative phosphoprotein analysis using Multiplexed Proteomics technology. *Proteomics.* **3**, 1128–1144.
2. Goodman T., Schulenberg B., Steinberg T. H., and Patton W. F. (2004). Detection of phospho-proteins on electroblot membranes using a small-molecule organic fluorophore. *Electrophoresis.* **25**, 2533–2538.
3. Schlessinger, J. (1993). Cellular signaling by receptor tyrosine kinases *Harvey Lect.* **89**, 105–123.
4. Reinders J. and Sickmann A. (2005). State-of-the-art in phosphoproteomics. *Proteomics.* **5**, 4052–4061.

3

Identification of Protein Phosphorylation Sites by Advanced LC-ESI-MS/MS Methods

Christoph Weise and Christof Lenz

Summary Phosphorylation, the process by which a phosphate group is attached to a pre-existing protein, is an evolutionarily and metabolically cheap way to change the protein's surface and properties. It is presumably for that reason that it is the most wide-spread protein modification: an estimated 10–30% of all proteins are subject to phosphorylation.

MS-based methods are the methods of choice for the identification of phosphorylation sites, however biochemical prefractionation and enrichment protocols will be needed to produce suitable samples in the case of low-stoichiometry phosphorylation. Using emerging MS-based technology, the elucidation of the "phosphoproteome," a comprehensive inventory of phosphorylation sites, will become a realistic goal. However, validating these findings in a cellular context and defining their biological meaning remains a daunting task, which will inevitably require extensive and time-consuming additional biological research.

Key Words Phosphorylation; LC-ESI-MS/MS; phosphoproteome.

1   Introduction

Phosphorylation, the process by which a phosphate group is attached to a pre-existing protein, is an evolutionarily and metabolically cheap way to change the protein's surface and properties. It is presumably for that reason that it is the most wide-spread protein modification: an estimated 10–30% of all proteins are subject to phosphorylation. The reaction is catalyzed by a set of enzymes called kinases that form one of the largest protein families of all. As the reaction is readily reversed by another group of enzymes, called phosphatases, phosphorylation turns out to be a pivotal regulatory mechanism that plays critical roles in the regulation of many metabolic pathways and cellular processes, including cell cycle, growth or differentiation (*1*). The determination of phosphorylation sites is the basis for a deeper understanding of cellular regulation and will allow conclusions about the enzymes involved in specific regulatory pathways. Because aberrant phosphorylation

From: *Post-translational Modifications of Proteins.*
Methods in Molecular Biology, Vol. 446.
Edited by: C. Kannicht © Humana Press, Totowa, NJ

events are known to occur in many diseases, including various types of cancer, this holds huge promise for the definition of new drug targets. Phosphorylation—from the history of its discovery to methodological advances and biological aspects—is covered by a number of excellent and exhaustive reviews (*1–3*).

Important recent technological advances have made mass spectrometry (MS) the method of choice for protein analysis and proteome research over the last decade, but despite the huge interest in protein phosphorylation, the determination of phosphorylation sites has remained analytically challenging. The classical chemical sequencing approach (Edman degradation) was hampered mainly by the insolubility of the phosphoamino acid products and the necessity to obtain highly purified phosphopeptides. Mass spectrometry, on the other hand, is well suited to deal even with complex peptide mixtures, but still suffers from the often low stoichiometry of phosphorylation leading to low signal intensities that tend to disappear into the background. The often cited low ionization efficiency of phosphopeptides relative to their nonmodified counterparts, though, appears to be a generalization that is not supported by experimental evidence (*4*).

Several approaches have been reported to deal with the stoichiometry challenge (*3*):

– chemical replacement of the phosphate group by other functionalities that enhance ionization efficiency and MS/MS fragmentation behavior, e.g., by β-elimination and subsequent Michael addition. Because of incomplete reaction and purification of the products this usually requires an increased amount of peptide.
– affinity enrichment of phosphorylated species, e.g., by immobilized metal-affinity chromatography (IMAC) on Fe^{3+} or Ga^{3+} matrices. ZrO_2 or TiO_2 have also been successfully used for this purpose, however the enrichment is rarely specific and acidic peptides are likely to be enriched as well.
– alternatively, peptides phosphorylated on tyrosine can be purified using anti-P-Tyr-antibodies. No antibodies with good specificity for P-Ser and P-Thr are available, however, although these form the bulk of cellular phosphorylation sites.

The phosphorylation-specific analytical method would have to introduce some sort of filter that will allow systematic screening for phosphorylated compounds.

In mass-spectrometric analysis, precursor ion scanning can be used to identify compounds from mixtures, such as proteolytic digests, that result in a common product ion. In this experiment a first mass analyzer is set to scan the entire mass range of possible precursor (peptide) ions. Through collision-induced dissociation (CID) these are fragmented to produce a marker product ion, which is selectively monitored using a second mass analyzer fixed on the m/z value of the marker ion. Under CID conditions in negative-ion mode phosphopeptides produce distinct marker ions at m/z 79 (PO_3^-) and m/z 63 (PO_2^-), which can be used for their selective detection (*5–7*). Various phosphorylation-specific aspects of precursor-ion-scanning methods are discussed in the Notes (**Section 4**).

In **Section 2.3** we describe a state-of-the-art LC-MS/MS method (direct coupling of a liquid-chromatography system to a mass spectrometer) for determining phosphorylation sites from a peptide mixture generated by in-gel digestion.

Phosphopeptides selectively detected by precursor ion scanning are subsequently fragmented by collision cell CID in a product-ion experiment to establish their sequence and the site of phosphorylation. Using this set-up, amounts of phosphorylated peptide as low as 5 fmol can be detected.

To summarize, MS-based methods are the methods of choice for the identification of phosphorylation sites, however biochemical prefractionation and enrichment protocols will be needed to produce suitable samples in the case of low-stoichiometry phosphorylation. Using emerging MS-based technology, the elucidation of the "phosphoproteome," a comprehensive inventory of phosphorylation sites, will become a realistic goal. However, validating these findings in a cellular context and defining their biological meaning remains a daunting task, which will inevitably require extensive and time-consuming additional biological research.

2 Materials

2.1 In-Gel Reduction, Alkylation and Tryptic Digestion of Phosphoproteins

1. Gel Washing Solution: 50 mM ammonium bicarbonate in water.
2. Coomassie Destaining Solution: 50 mM ammonium bicarbonate in acetonitrile/water (1:1, v:v).
3. Gel Dehydration Solution: acetonitrile.
4. Cystine Reduction Solution: 100 mM dithiothreitol in 100 mM aqueous ammonium bicarbonate.
5. Cysteine Alkylation Solution: 55 mM iodoacetamide in 100 mM aqueous ammonium bicarbonate.
6. Modified Sequencing Grade Porcine Trypsin (Promega): 10 µg/mL in 25 mM aqueous ammonium bicarbonate.
7. Gel Extraction Solvent: 0.5% formic acid in acetonitrile:water (2:8, v:v).
8. Acetonitrile and water, HPLC grade.
9. 0.5-mL Eppendorf tubes.
10. A temperature-controllable heater/shaker.
11. A SpeedVac concentrator.

The protocol supplied here describes the in-gel tryptic digestion of a protein detected by Coomassie staining on an SDS gel. One major challenge in the analysis of protein phosphorylation is the substoichiometric degree of this modification (3,4). The methodology described here is capable of detecting amounts as low as 5 fmol of phosphopeptide total. Assuming a 1% degree of phosphorylation of the protein and a digestion/extraction efficiency of 50%, this translates into an amount of 1 pmol protein loaded onto the gel. Protein amounts in the low picomole range are usually detectable by Coomassie staining.

Another challenge lies in the choice of the proper endopeptidase for digestion. The actual site of phosphorylation may lie in a region of the sequence where too many or too little trypsin cleavage sites are located, resulting in peptides not suitable for the LC-MS/MS analysis (MW range 700–3,000 Da). For a comprehensive phosphorylation analysis of unknown proteins additional analyses using digest agents with different specificities should be used, such as endopeptidase GluC or elastase *(8)*.

2.2   *LC-ESI-MS/MS Analysis of Tryptic Digests*

1. Loading Solvent: 0.5% formic acid in acetonitrile:water (2:98, v:v).
2. Mobile Phase A: 0.1% formic acid in acetonitrile:water (5:95, v:v).
3. Mobile Phase B: 0.1% formic acid in acetonitrile:water (95:5, v:v).
4. Make-up Solvent: 0.1% formic acid in acetonitrile:2-propanol:water (1:8:1, v:v:v).
5. A low-dead-volume T-junction (Upchurch Micro-Tee P775, Upchurch, Oak Harbor, WA/US) with fused silica capillary (20 µm ID) and Teflon sleeves for connection.
6. A hybrid triple-quadrupole/linear-ion-trap mass spectrometer (4000 Q TRAP LC-MS/MS system, Applied Biosystems, Foster City, CA/US) coupled on-line with nanoflow HPLC (Ultimate with Famos autosampler and Switchos column switching module, all Dionex, Idstein, Germany). A Micro-ion spray head (Applied Biosystems) fitted with a fused silica tapered tip sprayer needle (FS360-20-10-N, New Objective Inc, Woburn, MA/US) and zero grade air or nitrogen as Sheath Gas. A 75 µm × 15 cm PepMap RP-C18 column (3-µm particle size, 100-Å pore size) and a 300 µm × 5 mm precolumn of the same material (Dionex).
7. An additional pump capable of generating nanoliter flow rates (Harvard Apparatus Model 11 Pico Plus syringe Pump, Harvard Apparatus, Holliston, MA/US) with a 100-µL glass syringe (Hamilton).

The LC-MS/MS analysis of phosphopeptides consists of 2 steps: A) detection of phosphorylated peptides by a selective scan function, i.e., a precursor ion scan (Fig. 3.1) for m/z 79 (PO_3^-) in negative mode, and B) MS/MS analysis of the phosphopeptide sequence in positive mode (*see* **Notes 1** and **2**). On hybrid triple quadrupole/linear ion trap mass spectrometers these 2 steps can be carried out in a single integrated experiment *(9–11)*. If such an instrument is not available though, the 2 parts can also be carried out independently on other equipment. Both triple-quadrupole and hybrid-triple-quadrupole/time-of-flight mass spectrometers are capable of precursor ion scan experiments. The subsequent high-sensitivity MS/MS analysis can also be performed on e.g., hybrid triple quadrupole/time-of-flight or conventional ion trap mass spectrometers. The sample will need to be split for the 2 experimental steps *(5,6)*.

Special consideration should be given to the choice of solvents and organic modifiers that are used for the reverse-phase separation and LC-MS/MS analysis. For a single experiment, analysis conditions are needed that allow both for negative and positive mode electrospray ionization. A weak acidic modifier has proven to enable electrospray analysis of phosphopeptides in both polarities at comparable sensitivity. From a chromatographic point of view, trifluoroacetic (TFA) acid is the

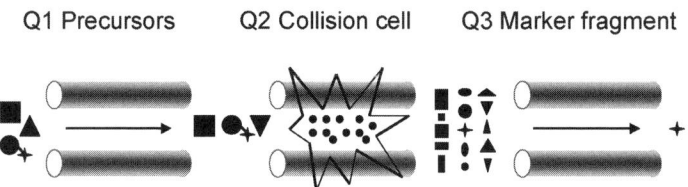

Q1 Precursors   Q2 Collision cell   Q3 Marker fragment

Fig. 3.1 Principle of a precursor ion scan experiment on a triple-quadrupole mass spectrometer. The first quadrupole Q1 scans the *m/z* range of possible intact phosphopeptide precursors; the second quadrupole Q2 serves as a collision cell where precursors are fragmented by collisionally induced dissociation (CID); Q3 is set to continually monitor production of the marker fragment *m/z* 79 (PO_3^-)

organic modifier of choice as it provides the best separation because of its strong ion-pairing properties. As TFA in practice is severely detrimental to negative-mode electrospray ionisation, formic acid is usually chosen as organic modifier.

Another challenge implied with negative-mode electrospray ionization is the possibility of high-voltage corona discharge, or "arching," leading to corrosion of the sprayer needle and irreproducible ionisation conditions. The post-column addition of isopropanol as a "make-up solvent" via a T-piece reduces the voltage necessary to achieve ionisation in negative mode, thus significantly reducing the danger of corona discharge. Isopropanol can be premixed with organic modifier and acetonitrile to achieve constant modifier concentration and reduce the otherwise high back-pressure of the more viscous isopropanol.

2.3   Data Interpretation

1. The amino-acid sequence of the protein.
2. Software for fragment mass matching of LC-ESI-MS/MS data against theoretical fragment patterns obtained from in silico digestion of protein sequences (MASCOT V2.1, Matrixscience Ltd., London, UK)
3. Software for annotation of raw ESI-MS/MS data with fragments generated from hypothetical sequences including modified residues (Bioanalyst 1.4, Applied Biosystems)

Using fragment mass matching software like MASCOT provides a good first screen for the detection of phosphorylated peptides from the LC-MS/MS data set. The probability score values obtained for phosphorylated peptides are generally lower compared to those obtained for nonmodified peptides though, resulting in a lower chance of detecting their presence. A manual evaluation of the data using mass lists of theoretically plausible phosphopeptides calculated from the protein sequence is therefore highly recommended.

Assignment by the algorithm of the site of phosphorylation to an individual S/T/Y residue in a phosphopeptide sequence is also frequently observed to be incorrect. The assignment suggested by the software should therefore be validated by annotating raw data MS/MS spectra with sets of theoretical fragments generated from putative sequences.

3   Methods

3.1   In-Gel Reduction, Alkylation and Tryptic Digestion of Phosphoproteins

1. After SDS-PAGE separation the Coomassie-stained gel band is excised and cut into smaller pieces using a scalpel. The pieces are transferred to a 0.6-mL Eppendorf tube, and washed twice with 0.5 mL of Gel Washing Solution for 10 min.
2. To destain the gel pieces 0.5 mL of the Destaining solution is added for 30 min, with occasional vortexing. The solution is then discarded. This procedure can be repeated until the gel piece is completely destained.
3. The gel pieces are dried by adding 0.1 mL of the Gel Dehydration solution. After 5 min the gel pieces shrink and turn white. The Gel Dehydration solution is pipeted off and discarded. Residual solvent is removed in a SpeedVac concentrator for 10 min.
4. Cystine bridges are reduced by adding 0.03 mL of the Cystine Reduction solution at 56°C for 30 min. Free cysteines are then alkylated by adding 0.03 mL of the Cysteine Alkylating Solution at room temperature in the dark for 20 min. The solvents are then pipeted off and discarded.
5. To remove residual reduction/alkylation agent, the gel piece is washed again with 0.5 mL of Gel Washing Solution for 10 min. Step 3 is then repeated to remove residual solvent.
6. Between 0.003 and 0.03 mL of trypsin solution are carefully added to the dried gel pieces until they are fully rehydrated. When the gel pieces do not take up any additional solution, 0.05 mL of gel washing solution is added, the solvents quickly spun down in a microcentrifuge, and the tube closed and sealed with Parafilm. Digestion is achieved by placing the tube in a heater/shaker combination at 37°C overnight.
7. The tube is removed from the heater/shaker and let cool to room temperature. The supernatant is pipeted off and set aside. To extract the majority of peptides 0.03 mL of the Gel Extraction Solvent is added to the gel piece, and the contents of the tube are sonicated twice for 15 min. The Extraction Solvent is now pipeted off, combined with the digestion supernatant in a 0.6-mL Eppendorf tube and dried down in a Speedvac concentrator.

3.2   LC-ESI-MS/MS Analysis of Tryptic Digests

1. Prepare the solvents and equilibrate the nanoflow LC system on 5% solvent B.

2. Set up a micro-T junction to split in the make-up solvent post-column at a flow ratio of 300:100 (LC eluent:make-up, nl/min). Measure the flow rates before and after the micro-T to ensure proper set-up of the junction.
3. Optimise source conditions for both negative mode and positive mode operation, and for rapid two-way switching between polarities.
4. Set up an LC-MS/MS acquisition method consisting of a precursor ion scan for m/z 79 in negative mode, a high resolution MS scan of detected signals in positive or negative mode, and of up to 3 MS/MS experiments of accepted precursors in positive mode (Fig. 3.2). The precursor ion scan should be set up using isolation widths (FWHH) of 2 Th in Q1 and 0.6 in Q3 to accommodate the isotopic patterns of the phosphopeptide precursors and the m/z 79 fragment ion, respectively. The chromatography method should include pre-column concentration and desalting. A linear gradient of 5–40% B across 45 min is often used.
5. Inject 100 fmol of a known phosphopeptide standard to evaluate the performance of the system with regard to stable polarity switching, sensitivity and MS/MS results. A tryptic digest of bovine casein α diluted down from a stock solution is usually used for this purpose (Fig. 3.3).

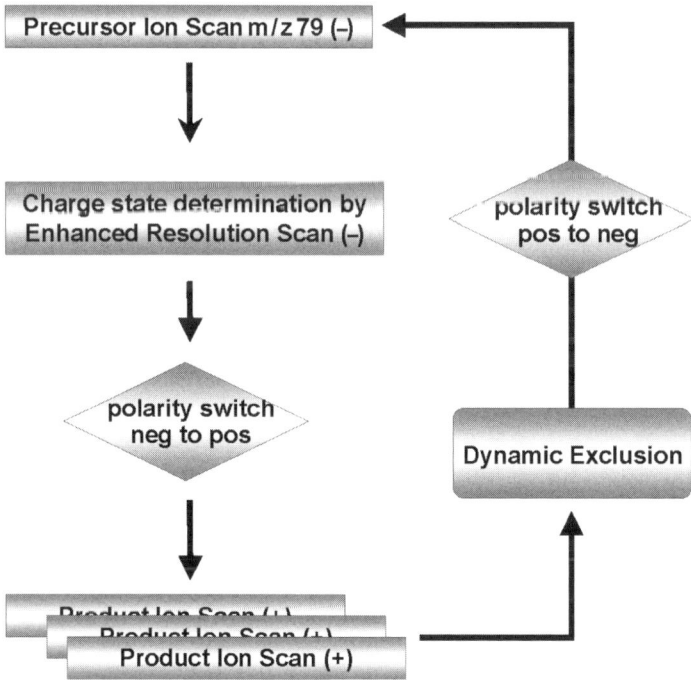

Fig. 3.2 Scan cycle for the selective detection and sequence determination of phosphopeptides on hybrid triple quadrupole/linear ion trap mass spectrometers. If a precursor has been selected for MS/MS in 2–3 consecutive cycles, it is excluded from further selection to allow the analysis to focus on other, lower-abundance precursors (dynamic exclusion)

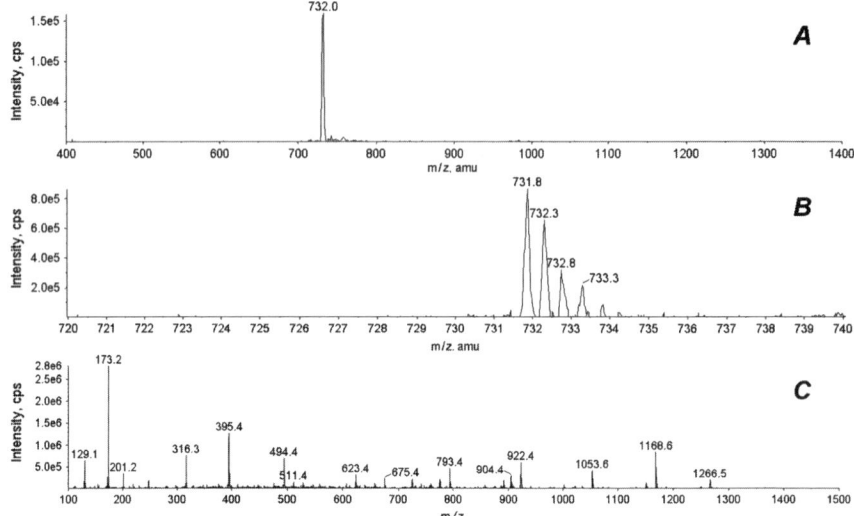

Fig. 3.3 Evaluation of the analytical system using a standard trypsin digest of bovine casein α at an amount of 100 fmol injected on column. The traces show the different experiments at the retention time of the phosphopeptide TVDMEpSTEVFTK (MW 1465.61 Da): (**A**) Precursor ion scan for *m/z* 79 in negative mode; (**B**) High resolution ion trap MS Scan of the detected [M-2H]²⁻ in negative mode; (**C**) MS/MS spectrum of [M+2H]²⁺ in positive mode

6. Once system performance has been established, inject approximately 1 pmol of the digested protein sample dissolved in 10 μL of the Loading solvent.

A hybrid triple quadrupole/linear ion trap mass spectrometer (4000 Q TRAP LC-MS/MS System, Applied Biosystems) coupled on-line with a nanoflow HPLC is used for the simultaneous detection and sequencing of phosphopeptides in complex mixtures.

Protein phosphorylation is usually substoichiometric. As a consequence an endopeptidase digest of a phosphoprotein will often contain only minor amounts of phosphopeptides. To increase the odds of detecting these compounds, an LC separation is employed to reduce the complexity of the mixture presented to the mass spectrometer at any given point in time. An additional simplification is achieved by the selective precursor ion scan for the phosphopeptide-specific fragment *m/z* 79 (PO₃⁻) in negative mode (Fig. 3.1).

As the precursor ion scan is a low-resolution experiment, it is usually not possible to determine the charge state of a detected peptide as its isotope pattern will not be resolved. The precursor ion scan is therefore followed by a higher-resolution ion trap MS experiment that allows for determination of the charge state of the precursor ion. From the *m/z* and the charge state the molecular weight of the phosphopeptide can be determined to an accuracy of 0.3 Da or better. This scan is usually carried out in negative mode (*see* **Note 3**).

If the two first scan events show one or multiple precursors that meet specific criteria (minimum signal intensity, MW range, charge state), these are selected for

MS/MS analysis to establish their sequence. After 2–3 occurrences, these precursor *m/z* values are excluded from selection for 60 s to focus on other, less abundant precursors eluting at the same time (Dynamic Exclusion, Fig. 3.2).

As MS/MS fragmentation of peptides in positive mode is better understood than in negative mode, the polarity has to be switched twice during the LC-MS/MS cycle: from negative to positive after precursors have been selected for MS/MS, and back to negative for the precursor ion scan of the following experiment. Switching polarity during a nanoLC-MS/MS experiment requires careful optimisation of the ion source parameters for both positive and negative mode. The performance of this experiment should always be tested using a known phosphopeptide standard before the first unknown sample is injected. Figure 3.3 shows a standard analysis of a 100 fmol injection of a casein α tryptic digest. At a retention time of 36.2 min, the precursor ion scan shows a single signal at *m/z* 732.0 in negative mode (Fig. 3.3A). The higher-resolution ion trap scan in negative mode (Fig. 3.3B) shows the isotope pattern of a doubly charged peptide precursor $[M-2H]^{2-}$, with the monoisotopic peak at *m/z* 731.8 indicating a molecular weight of 1,465.6 Da. The corresponding $[M+2H]^{2+}$ at *m/z* 733.8 is then selected for MS/MS in positive mode (Fig. 3.3C). Both the molecular weight and the MS/MS data unambiguously identify the peptide sequence as TVDMEpSTEVFTK from bovine casein α.

The results obtained on the standard sample should be examined for the following criteria: (i) intensity and signal-to-noise ratio in precursor ion scan mode, (ii) correct determination of the charge state from the linear ion trap scan, and (iii) high-quality MS/MS spectra that allow for the correct sequence assignment of the phosphopeptides at reasonably low amounts on column, preferably 100 fmol or less.

3.3   Data Interpretation

1. Open the LC-MS/MS data file generated from the sample. Select the Precursor Ion Scan data. Extract and print out a base peak chromatogram (BPC) annotated with retention time. Peaks in the precursor ion scan BPC will represent possible phosphopeptides.
2. Extract peak lists for all MS/MS spectra generated in the run and submit them for a combined database search against a protein database. If possible use a small dedicated database that just contains the protein sequence(s) of interest. Adjust mass tolerances for MS and MS/MS to reflect values typically achieved in routine analysis, e.g., 0.3 Da for the linear ion trap instrument used. Use phosphorylation of and neutral loss from S, T, and Y residues as variable modification filters. As phosphorylation of residues close to arginine or lysine frequently causes miscleavage of trypsin, allow for at least two missed cleavages in the database search.
3. Closely examine the database search results. Validate every tentative phosphopeptide assignment by matching theoretically calculated peptide fragment patterns to the raw MS/MS data. In case there are multiple possible phosphorylation sites in

a peptide sequence, compare the patterns for every possible site population as the database search engine can misassign the residue (*see* **Note 4**).

4. Manually look through the data to find high-quality MS/MS spectra that are not explained by the database search results. Look for MS/MS spectra that show neutral loss peaks at a distance of $-98/z$ (z = charge state) from the precursor. Compare the precursor m/z values of these MS/MS to a list of theoretically generated phosphopeptide m/z values. In case of a close match between theoretical and observed precursor m/z values, annotate the spectrum with the fragment pattern generated for this putative phosphopeptide sequence.

Figure 3.4 shows the sequence of the intracellular loop of acetylcholine receptor (δ-subunit from *Torpedo californica*) heterologously expressed in *Escherichia coli* *(12)*. The protein phosphorylated in vitro using protein kinase A was kindly provided by Dr.Viktoria Kukhtina, Berlin. After phoshorylation the molecular weight was determined by MALDI-ToF to be 17,633 Da (theor. 17,552 Da), indicating within experimental error at least one phosphorylation event (Δ 81 Da, theor. 80 Da).

After digestion and LC-MS/MS analysis with precursor ion scan detection, the base peak chromatogram depicted in Figure 3.5 (bottom trace) was obtained. A Mascot search against the SwissProt database indicated the presence of 5 phosphopeptides **A–E**, corresponding to 2 adjacent phosphorylation sites of serine residues 50 and 51, respectively. The higher number of peptides results from the observation of singly and doubly phosphorylated peptides, as well as missed cleavages from the trypsin digestion.

Figure 3.6 shows the MS/MS spectrum of the precursor m/z 1,119.0^{2+}, in chromatography peak **C**, corresponding to the sequence RSSSVGYISKAQEYFNIK doubly phosphorylated. In addition to fragment ions that can be assigned to neutral loss of the 2 phosphogroups ([M+2H-H$_3$PO$_4$]$^{2+}$ m/z 1070.3, [M+2H-2H$_3$PO$_4$]$^{2+}$ m/z 1021.6), a significant number of sequence-specific fragment ions is observed. The spectrum is labeled with the sequence RSpSpSVGYISKAQEYFNIK (aa 48–65) that gives the most comprehensive explanation of the fragments observed. While the b2 ion indicates that Serine 49 is not phosphorylated, both the b3 and b4 ions are observed in their phosphorylated and dephosphorylated states, indicating that the protein is indeed phosphorylated at residues 50 and 51. Alternative sequences like RpSSpSVGYISKAQEYFNIK or RpSpSSVGYISKAQEYFNIK do not explain the experimental data as consistently.

10	20	30	40	50
MHFRTPSTHV	LSTRVKQIFL	EKLPRILHMS	RADESEQPDW	QNDLKLRRSS

60	70	80	90	100
SVGYISKAQE	YFNIKSRSEL	MFEKQSERHG	LVPRVTPRIG	FGNNNENIAA

110	120	130	140	
SDQLHDEIKS	GIDSTNYIVK	QIKEKNAYDE	EVGNWNLVGQ	TIDR

Fig. 3.4 Amino-acid sequence of the protein analyzed here as an example (Figs. 3.5 and 3.6), the heterologously expressed intracellular loop of the δ-subunit of the acetylcholine receptor from *Torpedo californica (12)*

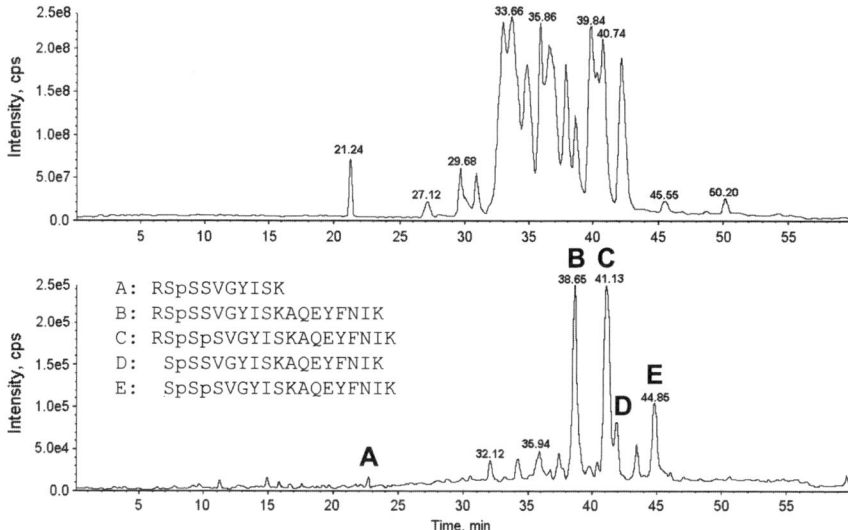

Fig. 3.5 The Precursor Ion Scan, a highly selective filter for the detection of phosphopeptides—comparison of basepeak chromatograms using regular linear ion trap MS detection (top trace) and precursor ion scan *m/z* 79 detection (bottom trace) of a tryptic digest of the acetylcholine receptor intracellular loop. The precursor ion scan base peak chromatogram is annotated with the phosphopeptide sequences identified by database searching

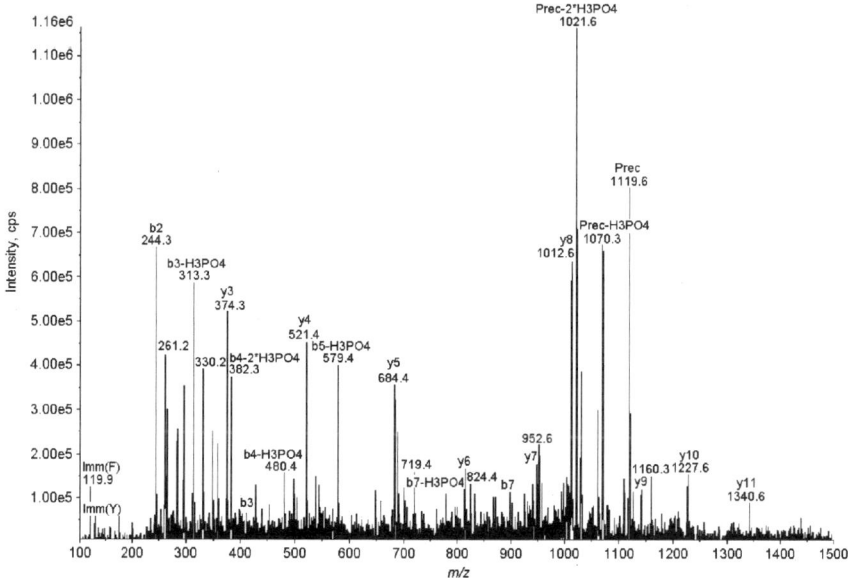

Fig. 3.6 MS/MS spectrum of the phosphopeptide precursor *m/z* 1,119.0^{2+} at a retention time of 41.1 min (Fig. 3.5, peak **C**). The spectrum is labeled with fragments calculated for the assigned sequence RSpSpSVGYISKAQEYFNIK. The ions of the lower b series indicate phosphorylation on S50 and S51, but not S49

The method described here is a significant improvement in the analysis of protein phosphorylation. It should be noted, though, that it will not always yield comprehensive results on all sample types. There are still multiple stages where problems can occur: (i) incomplete digestion of the protein around the site of modification, (ii) poor extraction of the phosphopeptide from the gel, (iii) loss of e.g., highly polar phosphopeptides during reverse-phase chromatography, (iv) failure of the phosphopeptide to produce sufficient signal response in MS and (v) inability of the database searching algorithm to assign the sequence and site of phosphorylation because of nonconclusive fragmentation. As a consequence, the method should be complemented by other strategies such as phosphopeptide enrichment *(3)*, off-line analysis or targeted LC-MS/MS analysis *(13,14)* wherever possible.

4 Notes

1. The method described here uses precursor ion scanning in negative-ion mode for the selective detection of phosphopeptides. Other methods have been described that use the neutral loss of phosphoric acid (H3PO4, −98 Da) in positive mode for this purpose *(15)*. Although, e.g., Constant Neutral Loss Scans on a triple quadrupole instrument in positive mode are easier to perform experimentally, they suffer from several shortcomings:

   – not all phosphopeptides exhibit a strong neutral loss fragmentation when collisionally activated in positive mode *(16)*. As a consequence, especially peptides containing phosphotyrosine are usually not detectable by neutral loss methods.
   – as the mass spectrometer analyses *m/z* (not MW), the actual neutral loss observed is dependent on the charge state, e.g., 98.0/2 = 49.0 or 98.0/3 = 32.7. This has to be accounted for when setting up the analysis.
   – Constant Neutral Loss experiments for, e.g., *m/z* 49 exhibit a high level of false positives, e.g., from sulfopeptides, iodoacetamide-methionine containing peptides *(17)* and to some degree from random tryptic peptides that show singly charged fragments at an *m/z* of [Prec-98/z]$^{z+}$.

   Another approach for the analysis of tyrosine phosphorylation is a precursor ion scan experiment targeted at the detection of the phosphotyrosine immonium ion at *m/z* 216 *(18)*. Although excellent selectivity can be achieved using high mass accuracy on the fragment ion, this experiment does not detect phosphoserine- and phosphothreonine-containing peptides.

   It should be noted in this context that the Neutral Loss/MS^3-based methods frequently described for ion trap mass spectrometers *(19)* are not selective experiments at all. In this approach peptides that are detected by regular MS and exhibit a strong neutral loss fragment at 98/z in MS/MS are selected for further analysis by MS^3. If the peptide has not been selected for MS/MS in the first place, however, the neutral loss will go undetected.

As a consequence, the experimentally more demanding Precursor Ion Scan method in negative-ion mode is the only approach that currently offers generic detection of different types of unknown phosphopeptides.

2. A more promising approach that utilizes neutral loss fragmentations in positive mode is the recently published MIDAS (MRM-initiated detection and sequencing) workflow *(13,14)*. As a targeted approach, however, it relies on accurate information about the protein sequence and the quality of the protein digestion.

3. On specific peptide sequences charge-state shifts have been observed between positive and negative mode, i.e., a phosphopeptide could have 2- as the most abundant charge state in negative mode, but 3+ as the most abundant charge state in positive mode *(4)*. This shift is dependent on the peptide sequence (number of basic residues), chromatography conditions (pH value) and instrument parameters (e.g., the declustering potential adjusted on the interface skimmer). As a consequence, it is advisable to perform the resolving ion trap MS scan in negative mode. Even if the positive mode MS/MS does not give conclusive results, the negative mode MS will at least allow to accurately determine the peptide's charge state and molecular weight. A second experiment targeted at different charge states of this peptide can then be used to obtain conclusive results.

4. A peptide sequence containing multiple possible sites of phosphorylation generates a set of multiple possible phosphopeptides that differ only in the residue actually modified. These regioisomers possess very similar physicochemical properties and are often not separated by reverse-phase chromatography. As a consequence, one should always account for the possibility of mixed MS/MS spectra, where the regiosisomer precursors are isolated and fragmented together as they do not separate in *m/z* or retention time. Each MS/MS spectrum should be carefully examined and all possible sequence hypotheses tested by annotating raw data to make a confident assignment of the residue actually modified.

References

1. Hunter, T. (1995) Protein kinases and phosphatases: the yin and yang of protein phosphorylation and signaling. *Cell* **80**, 225–236.
2. Cohen, P. (2002) The origins of protein phosphorylation. *Nat Cell Biol.* **4**, E127–30.
3. Reinders, J. and Sickmann, A. (2005) State-of-the-art in phosphoproteomics. *Proteomics* **5**, 4052–4061.
4. Steen, H., Jebanathirajah, J. A., Rush, J., Morrice, N. and Kirschner, M. W. (2006) Phosphorylation analysis by mass spectrometry: myths, facts and the consequences for qualitative and quantitative measurements. *Mol. Cell. Proteomics* **5**, 172–181.
5. Annan, R. S., Huddleston, M. J., Verma, R., Deshaies, R. J. and Carr, S. A. (2001) A multidimensional electrospray ms-based approach to phosphopeptide mapping. *Anal. Chem.* **73**, 393–404.
6. Zappacosta, F., Huddleston, M. J., Karcher, R. L., Gelfand, V. I., Carr, S. A. and Annan, R. S. (2002) Improved sensitivity for phosphopeptide mapping using capillary column hplc and microionspray mass spectrometry: comparative phosphorylation site mapping from gel-derived proteins. *Anal. Chem.* **74**, 3221–3231.

7. Steen, H., Küster, B. and Mann, M. (2001) Quadrupole time-of-flight versus triple-quadrupole mass spectrometry for the determination of phosphopeptides by precursor ion scanning. *J. Mass Spectrom.* **36**, 782–790.

8. Schlosser, A., Pipkorn, R., Bossemeyer, D. and Lehmann, W. D. (2001) Analysis of protein phosphorylation by a combination of elastase digestion and neutral loss tandem mass spectrometry. *Anal. Chem.* **73**, 170–176.

9. Hager, J. W. (2002) A new linear ion trap mass spectrometer. *Rapid Commun. Mass Spectrom.* **16**, 512–526.

10. Le Blanc, J. C. Y, Hager, J. W., Illisiu, A. M. P., Hunter, C., Zhong, F. and Chu, I. (2003) Unique scanning capabilities of a new hybrid linear ion trap mass spectrometer (Q TRAP) used for high sensitivity proteomics applications. *Proteomics* **3**, 859–869.

11. Williamson, B. F., Marchese, J. and Morrice, N. A., (2006) Automated identification and quantification of protein phosphorylation sites by lc/ms on a hybrid triple quadrupole linear ion trap mass spectrometer. *Mol. Cell. Proteomics* **5**, 337–346.

12. Kottwitz, D., Kukhtina, V., Dergousova, N., Alexeev, T., Utkin, Y., Tsetln, V. and Hucho, F. (2004) Intracellular domains of the δ-subunits of *Torpedo* and rat acetylcholine receptors— expression, purification, and characterization. *Protein Expr. Purif.* **38**, 237–247.

13. Cox, D. M., Zhong, F., Du, M., Duchoslav, E., Sakuma, T. and McDermott, J. C. (2005) Multiple reaction monitoring as a method for identifying protein posttranslational modifications. *J. Biomol. Tech.* **16**, 83–90.

14. Unwin, R. D., Griffiths, J. R., Leverentz, M. K. Grallert, A., Hagan, I. M. and Whetton, A. D. (2005) Multiple reaction monitoring to identify sites of protein phosphorylation with high sensitivity. *Mol. Cell. Proteomics* **4**, 1134–1144.

15. Covey, T., Shushan, B., Bonner, R., Schröder, W. and Hucho, F. (1991) In *Methods in Protein Sequence Analysis* (Jörnvall, H., Höög, J.-O. and Gustavsson, A.-M., eds.) Birkhäuser Verlag, Basel, Switzerland, pp. 249–256.

16. DeGnore, J. P. and Qin, J. (1998) Fragmentation of phosphopeptides in an ion trap mass spectrometer. *J. Am. Chem. Soc. Mass Spectrom.* **9**, 1175–1188.

17. Krüger, R., Hung, Ch.-W., Edelson-Averbukh, M. and Lehmann, W. D. (2005) Iodoacetamide-alkylated methionine can mimic neutral loss of phosphoric acid from phosphopeptides as exemplified by nano-electrospray ionisation quadrupole time-of-flight parent ion scanning. *Rapid Commun. Mass Spectrom.* **19**, 1709–1716.

18. Steen, H., Küster, B, Fernandez, M., Pandey, A. and Mann, M. (2001). Detection of tyrosine phosphorylated peptides by precursor ion scanning quadrupole tof mass spectrometry in positive ion mode. *Anal. Chem.* **73**, 1440–1448.

19. Schroeder, M. J., Shabanowitz, J., Schwartz, J. C., Hunt, D. F. and Coon, J. J. (2004) A neutral loss activation method for improved phosphopeptide sequence analysis by quadrupole ion trap mass spectrometry. *Anal. Chem.* **76**, 3590–3598.

4
Analysis of Tyrosine-*O*-Sulfation

Jens R. Bundgaard, Jette W. Sen, Anders H. Johnsen, and Jens F. Rehfeld

Summary Tyrosine *O*-sulfation was first described about 50 years ago as a post-translational modification of fibrinogen. In the following 30 years it was considered to be a rare modification affecting only a few proteins and peptides. However, in the beginning of the 1980s tyrosine (Tyr) sulfation was shown to be a common modification and since then an increasing number of proteins have been identified as sulfated. The target proteins belong to the classes of secretory, plasma membrane, and lysosomal proteins, which reflects the intracellular localization of the enzymes catalyzing Tyr sulfation, the tyrosylprotein sulfotransferases (TPSTs).

Traditionally, Tyr sulfation has been analyzed by incorporation of radiolabeled sulfate into target cells followed by purification of the target protein. Subsequently, the protein is degraded enzymatically or by alkaline hydrolysis followed by thin-layer electrophoresis to demonstrate the presence of radioactively labeled tyrosine. These techniques have been described in detail previously. The aim of this chapter is to present alternative analytical methods of Tyr sulfation than radioisotope incorporation before analysis.

Key Words Tyrosine *O*-sulfation; tyrosylprotein sulfotransferases; radio-immunoassay; MALDI-TOF.

1 Introduction

Tyrosine *O*-sulfation was first described about 50 years ago as a post-translational modification of fibrinogen *(1)*. In the following thirty years it was considered to be a rare modification affecting only a few proteins and peptides. However, in the beginning of the 1980s tyrosine (Tyr) sulfation was shown to be a common modification and since then an increasing number of proteins have been identified as sulfated. The target proteins belong to the classes of secretory, plasma membrane, and lysosomal proteins, which reflects the intracellular localization of the enzymes catalyzing Tyr sulfation, the tyrosylprotein sulfotransferases (TPSTs). TPSTs are

From: *Post-translational Modifications of Proteins.*
Methods in Molecular Biology, Vol. 446.
Edited by: C. Kannicht © Humana Press, Totowa, NJ

type II integral membrane glycoproteins that reside in the *trans* Golgi network and use adenosine 3'-phosphate 5'-phosphosulfate (PAPS) as a cosubstrate and sulfate donor. In higher organisms, 2 enzymes have been identified and cloned and they appear to be ubiquitously expressed in tissues *(2–4)*. Moreover, homologous genes are present in lower eukaryotes as well as in the nematode *Caenorhabditis elegans*, fruit fly, and zebrafish. Both TPST-1 and TPST-2 knock-out mice models have been established and both display distinct phenotypes. The TPST-1 knock out mice are slightly smaller than wild-type mice and the female produces smaller litters. The TPST-2 knock-out mouse has delayed growth but reaches normal weight at 10 wk. Strikingly, the TPST-2 knock-out male is infertile *(5,6)*.

The biological function of Tyr sulfation is modulation of protein-protein interactions. For instance, Tyr sulfation is necessary for binding of the peptide hormone cholecystokinin (CCK) to the CCK-1 receptor *(7)*; it is critical for interactions between p-selectin and PSGL-1 (P-Selectin Glycoprotein Ligand-l) *(8)*; between hirudin and thrombin *(9)*; and it modulates the ability of HIV −1 virus to enter cells via the chemokin receptor CCR5 and CD4 *(10)*. In addition, sulfation has been reported to modulate proteolytic processing *(11)* and secretion rates *(12)* of secretory proteins. From these functions, it follows that it is not only important to detect Tyr sulfation, but also to identify the sulfation site to establish effects on the activity on the protein. Attempts have been made to determine a consensus site for tyrosine sulfation and generate algorithms that predict sulfation sites *(13–15)*, one of these is the "sulfinator," which is available at www.expasy.org. However, it is known that the sulfinator at best gives an indication of possible sulfation sites, and it has been shown that nontypical structures, which are not identified by the sulfinator, can be sulfated *(16,17)*.

Traditionally, Tyr sulfation has been analyzed by incorporation of radio labeled sulfate into target cells followed by purification of the target protein. Subsequently, the protein is degraded enzymatically or by alkaline hydrolysis followed by thin-layer electrophoresis to demonstrate the presence of radio-actively labeled tyrosine. These techniques have been described in detail previously *(18)*. The aim of this chapter is to present alternative analytical methods of Tyr sulfation other than radioisotope incorporation before analysis. Thus, sulfation of protein or peptide fragments can be demonstrated by a combination of chromatography and specific assays. This has several advantages. First, depending on the assay used, the analysis can be quantitative. Second, problems with incorporation of isotope and the high degree of incorporation of radiolabeled sulfate into carbohydrate moieties that might interfere with detection, can be avoided. In addition, it allows two ways of identification of sulfated tyrosines depending on purity of the protein. One strategy demands purified protein using mass spectrometry or chromatography, whereas another uses Tyr sulfate sensitive antisera to distinguish between sulfated and nonsulfated forms in a crude extract. We provide protocols for establishment of sulfation specific antisera and give examples of chromatographic systems that are useful for analysis of Tyr sulfation.

Establishment of a specific immunoassay is costly and time-consuming and initiation of such a process demand an interest exceeding the mere confirmation of the modification. If such an assay is desirable, a number of considerations have to be taken into account. The size of an antibody binding site is complementary to that of a peptide ligand (or epitope) of 4–6 amino acid residues *(19)*. Within such an epitope the exact structure of each residue profoundly influences the binding of the antibody *(20)*. Moreover, the protein or peptide structures surrounding the particular epitope also modulate binding affinity considerably *(21)*. Accordingly, it is not surprising that attempts to raise antibodies that specifically bind a single *O*-sulfated Tyr residue as such—irrespective of neighboring structures—have failed (Hakanson, R. and Huttner, W. B. personal communications, and our own results). However, recently a promising monoclonal antibody was reported, which was raised against tyrosine sulfated P-selectin glycoprotein ligand-1. It has tyrosine sulfate specificity and show a wide tolerance to sequence variance around the sulfated tyrosine. Nevertheless, it does not bind sulfotyrosine alone and is thus not sequence independent *(22)*.

The discrepancy with the success to raise antibodies against the structurally similar phosphorylated Tyr residues *(23)* is at present unexplained. In accordance with the aforementioned consideration and with prevailing immunochemical theories, antibodies that recognize sulfation of a given Tyr residue consequently have to be raised against the entire epitope containing the sulfated tyrosine, i.e., a peptide sequence of approximately 5 or 6 amino acid residues. It has, however, proven difficult to raise high-titer and high-affinity antibodies or antisera against peptides of less than 10 residues *(24)*. Therefore, a peptide of approximately 10 residues containing the desired epitope in an otherwise expedient sequence should be designed, synthesized, and used for immunization. Hence, measurement of Tyr sulfation by immunochemical methods requires precise knowledge of the sequence and structure surrounding the O-sulfated residue in a peptide or protein. It is necessary to consider different strategies in the design of appropriate radioimmunoassays (RIA) (*see* **Note 1**).

Sulfation of a protein or peptide introduces a highly acidic group that alters its local physical and chemical properties such as conformation, hydrophilicity, and pK-value. This can be used to separate sulfated and nonsulfated forms, by various chromatographic techniques, e.g., ion-exchange chromatography and reverse-phase high-performance liquid chromatography (RP-HPLC) at near neutral pH under which the conditions for ionization of the sulfate group induces the maximum difference between the two forms (*see* **Note 2**). Also, removal of the sulfate group by digestion with arylsulfatase combined with chromatographic analysis of the peptide before and after the digestion can reveal the presence of Tyr sulfate. Often, identification of sulfated Tyr within a protein requires digestion of the protein into smaller peptides followed by separation of the fragments and isolation of relevant peptide. A number of proteinases may be used, for example trypsin or the endoproteinases Lys-C, Glu-C, and Asp- N, the choice being partly dependent on the individual protein. The details for these techniques are beyond the scope of the present chapter, but may be found in other volumes of this series. These analytical methods can profit from the availability of specific antisera or be carried out using pure proteins.

Having a purified protein offers alternative ways of analysis of Tyr sulfation. The sulfate ester bond is generally considered as labile and particularly susceptible to acidic hydrolysis. Despite of this, the tyrosine sulfate can generally withstand pH in the range of 1–3, often used during many standard analyses such as reversed phase chromatography and mass spectrometry, as long as the temperature is kept at room temperature or below and the exposure time is limited (25). However, sulfated Tyr can not be identified by protein sequence analysis because of the highly acidic environment and elevated temperatures during the analysis. Instead, non-derivatized Tyr will result. Likewise, the standard acidic hydrolysis used for amino acid analysis is not feasible; instead, demonstration of the presence of intact tyrosin sulfate in a peptide was originally obtained by basic hydrolysis before amino acid analysis (26). This technique has a limited sensitivity because of an inevitable background. Alternatively, nondestructive digestion can be obtained by aminopeptidase M or a combination of aminopeptidase M and prolidase followed by amino acid analysis (27). However, these techniques are not used much anymore. Mass spectrometry (MS) is now the preferred tool of analysis of many post-translational modifications. MS in combination with protein sequence analysis can disclose the presence of a sulfated Tyr. Thus, if the identified sequence contains Tyr and the measured mass is 80 Da higher than the calculated value, it is a good indication that the Tyr is sulfated. However, phosphorylation also results in an 80 Da increase of the molecular mass (see **Note 3**; Fig. 4.1). It should also be noted that the sulfate group is lost during ionization in both matrix assisted laser desorption/ionization (MALDI) and to some degree in electrospray ionization (ESI), resulting in a false-negative result. Using MALDI time-of-flight (TOF) MS, the loss of sulfate can be turned into a diagnostic advantage because the loss is often complete in the positive mode (31) and limited in negative mode. Hence, comparison of spectra obtained in positive and negative linear mode of the same sample will show a vast difference in the yield of the sulfated and nonsulfated species (see **Note 4**), whereas the picture will be more complex in the reflector mode (see **Note 5**; Fig. 4.2). In experiments using hybrid instruments of the quadrupole-TOF type it is possible to see fractions of the sulfated species in positive mode even though the desulfated species is generally dominating. The fact that tyrosine sulfate is unstable even at standard collision energy in a quadrupole-TOF instrument, can be used in identifying new sulfated targets by using neutral loss scan experiment detecting the loss of 79.957 (32). Though tyrosine sulfate and tyrosine phosphate has almost identical mass, it is possible to distinguish between the two as the tyrosine sulfate is generally much more unstable than the tyrosine phosphate, which is not nearly as prone to neutral loss as the tyrosine sulfate. Furthermore when performing MS/MS experiments the tyrosine phosphate gives rise to a stable immonium ion of 216.043 whereas the immonium ion of tyrosine sulfate is not detected. This can be used to further confirm the identification of a phosphate or sulfate. Another option is to investigate tyrosine sulfation by a mass spectrometer capable of electron capture dissociation (ECD). This is a fragmentation method that, to a higher degree, maintains the sulfate during peptide backbone fragmentation but requires specialized and expensive instrumentation (33,34).

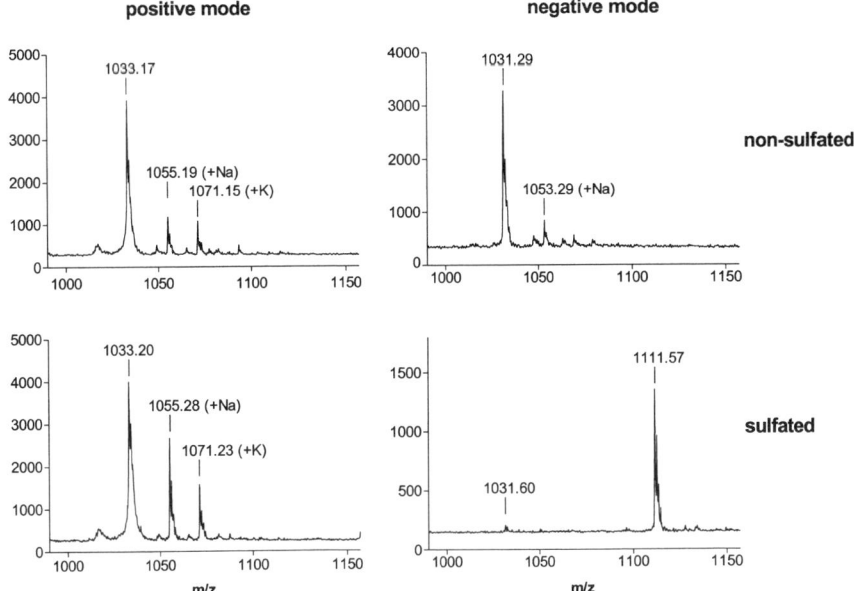

Fig. 4.1 Mass spectra obtained in the positive and negative mode in a mass spectrometer equipped with delayed extraction (Bruker Daltonik Biflex). As model peptides were used non-sulfated and sulfated shark gastrin-8 (Asp-Tyr(SO3)-Thr-Gly-Trp-Met-Asp-Phe-NH2) with the theoretical monoisotopic molecular masses 1032.40 and 1112.36 Da, respectively. In the positive mode the measured *m/z* values represent the protonated molecular ion (+1.01) and to a certain degree the Na$^+$ and K$^+$ adducts, whereas the deprotonated form (−1.01) is recorded in the negative mode. In this example the intact sulfated peptide is only observed in the negative mode together with a small fraction of the desulfated form. See however also **Note 3**. *y*-axis: number of collected ions

2   Materials

2.1   Generation of Tyr Sulfation Sensitive Radioimmunoassays

2.1.1   Preparation of Immunogen (Common for All 3 Coupling Methods; *see* Note 6)

1. 5 mg sulfated peptide.
2. *N, N*-dimethylformamide.
3. 25 mg bovine serum albumin.
4. 0.05*M* sodium phosphate buffer, pH 7.5
5. For carbodiimide-coupling: 1-ethyl-3-(3-dimethylaminopropyl) carbodiimide HCl.
6. For glutaraldehyde-coupling: 500 g/L glutaraldehyde.

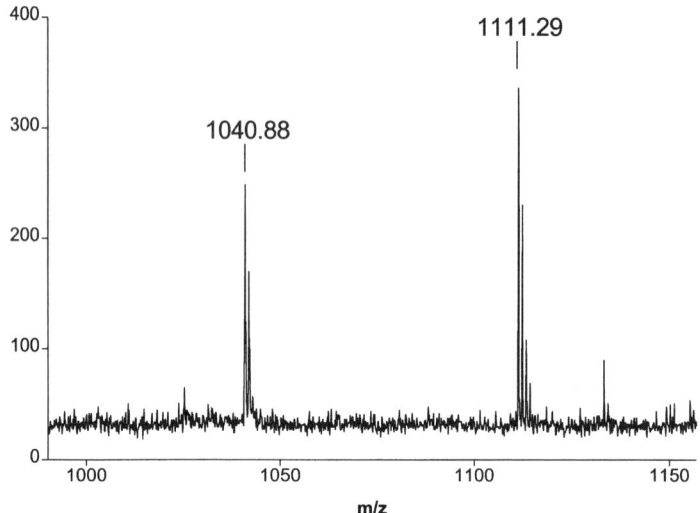

Fig. 4.2 Mass spectrum of sulfated shark gastrin-8 (*see* Fig. 4.1) obtained in the negative reflected mode using α-cyano-4-hydroxycinnamic acid as matrix. The peak at *m/z* 1111.29 represents the intact peptide (theoretical value 1112.36 −1.01) whereas the peak at 1040.88 represents the in-flight desulfated peptide. The difference between the two is considerably less than the expected 80 (*see* **Note 5**). Note that the resolution is higher in the reflected mode than in the linear mode (Fig. 4.1)

7. One Sephadex G-10 column with a fraction collector.
8. For maleimidobenzoyl-succinimede ester-coupling:
9. 25 mg *m*-maleimidobenzoyl-N-hydroxysuccinimide ester (MBS).
10. One Sephadex G-25 column with a fraction collector.

2.1.2 Immunization:

1. 6–8 rabbits.
2. 8.5 g/L saline buffer.
3. Freund's complete adjuvant.
4. Freund's incomplete adjuvant.

2.1.3 Preparation of Tracer (*see* Note 7; Fig. 4.3):

1. Synthetic peptide.
2. HPLC apparatus with reverse-phase column (Aquapore C-8 column, RP-300, 220 × 4.6 mm, 7-μm bead size) and fraction collector.
3. ethanol or acetonitrile (HPLC grade).
4. 1.0 mL/L trifluoroacetic acid (TFA; HPLC grade in water).

IMMUNOGEN

N-terminal tyrosyl sulfation

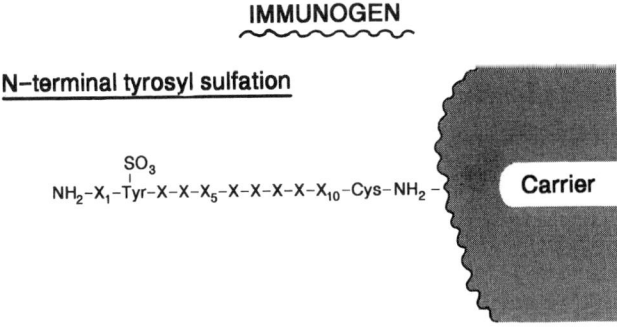

$$\overset{\displaystyle SO_3}{\underset{\displaystyle |}{}}$$
NH$_2$-X$_1$-Tyr-X-X-X$_5$-X-X-X-X-X$_{10}$-Cys-NH$_2$ –

Carrier

C-terminal tyrosylsulfation

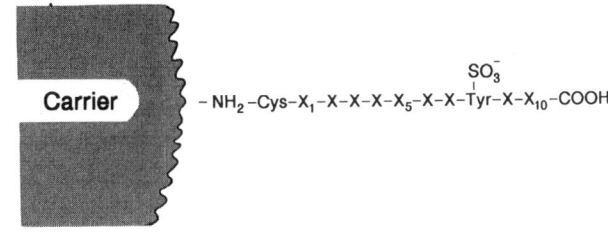

Carrier

$$\overset{\displaystyle SO_3^-}{\underset{\displaystyle |}{}}$$
– NH$_2$–Cys–X$_1$–X–X–X–X$_5$–X–X–Tyr–X–X$_{10}$–COOH

TRACER

N-terminal directed antibodies

$$\overset{\displaystyle SO_3^-}{\underset{\displaystyle |}{}} \qquad \overset{\displaystyle 125}{\underset{\displaystyle |}{}}$$
NH$_2$ X$_1$ Tyr X–X–X$_5$–X–X–X–X–X$_{10}$–Tyr

C-terminal directed antibodies

$$\overset{\displaystyle 125}{\underset{\displaystyle |}{}} \qquad\qquad\qquad \overset{\displaystyle SO_3^-}{\underset{\displaystyle |}{}}$$
Tyr–X$_1$–X–X–X–X$_5$–X–X–Tyr–X–X$_{10}$–COOH

Fig. 4.3 The structure of immunogens used for production of antibodies directed against tyrosyl *O*-sulfated peptides (upper part of the figure), and the structure of corresponding radioactive tracers (lower part of the figure)

5. 0.02*M* Barbital buffer, pH 8.4.
6. Iodination using chloramin T: I^{125} (Amersham IMS 30, specific activity 16.85 mCi/μg of iodine) 0.05*M* phosphate buffer, pH 7.5.
7. Disodium sulfite (Na$_2$SO$_3$).
8. Chloramin T.
9. Iodination using Bolton Hunter-reagent: Bolton and Hunter reagent for protein iodination (Amersham IM 5861).
10. 0.2*M* sodium borate buffer, pH 8.5 (mixture of 0.2*M* Na$_2$B$_4$O$_7$ and 0.2*M* H$_3$BO$_3$ with pH adjusted).

11. 0.1 *M* sodium borate buffer with 0.2*M* glycine, pH 8.5 (prepared from the buffer above).

2.1.4 RIA Procedure

1. Synthetic peptide for standard (*see* **Note 8**).
2. disposable plastic tubes.
3. 0.02*M* barbital buffer, pH 8.4, containing 1 g/L bovine serum albumin.
4. Activated charcoal.
5. Blood plasma (This is a mixture of buffer and outdated human plasma from Blood Banks in 0.02*M* sodium phosphate, pH 7.4. The optimal concentration normally varies between 10–50% plasma).
6. γ-scintillation counter.

2.2 Analysis of Tyr Sulfation Using FPLC Based Ion-Exchange Chromatography (see Note 9)

1. An FPLC system equipped with an ion-exchange column, e.g., an 5/5HR MonoQ anion-exchange column (Pharmacia) and a fraction collector.
2. Buffer A: 50 m*M* Tris-HCL, pH 8.2 added 10% (v/v) acetonitrile.
3. Buffer B: 50 m*M* Tris-HCL, pH 8.2 added 10% (v/v) acetonitrile and 1*M* NaCl. Both buffers are filtered through a 0.45 μm filter, degassed and stored at 4°C where it is stable for months.

2.3 Analysis of Tyr Sulfation Using Reverse Phase HPLC

1. An HPLC instrument with gradient formation capability and equipped with a suitable RP (reversed phase) column (C8 or C18).
2. Solvents: Stock solution of 0.5*M* ammonium acetate in water, filtered through 0.45 μm filter (can be kept for several months at 4°C). Solvents A and B are prepared by addition of 20 mL/L of H_2O and acetonitrile, respectively (both HPLC grade), to form 10*M* ammonium acetate.

2.4 Analysis of Tyr Sulfation Using Mass Spectrometry

2.4.1 For MALDI-TOF

1. Instrument: A mass spectrometer for MALDI-TOF MS.
2. Matrix solution: Prepare a 20 mg/mL solution of α-cyano-4-hydroxycinnamic acid in 30% (v/v) acetonitrile/0.1% (v/v) TFA in an Eppendorf tube and

centrifuge; use the supernatant as matrix solution. Alternatively 2,5-dihydroxy-benzoic acid (DHB) 20 mg/mL in 50% Acetonitrile/1% phosphoric acid can be used (*see* **Note 10**). For larger peptides 3,5-dimethoxy-4-hydroxy-*trans*-cinnamic acid (sinapinic acid) can be used. Premade matrix solutions ("MALDI-grade") are also commercially available (Hewlett-Packard).

2.4.2 Neutral loss scanning on quadrupole-TOF instrument

1. Instrument: A hybrid instrument of quadrupole time-TOF type. For complex samples connected to a HPLC instrument equipped with a reversed phased column for separation of peptides.
2. Desalted sample dissolved in 30–50% Acetonitrile 0.2% formic acid for direct infusion, or sample dissolved in aqueous solution for injection on reversed phase column.
3. Solution A: 95% Acetonitrile/0.2% formic acid, solution B: 5% Acetonitrile/0.2% formic acid.

2.5 Arylsulfatase Treatment

1. Arylsulfatase type VIII (Sigma S 9754, 20–40 U/mg).
2. Acetate buffer: 0.2*M* sodium acetate, pH 5.0.
3. 0.2% Sodium chloride.

3 Methods

3.1 Generation of Tyr Sulfation Sensitive Radioimmunoassays

3.1.1 Preparation of the Immunogen

3.1.1.1 Carbodiimide-Coupling

1. Dissolve 5 mg peptide hapten in 1 mL of *N,N*-dimethylformamide.
2. Dissolve 25 mg bovine serum albumin in 2.5 mL of 0.05*M* sodium phosphate, pH 7.5.
3. Conjugate by the addition of 125 mg of 1-ethyl-3-(3-dimethylaminopropyl) carbodiimide HCl to give molar ratios between the peptide, albumin and ethyl-carbodiimide of 1:0.1:40. Mix the reagents and incubate for 20 hours at 20°C.
4. Divide the mixture into six portions and store at −20°C until immunization (stable up to one year).

3.1.1.2 Glutaraldehyde Coupling

1. Dissolve 2.5 mg of sulfated peptide hapten together with 7.5 mg bovine serum albumin in 5 mL of 0.05M sodium phosphate, pH 7.5.
2. Conjugate by dropwise addition of 100 µL of 500 g/L glutaraldehyde.
3. Mix the solution and incubate for 4 h at 20°C.
4. Apply the mixture to a calibrated Sephadex G-10 column and elute at 20°C with 0.05M sodium phosphate, pH 7.5, in fractions of 1 mL.
5. The void volume fractions containing the conjugate are then pooled, divided into 6 portions, and stored at −20°C until immunization (stable for months).

3.1.1.3 Maleimidobenzoyl Succinimede ester Coupling

1. Dissolve 5 mg sulfated peptide hapten together with 25 mg of bovine serum albumin (BSA) in 10 mL 0.05M sodium phosphate buffer (pH 7.5).
2. Dissolve 25 mg of m-maleimidobenzoyl-N-hydroxysuccinimide ester (MBS) in 1 mL N, N-dimethylformamide.
3. Conjugate by dropwise addition of the MBS-solution to the sulfated peptide/BSA solution.
4. Stir at room temperature for 30 min.
5. Free MBS is separated from the activated albumin by Sephadex G-25 gel filtration (eluted at 20°C with 0.05M sodium phosphate, pH 7.5, in fractions of 1 mL).
6. Add the sulfated and cysteinylated peptide to the MBS-activated bovine serum albumin and stir the mixture stirred for 3 h at room temperature and a pH of 7.5.

3.1.2 Immunizations

1. The first portion of the antigen is suspended in 8.5 g/L saline to a volume of 5 mL and carefully emulsified with an equal volume of Freund's complete adjuvant. We recommend emulsification by pushing the saline-adjuvant forwards and backwards between two connected 10-mL syringes until a drop of the emulsion remains coherent and condensed on a larger water surface. Randomly bred rabbits are used for immunizations in series of 6–8 rabbits each.
2. Two subcutaneous injections of the mixture are given over the hips in amounts corresponding to 40 µg hapten per animal.
3. Five or more booster injections using Freund's incomplete adjuvant (prepared as described in 1) are then administered simultaneously in all rabbits at 8-wk intervals, using one-half of the initial dose of antigen per immunization.

4. The rabbits are bled from an ear vein 10 d after immunization. Sera from the bleedings are separated and stored at −20°C.
5. The antiserum obtained is evaluated (*see* **Note 11**).

3.1.3 Preparation of Tracer for Radioimmunoassay (*see* **Note 7**).

3.1.3.1 Iodination using Chloramin T

1. Dissolve 500 µg peptide in 1 mL 0.05*M* phosphate buffer, pH 7.5. Aliquot in 10 µL portions and store at −20°C.
2. Dilute 2 mCi I^{125} (~20 µL) with 85 µL 0.05*M* phosphate buffer, pH 7.5 (final activity about 200 µCi/10 µL).
3. Dissolve 5 mg chloramin T in 10 mL 0.05*M* phosphate buffer pH 7.5.
4. Dissolve 2 mg disodium sulfite in 10 mL 0.05*M* phosphate buffer pH 7.5.
5. Mix 10 µL peptide, 10 µL I^{125} and 10 µL chloramin T gently. After 30 s the reaction is arrested with 25 µL disodium sulfite. Shake gently. If more tracer is prepared, do iodination in batches and pool the preparations.
6. Purify the tracer by HPLC. Fractions with maximal counts of reactivity is diluted with 0.02*M* barbital buffer, pH 8.4 to a final activity of 10^6 cpm/mL. Store the tracer in 500 µL aliquots at −20°C.

3.1.3.2 Iodination using the Bolton Hunter Reagent

1. Evaporate the water from the I^{125} Bolton Hunter reagent under a slow flow of nitrogen.
2. Dissolve 5 µg of peptide in 10 µL water and add 10 µL 0.2*M* sodium borate buffer.
3. Add the peptide solution to the Bolton Hunter reagent and incubate the mixture at 4°C for about 20 h.
4. Add 500 µL 0.1*M* borate buffer with 0.2M glycine. Continue directly with HPLC purification of the tracer as described above, or store at 4°C.

3.1.4 RIA Procedure

1. At room temperature, prepare the following mixtures of total volumes of 2.4 mL (RIA is carried out as an equilibrium system at pH 8.4, using 0.02*M* barbital buffer, pH 8.4, containing 1 g/L bovine serum albumin) in disposable plastic tubes: 2.0 mL of antiserum dilution, 250 µL of tracer solution (giving 1000 cpm, corresponding to 0.5 fmol of freshly prepared ^{125}I-labeled tracer-peptide), and 150 µL of standard solution or sample. All samples should be assayed in duplicate.

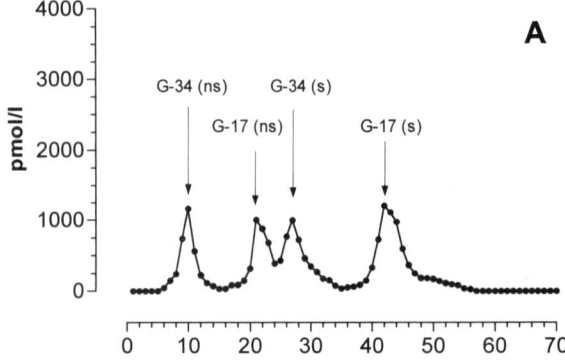

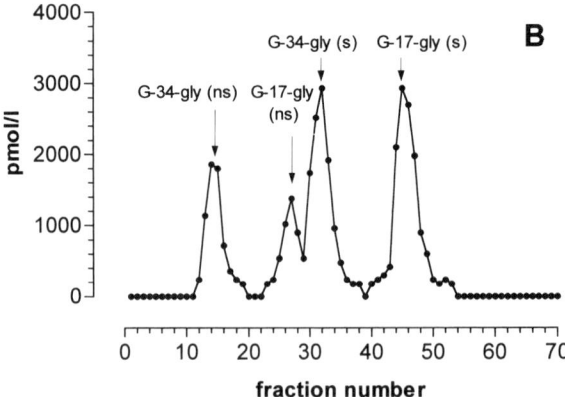

Fig. 4.4 Anion exchange chromatography using a FPLC system of the main processing forms of human gastrin heterologously expressed in a hamster β-cell line, HIT. Panel A shows carboxyamidated gastrin forms of 17 and 34 residues and demonstrates the separation of sulfated and nonsulfated forms. Panel B shows the immediate precursor of carboxyamidated gastrin, with the free C-terminus extended with a Gly. Fractions were eluted using a linear gradient of 20–55% buffer B over 60 min at a flow of 1 mL/min. Fractions were analyzed using specific radioimmunoassays. (Elution positions can be verified using synthetic peptides or by analysis of elution positions following arylsulfatase treatment.)

2. Incubate at 4°C for 2–5 days to reach equilibrium.
3. Antibody-bound (B) and free (F) tracers are separated by the addition of 0.5 mL of a suspension of 20 mg of activated charcoal and diluted blood plasma to each tube. The tubes are centrifuged for 10 min at 2000g.
4. Count the activities of the supernatant (B) and sedimented charcoal (F) in automatic γ-scintillation counters for 5 min.
5. Calculate the binding percentage as B − [(B + F) × D] / (B + F) − [(B + F) × D] × 100, where the "damage" (D) is defined as B × 100 / (B + F) in the absence of antiserum. The damage is usually 2–3%. The peptide concentration is determined from a standard curve based on known standards.

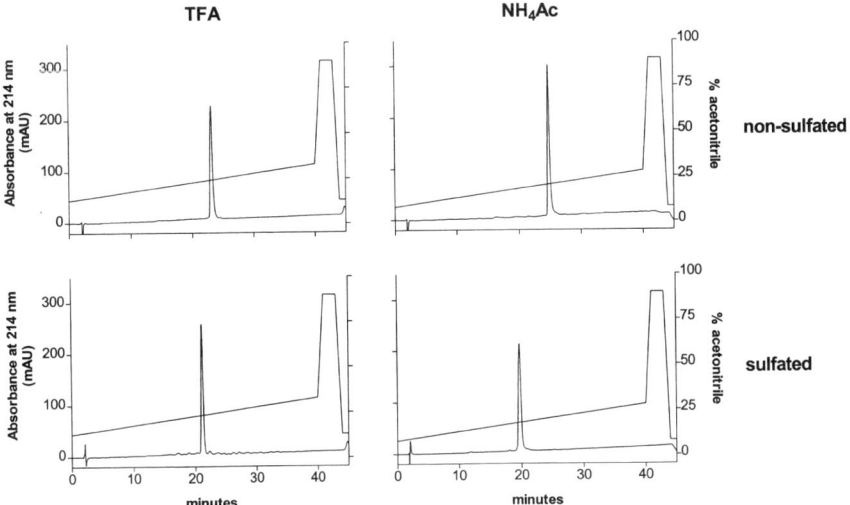

Fig. 4.5 HPLC separation of the model peptides, non-sulfated and sulfated shark gastrin-8 (*see* Fig. 4.1). The use of 0.1% trifluoroacetic acid (TFA) and 10 mM ammonium acetate (NH$_4$Ac) as buffering component are compared. In both cases 0.5%/min. gradients from H$_2$O to acetonitrile were employed, starting from 13% acetonitrile in the TFA system and from 8% in the NH$_4$Ac system (indicated by the straight lines). Although the two forms are indeed separated in the TFA system (by 1% acetonitrile) the separation is much more pronounced in the NH$_4$Ac system (by 2.5% acetonitrile). For bigger peptides, where sulfation influences the chromatographic behavior relatively less, use of the NH$_4$Ac system may be crucial for the separation of the non-sulfated and sulfated form

3.2 Analysis of Tyr Sulfation Using FPLC Based Ion-Exchange Chromatography

1. Equilibrate the column to start conditions and inject sample.
2. Elute with an appropriate gradient at the flow of 1 mL/min and collect fractions if necessary.
3. The eluting peptide is monitored by the UV signal recording or any appropriate specific assay. An example of an elution profile examined by radioimmunoassay is given in Figure 4.4.

3.3 Analysis of Tyr Sulfation Using Reverse Phase HPLC

1. Equilibrate column to start conditions and inject sample.
2. Elute with an appropriate gradient.
3. Record UV signal at 214 nm and—if further analyses are wanted—collect fractions automatically at regular intervals or collect peak manually. An example of an elution profile is given in Figure 4.5.

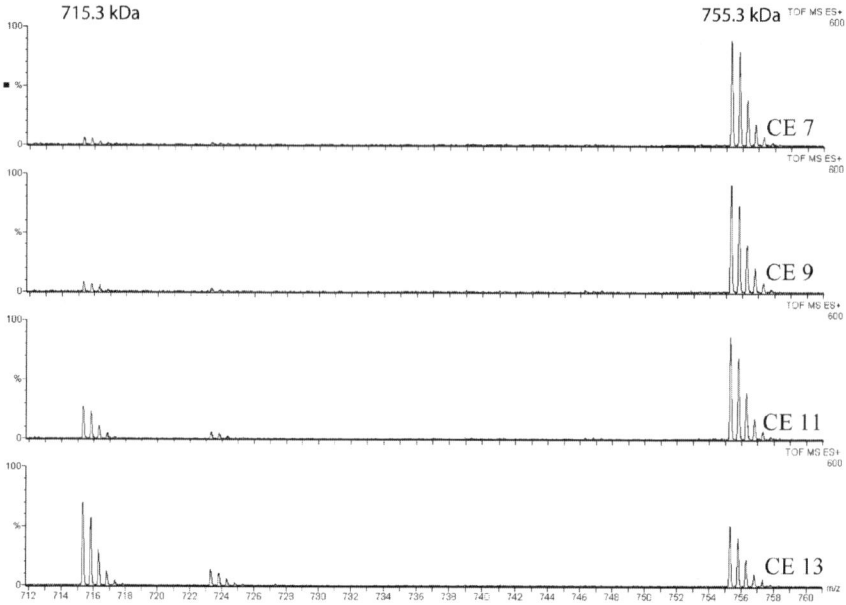

Fig. 4.6 Mass spectra of tyrosine sulfated drosulfokinin showing the increased neutral loss as the collision energy (CE) is increased in intervals of 2 from 7 to 13. The spectra were obtained on a Q-TOF2 using nanospray ionization. Sulfated drosulfokinin is seen at 755.3 Da and desulfated drosulfokinin at 715.3 Da. Both ions are doubly charged and the neutral loss is therefore observed as 40 Da

3.4 Analysis of Tyr Sulfation Using Mass Spectrometry

3.4.1 MALDI-TOF

1. Samples should be in a final concentration of 0.05–5 pmol/µL, sensitivity depends highly on the nature of the peptide (*see* **Note 12**).
2. Mix 0.5 µL of sample with 0.5 µL MALDI matrix solution. For manual preparation, this can be done in a microcentrifuge tube, or many samples can be prepared consecutively at different positions in a trough made from a cut-through polypropylene tube.
3. Apply immediately 0.5 µL of the mixture to the sample target (volume may depend on the target size) and let sample dry.
4. Insert sample(s) into the mass spectrometer and record the mass spectrum according to the instrument manual.

3.4.2 Neutral Loss Experiment on Quadrupole-TOF

3.4.2.1 Analyses of Complex Samples

1. The LC part of the system is programmed to run a gradient appropriate to the sample, e.g., 5–40% B over 120 min. (*see* **Note 13**).

2. The instrument is programmed to do neutral loss scanning for loss of 79.9568 Da and with alternating slightly higher and slightly lower collision energy than standard settings during the run of the gradient (e.g., on a Q-TOF2 instrument (Waters) collision energies 13 and 8 as opposed to a standard setting of 10), see Figure 4.6.
3. Peptides detected by neutral loss scanning are sequenced by MS/MS (*see* **Note 14**).
4. Tyrosine sulfated peptides are confirmed by the MS/MS data that should not contain the tyrosine phosphate immonium ion of 216.043 Da.

3.4.2.2   Direct Infusion of Simple Samples

1. For direct infusion, e.g., by nanospray without reversed phase separation the sample has to be clean from salts and only contain a limited number of peptides.
2. The experiment is run as described above but without the LC part. Alternatively collision energy can be changed manually and mass shift for the peptide suspected of being tyrosine sulfated monitored.

3.5   Arylsulfatase Treatment (see Note 15)

1. Prepare an enzyme solution of arylsulfatase VIII of 2 mg/mL in 0.2% sodium chloride.
2. Prepare the lyophilized sample by resuspension in 400 µL acetate buffer.
3. Add 100 µL enzyme solution to the sample preparation. The final concentration of arylsulfatase is 0 4 mg/mL (equivalent to 0.55 U/mL).
4. Incubate for 3 h at 37°C.
5. Terminate the reaction by boiling for 10 min.

4   Notes

1. The strategy depends on several factors: 1) The size of the protein or peptide; 2) the position of the sulfated Tyr within the protein; 3) the biology of the sulfated protein or peptide (a large membrane protein; a circulating hormone; or a small neurotransmitter peptide etc.); and 4) the concentration range in which to measure. If the sulfated Tyr is positioned at the N- or the C-terminus, production of antibodies against a synthetic analogue of the terminally sulfated decapeptide is straightforward. If, however, the sulfated Tyr is located in the middle of a protein or a long polypeptide chain, antibodies should be raised against a fragment containing the sulfated tyrosine, preferably a fragment that can be released from the protein by appropriate proteolytic cleavage. Finally, if the sulfated protein is available in substantial quantities for large immunization series, "shotgun" immunization with the entire protein and subsequent identification of epitope specificity can be attempted.

2. The presence of Met in the peptide presents a possible pitfall, because oxidation of Met to the sulfoxide (which may occur spontaneously in the presence of atmospheric oxygen) increases the hydrophilicity considerably, thus causing the peptide to elute significantly earlier than the reduced form in both a RP-HPLC and in an anion exchange chromatography system.

3. Both sulfation and phosphorylation add 80 Da to the molecular mass of the peptide (the exact monoisotopic values being 79.957 and 79.966, respectively). However, it is possible to distinguish the two by several criteria. Thus, under the standard conditions used for Edman sequencing peptides phosphorylated at Ser, Thr or Tyr give no signal (although for different reasons, [28–29]) resulting in a blank cycle, whereas sulfated Tyr is hydrolyzed and seen as Tyr in almost normal yield. Furthermore, in contrast to Tyr-sulfated peptides (Fig. 4.1), phosphorylated peptides are recorded as the intact species in MALDI-TOF linear mode (30). In the reflector mode a typical triplet is observed from phosphorylated peptides consisting of the intact molecular ion accompanied by a major and a much less abundant fragment because of the loss of H_3PO_4 (98 Da) and HPO_3 (80 Da), respectively (30). This is in contrast to the singular loss of 80 Da from a Tyr-sulfated peptide (see **Note 5**).

4. The relative yields of the sulfated and nonsulfated form depend on the composition of the peptide. Small acidic peptides like the model peptide give almost clean peaks representing the sulfated and nonsulfated form in the negative and positive (linear) mode, respectively (Fig. 4.1), whereas larger and more neutral peptides give a more mixed picture.

5. A MALDI-TOF mass spectrum obtained in the negative mode using reflector will show the correct molecular ion and an additional peak at a distance *less* than 80 Da below the value for the molecular ion (Fig. 4.2). This is caused by rupture of the sulfate ester bond in the linear flight tube after the peptide was accelerated but before the peptide reaches the reflector, and the calibration for the reflector requires ions with full accelerating energy. The decrease in molecular mass caused by sulfate loss observed in the reflected mode is instrument dependent, but once established it can also be used as an indicator of a sulfated peptide.

6. The tyrosine *O*-sulfated peptide, synthetic or purified has to be available in mg amounts. After dissolution in an appropriate buffer, the peptide is coupled to a protein carrier in amounts that ensure a coupling ratio of approximately 5–10 molecule haptens per carrier molecule. There are several useful carrier candidates. We recommend bovine serum albumin (BSA, approximately 5 mg/mg hapten decapeptide). BSA is easily available, inexpensive and in our experience as effective as any other protein carrier. When peptide hapten has to be synthesized, we recommend synthesis of an analogue of the genuine hapten equipped with a N- or C-terminal cysteinyl residue. Which terminus or end depends on the position of the sulfated tyrosyl residue (Fig. 4.3). The peptide hapten is then coupled to a free amino group in the carrier protein through the terminal cysteine using maleimidobenzoic acid N-hydroxysuccinimide ester as coupling reagent (35). Again, depending on the position of the sulfated tyrosyl residue,

conventional coupling using either ethylcarbodiimide or glutaraldehyde may also be used *(35)*.

7. With directional N- or C-terminal carrier-coupling (Fig. 4.3), it is important to design a tracer that corroborates and further advances the antibody specificity achieved. Because ^{125}I is the preferred RIA-isotope, and because iodine easiest is coupled to tyrosyl-residues, we recommend synthesis of a tyrosylated hapten decapeptide in which the additional tyrosyl-residue for iodination is placed in the same position as cysteine in the hapten analogue, i.e. at the terminus opposite to the epitopic part of the hapten containing the 0-sulfated tyrosine (Fig. 4.3). In this way, iodinated tyrosine will not interfere with the antibody binding. There is no risk for iodination of the *O*-sulfated tyrosine-residue. The sulfonate group blocks incorporation of iodine. Natural occurrence of an additional unsulfated tyrosyl in the short hapten sequence has not been experienced so far; but in such situation a new labeling strategy has to be delineated—based on the exact positions of the additional and the sulfated tyrosyl residue. The tyrosine-extended peptide analogue (Fig. 4.3) (4 nmoles) can be monoiodinated using the mild chloramine-T method, previously described *(36)*. If mild oxidation damages the peptide (containing for instance methionyl residues), it is possible to use the nonoxidative Bolton-Hunter iodination *(37)*. Subsequent purification on reversed-phase HPLC ensures a high specific radioactivity of the tracer. To evaluate the chromatographic separation of labeled and nonlabeled peptides, 1 mL of the monoiodinated peak fraction is mixed with 10 pmol of the relevant tyrosine-extended peptide and reapplied to the HPLC column as described. Both the radioactivity of the labeled peptides and the immunoreactivity are measured. Using the assumption that iodinated and unlabeled peptides are measured with identical affinity the specific radioactivity of the tracers can be determined by self-displacement *(38)*. The essence of the described procedure is to ensure monoiodination without oxidative damages to the peptide. High specific radioactivity is then achieved by efficient chromatographic purification.

8. As standard substance for RIA-measurements using antibodies specific for sequence containing a sulfated tyrosyl-residue, it is straightforward to use a peptide corresponding to the hapten decapeptide. If, however, this decapeptide in some way is partly or completely hidden within the structure or conformation of the genuine tyrosyl sulfated protein, or if the antibody requires that epitope has a free N- or C-terminus (Fig. 4.3), it may be necessary to release the hidden epitope by cleavage with proteases that expose the epitope for antibody binding *(39)*. If the natural occurring tyrosyl sulfated protein or polypeptide is recognized without cleavage, the protein may as well be used as standard—if available in sufficiently pure form.

9. In this example the peptide analyzed are acidic (the sequence of gastrin-17 is: pQGPWLEEEEEAYGWMDF-NH$_2$) for which reason an anion exchange column is used. Because Tyr sulfation predominantly occurs in acidic regions of proteins and peptides, similar conditions this will often be appropriate. However, it is also possible to use cation exchange chromatography to separate peptide forms or alternatively, if the net charges of the fragments are neutral, the pH of the buffers may be adjusted.

10. α-cyano-4-hydroxycinnamic acid gives a more homogeneous matrix than DHB and generally also a better signal. However, DHB is a cool matrix compared to α-cyano-4-hydroxycinnamic acid, and hence gives a reduced loss of sulfate in general and especially in reflector mode.

11. Four characteristics of sera from the immunized animals have to be examined: 1) The *titer* is defined as the antiserum dilution that binds 33% of the 0.5-fmol tracer at equilibrium. 2) *Affinity* is expressed by the "effective" equilibrium constant ($K°_{eff}$), determined as the slope of the curve at zero peptide concentration in a Scatchard plot *(40)*. 3) *Specificity* is determined in percentage as the molar ratio of the concentrations of the sulfated standard peptide, the unsulfated peptide and other related peptides that produce a 50% inhibition of the binding of the tracer. 4) *Homogeneity* of the antibodies with respect to binding kinetics is expressed by the Sips index *(41)*. An index of 1 indicates homogeneity of both the tracer and the antiserum in the binding, otherwise seen only for monoclonal antibodies *(21)*. The ability of the peptides to displace tracers from the antisera may be tested in peptide concentrations of 0, 3, 10, 30, 100, 1000, 10 000, and 100 000 pmol/L.

12. Samples in HPLC buffers are preferable (0.1% TFA in 30% acetonitrile is ideal for MALDI-TOF). Salt and buffer concentrations should be kept as low as possible. Volatile buffers may be removed in a vacuum centrifuge and the dried sample redissolved in 0.1% TFA/30% acetonitrile. Most detergents, especially SDS, will completely abolish the signal.

13. The gradient should be chosen to give a good separation of the peptides. The tyrosine sulfate containing peptides are increasingly suppressed as more peptides are eluted simultaneously. This results in higher risk of missing a tyrosine sulfated peptide in the neutral loss scanning procedure.

14. Note that the sulfate group will be completely lost during MS/MS and the spectrum will show the fragmentation pattern of the desulfated peptide, whereas the precursor ion is that of the sulfated peptide. This constitutes a problem to the general search engines, which are dependent on intact fragment ions of modified peptides to do correct annotation. Therefore manually annotation has to be applied or the parent ion mass must be altered before search.

15. It is our experience that the arylsulfatase treatment is tricky and sometimes is very inefficient. We therefore advice repetition of experiments with negative outcome and, if possible, the inclusion of a positive control peptide with the experiments.

References

1. Bettelheim FR. (1954) Tyrosine-O-sulfate in A peptide from fibrinogen. *J. Am. Chem. Soc.* **76**:2838–2839.
2. Ouyang Y., Lane W. S., and Moore KL. (1998) Tyrosylprotein sulfotransferase: purification and molecular cloning of an enzyme that catalyzes tyrosine O-sulfation, a common posttranslational modification of eukaryotic proteins. *Proc. Natl Acad Sci. U S A* **95**:2896–2901.

3. Beisswanger R., Corbeil D., Vannier C., Thiele C., Dohrmann U., Kellner R., et al. (1998) Existence of distinct tyrosylprotein sulfotransferase genes: molecular characterization of tyrosylprotein sulfotransferase-2. *Proc. Natl Acad. Sci. USA* **95**:11134–11139.
4. Ouyang Y. B. and Moore K. L. (1998) Molecular cloning and expression of human and mouse tyrosylprotein sulfotransferase-2 and a tyrosylprotein sulfotransferase homologue in Caenorhabditis elegans. *J. Biol. Chem.* **273**:24770–24774.
5. Ouyang Y. B., Crawley J. T., Aston C. E., and Moore K. L. (2002) Reduced body weight and increased postimplantation fetal death in tyrosylprotein sulfotransferase-1-deficient mice. *J. Biol. Chem.* **277**:23781–23787.
6. Moore K. L. (2003) The biology and enzymology of protein tyrosine O-sulfation. *J. Biol. Chem.* **278**:24243–24246.
7. Mutt V. (1980) Cholecystokinin: isolation, structure and functions, in *Gastrointestinal Hormones*. (Glass G. B. J., ed.).Raven Press, New York; pp. 169–221.
8. Pouyani T., Seed B. (1995) Psgl-1 recognition of p-selectin is controlled by a tyrosine sulfation consensus at the psgl-1 amino-terminus. *Cell* **83**:333–343.
9. Stone S. R., Betz A., Parry M. A., Jackman M. P., and Hofsteenge J. (1993) Molecular basis for the inhibition of thrombin by hirudin. *Adv. Exp. Med. Biol.* **340**:35–49.
10. Farzan M., Mirzabekov T., Kolchinsky P., Wyatt R., Cayabyab M., Gerard N. P. et al. (1999) Tyrosine sulfation of the amino terminus of CCR5 facilitates HIV-1 entry. *Cell* **96**:667–676.
11. Bundgaard J. R., Vuust J., Rehfeld J. F. (1995) Tyrosine O-sulfation promotes proteolytic processing of progastrin. *EMBO J* **14**:3073–3079.
12. Friederich E., Fritz H. J., and Huttner W. B. (1988) Inhibition of tyrosine sulfation in the trans-Golgi retards the transport of a constitutively secreted protein to the cell surface. *J. Cell Biol.* **107**:1655–1667.
13. Monigatti F., Gasteiger E., Bairoch A., and Jung E. (2002) The Sulfinator: predicting tyrosine sulfation sites in protein sequences. *Bioinformatics* **18**:769–770.
14. Rosenquist G. L. and Nicholas H. B., Jr. (1993) Analysis of sequence requirements for protein tyrosine sulfation. *Protein Sci.* **2**:215–222.
15. Nicholas H. B., Jr., Chan S. S., and Rosenquist G. L. (1999) Reevaluation of the determinants of tyrosine sulfation. *Endocrine* **11**:285–292.
16. Chen T. and Shaw C. (2003) Cloning of the (Thr6)-phyllokinin precursor from Phyllomedusa sauvagei skin confirms a non-consensus tyrosine O-sulfation motif. *Peptides* **24**:1123–1130.
17. Bundgaard J. R., Vuust J., and Rehfeld J. F. (1997) New consensus features for tyrosine O-sulfation determined by mutational analysis. *J. Biol. Chem.* **272**:21700–21705.
18. Huttner W. B. (1984) Determination and occurrence of tyrosine O-sulfate in proteins. *Methods Enzymol* **107**:200–223.
19. Schechter I. (1970) Mapping of the combining sites of antibodies specific for poly-L-alanine determinants. *Nature* **228**:639–641.
20. Rehfeld J. F. and Johnsen A. H. (1994) Residue-specific immunochemical sequence prediction. *J. Immunol. Methods* **171**:139–142.
21. Rehfeld J. F., Bardram L., Cantor P., Hilsted L., and Johnsen A. H. (1989) The unique specificity of antibodies in modern radioimmunochemistry–an essay on assays. *Scand J. Clin. Lab. Invest. Suppl.* **194**:41–44.
22. Adam J. Hoffhines, Eugen Damoc, Kristie G. Bridges Julie A. Leary, and Kevin L. Moore (2006) Detection and Purification of Tyrosine-sulfated Proteins Using a Novel Anti-sulfotyrosine Monoclonal Antibody *J. Biol. Chem.* **281**:37877–37887.
23. Ross A. H., Baltimore D., and Eisen H. N. (1981) Phosphotyrosine-containing proteins isolated by affinity chromatography with antibodies to a synthetic hapten. *Nature* **294**:654–656.
24. Rehfeld J. F. (1998) Accurate measurement of cholecystokinin in plasma. *Clin. Chem.* **44**:991–1001.
25. Balsved D., Bundgaard J. R., and Sen J. W. (2007) Stability of Tyrosine Sulfate in Acidic Solutions. *Anal. Biochem.* **363**:70–76.
26. Gregory H., Hardy P. M., Jones D. S., Kenner G. W., and Sheppard R. C. (1964) The antral hormone gastrin. Structure of gastrin. *Nature* **204**:931–933.

27. Johnsen A. H. (1991) Nondestructive amino acid analysis at the picomole level of proline-containing peptides using aminopeptidase M and prolidase: application to peptides containing tyrosine sulfate. *Anal. Biochem.* **197**:182–186.
28. Meyer H. E., Hoffmann-Posorske E., Donella-Deana A., and Korte H. (1991) Sequence analysis of phosphotyrosine-containing peptides. *Methods Enzymol.* **201**:206–224.
29. Meyer H. E., Hoffmann-Posorske E., and Heilmeyer L. M. Jr. (1991) Determination and location of phosphotyrosine in proteins and peptides by conversion to S-ethylcysteine. *Methods Enzymol.* **201**:169–185.
30. Annan R. S. and Carr S. A. (1996) Phosphopeptide analysis by matrix-assisted laser desorption time-of-fligh mass spectrometry. *Anal. Chem.* **68**:3413–3421.
31. Talbo G. and Roepstorff P. (1993) Determination of sulfated peptides via prompt fragmentation by UV matrix-assisted laser desorption/ionisation mass spectrometry. *Rapid. Commun. Mass. Spectrom.* **7**:201–204.
32. Salek M., Costagliola S., and Lehmann W. D. (2004) Protein tyrosine-O-sulfation analysis by exhaustive product ion scanning with minimum collision offset in a NanoESI Q-TOF tandem mass spectrometer. *Anal. Chem.* **76**:5136–5142.
33. Haselmann K. F., Budnik B. A., Olsen J. V., Nielsen M. L., Reis C. A., Clausen H., et al. (2001) Advantages of external accumulation for electron capture dissociation in Fourier transform mass spectrometry. *Anal. Chem.* **73**:2998–3005.
34. Monigatti F., Hekking B., and Steen H. (2006) Protein sulfation analysis-A primer, *Biochim. Biophys. Acta.* **1764**:1904–13.
35. Harlow E. and Lane D. (1988) *Antibodies: A laboratory manual.* Cold Spring Harbor Laboratory, Cold Spring Harbor, New York, pp 82–83.
36. Stadil F. and Rehfeld J. F. (1972) Preparation of ^{125}I-gastrin for radioimmunoanalysis. *Scand. J. Clin. Lab. Invest.* **30**:361–369.
37. Bolton A. E. and Hunter W. M. (1973) The labelling of proteins to high specific radioactivity by conjugation to a ^{125}I-containing acylating agent. *Biochem. J.* **133**:529–539.
38. Morris B. J. (1976) Specific radioactivity of radioimmunoassay tracer determined by self-displacement. *Clin. Chim. Acta* **73**:213–216.
39. Bardram L. and Rehfeld J. F. (1988) Processing-independent radioimmunoanalysis *Anal. Biochem.* **175**:537–543.
40. Ekins R. and Newman B. (1970) Theoretical aspects of saturation analysis. *Acta. Endocrinol.* **147**:11–36.
41. Sips R. (1948) On the structure of a catalyst. *J. Chem. Phys.* **16**:490–495.

5

α-Amidated Peptides: *Approaches for Analysis*

Gregory P. Mueller and William J. Driscoll

Summary α-Amidation is a terminal modification in peptide biosynthesis that can itself be rate-limiting in the overall production of bioactive α-amidated peptides. More than half of the known neural and endocrine peptides are α-amidated and in most cases, this structural feature is essential for receptor recognition, signal transduction, and thus, biologic function. This chapter describes methods for developing and using analytical tools to study the biology of α-amidated peptides. The principle analytical method used to quantify α-amidated peptides is the radioimmunoassay (RIA). Detailed protocols are provided for 1) primary antibody production and characterization; 2) radiolabeling of RIA peptides; 3) sample preparation; and 4) the performance of the RIA itself. Techniques are also described for the identification and verification of α-amidated peptides. Lastly, in vivo models used for studying the biology of α-amidation are discussed.

Key Words α-amidation; peptidylglycine α-amidating monooxygenase; secretory granules; radioimmunoassy; mass spectrometry; peptides; iodination.

1 Introduction

1.1 *Biologic Importance of α-Amidation*

α-Amidation is a terminal modification in peptide biosynthesis that can itself be rate limiting in the overall production of bioactive α-amidated peptides. More than half of the known neural and endocrine peptides are α -amidated and in most cases, this structural feature is essential for receptor recognition, signal transduction and thus, biologic function. Peptide α-amidation is catalyzed by a single protein (*see* **Section 1.2**) and genetic mutations resulting in loss of function are lethal *in utero (1)*. All twenty naturally occurring amino acids serve as terminal amides in the spectrum of α-amidated peptide messengers *(2)*. A list of selected α-amidated peptides is presented in Table 5.1.

From: *Post-translational Modifications of Proteins.*
Methods in Molecular Biology, Vol. 446.
Edited by: C. Kannicht © Humana Press, Totowa, NJ

Table 5.1 Representative α-Amidated Peptides of Diverse Biologic Systems

System/Peptide	Residue	Length	Functions/Properties
Nervous System:			
Gonadotrophin releasing hormone (GnRH)	Gly	10	Hypothalamic hormone/neurotransmitter
Corticotrophin releasing hormone (CRH)	Ile	42	Hypothalamic hormone/neurotransmitter
Growth hormone releasing hormone GH-RH	Leu	44	Hypothalamic hormone/neurotransmitter
Thyrotrophin releasing hormone	Pro	3	Hypothalamic hormone/neurotransmitter
Cholecystokinin Octapeptide (CCK-8)	Phe	8	Neurotransmitter
Neuropeptide Y (NPY)	Tyr	36	Neurotransmitter
Calcitonin gene related peptide (CGRP)	Phe	37	Neurotransmitter
Vasoactive intestinal polypeptide	Asn	28	Neurotransmitter
Substance P	Met	11	Neurotransmitter, representative tachykinin
Pituitary:			
Oxytocin	Gly	9	Parturition and nursing
Vasopressin	Gly	9	Renal water reabsorption and vasoconstriction
α-Melanotrophin (α-MSH)	Val	13	Melanocyte dispersion/skin coloration
Gastrointestinal Tract:			
Secretin	Val	27	Stimulates pancreatic exocrine secretion
Cholecystokinin (CCK-33)	Phe	33	Stimulates pancreatic exocrine secretion
Multiple proposed functions at many sites			
Adrenomedullin	Tyr	52	Neurotransmitter/hormone/autocrine
Amphibia peptides:			
Bombesin	Met	14	Promotes proliferation of cells and secretion
Dermenkephalin	Asn	7	Isolate from frog skin, opioid peptide
Sauvagine	Ile	40	Frog corticotrophin releasing hormone (CRH)
Insect Peptides:			
Adipokinetic hormones (several forms)	Thr,Gly,Trp	6–10	Regulate lipid use/neuropeptides
Pheromone biosynthesis activating neuropeptide 1	Leu	33	Reproduction pheromone
Melittin	Gln	26	Venom toxin
Chlorotoxin	Arg	36	Venom toxin
Scyllatoxin	His	31	Venom toxin

(continued)

Table 5.1 (continued)

System/Peptide	Residue	Length	Functions/Properties
Others			
Calcitonin	Pro	32	Calcium conservation in mammals
Egg laying hormone	Lys	36	Aplysia, reproduction
Conotoxins (several forms)	Cys, Ala, Tyr	13–26	Mulluscan toxins
Extendin-3	Ser	39	Ameoba, locomotion

1.2   Mechanism of Peptide α-Amidation

Peptide α-amidation is catalyzed by peptidylglycine α-amidating monooxygenase (PAM), a bifunctional enzyme localized to the trans-Golgi network and secretory granules of neural and endocrine tissues (Fig. 5.1). In a two step process PAM generates α-amidated peptides from their inactive glycine-extended precursors. Peptidylglycine α-hydroxylating monooxygenase (PHM; EC 1.14.17.3) catalyzes the formation of a peptidyl-α-hydroxyglycine intermediate, which is rapidly converted to α-amidated product and glyoxylate by peptidyl-α-hydroxyglycine-α-amidating lyase (PAL; EC 4.3.2.5) (Fig. 5.2). In this sequence, PHM is rate determining and requires ascorbate, molecular oxygen and copper for activity *(2–6)*. α-Amidation results in a dramatic change in the physicochemical properties of the peptide by removing the ionizable, free carboxyl group and leaving a nonionizable α-amide. These changes provide the basis for specific analytical procedures.

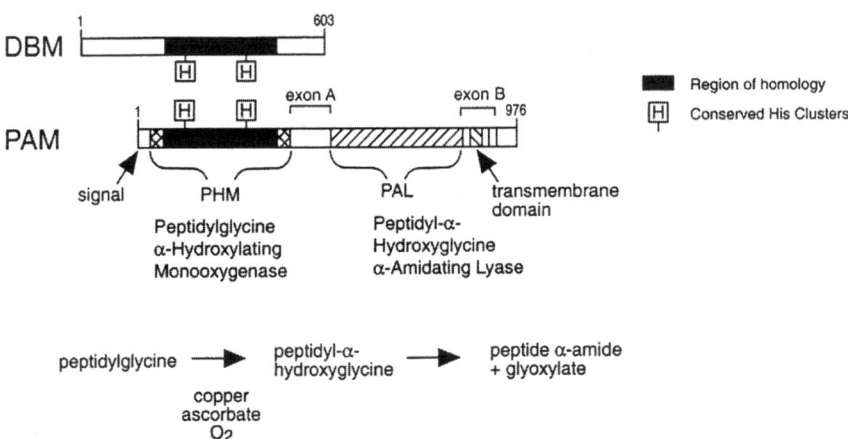

Fig. 5.1 Structural and functional organization of the rat PAM precursor protein. DBM = dopamine beta monooxygenase Both proteins are type II, copper, ascorbate and molecular oxygen dependent monooxygenases. Conserved, histidine rich, consensus metal binding sites [H] convey copper dependence to both enzymes *(1,5,6)*

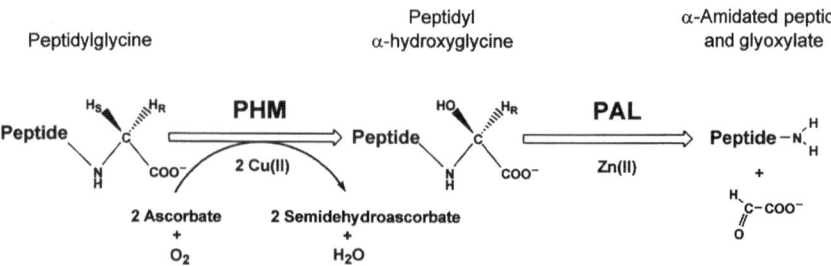

Fig. 5.2 Chemistry of α-amidation

1.3 Overview of the Procedures Used to Investigate α-Amidated Peptides

A flow diagram for the procedures used in studying α-amidated peptides is presented in Figure 5.3. The most powerful and versatile method for measuring α-amidated peptides is radioimmunoassay (RIA) (*see* **Note 1**). The distinctive structure of the carboxy terminal amide enables the development of remarkably specific and sensitive immunoassays. α-Amide-specific RIAs can generally detect 10–50 femtomoles of peptide per reaction with absolute specificity for the C-terminal α-amide. Because the antibody recognition epitope generally constitutes only 4–6 amino acids at the carboxy terminal, variations in the length and sequence of the N-terminal cannot be distinguished by the assay. Accordingly, structurally related peptides, which normally arise from the differential processing of a common precursor, can crossreact in the RIA. The peptide of interest can be separated from these crossreacting species by high pressure liquid chromatography (HPLC) before RIA analysis. Mass spectrometry (MS) provides the most definitive means for confirming the identity of an isolated peptide.

2 Materials

2.1 Radioimmunoassay (RIA) for an α-Amidated Peptide

2.1.1 Generation of a Primary Antibody

1. Synthetic α-amidated peptides for use as haptens, standards and radiolabeling.
2. Carrier protein: keyhole limpet hemocyanin or bovine thyroglobulin (Sigma, St. Louis, MO).
3. 1-ethyl-3-(3-dimethylaminopropyl)carbodiimide (EDC) (Pierce Chemical Co. Rockford, IL).

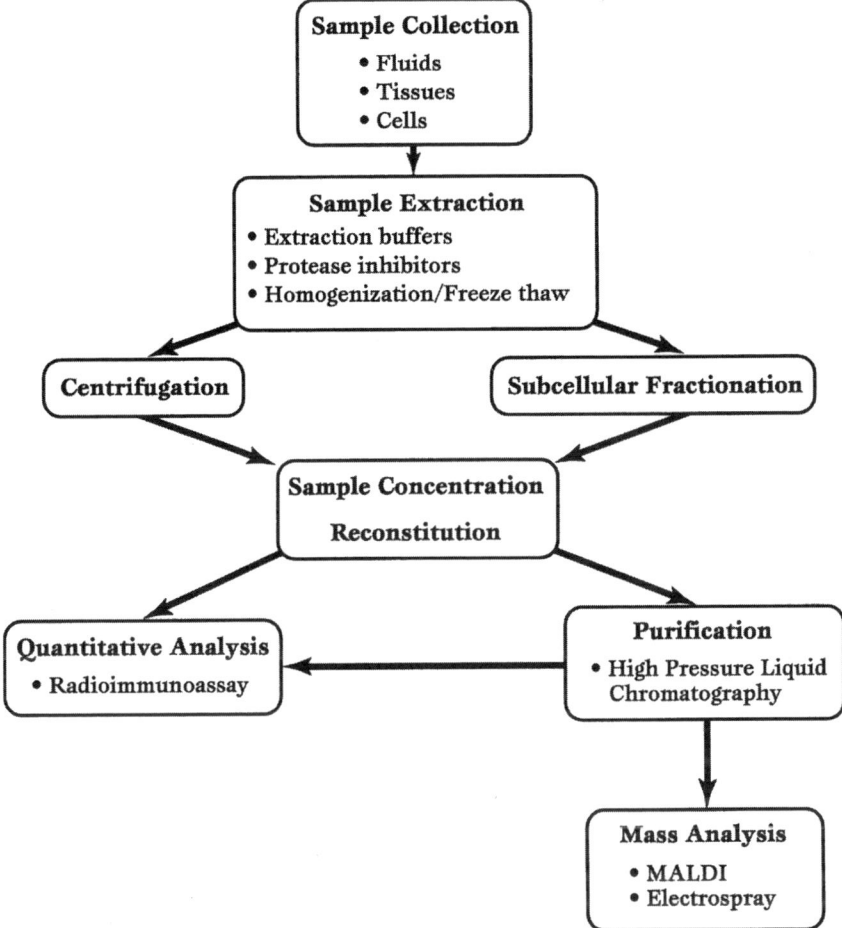

Fig. 5.3 Flow diagram showing procedures for studying an α-amidated peptide

2.1.2 Immunization

1. Freund's complete and incomplete adjuvants.

2.1.3 Preparation of Radiolabeled Peptide

1. Na[125]iodide.
2. 50 mM phosphate buffer, pH 7.0.
3. Reaction tubes, borosilicate glass, 12 × 75 mm.
4. Iodo-Beads (Pierce Chemical Co.).
5. Sep-Pak C$_{18}$ cartridges (Waters Chromatography, Division of Millipore, Marlboro, MA).

6. Sep-Pak mobile phase reagents, 0.1% trifluoroacetic acid (TFA) in water and acetonitrile.

2.1.4   Sample Preparation

1. 0.1N acetic acid (5.75 mL glacial acetic acid in 1 L).
2. Homogenization system: sonicator, Polytron or Potter-Elviehem.
3. Refrigerated centrifuge capable of generating 10,000g at 4°C.

2.1.5   The Working RIA

1. 50 mM phosphate buffer, pH 7.0.
2. Test tubes (disposable 12 × 75 mm borosilicate glass)
3. Bovine serum albumin solution, 0.05–0.1% in 50 mM phosphate buffer, pH 7.0.
4. Heat inactivated horse serum (GIBCO-BRL, Grand Island, NY).
5. Polyethylene glycol solution: 18% PEG (av. mol. wt 8,000) dissolved in 0.05M phosphate buffer, pH 7.0.
6. Refrigerated centrifuge capable of spinning assay tubes at 2,500g at 4°C.
7. Gamma counter

2.2   Procedures for the Identification/Verification of α-Amidated Peptides

2.2.1   High Pressure Liquid Chromatography (HPLC)

1. HPLC system equipped with a C_{18} column and UV absorbance detector (*see* **Note 2**).
2. HPLC mobile phase reagents (water containing 0.1% TFA and acetonitrile containing 0.1% TFA).

2.2.2   Dot-Blot Western Analysis

1. Nitrocellulose (Schleicher and Schuell, Keene, NH).
2. TBST: 50 mM Tris-HCl, pH 7.6, 0.8% NaCl, 0.1% Tween 20 (TBST)
3. Sodium azide.
4. Blocking solution: 1% nonfat dried milk + 1% heat inactivated horse serum in TBST.
5. Anti-rabbit/horse radish peroxidase-linked antibody (e.g., catalog # NA9340, Amersham Biosciences (GE Healthcare), Piscataway, NJ).
6. Enhanced chemiluminescent reagent system (ECL™, Amersham Biosciences [GE Healthcare]).

7. Radiographic film.
8. Dark room and film developing capabilities.
9. As an alternative to the use of radiographic film, western blots may be visualized by phosphoimaging using a variable mode imaging instrument such as those manufactured by FUJIFILM (Stamford, CT) or Amersham Biosciences (GE Healthcare).

2.2.3 Mass Spectral Analysis

1. Mass spectrometer: Matrix assisted laser desorption ionization-time-of-flight (MALDI-TOF) mass spectrometer and/or a HPLC-electrospray mass spectrometer (*see* **Note 3**).
2. Vacuum centrifuge.
3. 70% isopropanol/5%TFA for MALDI-TOF.
4. Matrix solution: α-cyano-4-hydroxy-trans-cinnamic acid (Aldrich, Milwaukee, WI), 10 mg/mL in 50% acetonitrile.

3   Methods

3.1   Radioimmunoassay (RIA) for an α-Amidated Peptide

A highly specific and sensitive RIA for an α-amidated peptide can generally be developed without significant problem. Ideally, 3 RIAs should be developed simultaneously: one for the α-amidated peptide, a second for its C-terminal glycine-extended precursor, and a third for its free acid. Together, these tools enable one to study precursor–product relationships, the peptide's metabolism and closely related crossreacting peptide species. The critical component of all immunoassays is the primary antibody. This reagent is usually developed in rabbits against the peptide conjugated to either keyhole limpet hemocyanin or bovine thyroglobulin. Polyclonal antisera have proven so effective in peptide RIAs that little attention has been given to exploring the utility of monoclonal antibodies.

3.1.1   Preparation of a Conjugate

Coupling with 1-ethyl-3-(3-dimethylaminopropyl)carbodiimide (EDC): Although a variety of crosslinking agents are commercially available for specialized uses in preparing conjugates, coupling with EDC has proven highly successful for α-amidated peptides. EDC mediates the linking of amino groups to carboxyl groups in a one-step procedure. The C-terminal amide ensures that this end of the peptide does not participate in the coupling reaction and is therefore available for immunologic recognition.

1. For a standard conjugation reaction, dissolve 5 mg of synthetic peptide (*see* **Note 4**) and 5 mg of carrier protein in 1 mL of water. It is convenient to use a flat bottom vial that will accommodate a small stirring bar.
2. Prepare a separate stock solution of EDC at 25 mg/mL in water and adjust pH to 4.0 to 5.0 with 0.1N HCl.
3. Working at room temperature with constant mixing, add 1 mL of the EDC stock solution slowly dropwise to the peptide/protein solution and allow to react for 60 min or more. Some cloudiness caused by aggregation is commonly observed.
4. Dialyze the mixture overnight against water (4°C, 2 L, 2 changes) to remove unreacted EDC (*see* **Note 5**).

The usefulness of an antiserum is dependent upon the quality of the immunogen. Accordingly, it is important to assess the effectiveness of the conjugation reaction before embarking upon the intensive and time consuming regimen of immunization, booster injections, sampling, titer checking, and antiserum characterization. Successful conjugation can be determined by (i) increased molecular weight of the carrier protein on sodium dodecyl sulfate-polyacrylamide gel electrophoresis (SDS-PAGE), (ii) altered retention time of the carrier protein on HPLC, or (iii) uptake of trace radioactive peptide into the conjugate product. The HPLC approach is the most convenient although aggregated material must be remove by either centrifugation or filtration (0.22-micron spin filter) before analysis. It can be assumed that the material in solution is representative of that in the aggregate. For highly insoluble conjugates, the product must be analyzed by SDS-PAGE.

3.1.2 Immunization

1. Dilute the dialyzed conjugate to 2 mg/mL and mix with an equal volume of Freund's complete adjuvant (*see* **Note 6**).
2. Prepare an emulsion by either sonication or repeated passages through a 20-gauge hypodermic needle using a glass syringe.
3. Administer the immunogen (1 mL/rabbit) in 10–20 subcutaneous injections. The innoculation of several animals *(2–4)* is recommended because individual immunoresponsiveness can vary greatly.
4. Perform subsequent booster injections with immunogen emulsified in Freund's *incomplete* adjuvant (1 mg/mL) similarly at 4–6 week intervals.
5. Take blood samples by ear vein cannulation for regular titer checks (1–2 mL; 10 d after the second and each subsequent boost) and periodic bulk antiserum harvesting (15–25 mL from a 2-kg rabbit).
6. Titer serum samples by serial dilution as described under **Section 3.1.9** *The Working RIA*. Maximal titers can be expected by 12–15 wk after the initial immunization. Antisera collected before this time may not be suitable for use in RIA because of high IgM content which is not stable through freeze-thawing.

3.1.3 Preparation of Radiolabeled Peptide

Radioiodination of peptides occurs on tyrosine and histidine residues. In cases where these amino acids are not present in the peptide of interest, an analog must be synthesized with a tyrosine in the 0 position. This addition should not interfere with immunologic recognition of the α-amide that occurs at the opposite end of the molecule. Standard iodination procedure:

1. Working at room temperature combine 1 μg of peptide with 1 mCi of $Na^{125}I$ in 100 uL of 50 mM phosphate buffer, pH 7.0.
2. Add 1 Iodo-Bead (a nonporous polystyrene bead containing immobilized N-chloro-benzsulfonamide) to mediate the iodination.
3. Allow the reaction to proceed for 10 min with periodic mixing (*see* **Note 7**).
4. Separate iodinated peptide from free iodine by chromatography on a SepPak C_{18} cartridge. Precondition the cartridge with 2 mL 100% acetonitrile/0.1% TFA followed by 2 mL of 0.1% TFA in water.
5. Dilute the iodination reaction mixture to 1–2 mL with 0.1% TFA in water and pass through the preconditioned SepPak.
6. Wash the cartridge with 5 mL of water containing 0.1% TFA to remove unincorporated iodine.
7. Elute the labeled peptide with stepwise, increasing concentrations of acetonitrile in water/0.1% TFA (1.5 mL each; 10% increments). α-Amidated peptides generally elute in the range of 20–40% CH_3CN/0.1% TFA. Iodinated peptides are stable for 4 or more weeks when stored at 4°C in SepPak mobile phase.

3.1.4 Conditions for Equilibrium Binding Reactions

The basis for an RIA is competitive binding between the sample peptide and radiolabeled peptide for the antibody recognition site. Although conditions for equilibrium binding assays vary greatly, a good starting point for establishing optimal conditions include the following:

1. buffer:	0.05 mM phosphate, pH 7.0
2. reaction tubes	borosilicate glass, 12 × 75 mm
3. reaction volume:	0.5 mL
4. sample volume	variable (e.g., 10–100 uL; depending upon peptide concentration)
5. range of standard curve:	1 femtomol–10 picomol
6. radioactivity/reaction:	20,000 cpm ("total counts")
7. antibody dilution:	primary antiserum diluted in buffer so that 20–35% of total counts are bound specifically; normally accomplished at a final dilution of 1:10,000 to 1:50,000
8. incubation conditions:	24 h, 4°C

9. inclusion of protein: bovine serum albumin (BSA) (0.05–0.1%)
                                    is often included in the reaction to decrease
                                    nonspecific binding (*see* **Section 3.1.6**)
10. separation of bound from free   polyethylene glycol (PEG) precipitation
                                    is recommended for establishing
                                    working conditions

3.1.5   Separation of Bound from Free by Polyethylene Glycol Precipitation

Separation of antibody-bound from free radioactive peptide is accomplished by
either precipitation with polyethylene glycol (PEG), charcoal absorption or second
antibody precipitation. All three approaches should be tested although the PEG
precipitation method usually works well for peptide RIAs (*see* **Note 8**). PEG precipitation works best when both the PEG reagent and assay tubes are precooled
(4°C).

1. Precipitate antibody-bound counts by adding of 1 mL of 18% PEG (av. mol. wt
   8,000 dissolved in $0.05M$ phosphate buffer, pH 7.0), vortexing and incubating at
   4°C for 30 min.
2. Centrifuge assay tubes for 45 min at $2,500g$, 4°C; higher force and longer run
   times should be tested for improved pellet formation.
3. Remove supernatant by vacuum aspiration using a Pasture pipet connected by
   flexible tubing to a low pressure collection vessel.
4. Count the tubes containing pellets.

3.1.6   Optimization of Conditions for Antibody Binding

The functional potential of a RIA is defined by the primary antibody's titer, affinity and specificity. Optimal conditions for antibody binding must be determined
empirically. The overall objective is to maximize specific binding and minimize
nonspecific binding (less than 5% of total counts). As a basis for comparison it is
recommended that initial conditions be established using the parameters defined
in **Section 3.1.4**. First, the antiserum is titered by serial dilution to achieve specific
binding equaling 20–30% of the total counts available. This level of binding
insures that the antibody is limiting in the reaction so that both increases and
decreases in specific binding can be readily detected. Second, room temperature
incubation should be tested for changes in specific binding. An increase in specific
binding (>5%) indicates that the antibody performs better under this condition.
Third, a time course for binding is performed to define the minimum period
required to reach equilibrium. Fourth, the source and amount of protein used in the
assay is optimized. Although BSA is widely used in peptide RIAs, we have often
found that its replacement with heat inactivated horse serum (HIHS) (final dilution
of 10%) is more effective than BSA in reducing nonspecific binding and actually

promotes antigen–antibody binding as well. In addition, the presence of whole serum enhances the effectiveness of PEG precipitation. It should be recognized that the beneficial effects of HIHS are not observed in all assays and therefore must be determine for each case. Finally, the alternative methods for separating bound from free (charcoal absorption and second antibody precipitation) should be evaluated for improvement in assay performance. It should be noted that the primary antibody may need to be periodically retitered as conditions are optimized.

3.1.7 Determination of Antibody Specificity

Specificity of the primary antibody is defined by evaluating crossreactivity of closely related peptides. Necessary determinations include those with the peptide free acid and its glycine extended precursor. Neither of these species should be detected by the antibody. By contrast, variations in the amino terminal that occur beyond the antigenic epitope generally do not alter antibody recognition. Accordingly, these peptides are equivalent to the authentic α-amidated peptide in their ability to interact with the antibody and thereby, reduce the binding of radiolabeled ligand. The extent to which N-terminally extended peptides contribute to the immunoreactivity measured in experimental samples must be determined by HPLC fractionation and subsequent RIA and mass analysis.

3.1.8 Sample Preparation

Samples to be analyzed for α-amidated peptides are usually prepared from whole tissues or cell cultures. Subcellular fractionation by differential centrifugation may be carried out before extraction to isolate a specific cellular organelle.

1. Extract tissue samples or cells in ten volumes of 0.1N acetic acid using either sonication, Polytron or Potter-Elviehem homogenization. Media from cell cultures may be acidified or assayed directly. A striking exception is CCK-8-NH$_2$ that does not solubilize well by this procedure and is effectively extracted into 90% methanol.
2. Freeze and thaw the homogenates 3 times to insure complete disruption and solubilization of all intracellular compartments
3. Centrifuge the homogenates at 10,000g, 4°C for 15 min to generate clarified supernatant.
4. Small volumes of the acid extracts can be added directly into RIA reactions where the buffering capacity of the assay is sufficient to neutralize the sample. Alternatively, the samples may be concentrated by lyophilization under vacuum and reconstituted in assay buffer containing complete protease inhibitor mix. Yields through an extraction protocol must be determined and shown to be reproducible. This may be performed by either RIA using known amounts of added peptide standard or by radioactive tracer.

3.1.9   The Working RIA

Each RIA consists of the following 5 sets of tubes performed in duplicate.

Set 1. Total Count Tubes (determine the total radioactivity in each reaction).
Set 2. Nonspecific Binding or Background Tubes (determine the amount of total
        radioactivity that is measured in the absence of primary antibody).
Set 3. Total Binding Tubes (determine the amount of radioactivity bound by the
        primary antibody in the absence of nonradioactive competing peptide).
Set 4. Standard Curve Tubes (determine the effect of increasing concentrations of
        nonradioactive competing peptide on antibody-bound counts).
Set 5. Unknown Tubes (determine the effect of experimental samples
        on antibody binding).

Specific binding is calculated by subtracting the nonspecific binding from the
total binding. As noted above, nonspecific binding should be less than 5% of total
counts and specific binding between 20% and 35% of total counts. Nonspecific
binding is also subtracted from all standard and unknown values as a step in data
analysis. Usually a standard curve entails six or more concentrations of competing
peptide. The concentrations used are selected to span the range required to reduce
specifically bound counts to a few percent to total specific binding. There is an
inverse relationship between the number of counts bound and the concentration of
standard peptide. Plotting the data obtained on semilog paper yields a sigmoid
curve having a linear segment between approximately 80% and 20% of total specific binding. This portion of the curve defines the working range of the assay.
Serial dilutions of experimental samples (unknowns) must parallel the standard
curve for the assay to be valid. Concentrations of peptide in the unknowns are
determined from the standard curve either by hand plotting or computer-based
analysis. Software programs for analyzing RIA data are standard features on
gamma counters. In some cases, a significant increase in assay sensitivity can be
achieved by allowing the standards and unknowns pre-equilibrate with the antibody before the radiolabeled peptide is introduced. The effectiveness of this
approach varies among assays.

3.2   Procedures for the Identification/Verification
of α-Amidated Peptides

3.2.1   High Pressure Liquid Chromatography (HPLC)

Reverse phase HPLC is indispensable in the purification and identification of peptide messengers. As a rule, peptide α-amides and their precursors and metabolites
exhibit distinctive retention/elution characteristics on C_{18} columns. Identification of
a sample peptide can be tentatively assigned on the basis of its having a retention
time identical to that of a known standard peptide.

1. Extract samples (tissue, cultured cells, or incubation medium) into 1% HCl; tissue and cell pellets are prepared in 10× volumes whereas media is acidified by the direct addition of concentrated acid.
2. Prepare clarified supernatants by high speed centrifugation (e.g., 10,000g, 15 min).
3. Inject an aliquot of the clarified supernatant onto a C$_{18}$ column (250×4.6 mm) equilibrated with 2.5% acetonitrile/0.1% TFA in water containing 0.1% TFA pumped at a flow rate of 1 mL/min (*see* **Note 9**).
4. Continue washing the column under these conditions for five or more min to elute material not retained on the C$_{18}$ column. Monitor optical absorbance at 214 nm (*see* **Note 2**).
5. Once a stable UV baseline is obtained, begin developing an increasing gradient of acetonitrile/0.1% TFA in water containing 0.1% TFA. A standard screening gradient involves a 1% per min change in acetonitrile concentration at a flow rate of 1 mL per min. Target peptides typically elute between 15% and 40% acetonitrile with recoveries that exceed 75%.

3.2.2 Detection of peptides by dot-blot Western analysis

HPLC profiles may be screened immunologically to identify peptides present in concentrations too low to be detected by UV absorbance. This method provides a convenient means for quickly analyzing an entire HPLC elution profile in a semiquantitative fashion. The effectiveness of this approach, of course, depends upon the ability of the peptide to adhere to the membrane and the ability of the primary antibody to interact with the bound peptide. Again, these factors must be determined empirically.

1. Collect 1 mL fractions of the entire HPLC elution profile including the initial flow-through period during which a stable baseline UV absorbance is established.
2. Apply a 10-μL aliquot of each fraction to nitrocellulose (Nytran, Schleicher and Schuell) that has been lined off in pencil to form a numbered grid. It may be necessary to concentrate the samples by lyophilization before spotting onto nitrocellulose. Allow the samples to dry completely and process the membrane following the procedures for standard Western-blot analysis *(7)* as follows.
3. Block the membrane at room temperature for 60 min by incubation with mixing in a solution of 50 m*M* Tris, pH 7.6, 0.8% NaCl, 0.1% Tween 20 (TBST) containing 0.1% sodium azide, 1% (wt/vol) nonfat dried milk and 1% heat inactivated horse serum (GIBCO-BRL) (blocking buffer).
4. Add anti-peptide antiserum directly to the blocking solution to achieve a final concentration of 1:1,000. Optimal working concentrations must be determined empirically for each antiserum.
5. Incubate the blot with mixing at either 4°C or room temperature for 4–18 h.
6. Wash the membrane 4 times, 15 min each, with TBST.
7. Incubate the membrane in blocking buffer containing anti-rabbit/horse radish peroxidase-linked antibody (1:10,000 dilution) and incubate with mixing at room temperature for 2–4 h. (Optimal antibody concentrations must be determined empirically.)

8. Wash the membrane 4 times, 15 min each, with TBST.
9. Immunoreactive peptides are visualized with an enhanced chemiluminescent reagent system according to the manufacturer's instructions.

3.2.3 Mass Spectral Analysis

Peptides purified by HPLC may be quantitated by RIA or analyzed by mass spectrometry (MS). The primary purpose for MS is to verify the identity of an isolated peptide. Two commonly used methods for the mass analysis of peptides and proteins employ different mechanisms of ionization: matrix assisted laser desorption ionization (MALDI)-MS and electrospray (ES)-MS. The reader is directed to several recent reviews for theoretical details on the procedures *(8–10)*. Additionally, tandem HPLC-ES-MS systems allow for direct analysis of the HPLC effluent without intervening sample manipulation. For both ES-MS and MALDI-MS, the limit of detection is defined, in large part, by the ability of the peptide to ionize. Unfortunately, MS is not quantitative due to differing ionization efficiencies that occur even under seemingly identical conditions. Nevertheless, the technique is highly sensitive, and the accuracy of mass assignment is routinely 0.01%. Accordingly, an α-amidated peptide can be readily differentiated from its glycine-extended precursor, free acid or related analog.

3.2.4 Sample Preparation

Because very limited amounts of α-amidated peptide are normally present in biologic samples, extracts routinely require concentration by drying (SpeedVac). Maximum concentration and recovery are achieved when samples are dried in conical vials that allow for reconstitution in 10 µL or less. Solubilization with 70% isopropanol/5%TFA (*see* **Note 10**) is appropriate for MALDI-MS analysis. Alternatively, 1% acetic acid is preferred for electrospray MS (*see* **Note 11**).

3.2.5 Slide Preparation for MALDI-Time of Flight (TOF)

Peptide sample is crystallized with matrix on a stainless steel sample slide. Matrix is a small, UV absorbing organic acid that transmits the energy from a laser pulse to the peptide and in the process, ionizes the sample by providing protons. The major ion product of a peptide is the +1 species. In the ionized state the peptide is able to "fly" down the instrument's sample path to the detector (electron multiplier). The time-of-flight is directly proportional to mass and inversely proportional to charge. Optimal conditions must be determined empirically for each specific peptide. Important variables include: chemical matrix, ratio of sample to matrix, and the energy of the laser pulse applied.

1. Resuspend HPLC fraction in 70% isopropanol/5% TFA.
2. Spot 0.5 μL of peptide sample on the sample slide followed by an equal volume of matrix. A generally effective matrix for most peptides is α-cyano-4-hydroxy-trans-cinnamic acid (10 mg/mL, in 50% acetonitrile) (*see* **Note 12**).
3. After all samples have been applied with matrix, dry slides thoroughly under vacuum.
4. Analyze samples by MALDI-TOF (*see* **Note 13**).

3.3 In Vivo Models for Investigating the Biology of α-Amidated Peptides

In vivo strategies for investigating the physiology of α-amidated peptides involve 3 approaches: nutritional, pharmacologic and genetic. All three approaches are designed to alter the function of PAM, either directly or via its essential cofactors, copper and ascorbate. Effective use of these experimental paradigms, however, requires the application of the methods described in this chapter. Copper and ascorbate deficient diets predictably decrease α-amidation in laboratory rodents and thus lower the concentrations and diminish the functions of α-amidated peptide messengers (*see* Reference 2). Several mechanism-based inhibitors have been developed for selectively inactivating PHM. Of these 4-phenyl-3-butenoic acid is the most effective at inhibiting α-amidation in vivo *(11)*. In addition, chelation of copper by treatment with disulfiram has been used to pharmacologically inhibit α-amidation *(12)*. Genetic models for investigating peptide α-amidation have been developed in both *Drosophila (13)* and mice *(1)*. Findings in PAM knockout mice demonstrate that PAM is essential for embryonic development and by inference, PAM is the sole mechanism for producing α-amidated peptides in mammalian embryos. Finally, several mouse strains bearing mutations in genes that encode critical copper transport proteins have been developed *(14,15)*. These animals exhibit reduced amounts of PAM activity and severe developmental abnormalities that closely resemble those observed in PAM knockout mice *(1,14,15)*.

4 Notes

1. Solid phase nonradioactive immunoassays offer a convenient alternative to solution based RIAs, however, at reduced sensitivity. Additionally, purified IgG is often required for optimal performance. In general, the requirement for detecting very low levels of α-amidated peptide in biologic samples has precluded the use of these nonisotopic assays.
2. Although not always available, a diode array detector provides the capability of continuously monitoring the spectral characteristics of the HPLC effluent. This information is useful for determining peak purity and can confirm the identity of peaks of interest by spectral characterization.

3. Because of the expense and complexity of mass spectrometry, most investigators carry out MS analyses in collaboration with a dedicated MS laboratory or core facility. MS laboratories commonly operate both MALDI-MS and HPLC-ES-MS instruments.

4. In the case of larger peptides, using the full length sequence for conjugation may be counter productive. This is because the resulting antiserum could recognize epitopes toward the N-terminus and not be α-amide specific. Accordingly, it is recommended that conjugations be carried out using haptens that are 8–12 residues in length.

5. The volume of the dialysate may increase by 50–100%. Loss of conjugate by binding to the dialysis tubing is not recognized as a problem in this procedure. The formation of a light colored precipitate of peptide/protein conjugate in the dialysate is not uncommon.

6. Considerable effort has been directed toward enhancing immunization procedures with the combined goal of improving antisera quality and limiting discomfort to the animals being inoculated. One development that has proven remarkably effective in this regard is the inclusion of colloidal gold with the immunogen. Colloidal gold alters antigen uptake and presentation in a manner that promotes the development of specific, high titer antisera against molecules with no intrinsic antigenicity (e.g., glutamate; *16*). Importantly, these effects are achieved with smaller amounts of immunogen. We have been successful with colloidal gold antigen using 100 μg as an immunization dose (in Freund's complete adjuvant) and 50 μg (in Freund's incomplete adjuvant) for subsequent boosts. Colloidal gold is commercially available from E-Y Laboratories (San Mateo, CA) or is easily prepared in house *(16)*. Although the specific application of colloidal gold with α-amidated peptides has not been reported, its proven effectiveness with weak antigens warrants its mention here.

7. Times and concentrations may be varied to increase or decrease labeling. In addition, the Na^{125}I can be reacted with the Iodo-Bead (5 min) before the addition of peptide. This enhances peptide labeling by increasing the conversion of iodide (I$^-$) to iodous ion (I$^+$), the species that actually mediates the iodination of tyrosine.

8. Perform charcoal absorption at room temperature by adding 1 mL of charcoal suspension (0.1% fine powder of activated charcoal/0.01% BSA in 0.05M phosphate buffer, pH 7.0) to each reaction tube, vortexing and centrifuging (2,500*g*, 45 min). Decant the supernatant containing antibody-bound counts into new tubes and count. It should be noted that charcoal suspensions must be stirred constantly during use and that activated charcoal can strip bound counts from antibodies of low avidity. Nonspecific binding can be reduced in some cases by increasing the concentration of activated charcoal/BSA. The second antibody precipitation method employs anti-rabbit IgG antiserum to form large immune complexes. In this case, normal rabbit serum is added to the reaction (1:100–1:300 final dilution) to generate complexes of sufficient mass for pelleting by centrifugation. Anti-rabbit IgG antisera prepared in goat or sheep are widely available from commercial suppliers. The second antibody must be appropriately

titered under assay conditions to insure that it is not limiting in complex formation and that maximal precipitation of antibody-bound counts is achieved.

9. Ideally, the volume injected is kept as small as reasonably possible (e.g., 50 µL) but may exceed 1 mL under conditions where the concentration of the peptide is low. Very dilute samples may be concentrated by lyophilization and then reconstituted in a small volume (e.g., 50 µL) of 2.5% acetonitrile containing 0.1% TFA. Samples are clarified by centrifugation (10,000g, 15 min) and injected onto the C_{18} HPLC column.

10. An alternative strong solvent worth testing for MALDI-MS is 50% formic acid. It should be noted that formylation of amino groups (+28 mass units) can occur spontaneously under neutral or basic conditions. However, under the acidic conditions of MALDI, this modification does not occur.

11. For HPLC-ES-MS acetic acid (1%) is substituted for TFA in the HPLC mobile phase because the strong ion pairing characteristics of TFA prevent efficient ionization.

12. Alternatives include 2,5-dihydroxybenzoic acid (10 mg/mL in 50% ethanol) and 3,5-dimethoxy-4-hydroxy-trans-cinnamic acid (10 mg/mL in 50% acetonitrile).

13. The inclusion of ammonium sulfate can significantly enhance ionization efficiency. This effect is assessed by adding 0.5 µL of saturated ammonium sulfate (in water) to the sample and matrix mixture. The order of addition to the slide is: sample, ammonium sulfate, matrix. Volumes can be reduced correspondingly (e.g., 0.3 µL each) to limit sample spreading.

References

1. Czyzyk, T. A., Ning, Y., Hsu, M. S., Peng, B., Mains, R. E., Eipper, B.A., and Pintar, J. E. (2005) Deletion of peptide amidation enzymatic activity leads to edema and embryonic lethality in the mouse. *Dev Biol.* **287**, 301–313.
2. Eipper, B., Stoffers, D., and Mains, R. (1992) The biosynthesis of neuropeptides: peptide _-amidation. *Annu. Rev. Neurosci.* **15**, 57–85.
3. Itoh, S. (2006) Mononuclear copper active-oxygen complexes. *Curr Opin Chem Biol.* **10**, 115–122.
4. Prigge, S. T., Mains, R. E., Eipper, B. A., Amzel, L. M. (2000) New insights into copper monooxygenases and peptide amidation: structure,mechanism and function. *Cell Mol. Life Sci.* **57**, 1236–1259.
5. Klinman, J. P. (2006) The copper-enzyme family of dopamine beta-monooxygenase and peptidylglycine alpha-hydroxylating monooxygenase: resolving the chemical pathway for substrate hydroxylation. *J Biol Chem.* **281**, 3013–3016.
6. Kulathilia, R., Merkler, K. A., and Merkler, D. J. (1999) Enzymatic formation of C-terminal amides. *Nat. Prod. Rep.* **16**, 145–154.
7. Driscoll, W. J., Mueller, S. A., Eipper, B. A., and Mueller, G. P. (1999) Differential regulation of peptide α-amidation by dexamethasone and disulfiram. *Mol. Pharm.* **55**: 1067–1076.
8. Domon, B. and Aebersold, R. (2006) Mass spectrometry and protein analysis. *Science* **312**, 212–217.
9. Reinders, J. and Sickmann, A. (2005) State-of-the-art in phosphoproteomics. *Proteomics* **5**, 4052–4061.

10. Baldwin, M. A. (2004) Protein identification by mass spectrometry: issues to be considered. *Mol. Cell Proteomics* **3**, 1–9.

11. Mueller, G. P., Driscoll, W. J. and Eipper, B. A. (1999) In vivo inhibition of peptidylglycine-α-hydroxylating monooxygenase by 4-phenyl-3-butenoic acid. *J. Pharmacol. Exp. Thera.* **290**, 1331–1336.

12. Mueller, G., Husten, E., Mains, R. and Eipper, B. (1993) Peptide α-amidation and peptidylglycine-α-hydroxylating monooxygenase: control by disulfiram. *Mol. Pharm.* **44**, 972–980.

13. Kolhekar, A. S., Roberts, M. S., Jiang, N., Johnson, R. C., Mains, R.E., and Eipper, B. A. (1997) Neuropeptide amidation in Drosophila: separate genes encode the two enzymes catalyzing amidation. *J. Neurosci.* **17**, 1363–1376.

14. Kuo, Y. M., Zhou, B., Cosco, D., and Gitschier, J. (2001) The copper transporter CTR1 provides an essential function in mammalian embryonic development. *Proc. Natl. Acad. Sci. U S A.* **98**, 6836–6841.

15. Steveson, T. C., Ciccotosto, G. D., Ma, X. M., Mueller, G. P., Mains, R. E., and Eipper, B. A. (2003) Menkes protein contributes to the function of peptidylglycine alpha-amidating monooxygenase. *Endocrinology* **144**, 188–200.

16. Shiosaka, S., Kiyama, H., Wanaka, A., and Toyama, M. (1986) A new method for producing a specific and high titer antibody against glutamate using colloidal gold as carrier. *Brain Res.* **382**, 399–403.

6

γ-Glutamate and β–Hydroxyaspartate in Proteins

Francis J. Castellino, Victoria A. Ploplis, and Li Zhang

Summary Vitamin K-dependent coagulation plasma proteins possess from 9–12 residues of γ-carboxyglutamic acid (Gla) distributed over a ca. 45 amino acid peptide sequence, i.e., the Gla domain, which encompasses the NH_2-terminal region. In addition, epidermal growth factor (EGF) homology units present in many of these same proteins contain β-hydroxyaspartate (Hya) residues, which is a modification decoupled from γ-carboxylation. The function of Gla residues in these proteins, viz., prothrombin, coagulation factors VII, IX, and X, along with anticoagulant protein C and protein S, is to coordinate Ca^{2+}. This results in a large conformational alteration in the proteins or peptides, which allows adsorption to membrane phospholipids (PL), an event that is critical is to their proper functions in the blood coagulation system. Less certain is the role of Hya in EGF domains, but it has been proposed that modification at this residue may negatively regulate fucosylation of these regions. In several proteins, these modules also interact with Ca^{2+}, but it has been shown that although the particular aspartate containing the β-OH group is critical to that interaction, β-hydroxylation of that Asp residue is not.

Because of their widespread distribution, quantitative detection protocols for both Gla and Hya are of importance. It is the purpose of this communication to detail a reliable method for these analyses that is employed in our laboratories.

Key Words γ-Carboxyglutamic acid (Gla); β-hydroxyaspartate (Hya); vitamin K-dependent proteins.

1 Introduction

Vitamin K-dependent coagulation plasma proteins possess from 9–12 residues of γ-carboxyglutamic acid (Gla) distributed over a approximately 45 amino acid peptide sequence, i.e., the Gla domain, which encompasses the NH_2-terminal region. In addition, epidermal growth factor (EGF) homology units present in many of

From: *Post-translational Modifications of Proteins.*
Methods in Molecular Biology, Vol. 446.
Edited by: C. Kannicht © Humana Press, Totowa, NJ

these same proteins contain β-hydroxyaspartate (Hya) residues, which is a modification decoupled from *gamma*-carboxylation *(1)*. The function of Gla residues in these proteins, viz., prothrombin, coagulation factors VII, IX, and X, along with anticoagulant protein C and protein S, is to coordinate Ca^{2+} and this results in a large conformational alteration in the proteins or peptides that allows adsorption to membrane phospholipids (PL), an event that is critical to their proper functions in the blood coagulation system *(2,3)*. Less certain is the role of Hya in EGF domains, but it has been proposed that modification at this residue may negatively regulate fucosylation of these regions *(4)*. In several proteins, these modules also interact with Ca^{2+}, but it has been shown that, although the particular aspartate containing the β-OH group is critical to that interaction, β-hydroxylation of that Asp residue is not.

A particularly novel class of naturally occurring peptides, the conantokins, also contain a wide variety of post-translational modifications, γ-carboxylation being one example. These small peptides, of up to 27 amino acids, are independent gene products found in predator snails of the species *Conus*, and function as noncompetitive inhibitors of Ca^{2+} flow through the ion channel of the *N*-methyl-D-aspartate (NMDA) receptor *(5–7)*. Because of this property, drugs based on these peptides have potential application in neuropathologies associated with NMDA receptor malfunction. The amino acid sequences of three of these peptides *viz.*, conantokin-G (con-G) *(8)*; conantokin-T (con-T) *(5)*, and conantokin-R (con-R) (unpublished), have been identified to date. These peptides contain 17, 21, and 27 amino acids, in which are present 5, 4, and 4 residues of Gla, respectively.

Other proteins and peptides contain Gla, such as matrix Gla protein and osteocalcin *(9–12)*, a protein from spermatozoa *(13)*, proline-rich transmembrane proteins *(14)*, and dentin *(15)*. Hya is not only found in EGF modules of proteins that are involved in coagulation *(16)*, but is also present in these same regions of thrombomodulin *(17)*, complement proteins *(18)*, and the LDL-receptor *(17)*.

Because of their widespread distribution, quantitative detection protocols for both Gla and Hya are of importance. It is the purpose of this communication to detail a reliable method for these analyses that is employed in our laboratories.

2   Materials

2.1   Hydrolysis of Samples

1. 2.5*M* KOH.
2. Saturated $KHCO_3$.
3. $HClO_4$ (60%; v/v).
4. 6*N* HCl.

2.2 HPLC Analysis of Hydrolysates by Ion Exchange HPLC

1. Preparation of OPA/ET reagent:
   Dissolve a total of 100 mg of o-phthalaldehyde (OPA, Sigma, St. Louis, MO) in 5 mL of methanol. Add a solution of 50 μL of ethanethiol (ET) in 10 mL of 0.15M sodium borate, pH 10.5, containing 0.2% Brij-35 (Pierce, Rockford, IL), and mix the solution thoroughly. Flush the mixture with N_2 and allow to stand for at least 16 h before analysis (*see* **Note 1**).
2. 0.1M KH_2PO_4 in 50% CH_3CN.
3. HPLC column (Nucleosil 5SB, 4 × 200 mm; 5 μm; Macherey-Nagel, Easton, PA) with Brownlee silica guard column (3.2 × 15 mm; 7 μ).
4. 20 mM sodium citrate, pH 4.2, 50% acetonitrile.
5. ANSFL(Gla)(Gla)RHSS (synthesized by standard Fmoc chemistry, *see* **Note 2**, *(19,20)*.
6. *erythro*- and *threo*-β-hydroxyaspartic acids (a gift from Dr. Marvin Miller, Dept. of Chemistry and Biochemistry, University of Notre Dame, Notre Dame, IN).

2.3 Gla Analysis by Reverse Phase HPLC

1. C8 column (2.0 × 250 mm, 5 μm).
2. Solvent A: 110 mL of 1M sodium acetate, pH 7.2, 95 mL methanol, 5 mL tetrahydrofuran, 790 mL H_2O.
3. Solvent B: 100% methanol.

2.4 Proteins and Peptides

1. Human plasma protein C (PC) (Enzyme Research Laboratories, South Bend, IN).
2. Recombinant human PC and mutants [Gla14D]-PC, [Gla20D]-PC and [D71A]-PC expressed in human kidney 293 cells *(21–28)*.
3. Bovine factors IX and X purified from bovine plasma *(29)*.
4. PC related peptide (residues 1–12 of human PC).
5. Conantokin-G and T *(19)*.

3   Methods

3.1 Hydrolysis of Samples

3.1.1 Gla Determination

1. Place lyophilized protein (20—100 μg) in a Pyrex glass tube (13 × 100 mm).
2. Add a solution of 220 μL of 2.5M KOH to the polypropylene tube, and then flush the entire system with N_2.

3. Seal the glass tube under vacuum (*see* **Note 3**) and insert in a deep-well heating block (Pierce, Rockford, IL) at 110°C. Allow the hydrolysis to proceed for 20 h.
4. After this time, place the glass tube on ice for 10 min, break the tube, and transfer the sample to an Eppendorf tube.
5. Combine the hydrolysate with a solution of 50 μL of saturated $KHCO_3$, and subject to centrifugation (3 min) with a table-top centrifuge to remove any precipitate.
6. Treat the solution with cold $HClO_4$ (60%; v/v) with frequent mixing until a pH of 7.0 (determined by spotting on pH paper), is attained.
7. After 30 min on ice, the $KClO_4$ precipitates. Remove the supernatant by aspiration, after centrifugation (800g) for 10 min, and store at −20°C until needed for amino acid analysis.

3.1.2   Determination of the Hya Content

1. Place the sample (100 μg) in a Pyrex tube (13 × 100 mm).
2. Add 300 μL of 6N HCl.
3. After flushing several times with N_2, seal the tube under vacuum and place the sample in the heating block at 110°C for 20 h for hydrolysis.
4. After this time, the HCl is removed by evaporation under N_2.
5. The hydrolysate is reconstituted in 50 μL of H_2O, and then employed for amino acid analysis.

3.2   HPLC Analysis of Hydrolysates by Ion Exchange HPLC

These methods represent operational modifications of previously described methodology (*19–23,30*). The derivation of amino acids in the hydrolysate was conducted with OPA/ET reagent because of the higher stability of the fluorescent derivative.

1. Add 10 μL of OPA/ET reagent to 10 μL of protein or peptide hydrolysate and mix the solution thoroughly by vortexing.
2. After 2 min, add 20 μL of 0.1M KH_2PO_4 in 50% CH_3CN.
3. A 20 μL aliquot of the sample is then injected into the HPLC column.

The separation of the OPA-conjugated amino acids is accomplished by isocratic elution on an ion exchange HPLC column (4 × 200 mm, Nucleosil 5SB) at 47°C, with detection at 340 nm. The elution buffer is 20 mM sodium citrate, pH 4.2, 50% acetonitrile. The elution positions of Glu, Asp, Hya, and Gla are shown in Figure 6.1. Because Glu and Asp are stable to both acid and alkaline hydrolysis, the ratios of Gla/Glu and Gla/Asp, combined with a standard curve, are used to obtain the number of Gla residues per mole of protein.

Two reference standards, a commercial amino acid standard mixture, as well as a peptide, ANSFL(Gla)(Gla)RHSS, are employed. Use of this peptide for this purpose allows for recovery values after alkaline hydrolysis to be obtained for

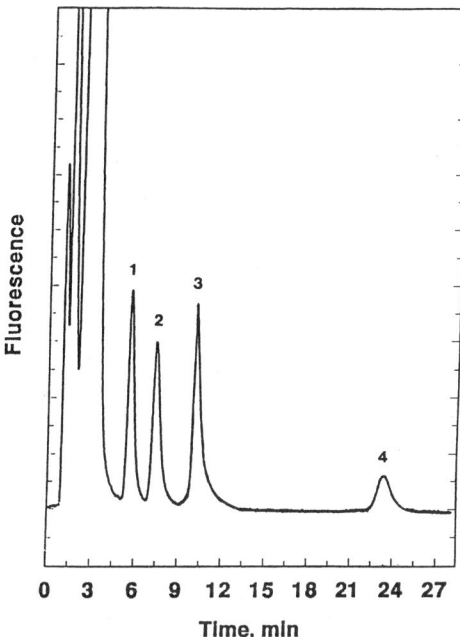

Fig. 6.1 The separation of acidic OPA-derivatized amino acids by anion-exchange HPLC. The separation was accomplished by isocratic elution on a Nucleosil 5SB column (4 × 100 mm; 5 μ) with a Brownlee silica guard column (3.2 × 15 mm; 7 μ). The elution buffer was 20 mM sodium citrate, pH 4.2, in 50% acetonitrile, at 47°C. Peak 1-Glu, peak 2-Asp, peak 3-*erythro*-β-hydroxyaspartic acid (Hya), peak 4-Gla

Gla and accurate conversion factors of peak area to concentrations of this amino acid. The Gla/Asp ratio of this peptide, of 2.0 (Asn is converted to Asp during hydrolysis), and the concentration response factor of Asp from the commercial standard mixture, are employed to obtain the concentration response for Gla.

For Hya determination, standard preparations of *erythro* and *threo* Hya, and the commercial amino acid standard mixture are used to obtain the concentration response factors of Hya and Asp. A ratio of Hya/Asp is used to calculate the amount of Hya per mole of protein in the hydrolysate.

3.3   Gla Analysis by Reverse Phase HPLC

Depending on the equipment available to an individual laboratory, an alternate method for quantitative Gla determinations is suggested using reverse-phase HPLC. After hydrolysis in alkali and precolumn derivatization with OPA/ET, as above, Gla analyses are carried out employing a C8 column (2.0 × 250 mm, 5 μ) column at 46°C with detection at 340 nm. The gradient used for elution consists of various combinations of two solvent systems. Solvent A contains 110 mL of 1M sodium acetate, pH 7.2/95 mL methanol/5 mL tetrahydrofuran/and 790 mL H_2O. Solvent B consists of 100% methanol. The column is equilibrated against 95% A:5% B. After application of the hydrolysate, the first gradient that is used immediately steps the solvent to 85% A:15% B (time = 0). From 0–10 min a linear gradient to a limit of 80% A:20% B is applied, followed by continued elution with 80% A:20% B for another 5 min. Under these conditions, Gla is eluted at 3.53 min with baseline separation from neighboring peaks, Asp at 4.98 min, and Glu at 5.95 min. Remaining materials are then batch eluted from the column with 2% A:98% B for 5 min, after which column start buffer (95% A:5% B) is applied for 5 min. The Gla content is determined using a standard synthetic Gla-containing peptide as described above.

3.4   Results and Conclusions

A representative example of the separation of standard OPA-derivatized acidic amino acids on a Nucleosil 5SB column is shown in Figure 6.1. It is seen that baseline separation of Glu, Asp, Hya, and Gla is achieved within a 26 min run time.

We have employed these analytical methods to determine Gla and Hya contents in recombinant and plasma proteins. Example chromatograms of an alkaline hydrolysate of recombinant PC and PC mutants, for Gla determinations, and an acid hydrolysate for Hya determinations, are provided in Figures 6.2 and 6.3. Using the approaches described above, quantitation of the Hya and Gla contents were determined for a variety of proteins and peptides, and a listing of the results of some of these analyses are provided in Table 6.1. The results point to the reliability of these analyses for all proteins and peptides studied.

In conclusion, reliable methodology is described for determination of Gla and Hya contents of a variety of proteins and peptides, which can be performed on μg quantities. Under the hydrolysis conditions described, the modified amino acids remain intact, but the exact hydrolysis conditions for any particular protein should probably be determined for each case.

4   Notes

1. The OPA/ET reagent can be stored in the dark at −20°C, and is usable for several months under these conditions.
2. N^α -Fmoc-γ,γ-ditBu-L-Gla-OH was used in the peptide synthesis for placement of γ in peptide *(19,20)*.
3. Using a Hamilton syringe, the protein solution is placed in a polypropylene tube fitted into the bottom of a pyrex 13 × 100 mm tube which was previously stretched into an hour-glass shape using an oxygen flame. The sample is then frozen in liquid nitrogen and evacuated 4 times. The tube is then sealed with a flame while under vacuum.

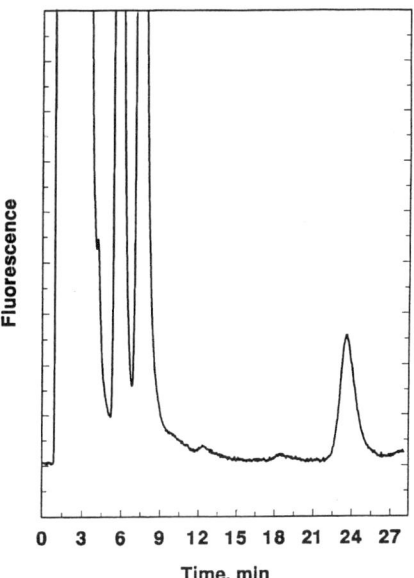

Fig. 6.2 HPLC chromatogram of an alkaline hydrolysate of wild-type recombinant protein C after OPA-derivatization. A quantity of 30 μg of protein was hydrolyzed with 2.5M KOH at 110°C for 20 h. After neutralization with 60% HClO$_4$, 5 μL of sample, corresponding to 0.5 μg of protein, was injected. The chromatography conditions are described in Figure 6.1. The peak at approximately 23 min represents Gla

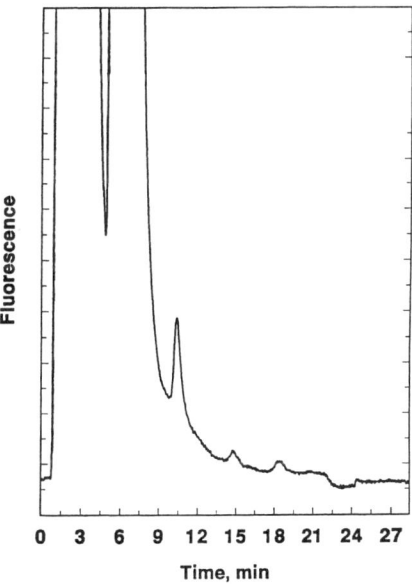

Fig. 6.3 HPLC chromatogram of an acid hydrolysate of wild-type recombinant protein C after OPA-derivatization. A quantity of 50 μg of protein was hydrolyzed with 6N HCl at 110°C for 20 h, and 5 μL, corresponding to 5 μg of protein, was applied to the column under the conditions of Figure 6.1. The peak at approximately 11 min represents Hya

Table 6.1 Gla and Hya Contents of Various Proteins

Protein/Peptide	Gla (mol/mol)		Hya (mol/mol)	
	expected	obtained	expected	obtained
PC-related peptide[a]	2.0	2.0	0	0
Bovine plasma FIX	12.0	12.0 ± 0.3	1.0	1.06 ± 0.05
Bovine plasma FX	12.0	12.5 ± 0.3	1.0	1.03 ± 0.04
Human plasma PC	9.0	8.7 ± 0.2	1.0	0.92 ± 0.04
Recombinant PC[b]	9.0	8.9 ± 0.3	1.0	1.04 ± 0.04
[Gla^{14}D]-PC[c]	8.0	7.8 ± 0.3	1.0	1.20 ± 0.13
[Gla^{20}D]-PC[c]	8.0	8.2 ± 0.2	1.0	0.78 ± 0.11
[D^{71}A]-PC[d]	9.0	8.7 ± 0.2	0	0
Conantokin-G[e]	5.0	4.7 ± 0.1	0	0
Conantokin-T[e]	4.0	4.1 ± 0.1	0	0

[a] Residues 1–12 of human protein C (PC)

[b] Wild-type recombinant PC, expressed in human kidney 293 cells.

[c] Human recombinant PC mutants. The mutation convention used is the [normal amino acid/ sequence position in PC/single letter code for the new amino acid placed in that sequence position by mutagenesis], followed by PC. Taken in part from (23).

[d] Taken from (28).

[e] Taken from (28).

References

1. Rabiet, M-J., Jorgensen, M. J., Furie B., and Furie, B. C. (1987) Effect of propeptide mutations on post-translational processing of factor IX. Evidence that β-hydroxylation and g-carboxylation are independent events. *J. Biol. Chem.* **262**, 14895–14898.

2. Nelsestuen, G. L. (1976) Role of γ-carboxyglutamic acid. An unusual protein transition required for the calcium-dependent binding of prothrombin to phospholipid. *J. Biol. Chem.* **251**, 5648–5656.

3. Nelsestuen, G. L., Broderius, M. and Martin, G. (1976) Role of γ-carboxyglutamic acid. Cation specificity of prothrombin and factor X-phospholipid binding. *J. Biol. Chem.* **251**, 6886–6993.

4. Harris, R. J., Ling, V. T., and Spellman, M. W. (1992) O-Linked fucose is present in the first epidermal growth factor domain of factor XII but not protein C. *J. Biol. Chem.* **267**, 5102–5107.

5. Haack, J. A., Rivier, J., Parks, T. N., Mena, E. E., Cruz, L. J., and Olivera, B. M. (1990) Conantokin-T. A γ-carboxyglutamate-containing peptide with N-methyl-D-aspartate antagonist activity. *J. Biol. Chem.* **265**, 6025–6029.

6. Skolnick, P., Boje, K., Miller, R., Pennington, M., and Maccecchini, M.-L. (1992) Noncompetitive inhibition of N-methyl-D-aspartate by conantokin-G: evidence for an allosteric interaction at polyamine sites. *J. Neurochem.* **59**, 1516–1521.

7. Blandl, T., Prorok, M., and Castellino, F. J. (1998) NMDA-receptor antagonist requirements in conantokin-G. *FEBS Lett.* **435**, 257–262.

8. McIntosh, J., Olivera, B. M., Cruz, L., and Gray, W. (1984) γ-Carboxyglutamate in a neuroactive toxin. *J. Biol. Chem.* **259**, 14343–14346.

9. Price, P. A., Poser, J. W., and Raman, N. (1976) Primary structure of the *gamma*-carboxyglutamic acid-containing protein from bovine bone. *Proc. Natl. Acad. Sci. USA* **73**, 3374–3375.

10. Hauschka, P. V., Frenkel, J., DeMuth, R., and Gundberg, C. M. (1983) Presence of osteocalcin and related higher molecular weight 4-carboxyglutamic acid-containing proteins in developing bone. *J. Biol. Chem.* **258**, 176–182.

11. Price, P. A., Urist, M. R., and Otawara, Y. (1983) Matrix Gla protein, a new *gamma*-carboxyglutamic acid-containing protein which is associated with the organic matrix of bone. *Biochem. Biophys. Res. Comm.* **117**, 765–771.

12. Price, P. A., Williamson, M. K. (1985) Primary structure of bovine matrix Gla protein, a new vitamin K-dependent bone protein. *J Biol Chem* **260**, 14971–14975.

13. Soute, B. A., Muller-Esterl, W., de Boer-van den Berg, M. A., Ulrich, M., and Vermeer, C. (1985) Discovery of a *gamma*-carboxyglutamic acid-containing protein in human spermatozoa. *FEBS Lett.* **190**, 137–141.

14. Kulman, J. D., Harris, J. E., Haldeman, B. A., and Davie, E. W. (1997) Primary structure and tissue distribution of two novel proline-rich *gamma*-carboxyglutamic acid proteins. *Proc. Natl. Acad. Sci. USA* **94**, 9058–9062.

15. Linde, A., Bhown, M., Cothran, W. C., Hoglund, A., and Butler, W. T. (1982) Evidence for several *gamma*-carboxyglutamic acid-containing proteins in dentin. *Biochim. Biophys. Acta* **704**, 235–239.

16. Drakenberg, T., Fernlund, P., Roepstorff, P., and Stenflo, J. (1983) β-Hydroxyaspartic acid in vitamin K-dependent protein C. *Proc. Natl. Acad. Sci. USA* **80**, 1802–1806.

17. Stenflo, J., Ohlin, A.-K., Owen, W. G., and Schneider, W. J. (1988) β-Hydroxyaspartic acid or β-hydroxyasparagine in bovine low density lipoprotein receptor and in bovine thrombomodulin. *J. Biol. Chem.* **263**, 21–24.

18. Thielens, N. M., Van Dorsselaer, A., Gagnon, J., and Arlaud, G. J. (1990) Chemical and functional characterization of a fragment of C1-s containing the epidermal growth factor homology region. *Biochemistry* **29**, 3570–3578.

19. Prorok, M., Warder, S. E., Blandl, T., and Castellino, F. J. (1996) Calcium binding properties of synthetic g-carboxyglutamic acid containing marine cone snail "sleeper" peptides, conantokin-G and conantokin-T. *Biochemistry* **35**, 16528–16534.

20. Colpitts, T. L. and Castellino, F. J. (1994) Calcium and phospholipid binding properties of synthetic *gamma*-carboxyglutamic acid-containing peptides with sequence counterparts in human protein C. *Biochemistry* **33**, 3501–3508

21. Zhang, L. and Castellino, F. J. (1990) A γ-carboxyglutamic acid variant (γ⁶D, γ⁷D) of human activated protein C displays greatly reduced activity as an anticoagulant. *Biochemistry* **29**, 10828–10834.

22. Zhang, L. and Castellino, F. J. (1991) Role of the hexapeptide disulfide loop present in the γ-carboxyglutamic acid domain of protein C in its activation properties and in the in vitro anticoagulant activity of activated protein C. *Biochemistry* **30**, 6696–6704.

23. Zhang, L., Jhingan, A., and Castellino, F. J. (1992) Role of individual *gamma*-carboxyglutamic acid residues of activated human protein C in defining its *in vitro* anticoagulant activity. *Blood* **80**, 942–952.

24. Zhang, L. and Castellino, F. J. (1992) Influence of specific γ-carboxyglutamic acid residues on the integrity of the calcium-dependent conformation of human protein C. *J. Biol. Chem.* **267**, 26078–26084.

25. Zhang, L. and Castellino, F. J. (1993) The contributions of individual *gamma*-carboxyglutamic acid residues in the calcium-dependent binding of recombinant human protein C to acidic phospholipid vesicles. *J. Biol. Chem.* **268**, 12040–12045.

26. Yu, S., Zhang, L., Jhingan, A., Christiansen, W. T., and Castellino, F. J. (1994) Construction, expression, and properties of a recombinant human protein C with replacement of its growth factor-like domains by those of human coagulation factor IX. *Biochemistry* **33**, 823–831.

27. Zhang, L. and Castellino, F. J. (1994) The binding energy of human coagulation protein C to acidic phospholipid vesicles contains a major contribution from leucine-5 in the *gamma*-carboxyglutamic acid domain. *J. Biol. Chem.* **269**, 3590–3595.

28. Cheng, C.-H., Geng, J.-P., and Castellino, F. J. (1997) The functions of the first epidermal growth factor homology region of human protein C as revealed by a charge-to-alanine scanning mutagenesis investigation. *Biol. Chem.* **378**, 1491–1500.
29. Stenflo, J. (1976) A new vitamin K-dependent protein. Purification from bovine plasma and preliminary characterization. *J. Biol. Chem.* **251**, 355–363.
30. Kuwada, M. and Katayama, K. (1983) An improved method for determination of *gamma*-carboxyglutamic acid in proteins, bone and urine. *Anal. Biochem.* **131**, 173–179.

7

Lysine Hydroxylation and Cross-linking of Collagen

Mitsuo Yamauchi and Masashi Shiiba

Summary Collagens represent a large family of structurally related extracellular matrix proteins containing unique triple helical structure. One of the characteristics of this structural protein is its extensive post-translational modifications that have major effects on molecular assembly, stability, and metabolism. Hydroxylation of specific lysine residues is one of such unique modifications found in collagen, and the pattern/extent of this modification influences fibrillogenesis, cross-linking, and matrix mineralization. The formation of covalent intermolecular cross-linking is the final modification in collagen biosynthesis and is critical for the stability of collagen. The process of cross-linking is dynamic and the pathways vary depending on the tissues and tissue's physiological state. This tissue specificity of cross-linking pattern may in part be the results of differential expression of various isoforms of lysyl hydroxylases and lysyl oxidases that have been recently identified and partially characterized. This chapter concentrates on recent research progress on these two modifications and the methods for analysis we have developed.

Key Words Collagen post-translational modifications; collagen cross-linking; lysine hydroxylation; amino acid analysis.

1 Introduction

Collagen is a large family of extracellular matrix proteins that contain at least one triple helical domain formed by repeating (Gly-X-Y) sequences *(1)*, and is the most abundant molecule in vertebrates. Type I collagen, the predominant genetic type in the collagen family, is a heterotrimeric molecule composed of 2 $\alpha1$ chains and 1 $\alpha2$ chain, and is the major fibrillar component in most connective tissues. This molecule consists of 3 domains: amino-terminal nontriple helical (*N*-telopeptide), central triple helical (helical), and carboxy-terminal nontriple helical (*C*-telopeptide) domains (Fig. 7.1). The central helical domain of each chain contains more than 300 repeats of (Gly-X-Y) sequence and represents more than 95% of the polypeptide. One of the characteristic features of collagen is its extensive post-translational

From: *Post-translational Modifications of Proteins.*
Methods in Molecular Biology, Vol. 446.
Edited by: C. Kannicht © Humana Press, Totowa, NJ

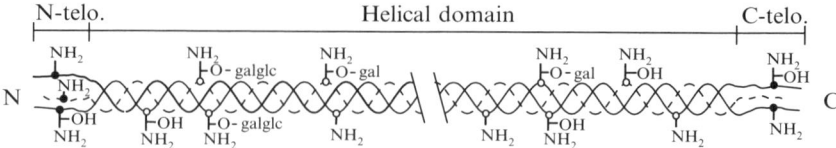

Fig. 7.1 Structure of type I collagen molecule. Type I collagen molecule is composed of two α1 chains (solid line) and one α2 chain (dotted line). The molecule consists of three domains: amino-terminal non-helical (N-telo.), central triple helical (helical) and carboxy-terminal non-helical (C-telo.) domains. Certain lysine residues in these domains are hydroxylated. Specific hydroxylysine residues in the helical domain can be o-glycosylated to form galactosylhydroxylysine or glucosylgalactosylhydroxylysine residues. In general, the dissaccharide form is more prevalent (*30*) although the monosaccharide derivartive is relatively abundant in certain tissues

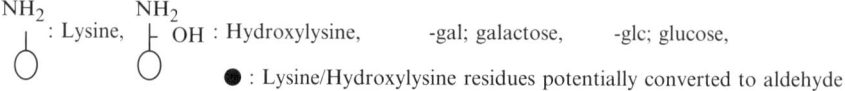

● : Lysine/Hydroxylysine residues potentially converted to aldehyde

modifications most of which are unique to collagen protein. Such modifications include hydroxylation of proline (Pro) and lysine (Lys) residues, glycosylation of specific hydroxylysine (Hyl) residues, oxidative deamination of the ε-amino groups of peptidyl Lys/Hyl in the telopeptide domains of the molecule, and subsequent intra/intermolecular covalent cross-linking *(2)* (Fig. 7.1). The hydroxylation of Pro is catalyzed by prolyl 4-hydroxylase and prolyl 3-hydroxylase. The former reacts on Pro with the minimum sequence X-Pro-Gly and the latter appears to require a Pro-4-Hyp-Gly sequence *(3)*. The presence of 4-Hyp (predominant form) is critical in stabilizing the triple helical conformation of collagen providing hydrogen bonds and water bridges *(4)*. The content of Pro/Hyp in type I collagen is relatively high representing about 25% of the total amino acids and 40–45% of Pro is hydroxylated (mostly 4-Hyp).

The hydroxylation of Lys is important for 2 further critical modifications of collagen, i.e., glycosylation and covalent cross-linking. In type I collagen, glyco-sylation (either by a single galactose or a disaccharide glucosyl-galactose residue) occurs only at the hydroxyl group of certain Hyl residues in the helical domain. Although the function of the collagen glycosylation is still unclear, several functions have been proposed *(2)*. In human type I collagen, there are 38 residues of Lys in an α1 chain (36 in the helical, 1 in the C-, and 1 in the N-telopeptide domains) and 31 in an α2 chain (30 in the helical, 1 in the N-telopeptide, and none in the C-tel-opeptide domains) (Sequence accession #P02452, P02464 and P08123) (Fig. 7.2A). The extent of Lys hydroxylation of collagen is highly variable in comparison to that of Pro. It varies from one genetic type to another *(5)* and, even within the same genetic type I collagen, it significantly varies depending on the tissues and the tis-sue's physiological/pathological conditions. Furthermore, a difference in the extent of Lys hydroxylation exists between the helical and telopeptide domains of a type I collagen molecule. The latter observation, together with the fact that a purified lysyl hydroxylase fails to hydroxylate the Lys residues in the telopeptide domains, led investigators to speculate that there is more than one mechanism for Lys

(A)

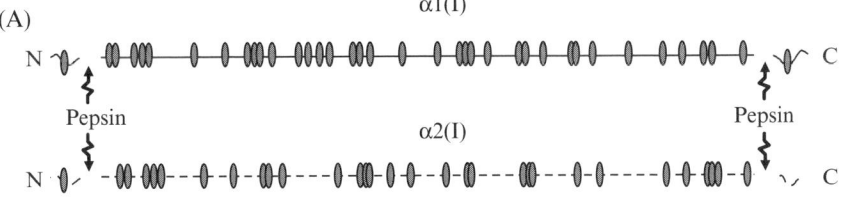

(B)

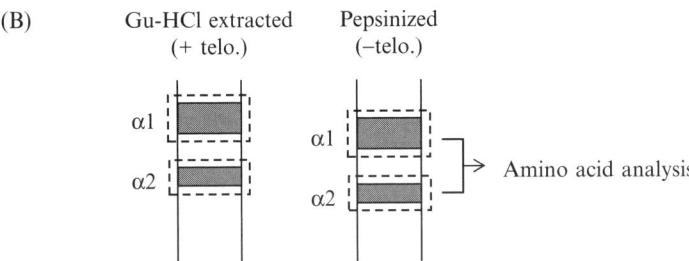

Fig. 7.2 (**A**) Relative distribution of lysine residues on α1 (solid line) and α2 (dotted line) chain of human type I collagen. The central straight line represents the helical and the curved lines at both ends telopeptide domains. The lysine residues are indicated by closed grey ovals. The helical domains of α1 and α2 chains contain 36 and 30 residues of lysine, respectively. Note that each telopeptide contains one lysine residue except for the C-telo peptide domain of an α2 chain. The pepsin treatment of type I collagen removes the telopeptide domains. When a collagen molecule is treated with pepsin, the telopeptide domains are cleaved. (**B**) Typical SDS-PAGE pattern of type I collagen with (left panel) and without (right panel) telopeptides. Note that α1 and α2 chains of pepsinized collagen migrate slightly faster than those of guanidine-HCl extracted collagen because of the lack of the telopeptides. Each band is excised and subjected to amino acid analysis. Based on the results, the extent of Lys hydroxylation in the helical and teloprptide domains can be calculated

hydroxylation of collagen *(6,7)*. Recently, 3 genes encoding for isoforms of lysyl hydroxylase (Procollagen-lysine, 2 oxyglutarate, 5 dioxygenase, *PLOD 1-3/LH1-3*) have been cloned and characterized *(8, 9)*. In addition, an alternatively spliced variant of LH2, LH2b (also described as LH2alt), with an additional 63 bp-exon 13A has been reported *(10)*. Our study on the expression pattern of these genes during human osteoblastic cell differentiation indicated that *LH2* gene (PLOD2) expression was associated with Lys hydroxylation in the telopeptide domains of the type I collagen molecule (*see* Fig 2A, B for the method) *(11)*. More recently, we have reported that the major form of LH2 in osteoblastic cells is LH2b and that LH2b indeed directs the collagen cross-linking pathway by catalyzing Lys hydroxylation in the telopeptides *(12)*. In this report, we demonstrated that when LH2b was overexpressed in osteoblasts, Hyl[ald]-derived collagen cross-links became predominant, but when underexpressed, Lys[ald] and its derived cross-links were significantly increased (Figs. 7.3, 7.4). A deficiency of pyridinoline cross-link (i.e., Hyl[ald]-derived cross-link, see Fig. 7.3) in patients with Bruck syndrome, a rare autosomal recessive disorder associated with mutations in the LH2 gene *(13)*, and other recent studies *(14,15)* also support the involvement of LH2b in the hydroxylation of Lys residues in the telopeptides of collagen. The LH2b induced modification of collagen and its consequences, i.e., telopeptidyl Lys hydroxylation and cross-linking pattern, has significant effects on collagen fibrillogenesis and matrix mineralization *(16)*. Interestingly, the effect of LH overexpression on collagen matrix mineralization in vitro varies significantly among the different isoforms *(17)*. These results indicate a critical role of domain-specific Lys hydroxylation of collagen in connective tissue functions.

Although Lys hydroxylation of type I collagen is important to determine the pattern of cross-linking, the process of cross-linking does not begin without the action of an amine oxidase, lysyl oxidase (LOX), on Lysl/Hyl located in the telopeptides of collagen. The formation of covalent intra/intermolecular cross-links, the final step of collagen biosynthesis, is critical in providing the fibrils with stability. This process is initiated with conversion of the ε-amino groups of specific Lys and Hyl residues located in the telopeptide domains of the molecule to aldehyde (Lys^{ald} and Hyl^{ald}, respectively) that is catalyzed by LOX *(18)*. Recently, as in the case of LH, several isoforms of LOX that possess amine oxidase activity have been identified and partially characterized. At present, 4 LOX-like proteins (LOXL1–4) are known to be present in multiple tissues and each LOX isoform appears to be expressed and localized in a cell/tissue specific manner *(19)*. It is likely that the cross-linking pattern/quantity is also regulated by the differential expression of these isoforms *(20)*. The aldehyde produced, i.e., Lys^{ald} and Hyl^{ald}, then undergoes a series of condensation reactions involving the juxtaposed Lys/Hyl or histidine (His) residues on the neighboring molecules (Fig. 7.3). Obviously, the extent of Lys hydroxylation in the telopeptide domain and that in the juxtaposed helical domain of a neighboring molecule are a key determinant for the subsequent condensation reaction products. Lys^{ald} in the N-telopeptide domain, for instance, can react with another Lys^{ald} within the same molecule to form an intramolecular aldol condensation product (ACP), then matures into a tetravalent intermolecular cross-link, dehydrohistidinohydroxymerodesmosine (deH-HHMD), by involving the juxtaposed His and Hyl residues on a neighboring molecule *(18)*. In skin and cornea, Lys^{ald} in the C-telopeptide domain reacts with the specific Hyl of an $\alpha1$ chain on a neighboring molecule and the resulting iminium bifunctional cross-link, dehydro-hydroxylysinonorleucine (deH-HLNL), matures into a trifunctional stable cross-link, histidinohydroxylysinonorleucine (HHL) by involving His of an $\alpha2$ chain on another molecule *(21–23)*. The major Hyl^{ald}-derived pathway leads to the trifunctional cross-links, pyridinoline (Pyr) and/or its lysyl analog deoxypyridinoline (d-Pyr) via the intermediate bifunctional cross-link, dehydrodihydroxylysinonorleucine (deH-DHLNL) /its ketoamine *(24)*. A pyrrole compound is also a mature trifunctional cross-link of collagen *(25,26)*. Although the mechanism of its formation is still not clear, it likely involves both Hyl^{ald} and Lys^{ald}. Detailed chemistries of these LOX mediated cross-linking pathways are described in several review articles *(24, 27–29)*. The stereospecificity and stoichiometry of these cross-links at the specific molecular loci (e.g., the frequency and the relative involvement of Hyl of $\alpha1$ and $\alpha2$ chains) provide valuable information concerning the molecular packing structure of a fibril that seems to be tissue specific *(30, 31)* and the substrate specificity of various isoforms of LH and LOX.

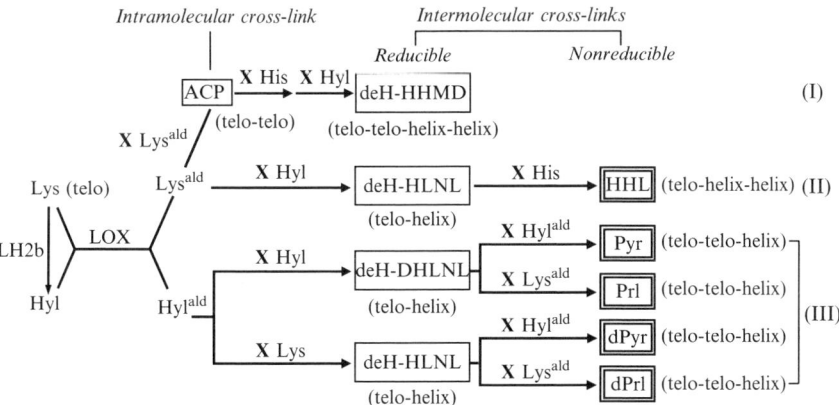

Fig. 7.3 Major cross-linking pathways of collagen. Predominant in (**I**): soft tissues in general, (**II**): skin and cornea, (**III**): skeletal tissues LH2b: Lysyl hydroxylase-2b, LOX: Lysyl oxidase, *ald*: aldehyde, ACP: Aldol Condensation Product (*Intra*molecular cross-link), deH: dehydro, HLNL: hydroxylysinonorleucine, DHLNL: dihydroxylysinonorleucine, HHMD: histidinohydroxymerodesmosine, HHL: histidinohydroxylysinonorleucine, Pyr: pyridinoline, d: deoxy, Prl: pyrrole Cross-linking compounds are indicated in the rectangular boxes with a single (reducible with borohydride under mild conditions) or double (nonreducible) line. All compounds are intermolecular cross-links except ACP. The loci of each cross-link within the molecular domains (telopeptides or helix) are described in a bracket

Thus, the analysis for the extent of Lys hydroxylation in different domains of each α chain provides important information about the cell phenotypes *(32)*, functions of various LH encoding genes and the tissue specific cross-linking chemistries. In the past, because of the difficulties in solubilizing and purifying the α chains of tissue collagen, Lys hydroxylation was analyzed, in most cases, on a selective pool of tissue collagen or selective domains of α chains. In this chapter, methods we have developed for characterization of the Lys hydroxylation of the individual α chains of type I collagen with and without telopeptides and for collagen cross-links using a cell culture system are described. By using a cell culture system, the matrix type I collagen synthesized by specific cells and deposited in the cultures can be readily solubilized/extracted by pepsin digestion (without telopeptides) or by guanidine-HCl (with telopeptides). Based on the amino acid analysis of the individual α chains obtained from these two methods, the extent of Lys hydroxylation in helical and telopeptide domains can be calculated. Furthermore, the cross-linking profile (mostly reducible cross-links) can be also analyzed for the same collagen matrix.

2 Materials

2.1 Cell Culture/Extraction of Type I Collagen

1. Cells: Cell line or those obtained from human tissues.
2. Lysis buffer: $0.1M$ Tris-HCl, $0.125M$ NaCl buffer containing 1% Triton X-100, 0.1% sodium dodecyl sulfate (SDS) and 1% deoxycholate.

3. Protease inhibitors: 0.5 mM phenylmethanesulfonylfluoride, 5 mM benzamidine, 2 mM pepstatin A and 1 mM leupeptin in dimethylsulfoxide.
4. Phosphate buffered saline (PBS): 6.7 mM potassium phosphate, pH 7.4, containing 150 mM NaCl.

2.2   Isolation of Type I Collagen without Telopeptide Domains

1. Pepsin (Worthington).
2. Buffer for pepsin digestion: cold 0.5N acetic acid.
3. Buffer for collagen precipitation: cold 0.5N acetic acid containing 0.7M NaCl.
4. Instrument: Ultracentrifuge.

2.3   Isolation of Type I Collagen with Telopeptide Domains

1. Extraction buffer: 0.05M Tris-HCl, pH7.4, containing 6M guanidine-HCl.
2. Instrument: Ultracentrifuge.

2.4   Separation and Blotting of α1 and α2 Chains of Type I Collagen

1. SDS gel-loading (sample) buffer: 63 mM Tris-HCl, pH6.8, 10% glycerol, 2% SDS, 0.0025% bromophenol blue.
2. Acrylamide gel: 5% acrylamide gel (acrylamide:N, N'-methylenebisacrylamide; 29:1).
3. Transfer buffer: 10 mM 3-1-propanesulfonic acid, pH 11.0 and 10% methanol.
4. Polyvinylidene fluoride (PVDF) membrane.
5. Gel staining solution: 0.25% coomassie blue and 10% glacial acetic acid.
6. Gel destaining solution: 30% methanol and 10% acetic acid.
7. Instrument: Transfer cell (Trans-Blot SD, Bio-Rad).

2.5   Acid Hydrolysis and Amino Acid Analysis

1. 6N HCl (sequencing grade).
2. N$_2$ gas.
3. Centrifuge tube filter: 0.22 μm; cellulose acetate membrane.
4. Instruments:

    a. Oven (115°C) for acid hydrolysis.
    b. Speed vacuum centrifuge system to evaporate acid.

  c. Desktop microfuge for filtration of the hydrolysates.
  d. HPLC system equipped with a strong cation exchange column (AMINOSep #AA-911, Transgenomic) for amino acid analysis.

5. HPLC buffer:

  a. Buffer A: 12.5 mM tartaric acid disodium, 17.2 mM maleic acid, 2.5% isopropyl alchohol, pH 2.78.
  b. Buffer B: 0.3 M NaOH, 107.7 mM maleic acid, 48.5 mM boric acid, pH 9.91.

6. Ninhydrin, ninhydrin reactor (135°C).

2.6 Collagen Cross-Link Analysis

1. Reducing buffer: 0.15M N-trismethyl-2-aminoethanesulfonic acid (TES), 0.05M Tris-HCl, pH7.4.
2. Antifoam solution.
3. NaB^{3}H$_4$.
4. Acid hydrolysis: same as **Section 2.5**.
5. Instruments: HPLC system linked to an on-line fluorescence flow monitor and a liquid scintillation monitor.
6. Column: same as **Section 2.5.** item **4.d**.
7. HPLC buffer: same as **Section 2.5**.

3 Methods

3.1 Cell Culture/Extraction of Type I Collagen (see Note 1 as well)

1. Culture the cells of interest with appropriate medium.
2. Maintain the cells for 2–4 weeks with replacement of medium twice weekly.
3. Collect the cells and insoluble matrices by scraping the culture dishes.
4. Centrifuge at 2,000g for 30 min at 4°C.
5. Remove the supernatant and, if necessary, add 15 mL of lysis buffer including a cocktail of protease inhibitors (for cross-link analysis, steps 5–7 can be omitted).
6. Incubate for 30 min at room temperature and then incubate overnight at 4°C on a rotating platform.
7. Centrifuge at 2,000g for 30 min at 4°C.
8. Wash the residue with PBS (×3) and with cold distilled water (×1).
9. Lyophilize and store at −80°C.

3.2 Isolation of Type I Collagen α Chains Without Telopeptide Domains (see Note 2 as well) (Fig. 7.3)

1. Dissolve 2 mg of lyophilized samples (*see* **Section 3.1.**) in 2 mL of cold 0.5N acetic acid.
2. Add pepsin (20% W/W) and incubate for 24 h at 4°C with constant stirring.
3. Centrifuge at 40,000g for an hour at 4°C.
4. Remove the residue and add NaCl gradually to the supernatants to a final concentration of 0.7M while stirring.
5. Stir the solution gently for 24 h at 4°C.
6. Centrifuge at 40,000g for an hour at 4°C.
7. Discard the supernatant and dissolve the precipitate in 2–3 mL of 0.5N acetic acid.
8. Dialyze against cold distilled water overnight at 4°C, lyophilize and store at −80°C.

3.3 Isolation of Type I Collagen α Chains with Telopeptide Domains (see Note 2 as well) (Fig. 7.3)

1. Dissolve 2 mg of lyophilized samples (*see* **Section 3.1.**) in extraction buffer (6M guanidine-HCl, see **Section 2.3.**).
2. Stir for 3–4 days at 4°C.
3. Centrifuge at 40,000g for an hour at 4°C.
4. Dialyze the supernatant against cold distilled water for 2 d at 4°C.
5. Lyophilize and store at −80°C

3.4 Separation and Blotting of α1 and α2 Chains of Type I Collagen (see Note 3) (Fig. 7.4)

1. Dissolve 100 µg of the dried samples (*see* **Section 3.2. and 3.3.**) in SDS gel-loading buffer.
2. Boil for 5 min.
3. Apply the samples to SDS-PAGE using a 5% acrylamide gel under non-reducing condition.
4. Transfer onto a PVDF membrane with transfer buffer.
5. Stain the membrane lightly with the gel staining solution and destain with the destaining solution.
6. Excise the bands corresponding to α1 and α2 chains from the PVDF membrane.

3.5 Acid hydrolysis and Amino Acid Analysis (see Note 4)

1. Hydrolyze the excised membrane (*see* **Section 3.4., step 6**) with 6N HCl *in vacuo*, after flushing with N_2, for 22 h at 115°C.

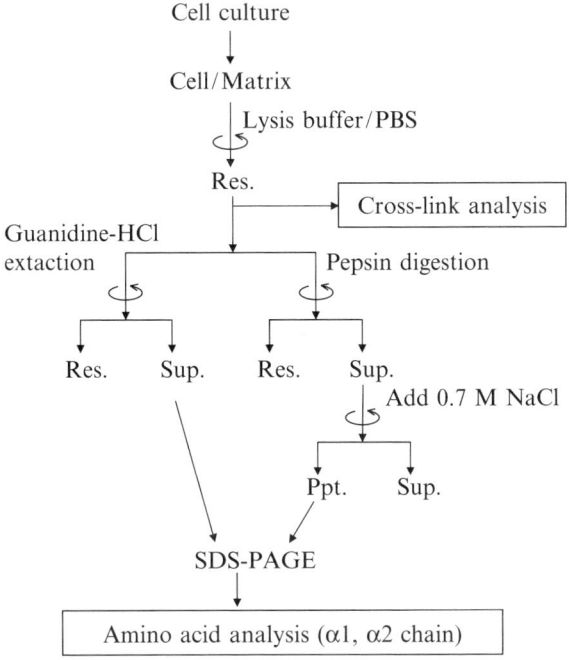

Fig. 7.4 Flow diagram of sample preparation for lysine hydroxylation and cross-link analysis of collagen from cell cultures

2. Dry the hydrolysate under reduced pressure by a speed vacuum centrifuge.
3. Dissolve the hydrolysate in distilled water or HPLC buffer A (*see* **Section 2.5.5.a.**) and filter with a 0.22-µm centrifuge tube filter.
4. Apply the hydrolysates to the HPLC system configured as an amino acid analyzer.
5. Elute amino acids with a combination of isocratic and linear gradients at a flow rate of 0.5 mL/min and detect the amino acids using ninhydrin for color development.
6. Analyze the Lys hydroxylation of collagen α chains based on the values of Hyl and Lys (*see* **Note 5**) (Fig. 7.5).

3.6 Collagen Cross-Link Analysis (see Note 6)

1. Dissolve 1–2 mg of dried samples (*see* **Section 3.1.**) in 3 mL of reduction buffer (*see* **Section 2.6.1.**). Add a few drops of antifoam solution. Ensure that pH of the solution is close to 7.5.
2. Add a total of 50–100-fold molar excess of NaB^{3}H$_4$ with three 5 min intervals with constant stirring and, after the third addition, continue to stir for 10 min (*see* **Note 7**).

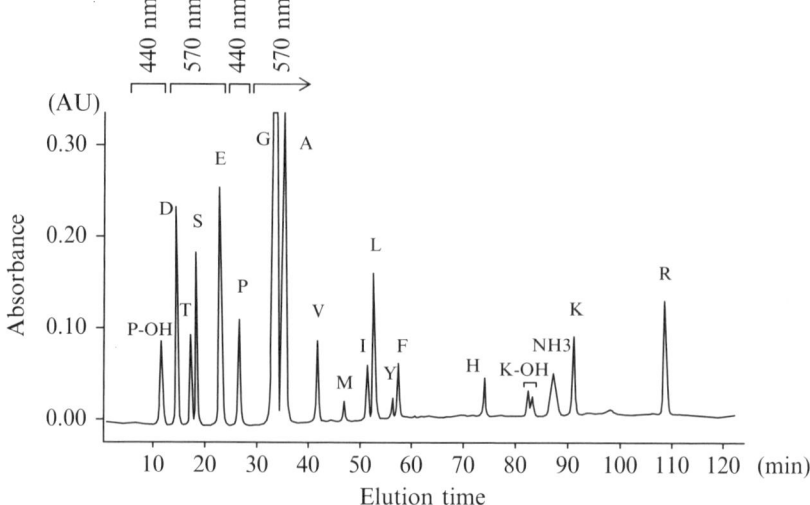

Fig. 7.5 An example of amino acid analysis of an α chain of type I collagen. One-letter code is used to designate amino acids. P-OH and P are measured by absorbance at 440 nm and all others at 570 nm. The automatic wavelength change (from 440 nm to 570 nm or vice versa) is programmed at appropriate points. The extent of Lys and Pro hydroxylation can be calculated by Hyl/(Lys+Hyl) and Hyp/(Pro+Hyp), respectively
P-OH: hydroxyproline, K-OH: hydroxylysine

3. Add a few drops of 50% acetic acid to drop the pH to 3–4.
4. Wash extensively with distilled water by centrifugation or dialysis to remove free NaB^3H_4 (*see* **Note 8**).
5. Lyophilize the sample and store at −80°C.
6. Hydrolyze with 6N HCl as in **Section 3.5., step 1–3**.
7. Determine the amount of collagen in the hydrolysate by Hyp content measured by amino acid analysis as in **Section 3.5., step 4–5**.
8. Apply the hydrolysates with known amount of Hyp to the HPLC system fitted with an ion-exchange column linked to a fluorescence flow monitor and a liquid scintillation monitor (*see* **Note 9**) (Fig. 7.6).
9. Measure the contents of reducible cross-links and aldehydes as their reduced forms based on the specific activity of NaB^3H_4 (*see* **Note 6**). The fluorescent cross-links (pyridinoline and deoxypyridinoline) are quantified by integrating the areas of the respective fluorescent peaks standardized by the hydrolysate of an apparently pure pyridinoline containing peptides *(30)*.
10. The cross-links and cross-link precursor aldehydes are calculated as a mole/mole of collagen basis based on the value of 300 residues of Hyp per collagen molecule.

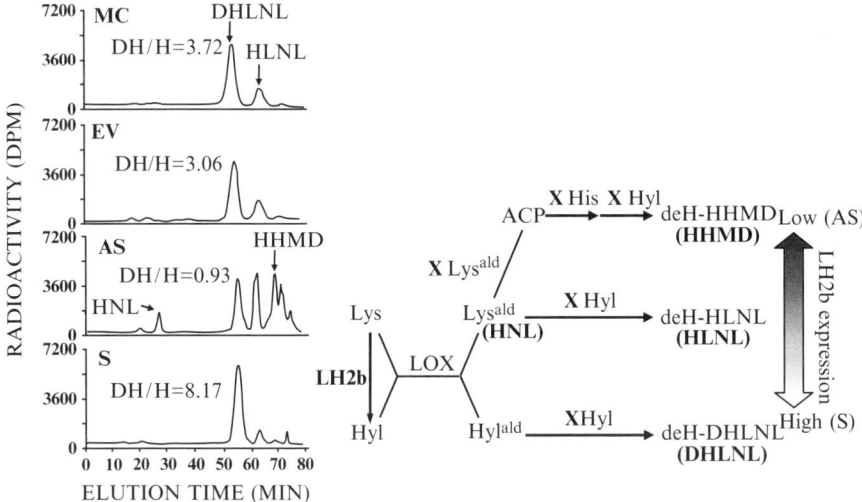

Fig. 7.6 Examples of cross-link analysis by HPLC on collagen obtained from matrices deposited by preosteoblastic MC3T3-E1 (MC) cells (subclone 4) and MC-derived clones (left panel). On the right panel, the cross-link pathway related to this analysis is shown. All cross-links and a precursor aldehyde shown in the chromatograph are NaB³H₄-reduced forms and they are indicated by bold letters in parentheses in the right panel. EV: a clone transfected with an empty pcDNA3.1/V5-His A vector (Invitrogen), AS/S: clones transfected with the vectors containing LH2b cDNA in an antisense (AS) or sense (S) direction. For details see ref 12. The major collagen cross-links in controls (MC and EV) are DHLNL (reduced deH-DHLNL/its ketoamine) and HLNL (reduced deH-HLNL/its ketoamine), and the ratio of the former to the latter (shown as DH/H in the figure) was 3-4 in these controls. When LH2b expression is low (AS), the ratio became lower than 1, and HNL (reduced form of Lys^ald) and a Lys^ald-derived cross-link, HHMD (reduced deH-HHMD), are formed in the clone. However, when the expression is high (S), the ratio went up to over 8, and neither Lys^ald nor its derived cross-links was formed. This provides evidence that LH2b in osteoblastic cells does function at least in part as a telopeptidyl LH

4   Notes

1. Collagen samples can be stored at −80°C. Try to avoid an exposure to UV light (e.g., store the samples in an amber bottle) because of the potential photolysis of pyridinium cross-links though they are relatively stable in acid and when they are bound to peptide chains. Most intra- and extracellular noncollagenous soluble proteins and procollagen molecules are removed by these extraction and washing procedures.

2. Under the conditions employed, pepsin cleaves only telopeptide domains but not the helical domain. Therefore, α chains purified from a pepsin digest lack the telopeptides whereas those purified from the guanidine-HCl extract do retain those. Thus, the extent of Lys hydroxylation in each domain of each α chain can be calculated based on the ratio of Hyl to Lys of these α chains because the number of Lys residues in these domains are known (Sequence accession #P02452, P02464 and P08123) (Fig. 7.4A). This information is useful to correlate the Lys hydroxylation in these domains to the expression of various LH

encoding genes *(11, 33)*. By this method, however, the Lys hydroxylation between the C- and N-telopeptides of an $\alpha 1$ chain are indistinguishable.

3. For this procedure, $\alpha 1$ and $\alpha 2$ chains need to be well separated from each other on the gel. Five % acrylamide gel is an excellent concentration for this purpose *(11)*. Because of the loss of the samples that occurs during blotting, a relatively large quantity of the sample (about 100 µg) should be loaded on a large gel (18 × 16cm). Because type III collagen that contains inter-chain disulfide bonds can be also present in the precipitate/extract, it is important to dissolve the sample with the non-reducing sample buffer for SDS-PAGE. Otherwise the type III α chain co-migrates with an $\alpha 1$ chain of type I collagen.

4. The pH of the HPLC buffer should be adjusted by adding maleic acid to the solution. All collagen associated amino acids can be separated by a combination of isocratic and linear gradients from 0 to 98% of buffer B over a period of 2h. The column temperature is set at 60°C (0–60min) and 90°C (60–120min). The color development with ninhydrin is performed at 135°C in a stainless steel reaction coil. The temperature is maintained to ±0.1°C with a thermostated silicone oil bath. The eluent is constantly monitored at 570nm except for Hyp and Pro that are monitored at 440nm. The wavelength change between 570 and 440nm is precisely set before and after these amino acids. By this system, Hyl and Lys elute before and after NH_3, respectively (Fig. 7.5). The quantity of each amino acid is calculated based on the respective amino acid standards. Using this system, many cross-linking amino acids including Pyr, d-Pyr, DHLNL, HLNL, HHL and HHMD can be separated from other amino acids and from each other. The extent of Pro hydroxylation in each α chain can also be calculated by Hyp/(Pro+Hyp) *(34)*.

5. The number of Hyl residues in an α chain with or without telopeptides can be calculated based on the ratio of Hyl to Lys and the number of Lys residues in each domain (Fig. 7.4A). The number of Hyl residues in the telopeptide domains then can be calculated by subtracting the number of Hyl residues in an α chain with telopeptides from that in the α chain without.

6. Reducing procedure must be carried out in a hood. NaB^3H_4 is standardized as follows: demineralized dentin collagen obtained from unerupted bovine teeth, which has a very simple cross-link pattern (mostly DHLNL with a small amount of HLNL), is reduced with the NaB^3H_4 and hydrolyzed with constant boiling 6N HCl for 22h at 115°C. The hydrolysate will be dried, dissolved in water and subjected to standardized P-2 column chromatography *(21)*. Radioactive fractions that encompass these cross-links (DHLNL and HLNL) are pooled and lyophilized. After dissolving in water or HPLC buffer A, an equal amount of the sample solution is subjected to amino acid analysis (for their quantities) and cross-link analysis (for their radioactivity counts), respectively. Elution positions of these cross-links are identified by cross-link standards *(21)*. The ninhydrin color factor for cross-links are obtained from amino acid analysis of apparently pure cross-linked peptides isolated from various connective tissues *(31)*. In this manner, the specific activity of the NaB^3H_4 can be calculated and expressed in DPM/µmole.

7. This wash process can be done by repeated centrifugation or exhaustive dialysis. Either way, it is important to remove most of the free NaB^3H_4 to obtain a clean analysis pattern.

8. For cross-link analysis, a linear gradient of buffer B from 0 to 93% in 57 min (0 to 55% in 15min, 55–60% in 25 min and 60–93% in 17 min) and an isocratic gradient at 93% for the next 40 min (including wash process) at a flow rate of 0.5 mL/min are used. The column temperature is set at 58°C for the cross-link analysis and then 88°C for a column wash. The pyridinium cross-links (Pyr and d-Pyr) are identified and quantified by a fluorescence flow monitor at excitation 330 nm and emission 390nm. Then the eluate is mixed with scintillation fluid and the reducible compounds (i.e. reduced aldehydes, DHLNL, HLNL and HHMD) are measured by a calibrated radioactivity flow monitor. Thus fluorescence and radioactivity are continuously monitored, recorded and the areas of the respective peaks are integrated and calculated.

Acknowledgement This study was supported by NIH grants DE10489, AR052824 and NASA grant NAG2-1596.

References

1. Myllyharju, K. and Kivirikko, K. I. (2004) Collagens, modifying enzymes and their mutations in humans, flies and worms *Trends Genetics* **20**, 33–43.
2. Yamauchi, M. (2003) Collagen Biochemistry: An Overview, in *Bone Morhogenetic Protein and Collagen* (Phillips, G. O., ed.), World Scientific, Singapore, pp. 93–148.
3. Kivirikko, K. I. and Myllyla, R. (1982) Posttranslational enzymes in the biosynthesis of collagen: intracellular enzymes. *Methods Enzymol.* **82**, 245–304.
4. Berg, R. A. and Prockop, D. J., (1973) The thermal transition of a non-hydroxylated form of collagen. Evidence for a role for hydroxyproline in stabilizing the triple-helix of collagen. *Biochem. Biophys. Res. Commun.* **52**, 115–120.
5. Miller, E. J. (1984) Chemistry of the collagens and their distribution, in *Extracellular Matrix Biochemistry* (Piez, K. A. and Reddi, A. H., ed.), Elsevier, New York, NY.
6. Royce, P. M. and Barnes, M. J. (1985) Failure of highly purified lysyl hydroxylase to hydroxylate lysyl residues in the non-helical regions of collagen. *Biochem. J.* **230**, 475–480.
7. Gerriets, J. E. and Curwin, S. L., and Last, J. A. (1993) Tendon hypertrophy is associated with increased hydroxylation of nonhelical lysine residues at two specific cross-linking sites in type I collagen. *J. Biol. Chem.* **268**, 25553–25560.
8. Valtavaara, M., Papponen, H., Pirttila, A. M., Hiltunen, K., Helander, H., and Myllyla R. (1997) Cloning and characterization of a novel human lysyl hydroxylase isoform highly expressed in pancreas and muscle. *J. Biol. Chem.* **272**, 6831–6834.
9. Valtavaara, M., Szpirer, C., Szpirer, J., and Myllyla, R. (1998) Primary structure, tissue distribution, and chromosomal localization of a novel isoform of lysyl hydroxylase (lysyl hydroxylase 3). *J. Biol. Chem.* **273**, 12881–12886.
10. Yeowell, H. N. and Walker, L. C. (1999). Tissue specificity of a new splice form of the human lysyl hydroxylase 2 gene. *Matrix Biol.* 18, 179–187.
11. Uzawa, K., Grzesik, W. J., Nishiura, T., Kuznetsov, S. A., Robey, P. G., Brenner, D. A., and Yamauchi, M. (1999) Differential expression of human lysyl hydroxylase genes, lysine hydroxylation, and cross-linking of type I collagen during osteoblastic differentiation in vitro. *J. Bone Miner. Res.* **14**, 1272–1280.
12. Pornprasertsuk, S., Duarte, W. R., Mochida, Y., and Yamauchi, M. (2004) Lysyl hydroxylase-2b directs collagen cross-linking pathways in MC3T3-E1cells. *J. Bone. Miner. Res.* 19, 1349–1355.
13. Van Der Slot, A. J., Zuurmond, A. M., Bardoel, A. F., Wijmenga, C., Pruijs, H. E., Sillence, D. O., Brinckmann, J., Abraham, D. J., Black, C. M., Verzijl, N., DeGroot, J., Hanemaaijer, R., TeKoppele, J. M., Huizinga, T. W., and Bank, R. A. (2003) Identification of PLOD2 as telopeptide lysyl hydroxylase, an important enzyme in fibrosis. *J. Biol. Chem.* 278, 40967–40972.

14. Mercer D. K., Nicol P. F., Kimbembe C., and Robins S. P. (2003) Identification, expression, and tissue distribution of the three rat lysyl hydroxylase isoforms. *Biochem Biophys Res Commun* **307**, 803–809.

15. van der Slot, A. J., Zuurmond, A. M., van den Bogaerdt, A. J., Ulrich, M. M. W., Middelkoop, E., Boers, W., Ronday, H. K., DeGroot, J., Huizinga, T. W. J., and Bank, R. A. (2004) Increased formation of pyridinoline cross-links because of higher telopeptide lysyl hydroxylase levels is a general fibrotic phenomenon. *Matrix Biol* **23**, 251–257.

16. Pornprasertsuk, S., Duarte, W. R., Mochida, Y., and Yamauchi, M. (2005) Overexpression of lysyl hydroxylase-2b leads to defective collagen fibrillogenesis and matrix mineralization. *J. Bone. Miner. Res.* **20**, 81–87.

17. Pornprasertsuk-Damrongsri, S., Duarte, W. R., Miguez, P. A., Mochida, Y., and Yamauchi, M. (2005) Differential roles of lysyl hydroxylase isoforms in collagen matrix mineralization. *8th Int Conf Chem Biol Min Tis*, Univ Toronto Press, pp.184–187.

18. Tanzer, M. L. (1973) Cross-linking of collagen. *Science.* **180**, 561–566.

19. Molnar, J., Fong, K. S. K., He, Q. P., Hayashi, K., Kim, Y., Fong, S. F. T., Fogelgren, B., Molnarne Szauter, K., Mink, M., and Csiszar, K. (2003) Structural and functional diversity of lysyl oxidase and LOX-like proteins" *Biochim. Biophys. Acta* **1647**, 220–224

20. Atsawasuwan, P., Mochida, Y., Parisuthiman, D., and Yamauchi, M. (2005) Expression of lysyl oxidase isoforms in MC3T3-E1 osteoblastic cells. *Biochem Biophys Res Commun.* **327**, 1042–1046.

21. Yamauchi, M., London, R. E., Guenat, C., Hashimoto, F., and Mechanic, G. L. (1987) Structure and formation of a stable histidine-based trifunctional cross-link in skin collagen. *J. Biol. Chem.* **262**, 11428–11434.

22. Mechanic, G. L., Katz, E. P., Henmi, M., Noyes, C., and Yamauchi, M. (1987) Locus of a histidine-based, stable trifunctional, helix to helix collagen cross-link: stereospecific collagen structure of Type I skin fibrils. *Biochemistry* **26**, 3500–3509.

23. Yamauchi, M., Chandler, G. S., Tanzawa, H., and Katz, E. P. (1996) Cross-linking and the molecular packing of corneal collagen. *Biochem. Biophys. Res. Commun.* **219**, 311–315.

24. Eyre, D. R., Patz, M. A., and Gallop, P. M. (1984) Cross-linking in collagen and elastin. *Ann. Rev. Biochem.* **53**, 717–748.

25. Kuypers, R., Tyler, M., Kurth, L. B., Jenkins, I. D., and Horgan, D. J. (1992) Identification of the loci of the collagen-associated Ehrlich chromogen in type I collagen confirms its role as a trivalent cross-link. *Biochem. J.* **283**, 129–136.

26. Hanson, D. A. and Eyre, D. R. (1996) Molecular site specificity of pyridinoline and pyrrole cross-links in type I collagen of human bone. *J. Biol. Chem.* **271**, 26508–26516.

27. Bailey, A. J., Paul, R. G., and Knott, L. (1998) Mechanisms of maturation and ageing of collagen. *Mech. Ageing Dev.* **106**, 1–56.

28. Yamauchi, M. and Mechanic, G. L. (1988) Cross-linking of Collagen, in *Collagen, vol I* (Nimni, M. E., ed.), CRC Press, Boca Raton, FL, pp. 157–186.

29. Robins, S. P. (1982) Analysis of the crosslinking components in collagen and elastin. *Methods Biochem. Anal.* **28**, 329–379.

30. Yamauchi, M., Katz, E. P., and Mechanic, G. L. (1986) Intermolecular cross-linking and stereospecific molecular packing in type I collagen fibrils of the periodontal ligament. *Biochemistry.* **25**, 4907–4913.

31. Yamauchi, M. and Katz, E. P. (1993) The post-translational chemistry and molecular packing of mineralizing tendon collagens. *Connect. Tissue Res.* **29**, 81–98.

32. Uzawa, K., Yeowell, H. N., Yamamoto, K., Mochida, Y., Tanzawa, H., and Yamauchi, M. (2003) Lysine hydroxylation of collagen in a fibroblast cell culture system. *Biochem. Biophys. Res. Commun.* **305**, 484–487

33. Saito, M., Soshi, S., Tanaka, T., and Fujii, K. (2004) Intensity-related differences in collagen post-translational modification in mc3t3-e1 osteoblasts after exposure to low- and high-intensity pulsed ultrasound. *Bone* **35**, 644–655

34. Yamamoto, K. and Yamauchi, M. (1999) Characterization of dermal type I collagen of C3H mouse at different stages of the hair cycle. *Br. J. Dermatol.* **141**, 667–675.

8

Mass Spectrometric Determination of Protein Ubiquitination

Carol E. Parker, Maria R. E. Warren, Viorel Mocanu, Susanna F. Greer, and Christoph H. Borchers

Summary Mass spectrometric methods of determining protein ubiquitination are described. Characteristic mass shifts and fragment ions indicating ubiquitinated lysine residues in tryptic and gluC digests are discussed. When a ubiquitinated protein is enzymatically digested, a portion of the ubiquitin side chain remains attached to the modified lysine. The ubiquitinated peptide thus has two N-termini – one from the original peptide and one from the ubiquitin side chain. Thus, it is possible to have two series of b ions and y ions, the additional series is the one that includes fragments containing portions of the ubiquitin side chain. Any diagnostic ions for the modification must include portions of this side chain. Fragment ions involving any part of the "normal" peptide will vary in mass according to the peptide being modified and will therefore not be of general diagnostic use. These diagnostic ions, found through examination of the MS/MS spectra of model ubiquitinated tryptic and gluC peptides, have not previously been reported. These ions can be used to trigger precursor ion scanning in automated MS/MS data acquisition scanning modes.

Keywords Mass spectrometry; ubiquitination; diagnostic ions; MS/MS.

1 Introduction

Cellular homeostasis requires a delicate balance of protein synthesis and degradation, a balance which is maintained by the actions of regulatory proteins. In turn, proteins which are no longer required must be degraded in a rapid and selective manner. The selective degradation of proteins involved in diverse regulatory mechanisms such as signal transduction *(1)*, transcriptional regulation *(2)*, cell cycle regulation *(3)*, and stress response *(4)* have been linked to the covalent attachment of ubiquitin to proteins. Ubiquitin is a 76-residue polypeptide, which attaches via its carboxy-terminus to lysine ε amino groups of the target protein. Ubiquitin-protein conjugates are short-lived, primarily due to proteolysis by the 26S proteosome or, in some cases, dissociation of the complex with removal of the ubiquitin by ubiquitin isopeptidases *(5)*. Other roles for monoubiquitination

From: *Post-translational Modifications of Proteins.*
Methods in Molecular Biology, Vol. 446.
Edited by: C. Kannicht © Humana Press, Totowa, NJ

have recently been uncovered, including acting as a signal for protein trafficking, cell division, targeting proteins to subnuclear structures, endocytosis, signal transduction, and kinase activation *(6–11)*. Both the dynamic nature·of these important regulatory proteins, and the low protein levels *in vivo*, make analysis of protein ubiquitination inherently difficult.

The use of mass spectrometry to identify sites of ubiquitination within target proteins *in vivo* will allow greater understanding of the ways in which ubiquitination alters protein function. Therefore, the development of broadly applicable identification approaches is critical. The importance of ubiquitination and the role of mass spectrometry in the study of ubiquitination has recently been reviewed by Kirkpatrick, Denison, and Gygi *(12,13)*.

2 Materials

2.1 Mass Spectrometry

2.1.1 MALDI Analysis Of Intact Proteins

1. Bruker Reflex III matrix-assisted laser desorption (MALDI-MS, (Bruker Daltonics, Billerica, MA).
2. α-cyano 2-hydroxycinnamic acid (Aldrich; St. Louis, MO).
3. Ethanol (AAPER Alcohol and Chemical Co., Shelbyville, KY).
4. Water, HPLC-grade (Pierce, Rockford, IL).
5. Formic acid (Fisher, Pittsburgh, PA).

2.1.2 LC/MS/MS

1. Waters/Micromass Q-TOF (Waters/Micromass Corp., Milford, MA).
2. PepMap C_{18} 15 cm × 75 µm id capillary column (Dionex; Sunnyvale, CA).
3. Trapping column 5 mm × 800 Å id C_{18} P3 (Dionex).
4. Water, HPLC-grade (Pierce).
5. Acetonitrile, HPLC-grade (Pierce).
6. Formic acid (Fisher).

2.1.3 Affinity Purification

1. Anti-HA antibody beads (Sigma, St. Louis, MO).
2. Anti-FLAG antibody beads (Sigma).
3. Ammonium bicarbonate (Fluka; Milwaukee, WI).
4. Ethanol (AAPER).
5. Water (Pierce).
6. Formic acid (Fisher).

2.1.4   In-Solution Digestion

1. Water, deionized or HPLC-grade (Pierce).
2. Trypsin, sequencing-grade (Promega; Madison, WI).
3. GluC (Sigma).
4. Ammonium bicarbonate (Fluka).
5. Low-retention Eppendorf tubes (Axygen; Union City, CA).
6. Thermomixer (Eppendorf; Hamburg, Germany).

2.1.5   Lyophilization and Reconstitution

1. Freeze dryer (Labconco; Kansas City, MO).
2. Water (Pierce).

3   Methods

3.1   Evidence from Gels

Most direct evidence for ubiquitination comes from gel electrophoresis (Fig. 8.1), where a series of higher molecular weight bands are observed above the molecular weight of the protein, or from Western blot analysis using anti-ubiquitin antibodies (Fig. 8.2).

3.2   Mass Spectrometric Evidence for Protein Ubiquitination

While ubiquitination is clearly important for protein degradation, most mass spectrometric studies on ubiquitination have focused on protein phosphorylation, rather than direct mass spectrometric studies of protein ubiquitination. There are three types of mass spectrometric evidence for ubiquitination: the first is ubiquitination of the intact protein, the second is co-electrophoretic migration of the target protein and the attached ubiquitin and, the third is the mass shift of a ubiquitinated peptide relative to the non-ubiquitinated peptide.

3.2.1   Evidence from MS of the Intact Protein

Direct mass spectrometric evidence of intact ubiquitinated HSP70 is shown in Fig. 8.3. This spectrum was obtained by direct MALDI-MS analysis of ubiquitinated HSP70, affinity-bound to anti-HSP70 affinity beads.
Direct MALDI Analysis of Proteins bound to Affinity Beads (*see* **Note 1**)

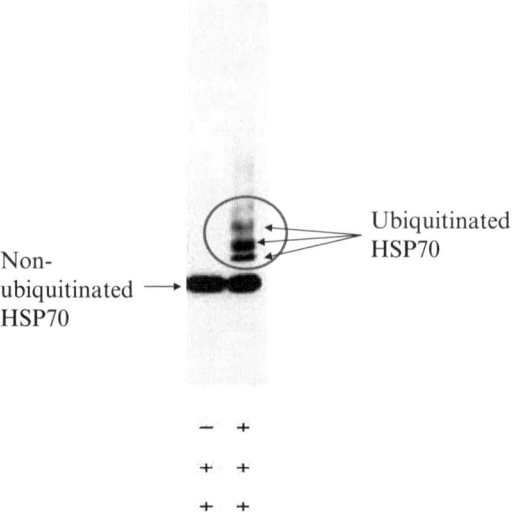

Fig. 8.1 Ubiquitination evidence from Gels. PAGE gel showing a series of bands above the molecular weight of the non-ubiquitinated protein. (Collaborator: W. C. Patterson)

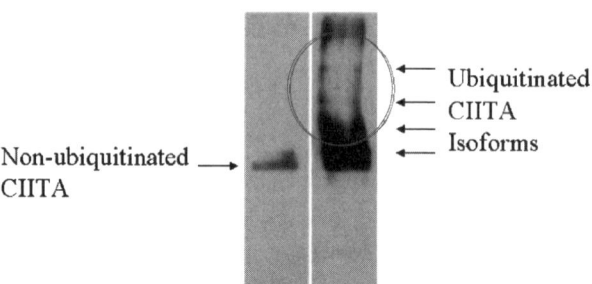

Fig. 8.2 Ubiquitination evidence from Western blot analyses. PAGE gel showing a series of bands above the molecular weight of the non-ubiquitinated protein: Western blot analysis of CIITA, showing ubiquitination

1. Use antibodies *covalently* bound to the affinity beads (*see* **Note 2**)
2. Bind the antibody according to the bead manufacturers' instructions (*see* **Note 3**, **Note 4**)
3. Rinse beads 3 times with 2–3 bead volumes of 100 m*M* ammonium bicarbonate (*see* **Note 5**).

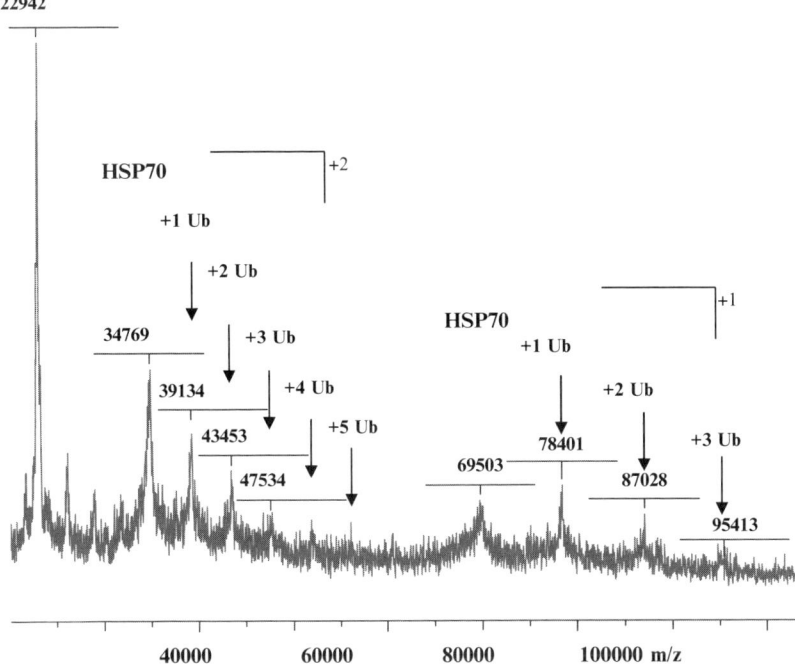

Fig. 8.3 Ubiquitination evidence from MALDI of intact proteins. MALDI-MS of HSP70 with a series of ubiquitin attached moieties

Spotting the MALDI target

1. Prepare the MALDI matrix solution – a saturated solution of recrystallized (*see* **Note 6**) α-cyano 2-hydroxycinnamic acid in 45:45:10 ethanol:water:formic acid.(*see* **Note 7**)
2. Pipet 0.5 μL of settled beads onto the MALDI target, followed by 0.5 μL of MALDI matrix solution.

3.2.2 Evidence from MS and MS/MS of Peptides

The second type of mass spectrometric evidence for ubiquitination comes from in-gel digestion and protein identification studies. In-gel digestion of the higher molecular weight bands followed by protein identification by MALDI-MS or LC/MS/MS on a gel can sometimes provide evidence for ubiquitination – evidence comes from peptides rather than proteins. In the example shown in Figure 8.4, peptides from the target protein CIITA and peptides from ubiquitin were found in a gel band of approximately 200Kda, clearly higher than the unmodified 150 kDa CIITA or the 8.7 kDa unbound modified monoubiquitin (see **Section 3.2.4**).

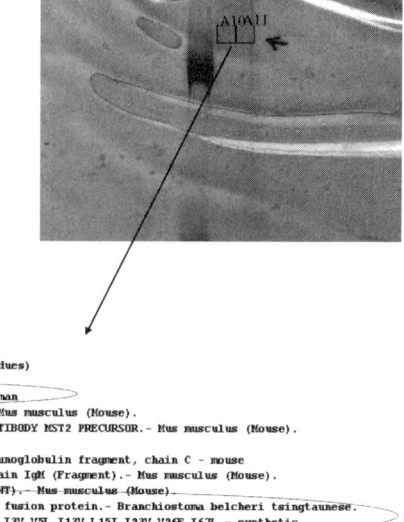

User   : Carol Parker
Email   : carol_parker@med.unc.edu
Search title : 070903_Greer_2+_QA50
MS data file : D:\MethodD.Pro\pkl_files\071703_Greer_06b.pkl
Database   : MSDB 20030710 (1206286 sequences; 383022796 residues)
Timestamp   : 21 Jul 2003 at 02:47:12 GMT

{*MATRIX*}
{*SCIENCE*} **Mascot Search Results**

Significant hits: A40843   MHC class II transactivator - human
                  Q8R3V9   Hypothetical 52.0 kDa protein.- Mus musculus (Mouse).
                  CAA56280 HEAVY CHAIN OF THE MONOCLONAL ANTIBODY MST2 PRECURSOR.- Mus musculus (Mouse).
                  2H1PL   2h1 fab, chain L - mouse
                  1CICC   ig heavy chain v regions fab immunoglobulin fragment, chain C - mouse
                  AAB48774 Anti-DNA immunoglobulin light chain IgM (Fragment).- Mus musculus (Mouse).
                  AAB70775 MAB544 IGG1/K VLJ REGION (FRAGMENT).- Mus musculus (Mouse).
                  Q8VQK2   Ubiquitin/ribosomal protein S27a fusion protein.- Branchiostoma belcheri tsingtaunese.
                  1UD7A   ubiquitin core mutant 1d7 mutant I3V,V5L,I13V,L15I,I23V,V26F,I67L - synthetic
                  1AVHA   trypsin (EC 3.4.21.4), chain A - pig
                  I61770   keratin 6e, type II - human
                  CAA82617 SINGLE CHAIN FV ANTIBODY (FRAGMENT).- Mus musculus (Mouse).
                  AAP21111 AY263789 NID: - synthetic construct
                  AAC37617 IMMUNOGLOBULIN HEAVY CHAIN (FRAGMENT).- Mus musculus (Mouse).
                  G87446   potassium-transporting ATPase, B subunit CC1592 [imported] - Caulobacter crescentus
                  Q9EQD7   Keratin intermediate filament 16b.- Mus musculus (Mouse).

Fig. 8.4 Peptides from the target protein CIITA and peptides from ubiquitin from a gel band at mw ~200 kDa

3.2.3   MS and MS/MS Spectra of Ubiquitinated Peptides

The third type of mass spectrometric evidence comes from the modified peptide itself, either from a shift in peptide molecular weight, or from MS/MS data. Ubiquitin is a 76-amino acid protein. An E3 ligase attaches ubiquitin to the ε amino group of a lysine residue in the target protein. This covalent linkage is formed at C-terminal glycine residue of the ubiquitin, with loss of the elements of water. Cleavage with trypsin or gluC (**Note 8**) leaves characteristic "tails" on the modified lysine. These "tails" cause a shift in the molecular weight of the peptide, which can be used to distinguish these modified peptides from the unmodified peptides (Fig. 8.5). The actual site of ubiquitination can then be determined by MS/MS sequencing.

These peptide mass shifts can be used to find and sequence ubiquitinated peptides. Mascot, for example, allows the user to create a modification of a particular mass, which can then be used to search the peptide data from a given digest. Due to their transient nature and low natural abundances, ubiquitinated peptides are difficult to detect. A study by Gururaja, et al. *(14)* on Hela cell lysates, used a 6 x his-tagged ubiquitin, IMAC purification, digested by Lys-C and trypsin, and protein identification by 2D-LC/MS/MS using strong cation exchange and C18 reversed phase. Ubiquitination was confirmed by anti-his tag and anti-ubiquitin Western

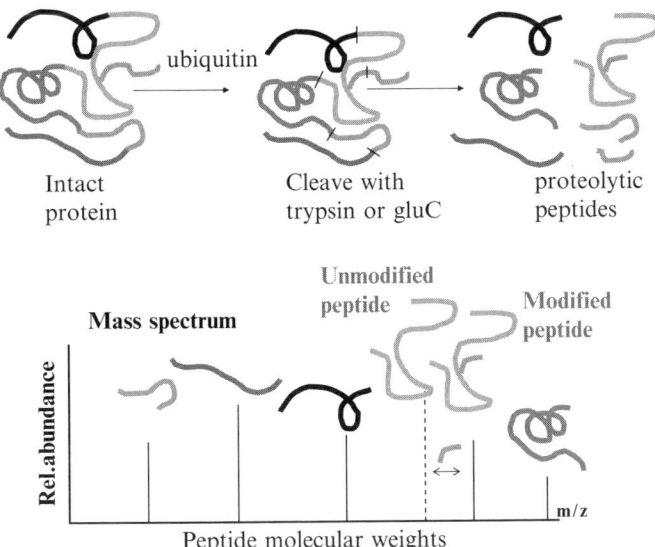

Intact   Cleave with   proteolytic
protein   trypsin or gluC   peptides

Mass spectrum

Unmodified
peptide   Modified
peptide

Rel.abundance

m/z

Peptide molecular weights

Fig. 8.5 Schematic showing the expected peptide molecular weight shift caused by ubiquitination

blotting on the undigested lysates. A total of 244 proteins were found, which the authors categorized into functional groups, but determination of the exact ubiquitination sites was not the focus of this study.

Using a similar approach, Kirkpatrick, et al., *(15)* also studied the "ubiquitinome" of human cells, this time HEK293 cells. As in the above studies, 6 x his-tagged ubiquitin was used, followed by Ni-IMAC purification. Lysates from as many as 80 plates were pooled, and the eluted proteins were digested with trypsin, followed by LC/MS/MS. Twenty-two ubiquitinated proteins were identified, along with 19 additional proteins non-specifically-bound to the IMAC beads. An attempt was made to find the ubiquitinated peptides, but only branched peptides from polyubiquitin were identified in this study – by the presence of the KGG modification. No consistent fragmentation (i.e., no loss of 114 Da) was observed in these KGG-modified peptides. Confirmatory evidence of the ubiquitination at these specific sites was the lack of tryptic cleavage at the modified lysine.

An alternative, but still "shotgun" approach is to use an anti-ubiquitin antibody to accomplish the enrichment. This approach was used by Figeys' group, and the digestion of 30 bands resulted in the identification of 70 proteins *(16)*. Several GGK-containing "signature" peptides were located by their mass shifts, using a Mascot database search.

Surprisingly few ubiquitinated peptides were found by mass spectrometry until the work by Gygi and coworkers *(17)*. As in the studies described above, a 6 x his-tagged ubiquitin was also used, and ubiquitinated proteins from yeast were purified by IMAC. The ubiquitin-enriched fraction (0.2 mg out of each original 100 mg of yeast cell lysate) was then digested with trypsin, fractionated into 80 fractions on a

SCX column, and the 80 fractions were analyzed by capillary LC/MS/MS. Peptide MS/MS data was searched using Sequest software, allowing for a ubiquitin-modified lysine with a mass shift of +114 Da. In this manner, 110 ubiquitination sites were determined. In another study, using 6× his-myc-Ub, 211 proteins were identified. Ubiquitination sites were identified on 15 of these proteins *(18)*.

3.2.3.1   MS/MS Fragmentation of ubiquitinated peptides

We could find only two reports of ubiquitination sites prior to the work of Gygi's group *(17)*, and these were on specifically-targeted proteins of interest. The first was the work of Laub, et al. *(19)*, who used gluC for proteolytic digestion of the protein, and the second was the work of Dohlman's group *(20)*, who used trypsin. Both of these papers showed MS/MS spectra of the ubiquitinated peptides, and we decided to use these peptides as the starting point for a detailed study of the fragmentation of ubiquitinated peptides with the goal of finding specific diagnostic fragment ions to aid in detecting low levels of ubiquitinated peptides at lower levels in biological materials *(21)*.

When a ubiquitinated protein is enzymatically digested, a portion of the ubiquitin side chain remains attached to the modified lysine (Fig. 8.6). A ubiquitinated peptide therefore has 2 N-termini—one from the original peptide and one from the ubiquitin side chain. Thus, it is possible to have two series of b ions and y ions. For the sake of clarity, we have chosen to refer to those b and y ions involving the ubiquitin side chain as *b* and *y* ions. Obviously, diagnostic ions for the modification

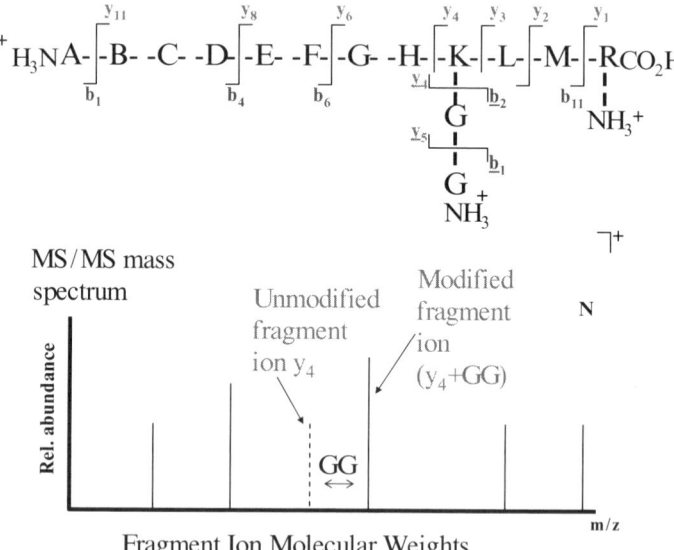

Fig. 8.6 Schematic showing fragmentation of ubiquitinated peptides

must come from fragmentation of this side chain. Fragment ions involving any part of the "normal" peptide will vary in mass according to the peptide being modified and will therefore not be of general diagnostic use.

3.2.3.2 Examination of the literature tryptic peptide spectra

MS/MS spectra of GG-tagged tryptic peptides are shown in both the paper from Dohmann's group, and that of the Gygi group. Examination of these spectra show that b and y ions are produced by dissociation of the peptide from the target protein, but there do not appear to be any *b* ions from the GG side chain, or the GGK portion of the peptide.

3.2.3.3 Examination of the literature gluC peptide spectra

The MS/MS spectrum of the gluC ubiquitinated peptide from rXL-calmodulin which was shown in the Laub, *et al.* paper **(19)**, reveals b and y ions from the calmodulin portion of the peptide (Fig. 8.7). Interestingly, it also shows doubly-charged $(b_7-H_2O)^{2+}$, $(b_{14}-H_2O)^{2+}$, and $(b_{12}-H_2O)^{2+}$ ions which contain both of the N-termini (one from the calmodulin peptide and one from the ubiquitin side chain) of the original branched peptide. These two ions (athough of low relative abundance) can be seen in the MS/MS spectrum of the ubiquitinated peptide from natural BT-calmodulin. Careful examination of the both spectra reveals ions which might be from the ubiquitin side chain, and therefore of possible diagnostic utility. The y_{14} ion includes ions from both the side chain and the C-terminus of the calmodulin-portion of the peptide, so it cannot be used as a ubiquitination marker ion. However, there are also ions from the side chain, b_4 at m/z 478.4 (obs) and, possibly, b_2, at m/z ~191.2 (obs),, which appear to be diagnostic ions of ubiquitination, after digestion with gluC.

3.2.3.4 Preparation of model ubiquitinated peptides

In order to further study the fragmentation of ubiquitinated peptides, we had a model peptide synthesized. A peptide was synthesized with the structure shown in Fig. 8.7. These modified peptides should show a characteristic mass shift from their

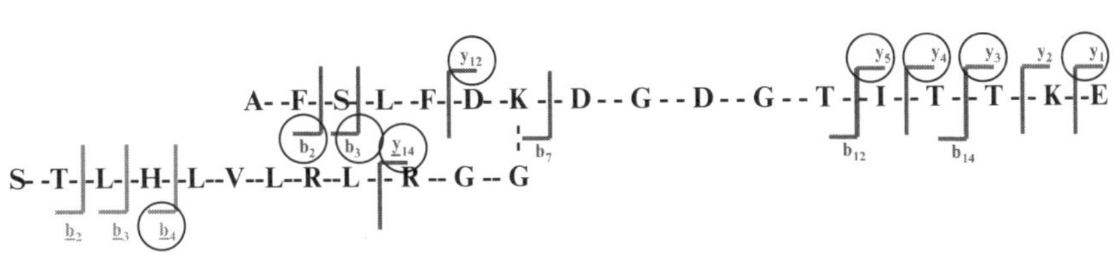

Fig. 8.7 Calmodulin ubiquitinated peptide structure after cleavage with gluC. Circled peptide fragments were not noted in the original publication by Laub, et al.(*19*), but appear to have the appropriate masses

unmodified analogues: in the case of trypsin, a mass shift of 114.0428 Da from the GG "tag" left on the modified lysine. In the case of gluC, a mass shift of 1302.7883 Da, a much larger mass shift resulting from the much longer "tail" (STLHLVLRLRGG-) on the modified lysine, is expected. Unfortunately, the model peptide was synthesized with an amide at the C-terminus, and an acetyl group at the N-terminus, so it is not an exact model of the peptide found by Laub, et al. *(19)*. Fortunately, the acetyl group was in the "calmodulin" portion of the peptide, so we expected that diagnostic fragments from the ubiquitin side chain would still be present in the MS/MS spectrum.

ESI-MS/MS spectra of the model peptides were obtained by LC/MS/MS analysis using a Waters/Micromass Q-tof API US, equipped with a Waters capLC system. An aliquot of the sample was injected first onto a Dionex trapping column, which was then switched so that it was connected on-line to a 75 µm Dionex Pep-map analytical column and the ESI source. The original synthetic peptide, the model gluC peptide, was dissolved in water, and injected without further purification. It proved to be a mixture of acetylated and methylated forms, whose MS/MS spectra could be analyzed separately after separation by LC/MS/MS. The actual methylation sites can be deduced from the MS/MS spectrum: one is on the S in the calmodulin part of the peptide, the other is on one of the first two residues (S or T) of the ubiquitin tail.

To create a model tryptic peptide, the tryptic peptide was prepared from the synthetic peptide by means of an in-solution tryptic digest (see **Section 3.2.5.3.**), using 2 µg of trypsin and 1 µg of the original peptide in 100 µL of 100 mM ammonium bicarbonate. As stated earlier, an aliquot of the digest was injected into the LC/MS/MS system, and went first onto a Dionex trapping column, which was then switched so that the column was connected on-line to a 75 µm i.d. Dionex Pep-map analytical column and the ESI source.

Two main products were formed: a peptide with the expected "GG" tail on the ubiquitinated lysine, and a second peptide with an "LRGG" tail, resulting from a missed cleavage. Even though this was an in-solution tryptic digest with a large amount of trypsin, a significant amount of a peptide was produced with a missed cleavage site on the side chain. Although this missed cleavage was unexpected, it is not unreasonable, since this cleave site is close to the branch point so cleavage at this site is likely to be sterically hindered. MS/MS spectra were obtained for both of these products. The formation of this peptide is of significant analytical interest as it provides a second characteristic molecular weight shift for ubiquitinated peptides after tryptic digestion.

3.2.3.5   Fragmentation of model gluC ubiquitinated peptide

Because of the various possible dimethylated isoforms, the first model gluC peptide examined was the dimethylated version. The resulting spectrum is shown in Fig. 8.8. Both b and y fragment ions are found from the "normal" part of the peptide. Most interestingly, several fragments are found which only involve the side chain. As were

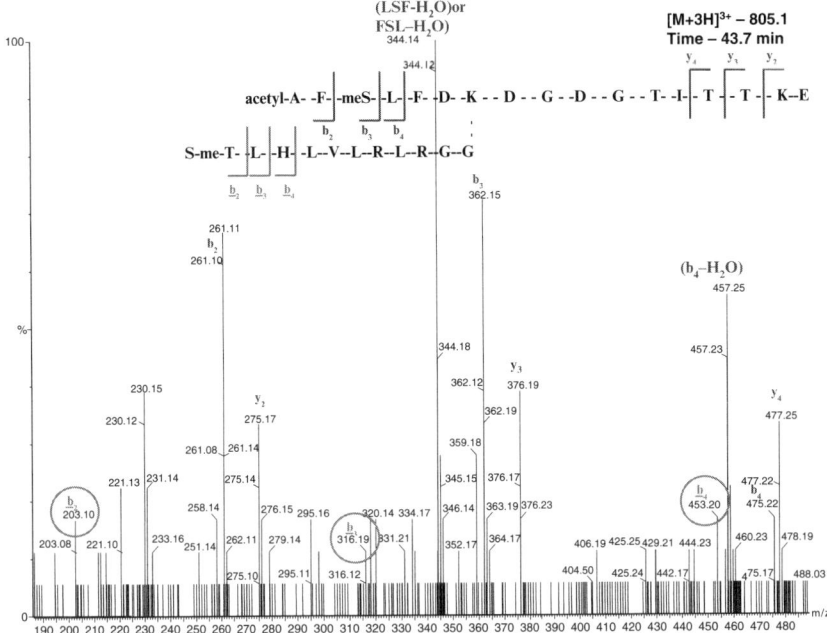

Fig. 8.8 MS/MS Fragmentation of synthetic model calmodulin gluC peptide, dimethyl form. Reproduced with permission from *(21)*, copyright John Wiley & Sons, Limited

observed in the literature spectrum, three characteristic ions were detected from the "ubiquitin" side chain: these are b_2, b_3, and b_4. These ions, in their un-methylated forms, (m/z 189.088, 302.172, 439.231, 555.315, and 651.383) should thus be diagnostic ions for peptides with ubiquitin side chains that have been cleaved with gluC.

3.2.3.6 Fragmentation of model tryptic peptides

Cleavage of the synthetic model peptide with trypsin resulted in a GG-tagged peptide whose MS/MS spectrum is shown in Fig. 8.9. Unfortunately, as in the literature spectra, no characteristic fragment ions could be found which were diagnostic of the critical GGK portion of the molecule.

The MS/MS spectrum of the model tryptic peptide which had the LRGG tag (resulting from a missed cleavge) was more analytically useful (Fig. 8.10). Diagnostic ions of the ubiquitin tag were found. These include b_2, b_4, and the internal fragment ion (LRGGK-28). There is also a LRGGKD ion fragment ion, but since this includes the "D" from the calmodulin peptide, it cannot be used as a diagnostic ion for modification by ubiquitin. The assignment of these peptide fragments is confirmed by the MS/MS spectrum of the S-methylated calmodulin peptide is shown in Fig. 8.11. The same diagnostic ions b_2, b_4, (LRGGK-28) ions are observed in this spectrum.

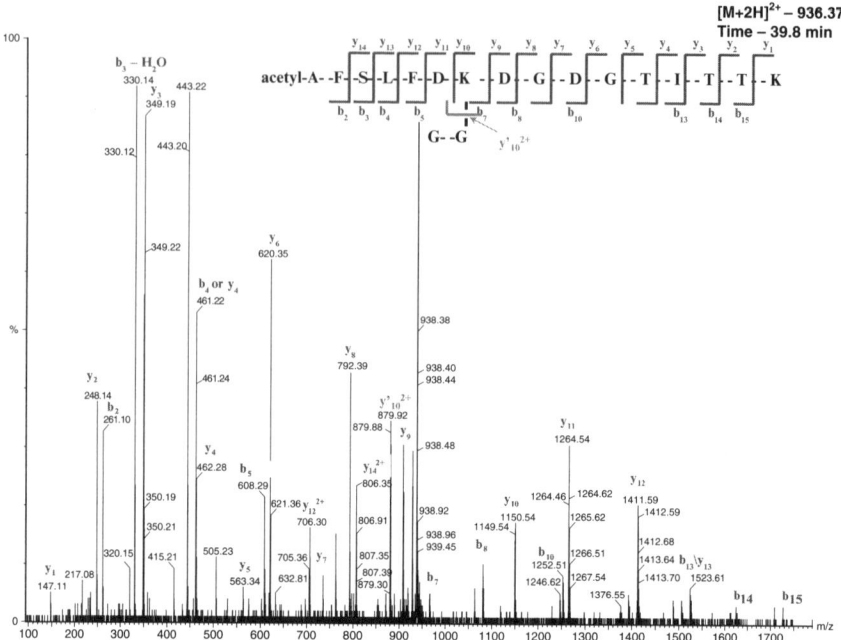

Fig. 8.9 Fragmentation of model "GG-" ubiquitinated tryptic peptide. Reproduced with permission from *(21)*, copyright John Wiley & Sons Limited

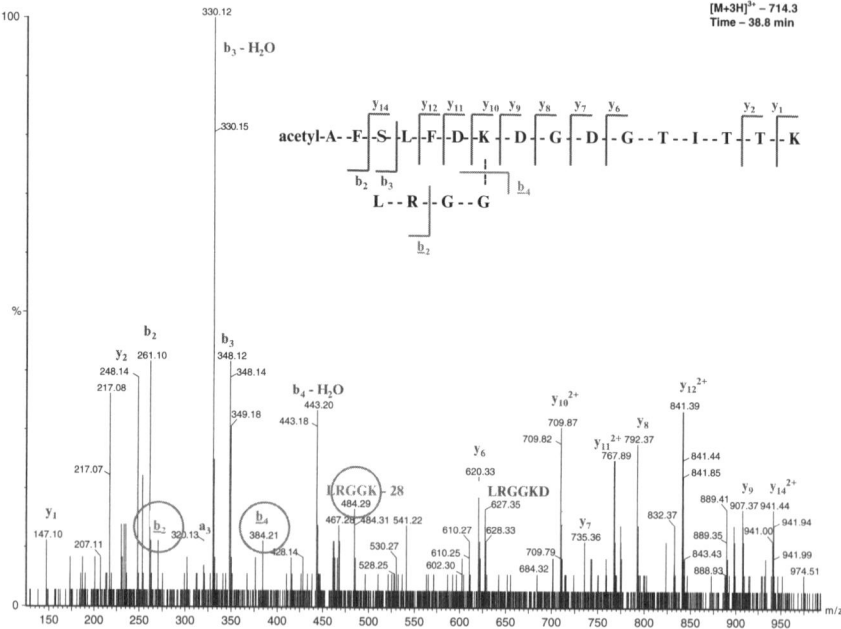

Fig. 8.10 Fragmentation of model "LRGG-" ubiquitinated tryptic peptide. Reproduced with permission from *(21)*, copyright John Wiley & Sons

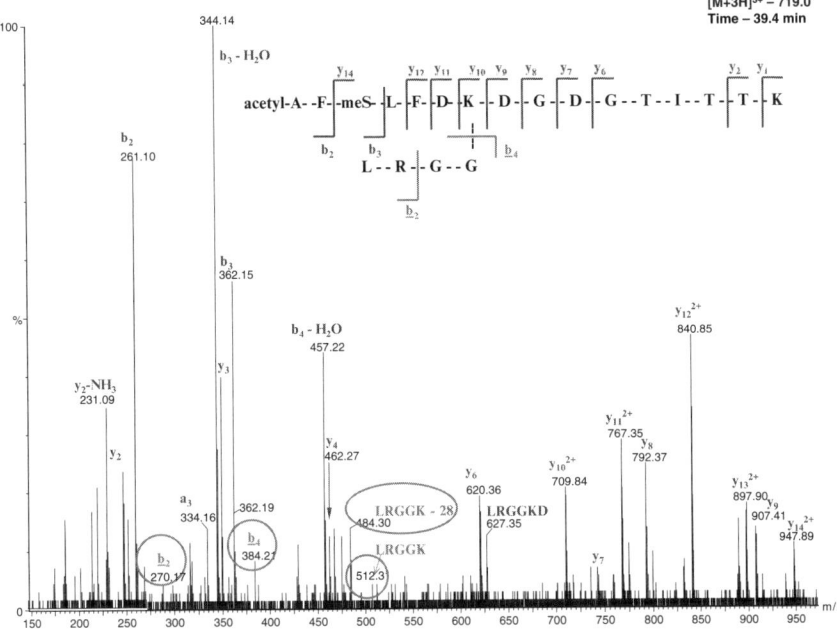

Fig. 8.11 Fragmentation of model "LRGG-" ubiquitinated tryptic peptide, methylated form. Reproduced with permission from *(21)*, copyright John Wiley & Sons Limited

3.2.4 Enrichment of Samples for Specific Ubiquitin-Modified Proteins

As described above (**Section 3.2.3**), the work of Gururaja *(14)*, Kirkpatrick *(15)*, and Gygi *(17)*, depended upon enriching the sample in ubiquitinated peptides. Their goals were to find as many ubiquitination sites as possible in *any* protein, so their approach was to use a 6 x his-tagged ubiquitin, and enrich the sample in ubiquitinated proteins through the use of IMAC or 6 x his – myc-Ub *(18)*. Other researchers have used anti-ubiquitin for "shotgun" enrichment *(16)*.

To find ubiquitination sites in a *particular* target protein requires a different approach. One method is the use of a GST-tagged substrate, and anti-GSH affinity beads. This method was used by the Marshall group for an FT-MS study of polyubiquitinated GST-Ubc5, ubiquitinated *in vitro*. FT-MS identified fifteen ubiquitination sites in GST-Ubc5, and four sites in ubiquitin, although large quantities of material were used (*e.g.*, 700 µg GST-Ubc5).

If an antibody is available against the target protein, then one strategy can be to affinity purify the target protein from the cell lysate. If an antibody against the protein of interest is *not* available, an affinity tag can be incorporated into the target protein sequence. In our case, we used a FLAG tag added to the amino-terminus of the CIITA target protein so that it could be immunoprecipitated with anti-FLAG antibody beads.

As a second strategy for increasing the proportion of ubiquinated vs. nonubiquitinated protein, a plasmid was used which coded for a modified ubiquitin which had all of the lysines modified to arginines and a HA tag on the C-terminus. This construct was designed to prevent the formation of polyubiquitin chains and thus to inhibit degradation of the target protein *(22)*. The HA tag allows a second affinity purification step, either before or after proteolysis, this time on anti-HA beads. To avoid proteolysis of the ubiquitin (and loss of the HA tag), LysC was used for the initial digestion of the protein, instead of trypsin.

The affinity-bound protein can them be digested overnight with trypsin or with gluC (**Note 8**) while still attached to the beads, using the protocols described in the next subheading. Alternatively, the protein can be eluted from the affinity beads using eluted with 1:1:8 ethanol:formic acid:water, lyophilized and digested in solution.

3.2.5   Elution and Enzymatic Digestion Procedures

3.2.5.1   Elution of proteins from beads

1. Place ~50–100 µL of beads in an Eppendorf tube.
2. Wash beads 3×with 200 µL of 100 mM ammonium bicarbonate, and discard wash solutions.
3. Add 100 µL 1:1:8 Ethanol:formic acid:water.
4. Vortex, let settle.
5. Remove and save eluates.
6. Repeat extraction 2 more times.
7. Save and combine eluates.
8. Lyophilize.
9. Store at −80°C.

3.2.5.2   On-bead digestion tryptic procedure(**23**) (see **Note 9**)

1. Place ~50–100 µL of antibody beads (with the attached affinity-bound protein) in an Eppendorf tube.
2. Wash 3× with 200 µL of 100 mM ammonium bicarbonate.
3. Add 100 µL of 100 mM ammonium bicarbonate.
4. Reconstitute Promega trypsin in 20 µL Promega resuspension buffer (0.015M acetic acid) (Promega trypsin comes in aliquots of 20 µg per vial).
5. Add 2 µL trypsin solution to each sample. (*see* **Note 10**)
6. Vortex/spin down at <1000 rpm to avoid breaking the beads.
7. Incubate overnight in sealed Eppendorf tubes, at 35°C with rotation (~400rpm). (*see* **Note 11**)
8. Remove supernatant from beads.
9. Lyophilize to reduce volume (if sample is concentrated enough, supernatant can be injected directly).
10. Store at −80°C.

3.2.5.3 In-solution tryptic digestion procedure

1. Calculate what a 1:50 Enzyme:Substrate ratio would be (*see* **Note 12**).
2. Reconstitute Promega trypsin in 20 μL Promega resuspension buffer (0.015 M acetic acid) (Promega trypsin comes in aliquots of 20 μg per vial.)
3. Dissolve sample in ~20 μL of 100 mM ammonium bicarbonate solution.(*see* **Note 13**)
4. Add calculated amount of trypsin solution to each sample.
5. Vortex/spin down.
6. Incubate for at least 4 h for a digest in solution, or overnight for beads, in sealed Eppendorf tubes, at 35°C with rotation (~400 rpm).

3.2.5.4 In-solution digestion with gluC (see **Note 14**)

1. Dissolve the enzyme (which comes lyophilized, 25 μg per ampule) in 25 μL sample in water.
2. Dissolve sample in 100 μL of 100 mM ammonium bicarbonate buffer.
3. Add 2 μL enzyme solution.
4. Incubate overnight at 35°C, at 35°C with rotation (~400 rpm).
5. Freeze at −80°C.

3.2.5.5 On-bead digestion with gluC (see **Note 14**)

1. Dissolve the enzyme (which comes lyophilized, 25 μg per ampule) in 25 μL sample in water.
2. Rinse ~50 μL of affinity beads 3 x with 100 μL of 100 mM ammonium bicarbonate buffer.
3. Add 100 μL of 100 mM ammonium bicarbonate to the beads.
3. Add 2 μL enzyme solution to the beads.
4. Incubate overnight at 35°C, at 35°C with rotation (~400 rpm).
5. Remove supernatant, and freeze at −80°C.

3.2.6 Mass Spectrometric Approaches for Detection of Ubiquitinated Peptides and Determination of Ubiquitination Sites

The first step is to simply perform LC/MS/MS on the peptide digest. Long runs with long linear gradients (>200 min) are preferred for separating complex mixtures of peptides in order to reduce suppression effects and to try to reduce the number of peptides that coelute, since selection of the various precursor ions is done on the basis of ion abundances. For an unknown modification site in a protein containing many potential sites, where the peptide molecular weight is therefore not known, automatic data-dependent triggering of MS/MS data collection (called "survey scan" mode in the Micromass MassLynx software) is the only feasible automatic scanning option.

The resulting MS/MS spectra are then analyzed by commercially-available software packages (such as Mascot or SEQUEST) which can be programmed (*see* **Note 15** and **Note 16**) to consider a lysine modified with a GG or, as we have learned from the above experiments on our model peptides, LRGG, for a tryptic digest, or with STLHLVLRLRGG for a gluC digest). Ideally, the ubiquitination will be found from this automated search routine (*see* **Note 17** and **18**). Fig. 8.12 shows an example of a database search of MALDI-MS data from a tryptic digest of a ubiquitinated protein. Here, a ubiquitinated peptide has been identified. Although promising, these results simply mean that there is a peak in the mass spectrum which has the mass of an expected tryptic peptide where a lysine has been modified with a GG tag. Since a peak at this mass could have come from another peptide in the mixture, this ubiquitination site cannot be confirmed without MS/MS sequence data of this peptide.

If MS/MS data has been acquired and searched, as is the case when LC/MS/MS has been used, the identity of the peptide can be confirmed and site of ubiquitination found from the database search results.(*see* **Note 19**) This was the method used to find the ubiquitination sites in the Gygi paper. Unfortunately, this approach often

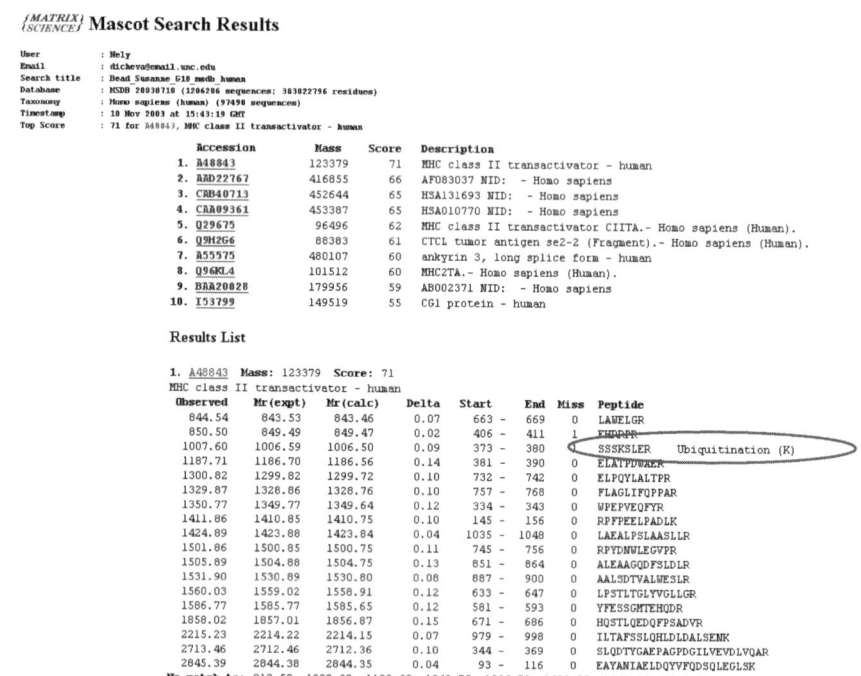

Fig. 8.12 Mascot database search results from a tryptic digest of ubiquitinated CIITA, showing a "hit" for a potential ubiquitinated peptide. This potential ubiquitination site has not yet been confirmed by MS/MS data

fails where lower amounts of biological material are available. Very low levels of the modified peptide mean that there may not be sufficient intensity of the modified peptide to trigger this automatic data-dependent MS/MS sequencing.

In this case, another much more time-consuming option is to examine the MS spectra to search for a peptide shifted by the masses corresponding these possible modifications. This can be done manually by creating a list of expected "normal" peptide masses, calculating the modified masses, and examining the MS spectra obtained during the LC/MS/MS run. Obviously, at 1 s per scan, thousands of spectra are obtained throughout the course of an extended LC/MS/MS run. Current software systems allow the combination of groups of spectra, and these groups of combined spectra can be examined. Most current software packages also allow the deconvolution of the spectra to singly-charged species, which reduces the complexity of the manual data analysis, since multiple charge states of the possible peptides are deconvoluted to singly-charged species.

A semi-automated approach to this task is to combine all of the spectra, perform a deconvolution on the MS data to generate a pseudo-singly charged spectrum (*see* **Note 20**), and then submitting this data to the database search software for searching as an MS data file. As above, the MS data can then be searched for modified peptides (*see* **Note 21**).

If a possible modified peptide is identified in this manner, it is useful to examine the original MS data to see if the calculated +2 and +3 charge states for this peptide co-chromatograph. Now that the masses are known, it is possible to perform a different type of MS scanning, where the precursor ion is specified – in Micromass software, it is called "include only" MS/MS. In this scan mode, MS/MS is only performed on the preselected precursor ions. In a previous study *(24)*, we have found that this can lead to an increase in sensitivity of a factor of 50–100 for these ions.

Although this scanning mode is dependent on intensity-based triggering, and thus has the same sensitivity limitations as the "survey scan" mode, knowing these characteristic fragment ions also allows the possibility of "precursor ion scanning." Here, when a preset characteristic ion is detected, the data system switches to the MS mode, detects the precursor, and collects the MS/MS data. Similar approaches are already commonly used, for example, in order to find peptides containing acetylated lysine from the acetylated lysine immonium ion *(25)*, or to find phosphotyrosine-containing peptides from its characteristic immonium ion *(26,27)*.

The identification of the specific and characteristic fragment ions which we have described above, provides a powerful new approach for finding ubiquitinated peptides. Searching the MS/MS chromatograms (*see* **Note 21**) for these characteristic ions (which should co-chromatograph because they are fragment ions from the same peptide) should allow one to find peptides containing ubiquitin side chains.

As mentioned above (Section **3.2.3.1**), ubiquitinated peptides differ from "standard" peptides in that they have 2 N-termini. Recently, Cotter and his group have developed an interesting new method which uses this feature *(28,29)*. This method is based on a derivatization procedure, similar to the chemistry behind Edman sequencing *(30)*. This reagent attaches an SPITC (4-sulfophenyl isothiocyanate) moiety to the N-termini – to *both* N-termini for ubiquitinated peptides. Under MS/MS conditions, ubiquitinated peptides can then be distinguished from non-ubiquitinated peptides by the loss of *two* SPITC groups forming a set of "signature" ions resulting from complete or partial loss of one or two of these tags ("normal" peptides can only have *one* loss of SPITC) (Fig. 8.13). This derivatization procedure, first developed by Keough, et al. *(31)*, produces a y series from a derivatized peptide, while reducing the b series (ions which contain the N-terminus). The Cotter group first modified this reagent for more efficient use in aqueous media,*(32)* and then used it to derivatize tryptic peptides prior to MALDI-MS and MALDI-MS/MS. They used this method for determination of ubiquitination sites in a synthetic peptide containing a KGG tag, and to tetraubiquitin *(28)*.

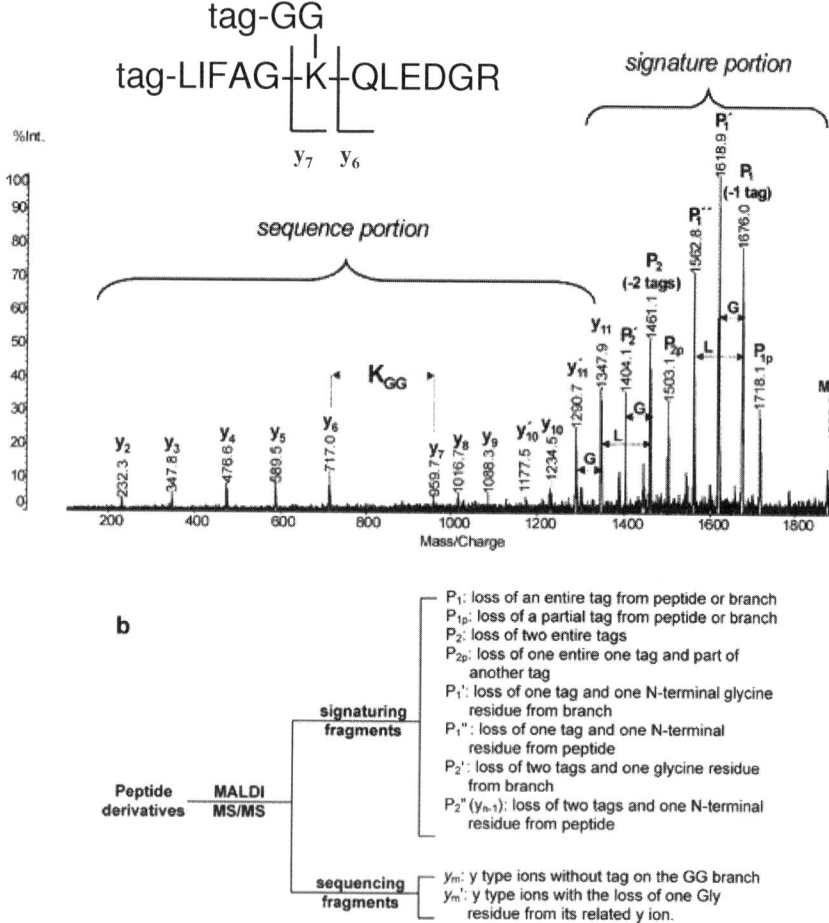

Fig. 8.13 Fragmentation of an SPITC-labeled ubiquitinated peptide. (Reprinted from *(28)*, with permission)

In a subsequent paper,(29) this procedure was used to find the ubiquitination sites on His-tagged CHIP (*C*-terminal HSP70-interacting protein), which was affinity purified using Cobalt beads, and IMAC. Three peptides were identified as doubly tagged based on their MS/MS spectra, which produced "signature" ions from the SPITC moieties and sequence ions which were used to identify the site of ubiquitination.

4   Notes

1. We call this "direct" analysis because the affinity beads are placed directly on the target, in contrast to methods were the affinity-bound proteins are eluted first, and the eluate is spotted on the target.
2. If binding has to be done through protein A, or the antibody is dissolved in ascites, the antibody can be crosslinked to the beads *(33)* before the target protein is affinity-bound.
3. Since only a small number of beads will be placed on the MALDI target, it is important to get much protein as possible on each bead, so use a small amount (~20 µL) of settled affinity bead slurry.
4. To avoid releasing parts of the antibody into the solution, try to avoid reducing agents such as β-mercaptoethanol or dithiothreitol (DTT) in the binding buffer.
5. Most salts will be removed during the wash steps. Other components not compatible with mass spectrometric analysis (such as glycerol or detergents) should be avoided or minimized Although Zwittergent is thought to be compatible with mass spectrometry, it seems to make agarose beads turn "gummy", so it should not be used.
6. To recrystallize α-cyano 2-hydroxycinnamic acid, make a saturated solution of α-cyano 2-hydroxycinnamic acid in boiling methanol. Pour off the solution and discard it. Add more methanol, and again make a saturated solution in boiling methanol. This time, pour off the methanol and save it. Evaporate the methanol to dryness in a hood, while protecting the solution and the crystals from light with aluminum foil. Store in the dark or in a vial wrapped with aluminum foil.
7. The matrix solvent must contain both organic and acid so that it dissociates the affinity bound protein from the antibody on the MALDI target.
8. Also known as *Staphloccus aureus* V-8 protease.
9. The antibody used should be covalently attached to the beads, and DTT should not be used (or used in a very low concentration, *(34)*) in the purification step. If the disulfide bridges in the antibody are reduced, the antibody can be enzymatically digested along with the attached protein. This will lead to a high background of IgG peptides along with the peptides from the target protein, and will make finding the modified peptide more difficult.
10. For proteins affinity-bound to antibody beads, much higher ratios of enzyme (e.g., 5:1) to substrate should be used than for proteins in solution *(23)*.

11. Be sure to add enough digestion buffer so that the beads can "slosh around" in the Eppendorf tube and don't dry out during the overnight digestion.
12. If you don't know the protein concentration, use ~2 μL of a 1 μg/μL solution.
13. If the sample to be digested is already dissolved in water or a buffer such as PBS, add enough ammonium bicarbonate to the solution to ensure that the pH will be ~7–8.
14. The specificity of gluC depends on buffer used and pH of the solution. Cleavage can be either C-terminal to glutamic acid (in ammonium acetate at pH 4.0) or ammonium bicarbonate at pH 7.8), or to *both* glutamic and aspartic acid (in PBS, pH 7.8). For a solution digest, you can add a second buffer to the original buffer in order to adjust the pH, but be sure to take both enzymes into account when calculating the expected of the peptide mw's.
15. This requires a site-license for Mascot, although the company says that if there is sufficient interest, they will add new modifications to their on-line website (www.matrixscience.com).
16. Programming in these modification means, in effect, telling the software that there are three additional types of lysines, with new masses: 242.1374 for GGK, 511.6294 for LRGGK, and 1430.8824 for STLHLVLRLRGGK.
17. In searching using Mascot, first search with no variable modifications, *then* select the target protein and search in the error tolerant mode, specifying the appropriate ubiquitination modifications you have previously entered.
18. Be sure to specify at least two missed cleavages. Cleavage is not expected to occur at the modified lysine.
19. The database search can also be "forced" to consider a specific target program. In MASCOT, this is done by adding the accession number from the appropriate database as the first line of the peak list being searched. (*e.g.*, accession=XXXXX, followed by a blank line)
20. For Micromass MassLynx software, this is MaxEnt 3, under "Tools".
21. If multiple levels of MS/MS spectra are produced, as in the Micromass MassLynx software, *all* of the MS/MS functions must be searched.

Acknowledgements This study was funded by a gift from an anonymous donor to support research in proteomics and cystic fibrosis, and grants from the Cystic Fibrosis Foundation (CFFTI STUTTS01U0) and from NIH (ES11997, 5U54HD035041-07, and P30 CA 16086-25).

References

1. Hochstrasser, M. (1996) Ubiquitin-dependent protein degradation. *Ann. Rev. Genet.* **30**, 405–439.
2. Levinger, L. and Varshavsky, A. (1982) Selective arrangement of ubiquitinated and D1 protein-containing nucleosomes within the Drosophila genome. *Cell* **28**, 375–385.
3. Glotzer, M., Murray, A. W., and Kirschner, M. W. (1991) Cyclin is degraded by the ubiquitin pathway. *Nature* **349**, 132–138.
4. Sasaki, A., Inagaki-Ohara, K., Yoshida, T., Yamanaka, A., Sasaki, M., Yasukawa, H., Koromilas, A., and Yoshimura, A. (2003) The N-terminal truncated isoform of SOCS3 translated from an

alternative initiation AUG codon under stress conditions is stable due to the lack of a major ubiquitination site, Lys-6. *Journal of Biological Chemistry* **278**, 2432–2436.

5. Hershko, A. and Ciechanover, A. (1992) The ubiquitin system for protein degradation. *Ann. Rev. Biochem.* **61**.

6. Pickart, C. M. (2001) Ubiquitin enters the new millennium. *Mol. Cell* **8**, 499–504.

7. Hicke, L. (2001) A new ticket for entry into budding vesicles—ubiquitin. *Cell* **106**, 527–530.

8. Johnson, E. S. (2002) Ubiquitin branches out. *Nature Cell Biology* **4**, E295–E298.

9. Sun, Z. W., and Allis, C. D. (2002) Ubiquitination of histone H2B regulates H3 methylation and gene silencing in yeast. *Nature* **418**, 104–108.

10. Conaway, R. C., Brower, C. S., and Conaway, J. W. (2002) Emerging roles of ubiquitin in transcription regulation. *Science* **296**, 1254–1258.

11. Hochstrasser, M. (2002) Evolution and function of ubiquitin-like protein-conjugation systems. *Nature Cell Biology* **2**, E153–E157.

12. Kirkpatrick, D. S., Denison, C., and Gygi, S. P. (2005) Weighing in on ubiquitin: the expanding role of mass-spectrometry-based proteomics. *Nature Cell Biology* **7**, 750–757.

13. Denison, C., Kirkpatrick, D. S., and Gygi, S. P. (2005) Proteomic insights into ubiquitin and ubiquitin-like proteins. *Current Opinion in Chemical Biology* **9**, 69–75.

14. Gururaja, T., Li, W., Noble, W. S., Payan, D. G., and Anderson, D. C. (2003) Multiple functional categories of proteins identified in an in vitro cellular ubiquitin affinity extract using shotgun peptide sequencing. *Journal of proteome research* **2**, 394–404.

15. Kirkpatrick, D. S., Weldon, S. F., Tsaprailis, G., Liebler, D. C., and Gandolfi, A. J. (2005) Proteomic identification of ubiquitinated proteins from human cells expressing His-tagged ubiquitin. *Proteomics* **5**, 2104–2111.

16. Vasilescu, J., Smith, J. C., Ethier, M., and Figeys, D. (2005) Proteomic Analysis of Ubiquitinated Proteins from Human MCF-7 Breast Cancer Cells by Immunoaffinity Purification and Mass Spectrometry. *Journal of Proteome Research* **4**, 2192–2200.

17. Peng, J., Schwartz, D. R., Elias, J. E., Thoreen, C. C., Cheng, D., Marsischky, G., Roelofs, J., Finley, D., and Gygi, S. P. (2003) A proteomics approach to understanding protein ubiquitination. *Nature Biotechnology* **21**, 921–926.

18. Hitchcock, A. L., Auld, K., Gygi, S. P., and Silver, P. A. (2003) A subset of membrane-associated proteins is ubiquitinated in response to mutations in the endoplasmic reticulum degradation machinery. *Proceedings of the National Academy of Sciences of the United States of America* **100**, 12735–12740.

19. Laub, M., Steppuhn, J. A., Bluggel, M., Immler, D., Meyer, H. E., and Jennissen, H. P. (1998) Modulation of calmodulin function by ubiquitin-calmodulin ligase and identification of the responsible ubiquitylation site in vertebrate calmodulin. *European Journal of Biochemistry* **255**, 422–431.

20. Marotti, L. A., Jr., Newitt, R., Wang, Y., Aebersold, R., and Dohlman, H. G. (2002) Direct Identification of a G Protein Ubiquitination Site by Mass Spectrometry. *Biochemistry* **41**, 5067–5074.

21. Warren, M. R. E., Parker, C. E., Mocanu, V., Klapper, D. G., and Borchers, C. H. (2005) Electrospray ionization tandem mass spectrometry of model peptides reveals diagnostic fragment ions for protein ubiquitination. *Rapid Communic. Mass Spectrom.* **19**, 429–437.

22. Greer, S. F., Zika, E., Conti, B., Zhu, X.-S., and Ting, J. P.-Y. (2003) Enhancement of CIITA transcriptional function by ubiquitin. *Nature Immunology* **4**, 1074–1082.

23. Parker, C. E. and Tomer, K. B. (2000) *in* Methods in Molecular Biology (Chapman, J. R., Ed.), Vol. 146 (Mass Spectrometry of Proteins and Peptides), pp. 185–201, Humana Press, Totowa, NJ.

24. Kast, J., Parker, C. E., van der Drift, K., Dial, J. M., Milgram, S. L., Wilm, M., Howell, M., and Borchers, C. H. (2003) MALDI-directed nano-ESI-MS/MS Analysis for Protein Identification. *Rapid Commun. in Mass Spectrom.* **17**, 1825–1834.

25. Borchers, C., Parker, C. E., Deterding, L. J., and Tomer, K. B. (1999) A preliminary comparison of precursor scans and LC/MS/MS on a hybrid quadrupole time-of-flight mass spectrometer. *J. Chromatogr. A* **854**, 119–130.

26. Steen, H., Kuester, B., Fernandez, M., Pandey, A., and Mann, M. (2001) Detection of tyrosine phosphorylated peptides by precursor ion scanning quadrupole TOF mass spectrometry in positive ion mode. *Anal. Chem.* **73**, 1440–1448.

27. Steen, H., Kuster, B., Fernandez, M., Pandey, A., and Mann, M. (2002) Tyrosine phosphorylation mapping of the epidermal growth factor receptor signaling pathway. *J. of Biol. Chem.* **277**, 1031–1039.

28. Wang, D. and Cotter, R. J. (2005) Approach for determining protein ubiquitination sites by MALDI-TOF mass spectrometry. *Analytical Chemistry* **77**, 1458–1466.

29. Wang, D., Xu, W., McGrath, S. C., Patterson, C., Neckers, L., and Cotter, R. J. (2005) Direct Identification of Ubiquitination Sites on Ubiquitin-Conjugated CHIP Using MALDI Mass Spectrometry. *Journal of Proteome Research* **4**, 1554–1560.

30. Edman, P. (1950) Preparation of phenyl thiohydantoins from some natural amino acids. *Acta Chem. Scand.* **4**, 277–282.

31. Keough, T., Youngquist, R. S., and Lacey, M. P. (2003) Sulfonic acid derivatives for peptide sequencing by MALDI MS. *Analytical Chemistry* **75**, 156A–165A.

32. Wang, D., Kalb, S. R., and Cotter, R. J. (2004) Improved procedures for N-terminal sulfonation of peptides for matrix-assisted laser desorption/ionization post-source decay peptide sequencing. *Rapid communications in mass spectrometry* **18**, 96–102.

33. Peter, J. F. and Tomer, K. B. (2001) A general strategy for epitope mapping by direct MALDI-TOF mass spectrometry using secondary antibodies and cross-linking. *Anal. Chem.* **73**, 4012–4019.

34. Hall, M. C., Torres, M. P., Schroeder, G. K., and Borchers, C. H. (2003) Mnd2 and Swm1 Are Core Subunits of the Saccharomyces cerevisiae Anaphase-promoting Complex. *J. Biol. Chem.* **278**, 16698–16705.

9
Analysis of Sumoylation

Andrea Pichler

Summary Reversible attachment of SUMO (*small ubiquitin related modifier*) regulates a large number of proteins and plays an important role in processes such as transcriptional regulation, nucleo-cytoplasmic transport, genome integrity, and cell cycle progression. The steady state level of most sumoylated proteins is very low, presumably caused by strictly regulated modification and/or rapid cycles of modification and de-modification. This often causes a detection problem of sumoylation in vivo. One approach to overcome this obstacle is described here and involves enrichment of sumoylated proteins under denaturing conditions. After sumoylation is verified, addressing its functional consequences is the logical next step. This will benefit significantly from the availability of large quantities of modified protein. A protocol for efficient in vitro sumoylation of target proteins is described here. It makes use of an E3 ligase fragment that functions without target discrimination.

Key Words SUMO; His-SUMO1; Ni2+ pull down; E3 ligase; RanBP2 IR1+M

1 Introduction

Sumoylation involves an enzymatic cascade to form an isopeptide bond between the C-terminal glycine residue of SUMO and the ε-amino group of the substrate lysine. SUMO is first activated by the heterodimeric E1 activating enzyme Aos1/Uba2 (SAE1/SAE2), which catalyses adenylation of the SUMO C-terminal glycine and subsequently transfers the adenylate to a conserved cysteine resulting in an E1-SUMO thioester linkage. SUMO is then transferred to the E2 conjugating enzyme Ubc9, again forming a thioester. In the final step the modifier is conjugated to its substrate, a reaction that often requires a third class of enzymes, the E3 ligases *(1–3)*.

Steady state levels of sumoylation are usually very low, in part because of the presence of cysteine proteases (members of the Ulp family) that specifically remove SUMO from its conjugates *(1,3,4)*. These enzymes are distributed throughout the cell and are highly active, which often causes a problem in detection of endogenous

From: *Post-translational Modifications of Proteins.*
Methods in Molecular Biology, Vol. 446.
Edited by: C. Kannicht © Humana Press, Totowa, NJ

sumoylation. Different methods to verify in vivo sumoylation have been described. One obvious approach combines immunoprecipitation of endogenous target protein with detection of a 15–20kDa slower migrating modified form with anti-SUMO antibodies and vice versa. However, even if excellent antibodies are available and SUMO protease inhibitors like iodoacetamide or N-ethyl maleimide are included, this often is not sufficient to detect the marginal amounts of modified protein. Overexpression of either SUMO and/or the target protein can increase the level of modification. The method described here is based on a technique originally described for ubiquitin substrate verification (5,6). Among others, Ron Hay's laboratory adapted this approach for sumoylation (7–9) and we successfully applied it in our lab (10). The experimental strategy involves the enrichment of SUMO1 conjugates from a HeLa cell line stably expressing polyhistidine-tagged SUMO1 (kindly provided by Ron Hay), (8) on Ni^{2+} beads under denaturing conditions. This effectively inhibits the SUMO protease activity immediately upon cell lysis. Detection is performed by immunoblot analysis against the endogenous target protein and does not require as excellent antibodies as immunoprecipitation. A further advantage of this method is the possibility to easily increase the amount of starting material.

Once in vivo sumoylation has been demonstrated, understanding its functional consequences is the next step. Investigation of SUMO function involves mapping of the sumoylation site, often addressed by mutagenesis of potential lysines within the SUMO consensus motif (ψKxE, ψ is a bulky aliphatic and x any residue). This motif is recognized by the SUMO E2 in unstructured regions but not in α-helices (11). However, the number of SUMO substrates modified on nonconsensus lysines is growing. A better way to identify modified lysines is by mass spectrometry. This requires relatively high amounts of modified protein but in vitro sumoylation of most substrates works only poorly in the absence of SUMO E3 ligases. Here, a tool is proposed that significantly enhances in vitro sumoylation as identified by analysing the SUMO E3 ligase RanBP2 (10). A small fragment of this ligase designated IR1+M significantly enhances the SUMO transfer from the E2 to the target protein retaining its specificity for SUMO site recognition but not for substrate discrimination (10). The obtained high yields of modified protein can be analysed by mass spectrometry but can also serve for further biochemical approaches to gain deeper insights into the SUMO function of a specific substrate.

2 Materials

2.1 Cell culture

1. Dulbecco's Modified Eagle's Medium (DMEM) (Gibco/BRL, Bethesda, MD) supplemented with 10% fetal bovine serum, puromycin (2µg/mL), penicillin/streptomycin.
2. Tissue culture dishes 150 × 25mm (Falcon)

2.2 Lysis and Ni²⁺-pull down

Unless stated otherwise all buffers are prepared fresh and supplemented with 1 mM β-mercaptoethanol and 10 mM iodoacetamide.

1. Lysis Buffer: 6M guanidine hydrochloride, 100 mM, NaH$_2$PO$_4$ 10 mM Tris-HCl, pH 8.0.
2. Wash buffer 1: 8M urea, 100mM NaH$_2$PO$_4$, 10 mM Tris-HCl, 10 mM imidazole, pH 8.0.
3. Wash buffer 2: 8M urea, 100 mM NaH$_2$PO$_4$, 10 mM Tris-HCl, 10 mM imidazole, pH 6.3.
4. Elution buffer: 8M urea, 100 mM NaH$_2$PO$_4$, 10 mM Tris-HCl, 10 mM imidazole, pH 4.5.
5. Ni²⁺- ProBond Resin (Invitrogen).
6. Laemmli buffer: 25 mM Tris, 192 mM glycine, and 0.1% SDS, bromophenol blue, and 100 mM DTT in aqueous solution.
7. Micro Bio-Spin Chromatography Columns (Biorad - 732–6204).
8. Standard SDS- Polyacrylamide Gel Electrophoresis (SDS-PAGE).

2.3 Purification of RanBP2 IR1+M

1. Competent bacterial strain BL 21 DE3.
2. RanBP2 IR1+M cloned in pGEX-2T (10).
3. Glutathione Sepharose TM 4B (Amersham Biosciences).
4. Thrombin Cleavage Capture Kit, Novagen.
5. Buffer 1: 50 mM Tris-HCl, pH 8, 300 mM NaCl supplemented with 1 mM PMSF, 1μg/mL each of aprotinin, leupeptin, pepstatin and 1 mM DTT.
6. Transport buffer (TB): 20 mM HEPES, 110 mM KOAc, 2 mM Mg(OAc)$_2$, 1 mM EGTA pH 7.3 supplemented with 1 μg/mL each of aprotinin, leupeptin, pepstatin and 1 mM DTT.
7. LB with ampicillin (100 μg/mL).
8. Lysozyme (SIGMA).
9. Micro Bio-Spin Chromatography Columns (Biorad - 732–6204).
10. Standard SDS-PAGE (15%).

2.4 In vitro SUMOylation with RanBP2 IR1+M

1. Recombinant proteins: SUMO, E1 and E2 (12), more detailed in (16), RanBP2 IR1+M, and target protein.
2. Shift buffer: Transport buffer (TB): 20 mM HEPES, 110 mM KOAc, 2 mM Mg(OAc)$_2$, 1 mM EGTA pH 7.3 supplemented with 1 μg/mL each of aprotinin,

leupeptin, pepstatin, 1 mM DTT, 0.05% (v/v) Tween, and 0.2 mg/mL ovalbumin grade VI (Sigma). Aliquots can be stored at −20°C.

3. ATP: 100 mM ATP, 100 mM Mg(OAc)$_2$, 20mM HEPES, titrate pH 7.4 with 10N NaOH.

4. Heat block.

5. Standard SDS-PAGE.

3   Methods

3.1   *Cell culture and lysis*

1. Grow three 150 × 25 mm dishes of His-SUMO1 expressing HeLa cells to 80% confluency (*see* **Note 1** and **2**). As negative control use the same amount of wt HeLa cells.

2. Wash the cells twice with ice cold 1 × PBS.

3. Lyse the cells in 6 mL lysis buffer (2 mL/plate) scrape and transfer them into an appropriate tube.

4. Incubate the lysate for 30 min at room temperature and sonicate the sample reduce viscosity.

5. Centrifuge the lysate for 1 h at 100,000g to remove the cellular debris that can cause unspecific binding.

6. Use the cleared supernatant for the pull down (*see* **Note 3**).

3.2   *Ni^{2+} -Pull Down*

After denaturing cell lysis, 2 subsequent rounds of pull down are performed to reduce unspecific binding.

1. During centrifugation of the lysate prepare the Ni^{2+}-bio spin column with each 200 µL (*see* **Note 4**) of Ni^{2+}-beads for the SUMO1 expressing cell line and the control cells, respectively.

2. Wash the beads 3 times with each 1 mL Lysis buffer.

3. Transfer the cell extracts carefully onto the column (flow through can be collected to control depletion of His-SUMO1 conjugates).

4. Wash the column 4 times with each 1 mL of Wash buffer 1.

5. Wash the column 4 times with each 1 mL of Wash buffer 2.

6. Elute His-SUMO1 with 1 mL Elution buffer into a new tube.

7. Adjust the eluate to pH 8 with NaOH and control pH with pH paper.

8. Add 50 µL washed Ni^{2+}-beads to the eluate and incubate the sample for 1 h at room temperature on a rotating wheel.

9. Apply the lysate/beads mixture onto new bio spin column.

10. Wash the beads 4 times with each 1 mL Wash buffer 1.

11. Elute His-SUMO1 conjugates with 100 μL hot 1 × Laemmli buffer into a new tube.
12. Separate the sample by standard SDS- polyacrylamide gel electrophoreses.
13. Detect the protein of interest by standard Western blotting (RanGAP1 sumoylation is demonstrated as example in Figure 9.1A, (right panel). An anti-SUMO1 (*see* **Note 5**) blot can be performed as control (Fig. 9.1A left panel).

3.3   Purification of RanBP2 IR1+M

1. Transform pGEX-RanBP2 IR1+M into the bacterial *Escherichia coli* strain BL21(DE3).
2. Use a single colony to inoculate 5-mL overnight culture in LB/Amp.
3. Dilute 2 mL of the overnight culture 1:50 in LB/Amp and grow for 2–3 h at 37°C to OD600 < 0.6.

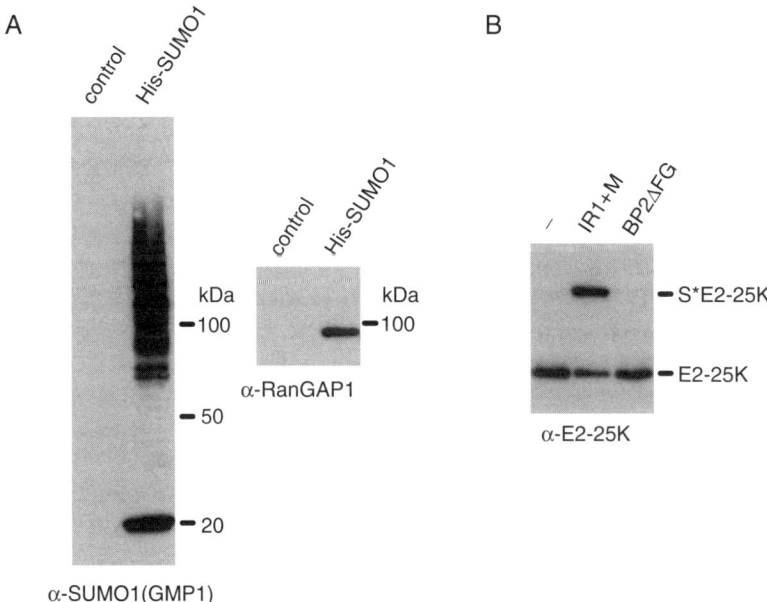

Fig. 9.1 (**A**) Ni^{2+} pull down from a His-SUMO1 expressing HeLa cell line as described here: Enrichment from His-SUMO1 conjugates in 2 rounds of Ni^{2+} pull down, elution with 1× Laemmli buffer and separation on 5–20% SDS – PAGE. Detection was performed with α-SUMO1 (GMP1) antibodies (left panel) and α-RanGAP1 antibodies (right panel), respectively. (**B**) In vitro sumoylation assay with E2-25K as SUMO substrate: A 20 μL reaction of E2-25K (450 nM, Boston Biochem), SUMO1 (5 μM), E1 (70 nM), E2 (25 nM) and 0.5 mM ATP with or without RanBP2 IR1+M (300 nM) or RanBP2ΔFG (8 nM) was incubated at 30°C for 30 min. Samples were stopped with 2× Laemmli buffer and separated on a 12.5% SDS-PAGE. Detection was performed by immunoblotting using α-E2-25K antibodies

4. Add 1 mM IPTG to induce protein expression and incubate for another 3 h at 37°C.
5. Harvest the cells by centrifugation at 3,000g.
6. Resuspend the cell pellet in 1.5 mL Buffer 1 supplemented with 1 mM PMSF, 1µg/mL each of aprotinin, leupeptin, pepstatin and 1 mM DTT.
7. Subject sample to one cycle of freeze-thawing in liquid nitrogen (or −80°C).
8. Transfer the solution to an appropriate ultracentrifuge tube, add 1mg/mL lysozyme and incubate the sample for 1 h on ice.
9. Centrifuge the sample at 100,000g for 1 h.
10. Meanwhile, add 200 µL glutathione beads into a Micro Bio-Spin Chromatography Column.
11. Wash the beads 3 times with 1 mL Buffer 1.
12. Apply the supernatant (9) carefully to the column.
13. Wash the column 3 times with each 1 mL Buffer 1.
14. Wash the column 3 times with each 1 mL TB.
15. Add 200 µL TB buffer and 1 µL biotinylated Thrombin and incubate overnight at 4°C.
16. Centrifuge the column and collect the flow through.
17. Incubate the flow through with 10 µL prewashed streptavidin beads to remove the biotinylated Thrombin.
18. Spin the sample to pellet the beads, aliquot the supernatant and shock-freeze the aliquots in liquid N2 and store them at −80°C.
19. Determine the protein concentration and analyse the sample on a 15% SDS-PAGE.

3.4   In vitro SUMOylation with RanBP2 IR1+M

1. Mix target protein (450 nM to 1 µM) (*see* **Note 6**) with SUMO1 (5 µM) (*see* **Note 7**), E1 (70 nM), E2 (25 nM), RanBP2 IR1+M (300 nM) and 0.5 mM ATP (*see* **Note 8**). As control include a sample without RanBP2 IR1+M (shows E3 independent reaction) and/or one without ATP (shows no shift).
2. Incubate the reactions for 30 min at 30°C.
3. Stop the reaction either by adding 2x Laemmli buffer (for SDS-PAGE) or by ATP depletion with 1 U Apyrase (SIGMA) or by blocking the E1 activity with 10 mM EDTA (for biochemical analysis) (*see* **Note 9**).
4. Separate the sample on an appropriate SDS-PAGE.
5. Detect the target protein and its sumoylated form either by Western blot (an example is demonstrated in Fig. 9.1B) or by Coomassie stain.

4   Notes

1. The amount of His-SUMO1 expressing HeLa cells to start with depends on the detection level of your antibodies and on the expected steady state level of sumoylation for the target protein and can therefore easily be scaled up- or down.

2. Usage of Ni2⁺-pull down has also been described for verification of substrate sumoylation in yeast (e.g., (13, 14)).

3. To determine the amount of modified versus unmodified protein, precipitate (TCA or Chlorophorm/Methanol-) an aliquot of the lysate.

4. High excess of Ni2⁺-beads over His-tagged protein increases the background (The QIAexpressionist handbook from Qiagen is highly recommended as trouble-shooting guide for Ni2⁺-pull downs).

5. Antibody to detect mammalian SUMO1: mouse monoclonal anti-GMP1 (anti-SUMO1) from Zymed

6. Usage of deletion fragments in the in vitro sumoylation assay is problematic because it can lead to wrong conclusions regarding the SUMO modification site (Lysines hidden in the full length protein can be exposed or small fragments become unstructured).

7. Recently, a SUMO mutant (SUMOT95R) that allows easier identification of sumoylated peptides by mass spectrometry analysis has been described (15). One may consider to use this mutant in the in vitro sumoylation assay for SUMO site identification.

8. Ovalbumin in the shift buffer is not required for assays using high protein concentration (but it is strongly recommended for dilutions to low concentrations, like for enzymes)

9. For studying the function of a sumoylated protein the modified protein has to be separated from the unmodified after the in vitro reaction. For small proteins this can be performed e.g. by separation on a Sephadex 75 FPLC column (11). Larger proteins can, e.g., be separated in a 2-step procedure by using a tagged SUMO that allows the separation of the modified from the unmodified form by pull down. In a second step the free SUMO is removed from the conjugate by gel-filtration.

Acknowledgements I would like to thank Frauke Melchior for her ongoing support, Annette Floto, Guillaume Bossis,and Guido Sauer for critical reading the manuscript and Ron Hay is kindly acknowledged for providing the His-SUMO1 HeLa cell line. This work was funded by Vienaa Science and Technology Fund WWTF LS05003 and FWF P18584–B12.

References

1. Melchior, F., Schergaut, M., and Pichler, A. (2003) SUMO: ligases, isopeptidases and nuclear pores. *Trends Biochem. Sci.* **28**, 612–618.
2. Johnson, E. S. (2004) Protein modification by SUMO. *Annu. Rev. Biochem.* **73**, 355–382.
3. Hay, R. T. (2005) SUMO: a history of modification. *Mol. Cell* **18**, 1–12.
4. Li, S. J. and Hochstrasser, M. (1999) A new protease required for cell-cycle progression in yeast. *Nature* **398**, 246–251.
5. Treier, M., Staszewski, L. M., and Bohmann, D. (1994) Ubiquitin-dependent c-Jun degradation in vivo is mediated by the delta domain. *Cell* **78**, 787–798.
6. Beers, E. P. and Callis, J. (1993) Utility of polyhistidine-tagged ubiquitin in the purification of ubiquitin-protein conjugates and as an affinity ligand for the purification of ubiquitin-specific hydrolases. *J. Biol. Chem.* **268**, 21645–21649.

7. Rodriguez, M. S., Desterro, J. M., Lain, S., Midgley, C. A., Lane, D. P., and Hay, R. T. (1999) SUMO-1 modification activates the transcriptional response of p53. *Embo J.* **18**, 6455–6461.

8. Girdwood, D., Bumpass, D., Vaughan, O. A., Thain, A., Anderson, L. A., Snowden, A. W., Garcia-Wilson, E., Perkins, N. D., and Hay, R. T. (2003) P300 transcriptional repression is mediated by SUMO modification. *Mol. Cell* **11**, 1043–1054.

9. Vertegaal, A. C., Ogg, S. C., Jaffray, E., Rodriguez, M. S., Hay, R. T., Andersen, J. S., Mann, M., and Lamond, A. I. (2004) A proteomic study of SUMO-2 target proteins. *J. Biol. Chem.* **279**, 33791–33798.

10. Pichler, A., Knipscheer, P., Saitoh, H., Sixma, T. K., and Melchior, F. (2004) The RanBP2 SUMO E3 ligase is neither HECT- nor RING-type. *Nat Struct. Mol. Biol.* **11**, 984–991.

11. Pichler, A., Knipscheer, P., Oberhofer, E., van Dijk, W. J., Korner, R., Olsen, J. V., Jentsch, S., Melchior, F., and Sixma, T. K. (2005) SUMO modification of the ubiquitin-conjugating enzyme E2-25K. *Nat. Struct. Mol. Biol.* **12**, 264–269.

12. Pichler, A., Gast, A., Seeler, J. S., Dejean, A., and Melchior, F. (2002) The nucleoporin RanBP2 has SUMO1 E3 ligase activity. *Cell* **108**, 109–120.

13. Johnson, E. S. and Blobel, G. (1999) Cell cycle-regulated attachment of the ubiquitin-related protein SUMO to the yeast septins. *J. Cell Biol.* **147**, 981–994.

14. Hoege, C., Pfander, B., Moldovan, G. L., Pyrowolakis, G., and Jentsch, S. (2002) RAD6-dependent DNA repair is linked to modification of PCNA by ubiquitin and SUMO. *Nature* **419**, 135–141.

15. Knuesel, M., Cheung, H. T., Hamady, M., Barthel, K. K., and Liu, X. (2005) A method of mapping protein sumoylation sites by mass spectrometry using a modified small ubiquitin-like modifier 1 (sumo-1) and a computational program. *Mol. Cell Proteomics* **4**, 1626–1636.

16. Bossis G., Chmielarska K., Gärtner U., Pichler A., Stieger E., and Melchior F. (2005) A FRET-based assay to study SUMO1 modification in solution. *Methods in Enzymol.* **398**, 20–32.

10
Detection and Analysis of Protein ISGylation

Tomoharu Takeuchi and Hideyoshi Yokosawa

Summary ISG15 is a ubiquitin-like modifier that is conjugated to target proteins by a sequential reaction catalyzed by E1/E2/E3 enzymes (protein ISGylation). ISG15 and protein ISGylation are upregulated by interferon stimuli. ISG15 functions as an antiviral protein against Sindbis virus and HIV-1, but the molecular mechanism remains unknown. Here we describe in detail methods for detecting and analyzing protein ISGylation. The methods consist of plasmid transfection and affinity purification of ISGylated proteins. In addition, we describe a method for detecting ISGylation of a target protein, Ubc13.

Key Words ISG15; ISGylation; interferon; ubiquitin; ubiquitin-like protein; post-translational modification.

1 Introduction

Intracellular events are regulated by various post-translational modifications of regulatory proteins, such as phosphorylation, methylation, and ubiquitination. Recently, many studies have focused on the mechanisms and roles of ubiquitination of regulatory proteins. Ubiquitination of proteins requires the sequential action of 3 enzymes, ubiquitin-activating enzyme (E1), ubiquitin-conjugating enzyme (E2), and ubiquitin-ligase (E3) (1) (Fig. 10.1). In addition to the ubiquitin modification system, cells have other protein modification systems mediated by ubiquitin-like proteins (2). ISG15 (an interferon-stimulated gene product with a molecular mass of 15 kDa), one of the type I ubiquitin-like protein family members, is composed of 2 tandem repeats of ubiquitin-like domains and is conjugated to various proteins (protein ISGylation) (3). Upon interferon stimuli, ISG15 and components of the ISGylation system are upregulated and protein ISGylation occurs via a pathway similar to ubiquitination (Fig. 10.1): E1 (UBE1L), E2 (UbcH8/UbcH6), and E3 (Efp/Herc5) for ISGylation and an ISG15-deconjugating enzyme (UBP43), all of which are interferon-inducible, have been identified

From: *Post-translational Modifications of Proteins.*
Methods in Molecular Biology, Vol. 446.
Edited by: C. Kannicht © Humana Press, Totowa, NJ

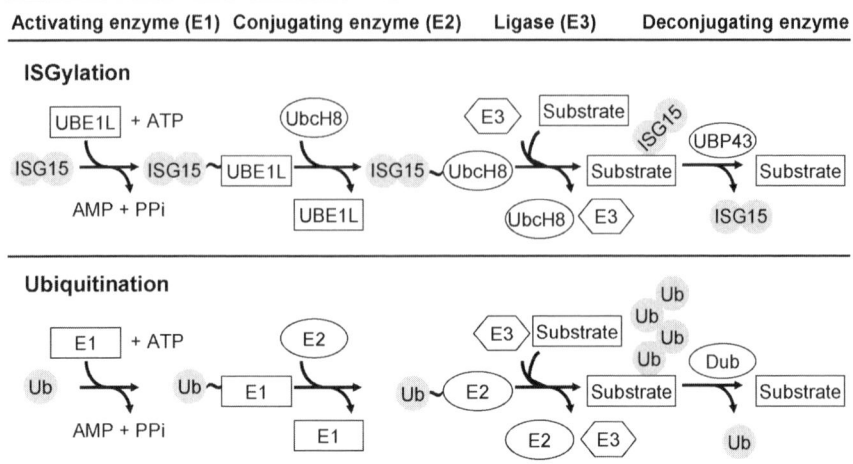

Fig. 10.1 Comparison of the ISGylation system and the ubiquitination system. Protein ISGylation occurs via a pathway similar to that for ubiquitination. Efp, a RING domain-containing protein, and Herc5, a HECT domain-containing protein, have been identified as E3 enzymes for ISGylation. Ub, ubiquitin; Dub, deubiquitinating enzyme

(4–7). Since ISG15 is an interferon-stimulated gene product, it is assumed that ISG15 plays a role in antiviral response. Although ISG15-knockout mice have been reported not to have any defect in antiviral response against VSV and LCMV infection *(8)*, it has recently been found that ISG15 functions as an antiviral protein against Sindbis virus and HIV-1 *(9,10)*, implying that the antiviral function of ISG15 is virus-specific.

To determine what kinds of proteins are ISGylated, we and others have performed affinity purification of ISGylated proteins and detected protein ISGylation by Western blotting *(5, 11–15)*. Here we describe in detail methods for transfection of plasmids for the ISGylation system and also the subsequent immunoprecipitation-based affinity purification of ISGylated proteins. Isolated ISGylated proteins can be subjected to proteomic analysis to identify target proteins for ISGylation. In addition, we describe a method for detecting ISGylation of a target protein, Ubc13 *(5,11)*.

2 Materials

2.1 Cell Culture and Transfection

1. Dulbecco's modified Eagle's medium (DMEM) (Sigma) supplemented with 10% heat-inactivated calf serum (Hyclone).

2. Solution consisting of trypsin (0.25%) and ethylenediaminetetraacetic acid (EDTA) (1 mM) (Invitrogen).
3. 2.5M CaCl$_2$, sterilized through a 0.22-μm filter.
4. 2× BBS: 50 mM N,N-bis(2-hydroxyethyl)-2-aminoethanesulfonic acid, 280 mM NaCl, and 1.5 mM Na$_2$HPO$_4$, pH adjusted to 6.93–6.97 with NaOH (*see* **Note 1**).
5. pCI-neo-3Flag-ISG15, pCI-neo-2S-UBE1L, and pCI-neo-2S-UbcH8 plasmids: These are mammalian expression plasmids that are transfected to mammalian cells to express Flag-tagged ISG15, S-tagged UBE1L and S-tagged UbcH8, respectively.

2.2 Cell Lysis and Immunoprecipitation

1. N-ethylmaleimide (NEM): prepare 0.5M stock solution in dimethylsulfoxide and immediately freeze in aliquots at −20°C (*see* **Note 2**).
2. Dithiothreitol (DTT): prepare 1M stock solution in water and immediately freeze in aliquots at −20°C.
3. RIPA buffer for cell lysis: 50 mM Tris-HCl, pH 7.5, 150 mM NaCl, 0.1% sodium dodecyl sulfate (SDS), 0.5% sodium deoxycholate, 1% Nonidet P-40, 5 mM NEM, and 1 mM DTT (*see* **Note 3**). Store at 4°C. Add DTT and NEM just before use.
4. Washing buffer (buffer A): 20 mM Tris-HCl, pH 7.5, 500 mM NaCl, 0.2% Nonidet P-40, and 10% glycerol.
5. Anti-Flag-tag M2 antibody-immobilized agarose beads from Sigma.
6. 3× Flag peptide (Sigma): prepare a stock solution (1 mg/mL) in 20 mM Tris-HCl, pH 7.5, and 150 mM NaCl and dilute 4-fold with buffer A before use.
7. MicroSpin empty columns from GE Healthcare.

2.3 SDS-Polyacrylamide Gel Electrophoresis (SDS-PAGE)

1. 5× SDS sample buffer: 250 mM Tris-HCl, pH 6.8, 5% SDS, 40% glycerol, 0.05% pyronin-Y, and 10% 2-mercaptoethanol. Store at room temperature. Add 2-mercaptoethanol just before use.
2. 30% acrylamide solution (29.2% acrylamide and 0.8% bis-acrylamide). Store at 4°C under light-blocking conditions.
3. Ammonium persulfate (APS): prepare 10% solution in water, store at 4°C, and use within 2 wk.
4. N,N,N',N'-tetramethylethylenediamine (TEMED) from Nakalai Tesque. Store at 4°C.
5. 1.5M Tris-HCl, pH 8.8, 0.5M Tris-HCl, pH 6.8, and 10% SDS. Store at room temperature.

6. Running buffer: 25 mM Tris, 192 mM glycine, and 0.1% SDS. Store at room temperature.
7. Molecular weight markers: Bench Mark Protein Ladder and Bench Mark Prestained Protein Ladder from Invitrogen.

2.4 Silver Staining

1. Silver stain KANTO III (Kanto Chemical), a kit for silver staining.
2. Fixation buffer for silver staining: 50% methanol and 10% acetic acid.

2.5 Western Blotting

1. Transfer buffer: 25 mM Tris, 20 mM glycine, and 10% methanol. Store at room temperature.
2. Nitrocellulose membrane from Bio-Rad and chromatography paper (85 × 90 mm) from Bio-Craft.
3. Ponceau S solution: 1% Ponceau S and 1% acetic acid.
4. Phosphate-buffered saline (PBS): prepare a 10-fold concentrated stock solution containing 1.37 M NaCl, 27 mM KCl, 81 mM Na_2HPO_4, and 15 mM KH_2PO_4 and dilute with H_2O just before use.
5. Phosphate-buffered saline with Tween (PBS-T): PBS containing 0.05% Tween 20.
6. Blocking buffer: 5% (w/v) nonfat dry milk in PBS-T.
7. Primary antibody: anti-Flag-tag antibody from Sigma.
8. Secondary antibody: anti-mouse immunoglobulin G (IgG) conjugated to horseradish peroxidase (GE Healthcare).
9. Enhanced chemiluminescent (ECL) reagent from GE Healthcare and X-ray film from Fuji Film.

2.6 Detection of ISGylation of Ubc13

1. pCI-neo-3T7-ISG15 and pCI-neo-3Flag-Ubc13 plasmids: These are mammalian expression plasmids that are transfected to mammalian cells to express T7-tagged ISG15 and Flag-tagged Ubc13, respectively.
2. Dulbecco's modified Eagle's medium (DMEM) (Sigma) supplemented with 10% heat-inactivated fetal bovine calf serum (Invitrogen).
3. OPTI-MEM (Invitrogen), a medium for transfection.
4. Metafectene (Biontex), a transfection reagent.
5. Buffer B for cell lysis: 50 mM Tris-HCl, pH 7.5, 150 mM NaCl, 0.1% SDS, 0.5% sodium deoxycholate, and 1% Nonidet P-40. Store at 4°C.
6. Anti-T7-tag antibody from Novagen.

3 Methods

To determine target proteins for ISGylation with Flag-tagged ISG15, it is important to discriminate target proteins that are covalently modified with ISG15 from other proteins that noncovalently bind with ISG15. Thus, immunoprecipitation with anti-Flag-tag M2 antibody-immobilized agarose beads is carried out under conditions using RIPA buffer containing detergents that disrupt noncovalent interaction. In addition, specific elution with Flag peptide of proteins covalently modified with Flag-tagged ISG15 from anti-Flag-tag M2 antibody-immobilized agarose beads has the advantage of avoidance of nonspecific interaction of proteins with the beads. The resulting target proteins covalently modified with Flag-tagged ISG15 are separated by SDS-PAGE, followed by silver staining and Western blotting with anti-Flag-tag M2 antibody. The coincidence between the silver-stained protein band and the Western blot provides information on the existence of ISGylated proteins. The ISGylated proteins can be subjected to proteomic analysis to determine the target candidate proteins. Subsequently, the candidate proteins are subjected to ISGylation in the presence of the ISGylation system to confirm the results of proteomic analysis.

3.1 Transfection of Plasmids to HeLa Cells by the Calcium Precipitation Protocol

1. HeLa cells that have been grown to confluence at 37°C in 5% CO_2 in an incubator are treated with trypsin/EDTA and transferred to a new medium in a 100-mm tissue dish. Ten milliliters of medium is used per 100-mm dish.
2. 50–60% confluent HeLa cells are used for transfection by the calcium precipitation protocol (16).
3. Prepare the following DNA-calcium phosphate complex for transfection: First, dissolve total 20 μg of plasmid DNA (10 μg of Flag-tagged ISG15 expression plasmid and 5 μg each of S-tagged UBE1L and S-tagged UbcH8 expression plasmids) in 450 μL of water (see **Note 4**) and mix by vortexing. Second, add 50 μL of 2.5 M $CaCl_2$ and mix by vortexing. Finally, add 500 μL of 2×BBS, mix gently, and incubate for 20 min at room temperature.
4. Add 1,000 μL of the above complex to a 100-mm dish in which HeLa cells have been grown. Mix gently by rocking the dish back and forth.
5. Incubate the cells for 12–16 h at 35°C in 3% CO_2 in an incubator.
6. Replace the medium and incubate the cells at 37°C in 5% CO_2 in an incubator.
7. Harvest after 36 h of incubation.

3.2 Cell Lysis and Immunoprecipitation of ISGylated Proteins

1. The transfected cells are washed with 3 mL of ice-cold PBS and lysed with 1 mL of ice-cold RIPA buffer. The cell lysate is sonicated for 3 s and the debris

is removed by centrifugation. The resulting supernatant is transferred to a centrifuge tube.

2. The supernatant is incubated with 20 μL of anti-Flag-tag M2 antibody-immobilized agarose beads for 2 h at 4°C. The immunoprecipitates are washed 3 times with 1 mL of RIPA buffer, followed by washing 3 times with 1 mL of buffer A. The beads are transferred to a MicroSpin empty column and incubated with 100 μL of 3× Flag peptide (200 μg/mL) in buffer A for 30 min at 4°C. The eluate is recovered by centrifugation and used as ISGylated proteins.

3. For preparation of samples for SDS-PAGE, add 400 μL of acetone to 100 μL of the above eluate in a microcentrifuge tube and mix by vortexing. After storage for 2 h at −80°C, the precipitates are recovered by centrifugation. The resulting precipitated proteins are air-dried, suspended in 40 μL of 1× SDS sample buffer, and boiled for 2 min to give a sample for SDS-PAGE.

3.3   SDS-PAGE of ISGylated Proteins

1. SDS-PAGE is performed using a BE-230 or BE-240 gel system (Bio-Craft) (*see* **Note 5**).

2. Prepare a 7.5% gel by mixing 2 mL of 30% acrylamide solution, 2 mL of 1.5 M Tris-HCl, pH 8.8, 3.9 mL of water, 80 μL of 10% SDS, 30 μL of 10% APS, and 15 μL of TEMED. Pour the gel, leaving space for a stacking gel, and gently overlay with 1 mL of water. The gel should polymerize within 30 min.

3. Pour off the water.

4. Prepare the stacking gel by mixing 335 μL of 30% acrylamide solution, 625 μL of 0.5 M Tris-HCl, pH 6.8, 1.5 mL of water, 25 μL of 10% SDS, 10 μL of APS, and 5 μL of TEMED. After rinsing the top of the gel with about 150 μL of the stacking gel solution without APS and TEMED, pour the stacking gel solution and insert the comb. The stacking gel should polymerize within 30 min.

5. Once the stacking gel has set, remove the comb carefully and wash the wells with water.

6. Add the running buffer to the chamber of the gel unit and load samples and molecular weight markers in the respective wells (*see* **Note 6**).

7. After assembly of the gel unit, the gel can be run at 20 mA through the stacking gel and at 30 mA through the separating gel for an appropriate time (*see* **Note 7**).

3.4   Silver Staining of ISGylated Proteins

1. The gel in which proteins have been separated by SDS-PAGE is incubated with the fixation buffer for 20 min.

2. Silver staining of proteins is performed using Silver stain KANTO III according to the manufacturer's instructions. A representative result is shown in Fig. 10.2A.

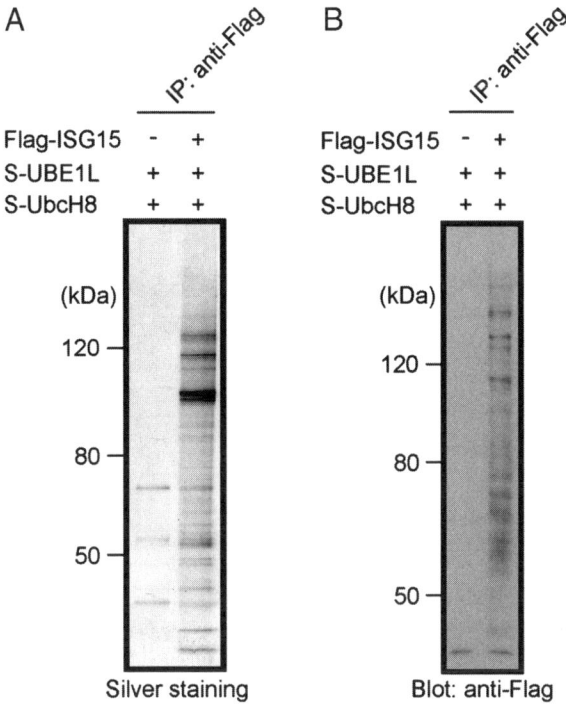

Fig. 10.2 Separation and analysis of ISGylated proteins isolated from HeLa cells. The extract of HeLa cells expressing Flag-tagged ISG15 together with S-tagged UBE1L and S-tagged UbcH8 was subjected to immunoprecipitation (IP) with anti-Flag-tag antibody-immobilized agarose beads, and isolated ISGylated proteins were separated by SDS-PAGE. The separated proteins were detected by silver staining (**A**, right lane) and ISGylated proteins were detected by Western blotting with anti-Flag-tag antibody (**B**, right lane). In the control, the extract of HeLa cells expressing S-tagged UBE1L and S-tagged UbcH8 without Flag-tagged ISG15 gave few signals (left lanes in **A** and **B**)

3.5 Western Blotting of ISGylated Proteins

1. Electrophoretic transfer to a nitrocellulose membrane of proteins that have been separated by SDS-PAGE is performed using a Bio-Craft semi-dry transfer system.
2. A sheet of the nitrocellulose membrane is cut to a size a little larger than that of the separating gel and is then submerged in the transfer buffer. Two sheets of chromatography papers are also submerged in the transfer buffer and pressed gently to avoid the generation of bubbles.
3. The stacking gel is removed.
4. A sheet of chromatography paper is put onto the lower panel of the transfer chamber and the separating gel is carefully laid on the paper to prevent bubbles from entering between the paper and the gel.

5. The membrane is carefully laid on the gel and then another sheet of chromatography paper is laid on the membrane. The upper panel of the transfer chamber is set and the chamber is pressed by placing a weight of about 1,000 g on the transfer chamber.

6. The transfer chamber is connected with a power supply and proteins are allowed to be transferred from the gel to the membrane at 80 V for 75 min.

7. Once the transfer is complete, the membrane is taken out from the transfer chamber and incubated with Ponceau S solution for 5 min at room temperature to stain the proteins on the membrane. The bands of Protein Ladder, detected by Ponceau S staining, are marked up using a ballpoint pen. Then the membrane is washed several times with water until the membrane is partially decolorized.

8. The membrane is incubated with 20 mL of blocking buffer for 30 min at room temperature on a rocking platform.

9. The blocking buffer is discarded and then the membrane is incubated with a 1:2,000 dilution of the primary antibody in blocking buffer for 1 h at room temperature on a rocking platform.

10. The primary antibody is discarded and the membrane is washed 3 times for 5 min each with 20 mL of PBS-T on a rocking platform.

11. The membrane is incubated with a 1:5,000 dilution of the secondary antibody in PBS-T for 1 h at room temperature on a rocking platform.

12. The secondary antibody is discarded and the membrane is washed 3 times for 5 min each with 20 mL of PBS-T, followed by washing twice for 5 min each with 20 mL of PBS.

13. During the final washing of the membrane, 600 μL each of the ECL reagents are mixed and the membrane is then incubated with a mixture of the ECL reagents for 1 min at room temperature.

14. After removal of the ECL reagent, the blotted membrane is placed onto a plastic sheet for an X-ray film cassette and is covered with Saran Rap (*see* **Note 8**).

15. The plastic sheet containing the membrane is placed in an X-ray film cassette with the film for a suitable exposure time. A representative result is shown in Fig. 10.2B.

3.6   Detection of ISGylation of Ubc13

1. Cell culture and transfection experiments are performed according to the procedure described in **Section 3.1** except for A549 cells cultured on 35-mm dishes (2 mL of medium used per 35-mm dish) and Metafectene-mediated transfection. Transfection is performed as follows: A total of 2 μg of plasmid DNA (0.67 μg each of T7-tagged ISG15, Flag-tagged Ubc13, and S-tagged UBE1L expression plasmids) is suspended in 200 μL of OPTI-MEM. Add 6 μL of Metafectene to the DNA solution and mix with tapping, followed by incubation at room temperature for 20 min. Then add 200 μL of the mixture to a 35-mm dish in which A549 cells have been grown. Cells are harvested after 30 h of incubation.

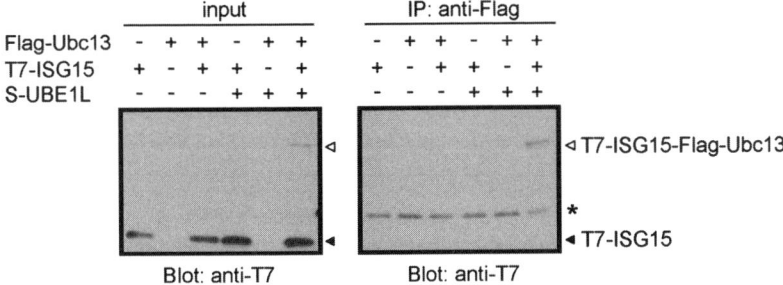

Fig. 10.3 Detection of ISGylation of Ubc13. The extract of A549 cells expressing Flag-tagged Ubc13, T7-tagged ISG15, and S-tagged UBE1L was subjected to immunoprecipitation (IP) with anti-Flag-tag antibody-immobilized agarose beads, followed by Western blotting with anti-T7-tag antibody. ISGylated Ubc13 (T7-ISG15-Flag-Ubc13), indicated by an open arrowhead, was detected in the case of expression of Flag-tagged Ubc13, T7-tagged ISG15 and S-tagged UBE1L but not in other combinations. Free T7-tagged ISG15 (T7-ISG15) is indicated by a closed arrowhead, and a non-specific band is indicated by an asterisk

2. Immunoprecipitation of ISGylated Ubc13 is performed according to the procedure described in **Section 3.2** except that the supernatant obtained from the cell lysate is incubated with 5 μL of anti-Flag-tag M2 antibody-immobilized agarose beads and the immunoprecipitates are washed 5 times with buffer B and incubated with 1 × SDS sample buffer in boiling water for 2 min for SDS-PAGE.

3. SDS-PAGE is performed according to the procedure described in **Section 3.3** except for the separation gel: The 12.5% gel is prepared by mixing 3.33 mL of 30% acrylamide solution, 2 mL of 1.5M Tris-HCl, pH 8.8, 2.54 mL of water, 80 μL of 10% SDS, 40 μL of 10% APS, and 15 μL of TEMED.

4. Western blotting of ISGylated Ubc13 is performed according to the procedure described in **Section 3.5** except for the use of anti-T7-tag antibody as a primary antibody. A representative result of Western blotting of ISGylated Ubc13 is shown in Fig. 10.3.

4 Notes

1. The pH of 2× BBS buffer is a crucial factor for efficient transformation of cells by the calcium precipitation protocol (*16*). We recommend that 2× BBS buffers at various values of pH, such as 6.93, 6.95, and 6.97, are prepared to check which pH is best for transfection. Calf serum included in the medium is another crucial factor because the DNA-calcium phosphate complex is formed gradually in the medium. When a lot of serum is checked, the rate of transfection efficiency should be measured.

2. All solutions, except for SDS-PAGE running buffer and PBS for Western blotting, should be prepared in water that has conductivity less than $18.2\,m\Omega$-cm.
3. Although DTT is not necessarily required for preparation of a sample, we add DTT to the buffer to avoid co-isolation of UBE1L or UbcH8 that has bound with ISG15 via a thioester bond.
4. The amount of total DNA can be made up to $30\,\mu g$ when a larger amount of proteins needs to be expressed.
5. In the case of silver staining of proteins, it is critical that the glass plate for SDS-PAGE is cleaned well to remove contaminating proteins because contaminating proteins are troublingly detected by silver staining. We recommend that the glass plate for silver staining be washed with alkaline detergent or a chromic acid solution.
6. For Western blotting, to easily distinguish right and left sides of the membrane, apply Protein Ladder and Prestained Protein Ladder, molecular weight markers, to the far left and the far right lanes, respectively.
7. For silver staining, the front of the dye, pyronin Y, can be run off, while the front of the dye should be retained for Western blotting because this dye can be transferred to the nitrocellulose membrane to confirm the respective lanes.
8. To confirm the orientation of the membrane under fluorescent light, the plastic sheet can be marked by putting a piece of fluorescent adhesive tape on the edge.

Acknowledgement This study was supported in part by grants-in-aid for scientific research from the Ministry of Education, Culture, Sports, Science and Technology of Japan.

References

1. Pickart, C. M. (2001) Mechanisms underlying ubiquitination. *Annu. Rev. Biochem.* **70**, 503–533.
2. Schwartz, D. C. and Hochstrasser, M. (2003) A superfamily of protein tags: ubiquitin, SUMO and related modifiers. *Trends Biochem. Sci.* **28**, 321–328.
3. Loeb, K. R. and Haas, A. L. (1992) The interferon-inducible 15-kDa ubiquitin homolog conjugates to intracellular proteins. *J. Biol. Chem.* **267**, 7806–7813.
4. Dao, C. T. and Zhang, D. E. (2005) ISG15: a ubiquitin-like enigma. *Front. Biosci.* **10**, 2701–2722.
5. Takeuchi, T., Iwahara, S., Saeki, Y., Sasajima, H., and Yokosawa, H. (2005) Link between the ubiquitin conjugation system and the ISG15 conjugation system: ISG15 conjugation to the UbcH6 ubiquitin E2 enzyme. *J. Biochem.* **138**, 711–719.
6. Zou, W. and Zhang, D. E. (2005) The interferon-inducible ubiquitin-protein isopeptide ligase (E3) EFP also functions as an ISG15 E3 ligase. *J. Biol. Chem.* **281**, 3989–3994.
7. Dastur, A., Beaudenon, S., Kelley, M., Krug, R. M., and Huibregtse, J. M. (2006) Herc5, an interferon-induced HECT E3 enzyme, is required for conjugation of ISG15 in human cells. *J. Biol. Chem.* **281**, 4334–4338.
8. Osiak, A., Utermohlen, O., Niendorf, S., Horak, I., and Knobeloch, K.-P. (2005) ISG15, an interferon-stimulated ubiquitin-like protein, is not essential for STAT1 signaling and responses against vesicular stomatitis and lymphocytic choriomeningitis virus. *Mol. Cell. Biol.* **25**, 6338–6345.

9. Lenschow, D. J., Giannakopoulos, N. V., Gunn, L. J., Johnston, C., O'Guin, A. K., Schmidt, R. E., Levine, B., and Virgin, H. W. (2005) Identification of interferon-stimulated gene 15 as an antiviral molecule during Sindbis virus infection in vivo. *J. Virol.* **79**, 13974–13983.

10. Okumura, A., Lu, G., Pitha-Rowe, I., and Pitha, P. M. (2006) Innate antiviral response targets HIV-1 release by the induction of ubiquitin-like protein ISG15. *Proc. Natl. Acad. Sci. USA* **103**, 1440–1445.

11. Takeuchi, T. and Yokosawa, H. (2005) ISG15 modification of Ubc13 suppresses its ubiquitin-conjugating activity. *Biochem. Biophys. Res. Commun.* **336**, 9–13.

12. Zou, W., Papov, V., Malakhova, O., Kim, K. I., Dao, C., Li, J., and Zhang, D. E. (2005) ISG15 modification of ubiquitin E2 Ubc13 disrupts its ability to form thioester bond with ubiquitin. *Biochem. Biophys. Res. Commun.* **336**, 61–68.

13. Malakhov, M. P., Kim, K. I., Malakhova, O. A., Jacobs, B. S., Borden, E. C., and Zhang, D. E. (2003) High-throughput immunoblotting. Ubiquitin-like protein ISG15 modifies key regulators of signal transduction. *J. Biol. Chem.* **278**, 16608–16613.

14. Zhao, C., Denison, C., Huibregtse, J. M., Gygi, S., and Krug, R. M. (2005) Human ISG15 conjugation targets both IFN-induced and constitutively expressed proteins functioning in diverse cellular pathways. *Proc. Natl. Acad. Sci. USA* **102**, 10200–10205.

15. Giannakopoulos, N. V., Luo, J. K., Papov, V., Zou, W., Lenschow, D. J., Jacobs, B. S., Borden, E. C., Li, J., Virgin, H. W., and Zhang, D. E. (2005) Proteomic identification of proteins conjugated to ISG15 in mouse and human cells. *Biochem. Biophys. Res. Commun.* **336**, 496–506.

16. Chen, C. and Okayama, H. (1987) High-efficiency transformation of mammalian cells by plasmid DNA. *Mol. Cell. Biol.* **7**, 2745–2752.

11

Analysis of Methylation, Acetylation, and other Modifications in Bacterial Ribosomal Proteins

Randy J. Arnold, William Running, and James P. Reilly

Summary A wide variety of post-translational modifications of expressed proteins are known to occur in living organisms *(1)*. Although their presence in an organism cannot be predicted from the genome, these modifications can play critical roles in protein structure and function. The identification of post-translational modifications can be critical in understanding the functions of proteins involved in important biological pathways and mass spectrometry offers a fast, accurate method for observing them. This chapter describes the procedure for analyzing ribosomal proteins of Escherichia coli by matrix-assisted laser desorption/ionization time-of-flight (MALDI-TOF) mass spectrometry and *Caulobacter crescentus* ribosomal proteins by electrospray quadrupole time-of-flight (ESI-QTOF) mass spectrometry.

Key Words Post-translational modifications; mass spectrometry; ribosomal proteins; MALDI; electrospray.

1 Introduction

A wide variety of post-translational modifications *(1)* of expressed proteins are known to occur in living organisms. Although their presence in an organism cannot be predicted from the genome, these modifications can play critical roles in protein structure and function. The identification of post-translational modifications can be critical in understanding the functions of proteins involved in important biological pathways and mass spectrometry offers a fast, accurate method for observing them. This chapter describes the procedure for analyzing ribosomal proteins of *Escherichia coli* by matrix-assisted laser desorption/ionization time-of-flight (MALDI-TOF) and *Caulobacter crescentus* ribosomal proteins by electrospray quadrupole time-of-flight (ESI-QTOF) mass spectrometry.

Because of their role as the center of protein synthesis in cells, ribosomes are a popular target of research. Many of the steps involved in the translation of messenger RNA into proteins have been elucidated. Nevertheless the ribosomes of prokaryotes,

From: *Post-translational Modifications of Proteins.*
Methods in Molecular Biology, Vol. 446.
Edited by: C. Kannicht © Humana Press, Totowa, NJ

which are somewhat simpler than those of eukaryotes, contain 3 ribonucleic acids and over 50 proteins intricately bound together into large and small subunits and much about their structure remains unknown. A molecular-level understanding of the role that ribosomal proteins play in the translation process remains to be attained *(2,3)*. Because they are present in all organisms, the RNA and protein components of ribosomes provide the material for fingerprinting and evolutionary studies *(4–7)*. The alteration or absence of ribosomal proteins conveys structural, genetic, and functional information *(7)*. Nine *E. coli* ribosomal proteins are reported to be post-translationally modified by acetylation or methylation *(7)*; others are believed to be carboxylated *(8)* and 1 thio-methylated *(9)*. *Caulobacter crescentus* appears to incorporate a comparable array of modifications. In general the roles of these modifications are not understood *(10)*. Despite its limited speed, 2-dimensional gel electrophoresis has been the preferred method for resolving ribosomal proteins derived from a single source *(2,11,12)*. Unfortunately, the resolution of this technique is not sufficient to distinguish most post-translational modifications.

Mass spectrometry is an optimal method for detecting mutations and modifications in biological macromolecules. Two methods developed in the 1980s, electrospray ionization *(13)* and matrix-assisted laser desorption/ionization (MALDI) *(14)* have made it possible to record mass spectra of large nucleic acids, proteins, and carbohydrates. Recent progress in time-of-flight instrumentation *(15–17)* has improved the quality of MALDI mass spectra to the point where the masses of molecules smaller than about 30,000 Da can typically be measured with accuracy on the order of 1 Da or better. Because up to 68% of the dry mass of rapidly growing *E. coli* cells corresponds to proteins, and up to 21% of the cellular protein content is ribosomal *(7,18)* these cells make an ideal system for studying ribosomal proteins. The methods presented in this chapter describe the analysis of *E. coli (19)* and *C. crescentus* ribosomes. However, they can be extended to ribosomes from other organisms *(20)* and to other well-characterized biological systems. Ribosomes were extracted from bacteria and their mass spectra were recorded, allowing for the detection and identification of 55 of the 56 *E. coli* ribosomal proteins and 53 of the 54 *C. crescentus* ribosomal proteins. Previously assigned post-translational modifications were easily observed by comparing measured masses to masses calculated from the amino acid sequences of known ribosomal proteins.

2   Materials

2.1   *Ribosome Extraction*

1. Luria broth (LB) growth medium:

   a. For growth of *E. coli* cells: Luria Broth (LB) growth medium containing 1% (w/v) tryptone, 0.5% (w/v) yeast extract, and 1% (w/v) sodium chloride in aqueous solution.

b. For growth of *Caulobacter crescentus* cells: Peptone Yeast Extract (PYE) medium containing 0.2% (w/v) peptone, 0.1% (w/v) yeast extract, 0.6 mM magnesium sulfate, 0.5 mM calcium chloride.

2. Refrigerated centrifuge capable of 100,000g centrifugal force.
3. Mortar, pestle, alumina or French press and pressure cell for cell lysis.
4. Buffer A: 20 mM Tris-HCl, 10.5 mM magnesium acetate, 100 mM ammonium chloride, 0.5 mM EDTA, 3 mM β-mercaptoethanol, pH 7.5.
5. Buffer B: same as buffer A, except 0.5M ammonium chloride instead of 100 mM.
6. Buffer E: 10 mM Tris-HCl, 5.25 mM magnesium acetate, 60 mM ammonium chloride, 0.25 mM EDTA, 3 mM β-mercaptoethanol.
7. DNase.
8. Sucrose.
9. 6–8 kDa MWCO dialysis tubing.

2.2   MALDI Sample Preparation

1. 0.1% aqueous trifluoroacetic acid (TFA).
2. 10% aqueous TFA.
3. Acetonitrile.
4. MALDI matrix: sinapinic acid (SA, 3,5-dimethoxy-4-hydroxycinnamic acid).
5. Calibration proteins: bovine ubiquitin, horse heart cytochrome c, bovine carbonic anhydrase.
6. 10- and 30-kDa MWCO centrifugal filters.
7. Time-of-flight mass spectrometer equipped with matrix-assisted laser desorption/ionization (MALDI) source.

2.3   LC-MS Sample Preparation

1. Glacial acetic acid.
2. 1M magnesium chloride.
3. 50 µg/mL Horse heart myoglobin in 50% aqueous acetonitrile with 0.1% formic acid (v/v) for tuning and calibration.
4. High performance liquid chromatograph capable of producing a 50 µL/min to 200 µL/min flow rate, and a flow splitter to decrease the flow rate into the mass spectrometer to 5–10 µL/min.
5. Chromatography media and mobile phases suitable for the separation of whole proteins with molecular weights between 4,000 and 25,000 Da.
6. Mass spectrometer with an electrospray ionization source capable of being coupled directly to the chromatograph.

3 Methods

3.1 Ribosome Isolation (see Note 1)

1. Bacteria growth

    a. Grow E. coli cells in ~1.8 L of LB growth medium to mid-log phase (~12.5 h, monitor by OD at 600 nm).
    b. Grow C. crescentus cells in ~1.5 L of PYE growth medium to mid-log phase (~16 h, monitor by OD at 600 nm).

2. Cell lysis

    For cell lysis by grinding:

    (a) Harvest cells by centrifugation at 10,000g and 4°C for 10 min (should yield about 5 g of wet cells).
    (b) Transfer cells to mortar and freeze cells at −20°C.
    (c) Break up cells by adding 12 g of alumina and grinding with pestle until a sticky paste is obtained.
    (d) Add 15 mL of buffer A and a few crystals of DNase.

    For cell lysis with a French Press:

    (a) Harvest cells by centrifugation at 6,000g and 4°C for 20 min.
    (b) Wash cells by resuspending pellets in Buffer A and centrifuging at 6,000g and 4°C for 20 min.
    (c) Resuspend cell pellets in a minimum volume (~30 mL) of Buffer A plus protease inhibitors.
    (d) Lyse cells using 5 passages through a French Press at 16,000 psi. After one passage, add a few crystals of DNase.

3. Transfer mixture to centrifuge tube and centrifuge at 10,000g and 4°C for 15 min to remove alumina and membranes. Retain the supernatant.
4. Centrifuge supernatant at 30,000g and 4°C for 45 min to remove more cell debris and alumina. Retain the supernatant.
5. Prepare 12 mL of 1.1M sucrose in buffer B. Add 3 mL to each of 4 clean centrifuge tubes. To each tube, layer ~4 mL of sample supernatant over sucrose cushion. Centrifuge at 100,000g and 4°C for 15 h. Remove and discard supernatant. Remove and discard the brown flocculent material (top portion of pellet), leaving clear, colorless, gel-like pellet.
6. Suspend each pellet in 1 mL of buffer A. Repeat step 5, this time layering resuspended pellet over sucrose cushion.
7. Suspend each pellet in 1 mL of buffer E. Dialyze (3×) entire sample (4 mL total) against 1-L volumes of buffer E using 6–8 kDa MWCO membrane tubing to remove excess sucrose.
8. Store samples frozen at −80°C (see Note 2) in buffer E prior to mass spectrometric analysis.

3.2 MALDI TOF Mass Spectrometry

1. Prepare 10 mg/mL sinapinic acid matrix solution by dissolving 3.0 mg of the solid in 200 μL of 0.1% TFA plus 100 μL of acetonitrile (*see* **Note 3**).
2. Acidify ribosomes by adding 1 μL of 10% aqueous TFA to 9 μL of ribosome solution (*see* **Note 4**).
3. Prepare calibration protein solutions by dissolving separately 1 mg of each protein in 1 mL of distilled, deionized water.
4. Mix 1 μL of TFA-treated ribosome solution with 9 μL of the matrix solution (*see* **Note 5**).
5. Prepare the calibration protein / matrix mixture by adding 1 μL of each calibration protein solution to 9 μL of matrix solution.
6. Apply 1 μL of the ribosome/matrix mixture to the sample probe and allow to air dry.
7. Apply 1 μL of the calibration protein / matrix mixture to the sample probe near the ribosome samples and allow to air dry.
8. Acquire MALDI mass spectra, externally calibrate spectra (*see* **Note 6**), and measure masses of observed peaks.
9. (Optional) Improved mass calibration may be achieved by using several observed peaks and their known sequence masses to internally calibrate the mass spectrum.
10. Compare masses measured by mass spectrometry to the masses calculated from the amino acid sequences of known proteins. Table 11.1 shows such a comparison for *E. coli* ribosomal proteins *(19)* (*see* **Note 7**).

3.3 LC ESI-TOF Mass Spectrometry

1. Mix the dialyzed ribosome suspension in Buffer E with 0.1 volume of 1*M* MgCl$_2$ and 2 volumes of glacial acetic acid to remove ribosomal RNA (*see* **Note 4**).
2. Vortex the mixture for ~30 s and allow to stand at room temperature for 15 min.

Table 11.1 Post-Translational Modifications Observed in *E. coli* Ribosomal Proteins by MALDI Mass Spectrometry

Subunit	Sequence mass (Da)	Measured mass (Da)	Difference (Da)	Modification
S5	17,472.3	17,514.8	+42.5	Acetylation
S11	13,713.8	13,727.7	+13.9	Methylation
S12	13,605.9	13,651.3	+45.4	β-methylthiolation
S18	8,855.3	8,897.0	+41.7	Acetylation
L3	22,243.6	22,257.2	+13.6	Methylation
L7	12,164.1	12,206.7	+42.6	Acetylation
L11	14,744.3	14,870.2	+125.9	9 methylations
L12	12,164.1	12,174.4	+10.3	Methylation
L31	7,871.1	6,971.1	−900.0	Cleavage of –RFNIPGSK from C-terminus
L33	6,240.4	6,254.1	+13.7	methylation

3. Pellet ribosomal RNA by centrifugation at 14,100*g* for 10 min at room temperature. Remove the protein-containing supernatant by aspiration.

4. Separate proteins using any of a number of LC modes (*see* **Note 8**). Reversed phase LC on C4 chromatography media using aqueous acetonitrile mobile phases containing 0.1% formic acid as an ion pairing reagent are most compatible with ESI MS.

5. Split the column effluent from the 50 μL/min flow rate used for separation to ~5 μL/min to enhance ionization efficiency.

6. Externally tune and calibrate the mass spectrometer using the +9 to +20 charge states of horse heart myoglobin directly infused at a flow rate of 5 μL/min in a solution of 50% aqueous acetonitrile with 0.1% (v/v) formic acid.

7. Generate a total ion chromatogram (TIC) by summing spectra for every 1 s of data collection and display total signal intensity as a function of elution time.

8. Calculate whole protein masses by summing blocks of spectra extracted from the TIC and deconvoluting the charge state distribution to derive a whole protein mass. Figure 11.1 shows a schematic of this process (*21*).

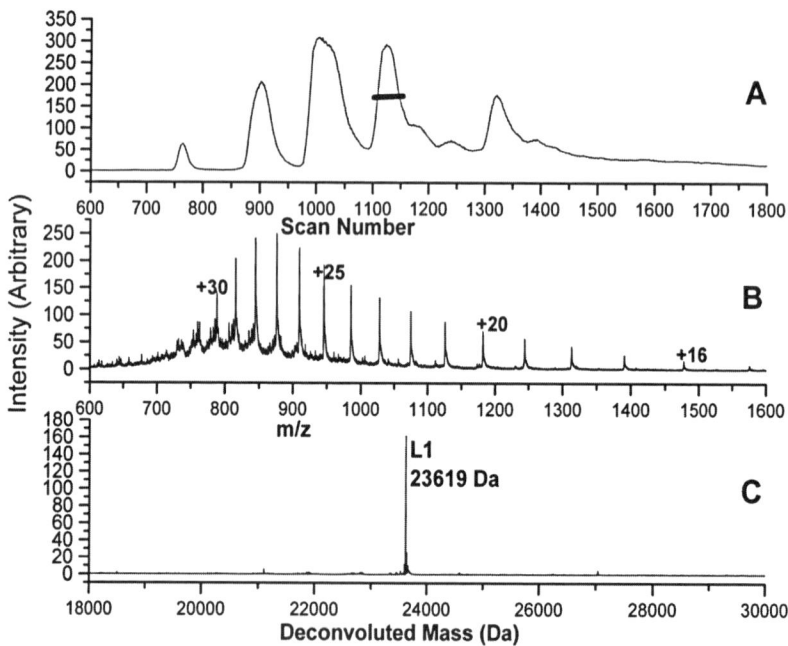

Fig. 11.1 Extraction of spectra from a TIC and deconvolution of a whole protein mass. **A** The total ion chromatogram of a fraction containing *Caulobacter* ribosomal proteins L1, L7/L12, L10, L11, L25, L29, L30, L31, and S6. The horizontal black bar indicates the spectra summed together to generate B. **B** The undeconvoluted spectrum of ribosomal protein L1, showing charge states from +16 to +30. **C** The deconvoluted spectrum calculated from B, showing ribosomal protein L1, with a mass that corresponds to removal of an N-terminal methionine and methylation at an amino group

Table 11.2 Post-Translational Modifications Observed in *Caulobacter crescentus* Ribosomal Proteins by LC/ESI-MS

Subunit	Measured mass (Da)	Sequence mass (Da)	Difference (Da)	Modifications
L1	23,618.8	23,735.3	−116.5	Methionine removal and methylation[a,b]
L3	26,666.7	28,079.9	−1,413.2	Removal of residues 1–13[d]
L7/L12	13,172.9	13,289.2	−116.3	Methionine removal and methylation[a,b]
L11	15,509.5	15,387.0	+122.5	Methionine removal and 18 +14 Da modifications
L16	15,883.8	15,851.6	+32.2	Methylation and a second +16 addition[b,c]
L18	12,327.4	12,443.2	−116.2	Methionine removal and methylation[a,b]
L27	9,111.2	9,370.7	−259.5	Methionine and C-terminal glutamate removal[a,f]
L32	6,794.7	6,910.9	−116.2	Methionine removal and methylation[a,b]
L33	6,297.5	6,413.6	−116.1	Methionine removal and methylation[a,b]
S5	21,496.6	21,611.7	−115.1	Methionine removal and methylation[a,b]
S9	17,099.6	17,186.7	−87.1	Methionine removal and acetylation[a,c]
S11	13,660.7	13,774.7	−114.0	Methionine removal and methylation[a,b]
S18	10,025.8	10,113.9	−88.1	Methionine removal and acetylation[a,c]
S21	9,062.4	10,174.0	−1,111.6	Removal of residues 1–10[d]

Notes:

[a] Methionine removal by the N-terminal methionine aminopeptidase reduces the average mass of the protein by 131.2 Da relative to the mass predicted from the sequence in the translated proteome.

[b] Addition of a methyl group, typically to an amino nitrogen, increases the average mass of the protein by 14.0 Da.

[c] Addition of an acetyl group, typically to the N-terminal amino group, increases the average mass of the protein by 42.0 Da.

[d] These extensive modifications have been confirmed by tryptic peptide mass mapping and C-terminal sequence determination using partial digestion with carboxypeptidases Y and P. The cause may be specific post-translational proteolysis or genome sequence errors.

[e] The site and identity of the second modification have not been confirmed.

[f] The C-terminal truncation has been confirmed using partial digestion with carboxypeptidases Y and P.

9. Deduce protein identities and proposed post-translational modifications by comparison of the deconvoluted masses and protein masses calculated from the proteome sequences. Table 11.2 shows such a comparison for *C. crescentus* proteins.

10. Confirm protein identities and proposed modifications by enzymatic analysis.

3.4   Recent Advances in Top-Down Proteomics

In addition to the methods described here for mass analysis of intact proteins, recent advances in instrumentation and protein fragmentation methods have enabled the development of "top-down" proteomics. Top-down proteomics involves the fragmentation of intact protein ions such that more detailed structural characterization of a protein is obtained while still providing full sequence coverage. With the additional information provided by the protein fragments, post-translational events including residue modifications, sequence truncation, and amino acid substitution can be located in the protein sequence *(22)*. The coupling of mass spectrometers such as linear ion traps (LIT) and quadrupoles (Q) with high-resolution instruments such as Fourier-transform ion cyclotron resonance mass spectrometers (FTICR or FTMS) and orbitraps (OT) have allowed large protein or protein fragment ions to be analyzed with high resolution and high mass accuracy *(23,24)*. At the same time, innovative ion fragmentation methods such as electron capture dissociation (ECD) *(25,26)*, electron transfer dissociation (ETD), infrared multiphoton dissociation (IRMPD), and sustained off-resonance collision-induced dissociation (SORI-CID) have been developed *(27)* and provide alternatives to the more conventional collision-induced dissociation (CID). This technology has been applied to a variety of biological systems, including yeast *(28)*, *Methanosarcina acetivorans (29)*, and *Arabidopsis* membrane proteins *(30)*. In addition, the top-down approach has been used in combination with bottom-up (protease digested) proteomics *(31,32)*, for measuring expression ratios *(33)*, or modified slightly to perform extended range proteomic analysis (ERPA) for increased sequence coverage *(34)*.

4   Notes

1. This method is optimized for the extraction of intact ribosomes from *E. coli* cells, although it has performed adequately for the isolation of intact ribosomes from *C. crescentus* and *Deinococcus radiodurans* with little modification. An extensive compilation of ribosome extraction procedures for other organisms can be found in *(2)*. Alternatively, other biological systems should also be amenable to MALDI mass spectral analysis, provided their protein amino acid sequences are well characterized. In all cases, special care should be taken to avoid adding chemicals to biological samples that either suppress ionization or create adducts

in the MALDI process *(35)*. The authors recommend avoiding large amounts of urea (>3M) and sodium dodecyl sulfate (>1%) and even small amounts of sodium, potassium, sulfate, and phosphate ions. Substituting ammonium, citrate, trifluoroacetate, and acetate ions (when appropriate) should lead to improved MALDI ion yield.

2. Samples may also be refrigerated at 4°C for short-term storage (hours to days) or frozen at −20°C for medium-term storage (days to weeks). Extended storage (months) should be done at −80°C. Best results are normally obtained by preparing MALDI samples immediately, prior to any intermediate storage.

3. For *E. coli* ribosomes, the most useful spectra have been obtained using sinapinic acid as the matrix. Nevertheless, other MALDI matrices may be used for analysis. Alternative matrices include ferulic acid (FA, 4-hydroxy-3-methoxycinnamic acid) and α-cyano-4-hydroxycinnamic acid (CHC). Varying the matrix can change the relative peak heights for a given sample. Thus the intensities of individual peaks may be enhanced (or diminished) simply by using a different matrix, without otherwise changing the sample.

4. Both of the acidification procedures described in this chapter disrupt the quaternary structure of the ribosome, denature the ribosomal proteins and remove ribosomal RNA by precipitation. Using aqueous TFA improves MALDI signal intensity for the intact proteins. Using 67% glacial acetic acid better solubilizes the intact denatured proteins for subsequent chromatography.

5. Since MALDI analysis of mixtures is commonly known to discriminate against the ionization of larger molecules in favor of smaller components, one may wish to enhance high-mass ion signal by removing low-mass components with either a 10- or 30- kDa MWCO centrifugal filter prior to mixing the TFA-treated ribosome solution with matrix solution.

6. Single- and double-charged ions of calibration proteins such as bovine ubiquitin, horse heart cytochrome c, and bovine carbonic anhydrase are normally observed. Masses of their singly-charged (M+H$^+$) ions are 8,565.88 Da, 12,361.15 Da, and 29,023.2 Da, respectively. Other well-characterized proteins may also be used for calibration.

7. In *E. coli*, *C. crescentus*, and most other organisms, methionine aminopeptidase *(36)* is known to cleave the N-terminal methionine residue when amino acids with small side chains (alanine, cysteine, glycine, proline, serine, threonine, and valine) are in position 2, next to methionine. This loss occurs in 34 of 56 observed *E. coli* and 38 of 54 *C. crescentus* ribosomal proteins. Mass shifts due solely to the loss of N-terminal methionine are not included in Tables 11.1 and 11.2.

8. The protein-containing supernatant from the acetic acid precipitation of ribosomal RNA has been fractionated both directly by chromatography on 4.6 × 300 mm C4 reversed phase columns and using strong cation exchange chromatography coupled directly to a second dimension of C4 reversed phase chromatography. Chromatography conditions can be found in *(37,38)*.

Acknowledgement This work has been supported by the National Science Foundation grant CHE0518234 and the Indiania Metacyte Initiative.

References

1. Krishna, R. and Wold, F. (1993) Post-translational modification of proteins in *Advances in Enzymology and Related Areas of Molecular Biology* (Meister A., ed.) pp. 265–296. John Wiley & Sons, New York.
2. Spedding, G. (1990) Isolation of ribosomes from prokaryotes, eukaryotes, and organelles in *Ribosomes and Protein Synthesis, A Practical Approach* (Rickwood, D and Hames, B.D. eds.) pp. 4–7, Oxford University Press, New York.
3. Ramakrishnan, V. and White, S. (1998) Ribosomal protein structures: insights into the architecture, machinery and evolution of the ribosome. *Trends in Biochem. Sci.* 23, 208–212.
4. Podzorski, R. P. and Persing, D. H. (1995) Molecular detection and identification of microorganisms in *Manual of Clinical Microbiology* (Murray R. D. et al., eds.) pp. 130–157, ASM Press, Washington, DC.
5. Woese, C. R. (1987) Bacterial evolution. *Microbiol. Rev.* **51**, 221–271.
6. Noller, H. F. and Nomura, M. (1996) Ribosomes in *Escherichia coli and Salmonella* (Neidhardt, F. D., ed.) Vol. 1, pp. 167–182, ASM Press, Washington, DC.
7. Wittmann, H. G. (1982) Components of bacterial ribosomes. *Ann. Rev. Biochem.* **51**, 155–183.
8. Van Buskirk, J. and Krisch, W. (1978) γ-carboxyglutamic acid in eukaryotic and prokaryotic ribosomes. *Biochem. Biophys. Res. Comm.* **82**, 1329–1331.
9. Kowalak, J. and Walsh, K. (1996) β-Methylthio-aspartic acid: identification of a novel post-translational modification in ribosomal protein S12 from *Escherichia coli. Protein Sci.* **5**, 1625–1632.
10. Neidhardt, F. C., Ingraham, J. L. and Schaechter, M. (1990) *Physiology of the Bacterial Cell*, Sinauer Associates, MA.
11. Geyl, D., Böck, A., and Isono, K. (1981) An improved method for two-dimensional gel-electrophoresis: analysis of mutationally altered ribosomal proteins of *Escherichia coli. Mol. Gen. Genet.* **181**, 309.
12. Datta, D. B., Changchien, L., Nierras, C. R., Strycharz, W. A. and Craven, G. R. (1988) Identification of *Escherichia coli* ribosomal proteins by an alternative two-dimensional electrophoresis system. *Anal. Biochem.* **173**, 241–245.
13. Fenn, J. B., Mann, M., C. K. Meng, Wong, S. K. and Whitehouse, C. M. (1989) Electrospray ionization for mass spectrometry of large molecules. *Science* **246**, 64–71.
14. Karas, M., Bachmann, D., Bahr U., and Hillenkamp, F. (1987) Matrix-assisted ultraviolet laser desorption of non-volatile compounds. *Int. J. Mass Spectrom. Ion. Proc.* **78**, 53–68.
15. Colby, S., King, T., and Reilly, J. (1994) Improving the resolution of matrix-assisted laser desorption/ionization time-of-flight mass spectrometry by exploiting the correlation between ion position and velocity. *Rapid Comm. Mass Spectrom.* **8**, 865–868.
16. Whittal, R. M. and Li, L. (1995) High-resolution matrix-assisted laser desorption/ionization in a linear time-of-flight mass spectrometer. *Anal. Chem.* **67**, 1950–1954.
17. Brown, R. S. and Lennon, J. J. (1995) Mass resolution improvement by incorporation of pulsed ion extraction in a matrix-assisted laser desorption/ionization linear time-of-flight mass spectrometer. *Anal. Chem.* **67**, 1998–2003.
18. Bremer, H. and Dennis, P. P. (**1996**) Modulation of chemical composition and other parameters of the cell by growth rate in *Escherichia coli and Salmonella* Vol. 2, (Neidhardt, F. D., ed.), pp. 1553–1569, ASM Press, Washington, DC.
19. Arnold, R. and Reilly, J. (1999) Observation of *Escherichia coli* ribosomal proteins and their post-translational modifications by mass spectrometry. *Anal. Biochem.* **269**, 105–112.
20. Arnold, R., Polevoda, B., Reilly, J., and Sherman, F. (1999) The action of *N*-terminal acetyl-transferases on yeast ribosomal proteins. *J. Biol. Chem.* **274**, 37035–37040.
21. McEwen, C. N. and Larsen, B. S. (1997) Electrospray ionization on quadrupole and magnetic-sector mass spectrometers in electrospray ionization mass spectrometry, (Cole R.B. ed.), Ch. 5, 171–202, John Wiley & Sons, Inc.

22. Roth, M. J., Forbes, A. J., Boyne II, M. T., Kim, Y. B., Robinson, D. E., and Kelleher, N. L. (2005) Precise and parallel characterization of coding polymorphisms, alternative splicing, and modifications in human proteins by mass spectrometry. *Mol. Cell. Proteomics* **4**, 1002–1008.

23. Jebanathirajah, J. A., Pittman, J. L., Thomson, B. A., Budnik, B. A., Kaur, P., Rape, M., Kirshner, M., Costello, C. E., and O'Connor, P. B. (2005) Characterization of a new qQq-FTICR mass spectrometer for post-translational modification analysis and top-down tandem mass spectrometry of whole proteins. *J. Am. Soc. Mass Spectrom.* **16**, 1985–1999.

24. Macek, B., Waanders, L. F., Olsen, J. V., and Mann, M. (2006) Top-down protein sequencing and MS3 on a hybrid linear quadrupole ion trap-orbitrap mass spectrometer. *Mol. Cell. Proteomics* **5**, 949–958.

25. Ge, Y., Lawhorn, B. G., ElNaggar, M., Strauss, E., Park, J. H., Begley, T. P., and McLafferty, F. W. (2002) Top down characterization of larger proteins (45 kDa) by electron capture dissociation mass spectrometry. *J. Am. Chem. Soc.* **124**, 672–678.

26. Zubarev, R. A. (2004) Electron-capture dissociation tandem mass spectrometry. *Curr. Opin. Biotechnol.* **15**, 12–16.

27. Cooper, H. J., Hakansson, K., and Marshall, A. G. (2005) The role of electron capture dissociation in biomolecular analysis. *Mass Spectrom. Rev.* **24**, 201–222.

28. Meng, F., Du, Y., Miller, L. M., Patrie, S. M., Robinson, D. E., and Kelleher, N. L. (2004) Molecular-level description of proteins from *Saccharomyces cerevisiae* using quadrupole FT hybrid mass spectrometry for top down proteomics. *Anal. Chem.* **76**, 2852–2858.

29. Patrie, S. M., Ferguson, J. T., Robinson, D. E., Whipple, D., Rother, M., Metcalf, W. W., and Kelleher, N. L. (2006) Top down mass spectrometry of <60-kDa proteins from *Methanosarcina acetivorans* using quadrupole FTMS with automated octopole collisionally activated dissociation. *Mol. Cell. Proteomics* **5**, 14–25.

30. Whitelegge, J. P., Laganowsky, A., Nishio, J. Souda, P., Zhang, H., and Cramer, W. A. (2006) Sequencing covalent modifications of membrane proteins. *J. Exp. Bot.* **57**, 1515–1522.

31. Strader, M. B., Verberkmoes, N. C., Tabb, D. L., Connelly, H. M., Barton, J. W., Bruce, B. D., Pelletier, D. A., Davison, B. H., Hettich, R. L., Larimer, F. W., and Hurst, G. B. (2004) Characterization of the 70S Ribosome from *Rhodopseudomonas palustris* using an integrated "top-down" and "bottom-up" mass spectrometric approach. *J. Proteome Res.* **3**, 965–978.

32. Millea, K. M., Krull, I. S., Cohen, S. A., Gebler, J. C., and Berger, S. J. (2006) Integration of multidimensional chromatographic protein separations with a combined "top-down" and "bottom-up" proteomic strategy. *J. Proteome Res.* **5**, 135–146.

33. Du, Y., Parks, B. A., Sohn, S., Kwast, K. E., and Kelleher, N. L. (2006) Top-down approaches for measuring expression ratios of intact yeast proteins using fourier transform mass spectrometry. *Anal. Chem.* **78**, 686–694.

34. Wu, S. L., Kim, J., Hancock, W. S., Karger, B. (2005) Extended Range Proteomic Analysis (ERPA): a new and sensitive LC-MS platform for high sequence coverage of complex proteins with extensive post-translational modifications – comprehensive analysis of beta-casein and epidermal growth factor receptor (EGFR). *J. Proteome Res.* **4**, 1155–1170.

35. Warren, M., Brockman, A., and Orlando, R. (1998) On-probe solid-phase extraction / MALDI-MS using ion-pairing interactions for the cleanup of peptides and proteins. *Anal. Chem.* **70**, 3757–3761.

36. Sherman, F., Stewart, J., and Tsunasawa, S. (1985) Methionine or not methionine at the beginning of a protein. *BioEssays*, **3**, 27–31.

37. Diedrich, G., Burkhardt, N., Nierhaus, K. H., (1997) Large-scale isolation of proteins of the large subunit from Escherichia coli ribosomes. *Protein Exp. Purif.* **10**, 42–50.

38. Champney, W. S, (1990) Reversed-phase chromatography of Escherichia coli ribosomal proteins. Correlation of retention time with chain length and hydrophobicity. *J. Chromatogr.* **522**, 163–170.

12
Analysis of S-Acylation of Proteins

Michael Veit, Evgeni Ponimaskin, and Michael F. G. Schmidt

Summary Palmitoylation or S-acylation is the post-translational attachment of fatty acids to cysteine residues and is common among integral and peripheral membrane proteins. Palmitoylated proteins have been found in every eukaryotic cell type examined (yeast, insect, and vertebrate cells), as well as in viruses grown in these cells. The exact functions of protein palmitoylation are not well understood. Intrinsically hydrophilic proteins, especially signaling molecules, are anchored by long-chain fatty acids to the cytoplasmic face of the plasma membrane. Palmitoylation may also promote targeting to membrane subdomains enriched in glycosphingolipids and cholesterol or affect protein–protein interactions.

This chapter describes (1) a standard protocol for metabolic labeling of palmitoylated proteins and also the procedures to prove a covalent and ester-type linkage of the fatty acids, (2) a simple method to analyze the fatty acid content of S-acylated proteins, (3) two methods to analyze dynamic palmitoylation for a given protein and (4) protocolls to study cell-free palmitoylation of proteins.

Key Words Hydrophobic modification; S-acylation; protein-palmitoylation; fatty acids; membrane proteins.

1 Introduction

Palmitoylation or S-acylation is the post-translational attachment of fatty acids to cysteine residues and is common among integral and peripheral membrane proteins. Palmitoylated proteins have been found in every eukaryotic cell type examined (yeast, insect, and vertebrate cells), as well as in viruses grown in these cells. The exact functions of protein palmitoylation are not well understood. Intrinsically hydrophilic proteins, especially signaling molecules, are anchored by long chain fatty acids to the cytoplasmic face of the plasma membrane. Palmitoylation may also promote targeting to membrane subdomains enriched in glycosphingolipids and cholesterol (lipid-rafts) or affect protein–protein interactions (see *1–5* for recent reviews).

From: *Post-translational Modifications of Proteins.*
Methods in Molecular Biology, Vol. 446.
Edited by: C. Kannicht © Humana Press, Totowa, NJ

Section 3.1. describes our standard protocol for metabolic labeling of palmitoylated proteins and also the procedures to prove a covalent and ester-type linkage of the fatty acids. Palmitic acid is a major fatty acid of cellular lipids. Thus, the vast majority (>99.5%) of the radioactivity is incorporated into lipids and only a tiny amount remains for the labeling of proteins. A protein must be fairly abundant in the cell type analyzed to detect palmitoylation. If DNA-clones of the potentially acylated protein are available, expression of the protein in vertebrate cells is often successful to increase its amount. **Section 3.1.** also describes our protocol for transient expression of recombinant proteins with the Vaccinia virus / T7-RNA-polymerase system.

Palmitate is usually found as the predominant fatty acid in S-acylated proteins, but other fatty acid species (myristic, stearic, oleic, and arachidonic acid) are often minor and sometimes even main components. In accordance, many S-acylated proteins can be labeled with more than one fatty acid and the palmitoyl-transferase shows no strict preference for palmitate in vitro *(6,7)*. **Section 3.2.** describes a simple method to analyze the fatty acid content of S-acylated proteins.

Palmitoylation is unique among hydrophobic modifications because the fatty acids may be subject to cycles of de- and reacylation. The turn-over of the fatty acids is often enhanced upon treatment of cells with physiologically active substances. It is supposed that reversible palmitoylation plays a role for the function of these proteins by controlling their membrane-binding and/or their protein-protein inter-actions *(8)*. An acyl-protein-thioesterase (APT), which cleaves fatty acids from a variety of palmitoylated proteins in vitro, has recently been purified and its cDNAs has been cloned *(9,10)*. However, the significance of this enzyme for deplamitoylation of proteins inside cells is not known. Furthermore, turn-over of fatty acids does not occur on every palmitoylated protein. **Section 3.3.** describes 2 methods to analyze dynamic palmitoylation for a given protein.

Integral membrane proteins are palmitoylated at cysteine residues located at the boundary between the transmembrane segment and the cytoplasmic tail *(11)*. Peripheral membrane proteins are often acylated at an N-terminal MGCXXS motif, which provides a dual signal for amide-myristoylation as well as S-palmitoylation *(12)*. Palmitoylation of peripheral membrane proteins can also occur at a cluster of cysteine residues located in the middle of the protein *(13)* or at C-terminal cysteines in the neigbourhood of an isoprenylated CAAX-box *(14)*. However, comparison of the amino acids surrounding the palmitoylated cysteine residues reveals no obvious consensus signal for palmitoylation. Thus, palmitoylation of a protein can not be predicted from its amino acid sequence and the palmitoylation site(s) have to be determined experimentally for each protein. This is usually done by site-specific mutagenesis of cysteine residues and subsequent expression of wild-type and mutant proteins in vertebrate cells. The most popular mutagenesis approach is the PCR overlap extension method, which we describe in **Section 3.4.** Once a non-palmitoylated mutant has been created, it can be used for functional studies to determine the role of the fatty acids in the life-cycle of the protein.

The enzymology of protein-palmitoylation is poorly understood. It was demonstrated early that cellular membranes contain an acylating-acitivity of proteinaceous nature operating on viral and cellular membrane proteins *(15)*. The enzymatic activity was extracted from the microsomes and characterized, but its purification to homogeneity failed *(7,16,17)*. Using a genetic approach diverse members of a protein-family containing a conserved Asp-His-His-Cys (DHHC) motif within a cysteine rich domain (CRD) were recently shown to palmitoylate cellular proteins *(18)*. On the other hand, in the test tube some cellular proteins can even be palmitoylated at authentic acylation sites in the absence of any enzyme source when incubated with palmitoyl-CoA *(19–24)*. It is currently not known whether this non-enzymatic or autocatalytic mechanism of palmitoylation occurs inside cells. In **Section 3.5.** we describe protocols to study cell-free palmitoylation of proteins.

2   Materials

2.1   Detection of Palmitoylated Proteins

1. Fetal calf serum (FCS).
2. Tissue culture medium without FCS.
3. Recombinant vaccinia virus vTF7-3 diluted to 200 µL with medium without FCS (*see* **Note 1**).
4. CO2 incubator.
5. Lipofectin solution: 10 µL of Lipofectin (Life Technologies) with 90 µL medium without FCS (*see* **Note 2**).
6. Vector DNA solution: Dilute 1–3 µg of pTM1 vector DNA with medium without FCS to a final volume of 100 µL (*see* **Note 2**).
7. Tritiated fatty acids, [9, 10-^{3}H (N)]-palmitic acid, and [9, 10-^{3}H (N)]-myristic acid, both at a specific activity of 30–60 Ci/mmol, are available from Amersham (Arlington Heights, IL), NEN-DuPont, (Boston, MA) or ARC (St. Louis, MO). [9, 10-^{3}H]-stearic acid, 10–30 Ci/mmol, is delivered by ARC only.
8. Phosphate-buffered saline (PBS): $0.14M$ NaCl, $27\,mM$ KCl, $1.5\,mM$ KH_2PO_4, $8.1\,mM$ Na_2HPO_4, pH 7.2.
9. RIPA-buffer: 0.1% SDS, 1% Triton-X-100, 1% deoxycholate. 0.15M NaCl, $20\,mM$ Tris-HCl, $10\,mM$ EDTA, $10\,mM$ Jodacetamide, $1\,mM$ PMSF, pH 7.2.
10. Protein A sepharose CL-4B is available from Sigma, (St. Louis, MO). Wash the beads 3 times with PBS. The packed beads are then resuspended in an equal volume of PBS and stored at 4°C.
11. SDS-PAGE sample buffer (non-reducing, 4 × concentrated): $0.1M$ Tris-HCl, pH 6.8, 4% SDS, 20% glycerol, 0.005% (w/v) bromophenole blue.
12. Gel-fixing solution: 20% methanol, 10% glacial acetic acid.
13. Scintillators for fluorography are available from Amersham (Amplify) or DuPont (Enlightening, En^{3}Hance). En^{3}Hance is also available as spray for fluorography of thin-layer plates.
14. 1M sodium salicylate, adjusted to pH 7.4.
15. Whatman 3MM filter paper.

16. Kodak X-OMAT AR film (Rochester, NY).
17. Phosphate-buffer (10 mM, pH 7.4; supplemented with 0.1% SDS).
18. Chloroform/methanol (2/1, by vol).
19. SDS-PAGE sample buffer (reducing, 4 × concentrated): 0.1M Tris-HCl, pH 6.8, 4% SDS, 20% glycerol, 0.005% (w/v) bromophenole blue, 10% mercaptoethanol.
20. 1M hydroxylamine, adjusted to pH 7.0 and pH 10, respectively (see **Note 3**).
21. 1 M Tris, adjusted to pH 7.0 and pH 10, respectively.
22. Dimethylsulfoxide (DMSO).
23. Mercaptoethanol.

2.2 Analysis of Protein-Bound Fatty Acids

1. Tritiated fatty acids, [9, 10-3H (N)]-palmitic acid, and [9, 10-3H (N)]-myristic acid, both at a specific activity of 30–60 Ci/mmol, are available from Amersham (Arlington Heights, IL), NEN-DuPont, (Boston, MA) or ARC (St. Louis, MO). [9, 10-3H]-stearic acid, 10–30 Ci/mmol, is delivered by ARC only.
2. Scintillators for fluorography are available from Amersham (Amplify) or DuPont (Enlightening, En3Hance). En3Hance is also available as spray for fluorography of thin-layer plates.
3. 1M sodium salicylate, adjusted to pH 7.4.
4. Dimethylsulfoxide (DMSO).
5. 6N HCl.
6. Glass ampoules for hydrolysis (e.g., Wheaton micro product V Vial, 3.0 mL with solid screw cap) are available from Aldrich (Milwaukee, WI).
7. Hexane.
8. HPTLC RP 18 thin-layer plates and hydroxylamine are available from Merck (Darmstadt, Germany).
9. Reference ³H-fatty acids (³H-myristate, ³H-palmitate and ³H-stearate).
10. TLC-solvent: acetonitrile/glacial acetic acid (1/1, by vol).
11. Kodak X-OMAT AR film (Rochester, NY).

2.3 Determination of a Possible Turnover of the Protein-Bound Fatty Acids

1. Tritiated [9, 10-³H (N)]-palmitic acid at a specific activity of 30–60 Ci/mmol, Amersham (Arlington Heights, IL), or NEN-DuPont, (Boston, MA) or ARC (St. Louis, MO).
2. Tissue culture medium containing 0.1% fatty acid free bovine serum albumin.
3. Palmitic acid stock solution: 100 mM palmitic acid in ethanol.
4. Cycloheximide stock solution: 50 mg/mL in ethanol.

2.4   PCR-Based Mutagenesis of DNA Sequences

1. GeneAmp PCR Reagents Kit (Perkin-Elmer) containing *Taq*-Polymerase, PCR-buffer and dNTPs
2. HPLC-purified synthetic oligonucleotides (primers)
3. JETsorb kit (Genomed, Germany)
4. plasmid containing the gene of interest

2.5   Cell-Free Palmitoylation of Proteins

1. Acyl-CoA Synthetase (from pseudomonas sp, >2 units/mg of protein, Sigma-Aldrich, A3352) stock solution: 10 mg/mL in phosphate-buffer.
2. Coenzyme-A stock solution: 10 mM in aqua distilled
3. ^{3}H-palmitic acid (125 µCi, 50–60 Ci/mmol, dissolved in 25 µL ethanol).
4. ATP stock solution: 100 mM in phosphate-buffer.
5. 100 mM MgSO$_4$ solution.
6. Phosphate-buffer 20 mM (pH 7.4).
7. Stop solution: acetonitrile/1M phosphoric acid (9/1 by vol).
8. Toluene.
9. TNE-buffer: 20 mM Tris-HCl, pH 7.4, 150 mM NaCl, 5 mM EDTA.
10. TLC-plates, Silica 60 (Merck, Darmstadt, Germany).
11. TLC solvent: butanol/acetic acid/aqua distilled (8/3/3, by vol).
12. PalCoA standard.
13. TNE-buffer with 2% Triton-X-100.
14. Hydroxylamine.
15. PD-10 column.
16. TNE-buffer containing 0.1% Triton-X-100.
17. Haemoglobin solution: 1% in TNE-buffer.
18. TNE-buffer with 1mM DTT.
19. Chloroform/methanol, 1:2 by volume.
20. Ethanol.

3   Methods

3.1   Detection of Palmitoylated Proteins

3.1.1   Vaccinia Virus-Based Expression of Foreign Genes for Acylated Proteins Mammalian Cells

The Vaccinia virus/T7-RNA-Polymerase expression system requires cloning of the gene of interest in an expression plasmid, which contains the promoter for the RNA-polymerase from bacteriophage T7 (e.g., pTM1). Mammalian cells do not express T7 RNA-polymerase and therefore they are infected with a recombinant vaccinia virus containing the gene for the polymerase (e.g., vTF7-3). Vaccinia virus has a broad host range and many different cell types can be used for expression. The T7 promoter is very strong causing a high rate of synthesis of the foreign protein. One disadvantage of this system is the severe and rapid cytopathic effect caused by the virus infection, which might prevent long-term functional studies with the expressed protein. Media and material in contact with the virus should be autoclaved or disinfected (e.g., with hypochlorite, 70% ethanol etc.). Gloves should be worn while handling virus and virus infected cells.

The time schedule given below is for monkey kidney cells (CV-1) and it might differ for other cell types. The quantities given in this and **Section 3.1.2.** are for labeling of one subconfluent cell monolayer grown in a plastic-dish with a diameter of 3.5 cm (approximately $1 \times 10^5 - 1 \times 10^6$ cells).

1. Wash the cells twice with medium without fetal calf serum (FCS).
2. Infect the cells at an multiplicity of infection (m.o.i.) of 10 plaque forming units (pfu)/cell with recombinant vaccinia virus vTF7-3 diluted to 200 μL with medium without FCS (*see* **Note 1**).
3. Incubate at 37°C in a CO2 incubator for 2 h.
4. Mix 10 μL of Lipofectin (Life Technologies) with 90 μL medium without FCS (*see* **Note 2**).
5. Dilute 1–3 μg of pTM1 vector DNA with medium without FCS to a final volume of 100 μL (*see* **Note 2**).
6. Carefully mix diluted Lipofectin and vector DNA solutions and incubate at room temperature for at least 15 min to allow DNA/Lipofectin complexes to form.
7. Meanwhile, remove the vaccinia virus inoculum from the dish and carefully wash the monolayer once with medium without FCS.
8. Add 400 μL medium without FCS to the Lipofectin/DNA mixture, swirl and carefully pipette the mixture onto the cells.
9. Place in CO2 incubator at 37°C for 2–4 h.
10. Replace medium containing the Lipofectin/DNA mixture with labeling medium containing ^{3}H-palmitic acid (*see* Section "Metabolic labeling of cells with ^{3}H-palmitate").

3.1.2   Metabolic Labeling of Cells with ^{3}H-Palmitate

1. Transfer 500 μCi 3H-palmitic acid to a polystyrene tube and evaporate the solvent (ethanol or toluene) in a speed vac centrifuge or with a gentle stream of nitrogen (*see* **Note 4**).
2. Redissolve 3H-palmitic acid in 2.5 μL ethanol by vortexing and pipeting the droplet several times along the wall of the tube. Collect the ethanol at the bottom of the tube with a brief spin.
3. Add 3H-palmitic acid to 500 μL tissue culture medium (*see* **Note 5**), vortex and add the labeling medium to the cell monolayer.

4. Label cells for 1–16 h (*see* **Note 6**) at 37°C in an incubator. During longer labeling times a slowly rocking platform is helpful to distribute the medium equally over the cell monolayer.

5. Place dishes on ice, remove labeling medium, wash cell-monolayer once with ice-cold PBS (1 mL) and lyse cells in 800 μL RIPA-buffer for 15 min on ice.

6. Transfer cell lysate to an Eppendorf-tube and pellet insoluble material for 30 min at 14,000 rpm in an Eppendorf-centrifuge to remove insoluble material (*see* **Note 7**).

7. Transfer supernatant to a fresh Eppendorf tube. Add antibody and protein-A-sepharose (30 μL) and rotate overnight at 4°C (*see* **Note 8**).

8. Pellet antigen-antibody-sepharose complex (5,000 rpm, 5 min), remove supernatant, add RIPA-buffer (800 μL) and vortex.

9. Repeat washing-step 8 at least twice.

10. Solubilize antigen–antibody–sepharose complex in 20 μL of nonreducing 1 × SDS-PAGE sample buffer (see **Note 9**). Heat samples 5 min at 95°C. Pellet sepharose beads (5,000 rpm, 5 min).

11. Load the supernatant on a discontinuous polyacrylamide-gel. SDS-PAGE should be stopped before the bromophenol blue has reached the bottom of the gel.

12. Agitate the gel for 30 min in fixing solution. Treat gel with scintillator as described by the manufacturer. All commercially available scintillators and PPO/DMSO are suitable for detection of palmitoylated proteins. We usually use the salicylate method as follows: Agitate the fixed gel for 30 min in distilled water and then for 30 min in 1M sodium salicylate, adjusted to pH

13. Dry the gel on Whatman 3MM filter paper and expose in a tightly fitting casset to X-ray film at −70°C (*see* **Note 10**). Kodak X-OMAT AR film (Rochester, NY) is supposed to be most sensitive.

3.1.3 Chloroform/Methanol Extraction of ³H-Palmitic Acid Labeled Proteins

Denaturing SDS-PAGE is usually sufficient to separate proteins from lipids, which run just below the dye front and appear as a huge spot at the bottom of the fluorogram. Some proteins have a strong affinity for phospholipids or other fatty acid containing lipids. If the binding of only a small amount of lipids were to resist SDS-PAGE, this would simulate palmitoylation. To exclude possible noncovalent lipid binding, immunoprecipitated samples should be extracted with chloroform/methanol before SDS-PAGE and the amount of chloroform/methanol resistant labeling should be compared with a nonextracted control.

1. Label cells with ³H-palmitate and immunoprecipitate protein as described.

2. Solubilize antigen–antibody–sepharose complex in 30 μL phosphate-buffer (10 m*M*, pH 7.4; supplemented with 0.1% SDS), pellet sepharose beads and dispense 2 × 15 μL of the supernatant into 2 eppendorf tubes.

3. Add 300 µL chloroform/methanol (2:1) to one tube. Vortex vigorously and extract lipids for 30 min on ice. The unextracted sample also remains on ice.
4. Pellet precipitated proteins for 30 min at 14,000 rpm in an Eppendorf centrifuge precooled at 4°C. Carefully remove the supernatant, air-dry the (barely visible) pellet and resuspend it in 1 × nonreducing SDS-PAGE sample buffer. Add 5 µL 4 × concentrated sample buffer to the unextracted sample.
5. Proceed with SDS-PAGE and fluorography as described.

If the fatty acids are non-covalently bound, the ³H-palmitic acid labeling of the extracted sample should be drastically reduced compared to the control sample.

3.1.4   Hydroxylamine and Mercaptoethanol Treatment

Two types of fatty acid linkages have been described in acylated proteins, an amide bond in myristoylated proteins and an ester-type linkage in palmitoylated proteins. Whereas amide-linked fatty acids are resistant to treatment with hydroxylamine, the esters are readily cleaved. Treatment with hydroxylamine, adjusted either to neutral or basic pH, can also be used to discriminate between thioesters to cysteine and oxygenesters to serine or threonine. Under alkaline conditions (pH 9–11) hydroxylamine cleaves both thio- and oxygenesters, whereas at neutral pH (ph 6.5–7.5) thioesters are selectively cleaved. A thioester type linkage can be further verified by its susceptibility to reducing agents, especially at high concentrations and temperatures. However, not all thioesters are equally sensitive to reducing agents (*see* **Note 9**).

Hydroxylamine treatment is usually done on gels containing ³H-palmitate labeled samples.

1. Run an SDS-PAGE with four samples of the ³H-palmitate labeled and immuno-precipitated protein. Each sample should be separated by two empty slots from its neighbours.
2. Fix the gel, wash out the fixing solution with destilled water (2 × 30 min).
3. Cut gel into four parts containing one lane each.
4. Treat 2 parts of the gel overnight under gentle agitation with 1M hydroxylamine, adjusted to pH 7.0 and pH 10, respectively (*see* **Note 3**). The remaining 2 gel parts are treated with 1*M* Tris, adjusted to the same pH values.
5. Wash out the salt solutions with distilled H_2O (2 × 30 min). Remove cleaved fatty acids by washing with dimethylsulfoxide (DMSO, 2 × 30 min) and wash out DMSO with distilled H_2O. Proceed with fluorography treatment. Reassemble the gel parts before drying and expose to X-ray film.

Treatment with reducing agents is done before SDS-PAGE.

1. Immunoprecipitate ³H-palmitate labeled protein.
2. Solubilize protein in 100 µL 1 × nonreducing sample buffer (2 min, 95°C).
3. Pellet sepharose beads. Make 5 aliquots of the supernatant, add mercaptoethanol to a final concentration of 5, 10, 15, and 20% (v/v). Mercaptoethanol is omitted from one sample. Dithiothreitol (DTT) can also be used at concentrations of 50, 100, 150, and 200 m*M*.

4. Heat samples for 10 min at 95°C. Centrifuge for 15 min at 14,000 rpm. Some proteins may precipitate after treatment with reducing agents and are pelleted. Exclude by analysis (e.g., by Western-blotting or ^{35}S-methionine labeling) that a decrease in the ^{3}H-palmitate labeling is not caused by its aggregation. Should this be the case, treatment with reducing agents needs to be done at lower temperatures (1 h, 50°C).
5. Proceed with SDS-PAGE and fluorography as described.

3.2   Analysis of Protein-Bound Fatty Acids

During metabolic labeling ^{3}H-palmitic acid is often converted into other ^{3}H-fatty acid species and even into ^{3}H-amino acids before incorporation into proteins. Thus, labeling of a protein with ^{3}H-palmitic acid does not necessarily prove that palmitate is its only or even its major fatty acid constituent and therefore its actual fatty acid content has to be analyzed. Furthermore, identification of the ^{3}H-palmitate derived labeling as a fatty acid is additional proof for its acylation. We describe a simple protocol feasible in laboratories without expensive equipment for lipid analysis. The method uses acid hydrolysis of ^{3}H-palmitate labeled proteins present in gel slices, extraction of the released fatty acids with hexane and thin-layer chromatography (TLC) to separate fatty acid species.

1. Label your protein with ^{3}H-palmitate or another ^{3}H-fatty acid as long as possible to allow its metabolism. Proceed with immunoprecipitation and SDS-PAGE as described.
2. Localize the protein by fluorography and cut out the band. Remove the scintillator by washing with distilled H20 (hydrophilic scintillators, e.g., salicylate, Enlightening) or DMSO (hydrophobic scintillators, e.g., En3Hance, PPO). 2 × 20 min are usually sufficient. Wash out the DMSO with distilled H20.
3. Cut the gel into small pieces, transfer them into glass ampoules and dry them in a desiccator.
4. Add 500 µL HCl (6N) and let the gel swell. The gel pieces should be completely covered with HCl after swelling.
5. Tightly seal glass ampoules and incubate at 110°C for at least 16 h. Polyacrylamide and HCl form a viscous fluid at high temperatures. Add an equal amount of hexane and vortex vigorously. Separate the 2 phases by gentle centrifugation (5 min, 2,000 rpm). Most of the polyacrylamide is sedimented to the bottom of the vessel.
7. Remove the upper organic phase containing the fatty acids with a Pasteur pipet and transfer it to conical glass vessels. Leave behind the traces of polyacrylamide that are present between the 2 phases. Repeat extraction of fatty acids with hexane and combine upper phases.
8. Concentrate pooled organic phases in a stream of nitrogen to a volume of 20 µL.
9. Draw a line with a soft pencil on the concentration zone of the TLC-plate (HPTLC RP 18 from Merck, Darmstadt), approximately 1 cm from the bottom. Apply your sample carefully in a spot as small as possible. Apply reference

³H-fatty acids (³H-myristate, ³H-palmitate, and ³H-stearate) on a parallel spot (*see* **Note 11**).

10. Put TLC-plate in an appropriate glass chamber containing the solvent system (acetonitrile/glacial acetic acid, 1:1). Take care that the samples do not dip into the solvent.
11. Develop chromatogram until the solvent front has reached the top of the plate (approximately 50 min).
12. Air-dry plate under a hood. Measure radioactivity on the plate with a radiochromatogram-scanner. Alternatively, spray plate with En3hance, air-dry completely and expose to X-ray film. Detection of ³H-fatty acids by fluorography of the TLC-plate requires long exposure times and is only feasible with a protein band easily visible in the SDS-gel after 3–5 d of film exposure.

3.3 Determination of a Possible Turn-Over of the Protein Bound Fatty Acids

3.3.1 Pulse-Chase Experiments with ³H-Palmitate

To show deacylation of a protein directly, pulse-chase experiments with ³H-palmitic acid have to be performed. Deacylation is visible as a decrease in the ³H-palmitate labeling with increasing chase time. The half-life of the fatty acid cleavage can also be determined from these experiments. However, ³H-palmitate labeling of protein can not be chased completely. Vast amounts of label are present in cellular lipids, which themselves show fatty acid turn-over, and a substantial fraction also as palmitoyl-coenzyme A, the acyl donor for palmitoylation. The following protocol is designed to minimize these problems.

1. Label several cell monolayers for 1 h with ³H-palmitic acid as described.
2. Remove labeling medium. Wash monolayer twice with 1 mL of medium containing fatty acid free bovine serum albumin (0.1%). Albumin will extract some of the remaining unbound fatty acids.
3. Add 1 mL of medium supplemented with 100 μM unlabeled palmitic acid. Palmitic acid is stored as a 100 mM stock solution in ethanol and is diluted 1:1,000 into the cell culture medium. Unlabeled palmitate will compete with ³H-palmitate for the incorporation into protein.
4. Lyse 1-cell monolayer immediately and chase remaining cells for different periods of time (e.g., 20, 40, 60 min up to 4 h) at 37°C.
5. Wash and lyse cells and proceed with immunoprecipitation as described.

3.3.2 Cycloheximide Treatment

Treatment of cells with appropriate concentrations of cycloheximide prevents protein synthesis immediately and nearly quantitatively, but has no obvious effect on palmitoylation per se. Thus, strong ^{3}H-palmitate labeling of a protein in the absence of ongoing protein synthesis is taken as an indication for reacylation of a previously deacylated protein. However, this issue is more complicated than it seems at first glance. Palmitoylation is a post-translational modification. Therefore, freshly synthesized proteins continue to incorporate ^{3}H-palmitate until all molecules have passed their intracellular site of palmitoylation. This takes approximately 10–20 min for proteins transported at a fast rate along the exocytotic pathway and their ^{3}H-palmitate incorporation decreases during this time. In contrast, ^{3}H-palmitate incorporation into a previously deacylated protein is not dependent on the labeling time after cycloheximide addition. It is therefore advisable to compare the ^{3}H-palmitate labeling of a protein at different time points after blocking protein synthesis.

1. Add 6 μL cycloheximide to 6 mL cell culture medium from a 50 mg/mL stock in ethanol to reach a final concentration of 50 μg/mL. Add 1 mL medium to cell monolayers and incubate at 37°C. One monolayer should not be treated with cycloheximide.
2. Label cells with 3H-palmitic acid for 1 hour, either immediately or 5, 10, 20, 30, and 60 min after cycloheximide addition. The labeling medium should also contain cycloheximide (50 μg/mL).
3. Proceed with immunoprecipitation, SDS-PAGE and fluorography as described.

3.4 PCR-Based Mutagenesis of DNA Sequences

The PCR-based mutagenesis approach is based on the fact that sequences added to the 5′-end of a PCR primer become incorporated into the end of the resulting molecule. By adding the appropriate sequences, a PCR amplified segment can be made to overlap sequences with another segment. In the second PCR this overlap serves as a primer for extension resulting in a recombinant molecule. A schematic overview of the method is shown in Figures 12.1 and 12.2.

1. Mix the following components for two different PCR reactions in separate autoclaved sample tubes to amplify products **A** and **B**.

	PCR Products	
Components	**A**	**B**
Template (*see* **Note 12**)	50 ng to 1 μg of appropriate gene	50 ng to 1 μg of appropriate gene
Primer *a1* (*see* **Note 13**)	0.1 μM to 1 μM	–
Primer *a2*	0.1 μM to 1 μM	–
Primer *b1*	–	0.1 μM to 1 μM
Primer *b2*	–	0.1 μM to 1 μM
10 × PCR buffer (*see* **Note 14**)	10 μL	10 μL
10 × dNTPs (*see* **Note 15**)	10 μL	10 μL
Taq Polymerase	0.5–2.5 Units	0.5–2.5 Units
H$_2$0 double distilled	to 100 μL	to 100 μL

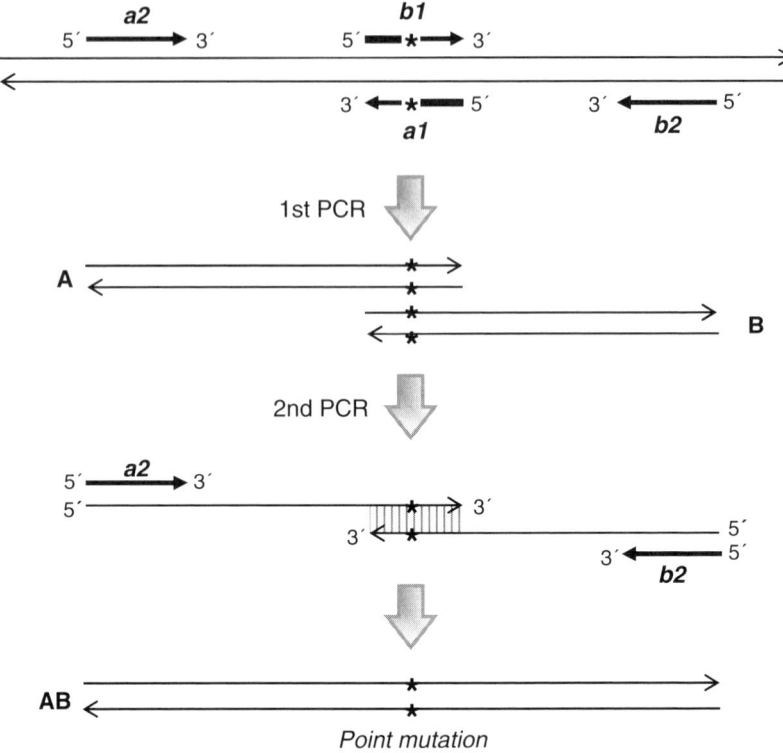

Fig. 12.1 Principle of introducing site-specific mutations into DNA fragment. In this case PCR is used to introduce a mutation near the overlapping ends of the targeted DNA fragments (first PCR). In the second PCR step the mutation is placed inside the resulting DNA molecule distant from either end. The asterisk indicates the mismatch in primer b1 and a1 resulting in a mutation in the amplified DNA-products

2. Vortex the mixtures and briefly spin down in a microcentrifuge.
3. Cycle for 25–35 rounds (initial denaturation at 94°C for 2 min, denaturation at 94°C for 1 min, primer annealing at 55°C for 1 min, primer extension at 72°C for 2 min, final extension at 72°C for 10 min).
4. Run the reaction on an agarose gel (1–1.5%) and cut out the bands that have the appropriate size for **A** and **B**.
5. Purify the DNA fragment from the gel slice. We routinely use a JETsorb kit (Genomed, Germany) for DNA extraction from agarose gel (*see* **Note 16**).
6. Pipet the reactants for the second PCR reaction into new sample tubes and subject to thermal cycling as in step 3 above:

Template **A**	50 ng to 1 μg
Template **B**	50 ng to 1 μg
Primer *a2*	0.1 μ*M* to 1 μ*M*

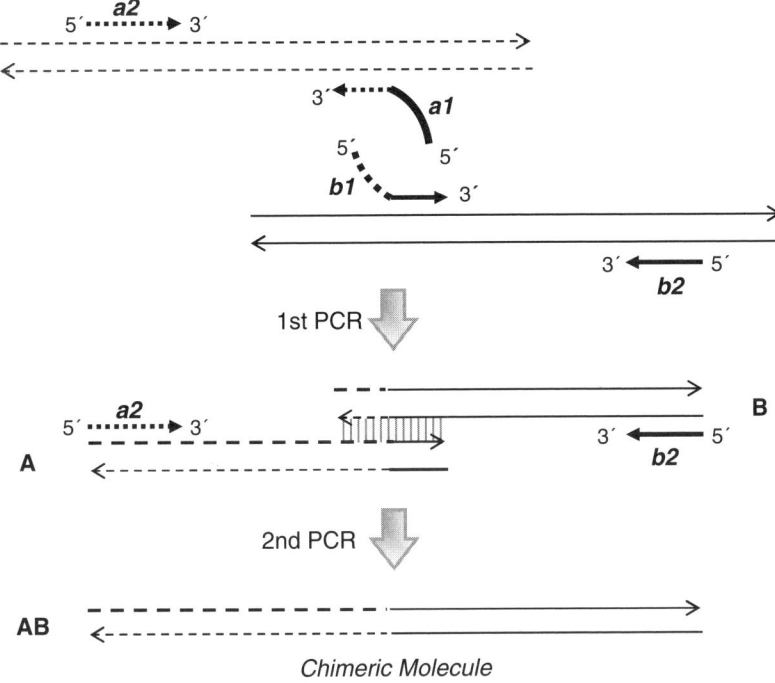

Fig. 12.2 Outline of overlap extension PCR application for the recombination of DNA molecules. Incorporating sequences complementary to the other gene into the 5′-end of a primer makes it possible to generate PCR products which overlap with the other gene segment. These segments can then be recombined during the second PCR step to produce the desired chimeric molecule

Primer *b2*	0.1 μ*M* to 1 μ*M*
10 × PCR buffer	10 μL
10XdNTPs	10 μL
Taq Polymerase	0.5–2.5 Units
H$_2$O bidest	to 100 μL

7. Run the reaction on an agarose gel (1–1.5%) and cut out the bands that have the appropriate size for **AB**. Purify the DNA fragment from the gel slice.

The purified DNA can be used for subcloning into a vector suitable for further applications.

3.5 Cell-free Palmitoylation of Proteins

Cell-free palmitoylation of a protein will give hints if acylation requires an enzyme. An effective palmitoylation assay might also be a starting point to characterize and purify the palmitoylating enzyme. The acceptor protein is incubated with the acyl donor ^{3}H-Pal-CoA in the presence or absence of an enzyme source. The samples are then subjected to SDS-PAGE and fluorography to check for incorporation of ^{3}H-palmitate. The protocols described in **Section 3.1.3.** and 3.1.4. can be used to verify covalent incorporation of the label into the protein. A mutant with the fatty acid binding site deleted or substituted is also as useful as a control to verify that the fatty acids are attached to the same site(s) that are palmitoylated in vivo. In the following section we describe our standard protocols to prepare ^{3}H-Pal-CoA and acceptor proteins. We mention also some considerations which have to be taken into account to set up an initial cell-free palmitoylation experiment.

3.5.1   Preparation of ^{3}H-Pal-CoA

[3]H-Pal-CoA is commercially available, but is expensive and has a low specific activity. We therefore prepare our own ^{3}H-Pal-CoA using acyl-CoA synthetase, Coenzyme A, ATP and ^{3}H-palmitic acid.

1. Mix into glass vessels:

   Acyl-CoA synthetase (10μL from a 10 mg/mL stock in phosphate-buffer). pH 7.0
   Coenzyme-A (15μL from a 10 mM stock in aqua distilled)
   ^{3}H-palmitic acid (125 μCi, 50–60 Ci/mmol, dissolved in 25 μL ethanol),
   ATP (2.5 μL from a 100 mM stock, 1 mM final concentration)
   $MgSO_4$ (5 μL of a 100 mM stock, 5 mM final concentration)
   phosphate-buffer (pH 7.4) to give a final volume of 250 μL.

2. Incubate for 10 min at 28°C.
3. Stop the reaction by adding 1 mL acetonitrile/1M phosphoric acid (9:1). Acyl-CoA synthetase precipitates and is pelleted with a centrifugation at 5,000g for 10 min.
4. Extract the supernatant with 1 mL of toluene. Let the glass vessels stand until the 2 phases are separated.
5. Discard the upper phase containing unreacted ^{3}H-palmitic acid. The toluene-extraction is repeated twice.
6. Remove the lower phase, which contains ^{3}H-Pal-CoA, but also traces of toluene. Wash the glass vessels with 200 μL TNE containing 0.1% Triton-X-100. This wash contains the highest concentration of ^{3}H-Pal-CoA and is used in the palmitoylation-assay.
7. Determine the radioactivity of the ^{3}H-Pal-CoA preparation by liquid-scintilation counting. 1 μL usually contains 50,000–200,000 counts/min.
8. Check the purity of the ^{3}H-Pal-CoA preparation using thin-layer chromatography with Silica 60 plates (Merck, Darmstadt) and butanol / acetic acid / aqua distilled (8:3:3) as solvent system. Run commercial Pal-CoA as a standard in parallel lane.

3.5.2 Preparation of Acceptor-Proteins

Proteins purified from several sources have been used successfully in a cell-free palmitoylation assay (1, 7, 15–23).

3.5.2.1 Purification of Proteins Using Overexpression of Fusion Proteins in *E. coli*

This is the method of choice to purify proteins which are not very abundant in their natural environment. Several expression and purification systems are commercially available (Qiagen, Pharmacia). The gene is expressed as a fusion protein equipped with a tag (e.g., 6 × histidine, glutathion transferase) that can be used in a rapid and simple 1-step affinity-chromatography purification. *E. coli* cells lack the machinery for protein-palmitoylation and therefore most proteins are expressed in a non-palmitoylated version which can be used directly in the palmitoylation assay. An exception is the cellular Bet3-protein which contains long chain fatty acids covalently bound to a cysteine residue even when purified from E. coli *(26)*. One disadvantage is the inability of *E. coli* cells to express transmembrane proteins in a soluble form. Therefore hydrophobic domains, which often contain putative fatty acid linkage sites or modulate protein-acylation, have to be deleted prior to expression. Also expression products from *E. coli* are never glycosylated which could be a critical modification in acylated glycoproteins for stabilizing conformation.

3.5.2.2 Purification of Proteins from Enveloped Viruses

We routinely purify palmitoylated glycoproteins from the membrane of enveloped viruses.

Semliki forest virus (SFV), which contains two palmitoylated glycoproteins (E1 and E2) is grown overnight in Baby hamster kidney (BHK) cells until a severe cytopathic effect becomes apparent. Cell debris are then removed by centrifugation of the culture medium at 3,000*g* for 30 min. Virus particles are pelleted from the supernatant at 30,000*g* for 3 h. The viral glycoprotein (E1 and E2) are then solubilized from the viral membrane and depalmitoylated with hydroxylamine.

1. Resuspend the virus pellet in 250 µL TNE-buffer.
2. Add an equal amount of TNE with 2% Triton-X-100 and incubate for 60 min with agitation at room temperture.
3. Pellet the viral core proteins at 100,000g for 60 min.
4. Add hydroxylamine to the glycoprotein containing supernatant to give a final concentration of 1M and incubate for 60 min with agitation.
4. Equilibrate a PD-10 column with TNE containing 0.1% Triton-X-100. Apply the sample to the column and collect 500 µL fractions.

5. Place 10 μL of the fraction in a microtiter plate well and add 50 μL of a hemo-globin solution. A color changes from red to brown indicates that the fractions contain hydroxylamine. These fractions are discarded because hydroxylamine interferes with the assay.

6. Check 10 μL of each fraction by SDS-PAGE and Coomassie-staining.

7. Pool the protein-containing, hydroxylamine-free fractions determine protein content with an aliquot in a protein assay (Bio Rad-assay) and store aliquots at −80°C.

3.5.3   Cell-free Palmitoylation-Assay

Obviously the conditions vary for each acceptor-protein and enzyme-source under investigation. Therefore only general considerations are mentioned which have to be taken into account for the set up of an initial palmitoylation-experiment.

Acceptor-protein: 1–10 μg of purified and deacylated protein (see above)

Lipid-Donor: 1×10^5–1×10^6 Counts per Minute of ^{3}H-Palmitoyl-Coenzyme A prepared as described in **Section 3.5.1.**

Enzyme source: approximately 10 μg of protein, either microsomes, detergent extracts of microsomes or fractions derived from ion-exchange chromatography.

Detergent: Diverse nondenaturing detergents, for example Triton-X-100 and Cholate are compatible with in-vitro palmitoylation. Octylglucosid is often inhibitory, probably because its carbon chain competes with fatty acid transfer. The concentration of the detergent varies with the experiment. Nonenzymatic palmitoylation is extremely sensitive to detergent and therefore its concentration should be as low as possible (e.g., 0.001%). Our partially purified enzyme preparation works optimal at a Triton-X-100 concentration of 0.025%, probably because it is an integral membrane protein requiring detergent to keep it soluble.

Buffer: We use TNE-buffer, but other buffers also work. No specific requirements for calcium or other ions have been reported. Low concentrations of reducing agents (e.g., 1mM DTT) often enhance palmitate incorporation, probably because they reduce disulfide-bonds thereby making additional cysteine residues available for fatty acid transfer. High concentrations of reducing agents are inhibitory, because they cleave thioester-linked fatty acids.

pH: The pH of the buffer affects the efficiency of the fatty acid transfer. Transfer of palmitate from Pal-CoA to cysteine residues is supposed to occur as a nucleophilic substitution reaction. To act as a good nucleophil the −SH group of the cysteine destined to be acylated must be deprotonated *(27)*. Obviously, a higher percentage of all cysteine residues are deprotonated when the pH of the buffer is basic. However, if palmitoylation in the test tube occurs only at very basic pH-values the reaction is unlikely to occur inside cells.

1. Incubate enzyme source, acceptor protein and 3H-Palmitoyl-Coenzyme A with buffer in a final volume of 50–250 μL for 30–60 min at 28°C or 37°C.

2. Precipitate samples with chloroform /methanol (1:2, 250 μL for a 50 μL reaction) to remove non-covalently bound fatty acids. Pellet precipitated proteins for 20

minutes at 15,000g. Remove the supernatant, wash the pellet once with ice-cold ethanol and centrifuge again. Air-dry the pellet before adding nonreducing SDS-PAGE sample buffer.

3. Subject samples to SDS-PAGE under nonreducing conditions and fluorography. Film-exposure times vary from overnight to several weeks.

4. Use appropriate control reactions, e.g., samples without enzyme source, samples with SDS- or heat-denatured enzyme and samples incubated on ice, to proof the enzymatic character of the reaction.

4 Notes

1. To amplify vaccinia virus, seed a maxi Petri dish (15-cm diameter) with 1×10^7 CV-1 cells in medium supplemented with 5% FCS. Incubate overnight at 37°C and at 5% CO_2. Remove media, wash cells twice with medium without FCS and then infect cells with vTF7-3 vaccinia virus at a low m.o.i., i.e., less than 1 pfu/cell for 1 h at 37°C. We routinely use 0.1 pfu/cell diluted with medium to a final volume of 1 mL. Discard inoculum into desinfectant and replace with 10 mL medium supplemented with 2% FCS. Incubate cells for 48 h at 37°C in a CO_2 incubator. After this incubation, discard the medium and resuspend the cells in 1 mL medium without FCS. Freeze/thaw the cells in liquid nitrogen and a 37°C water bath two times and centrifuge at 14,000 rpm for 1 min in Eppendorf centrifuge to remove cell debris. Store the virus-containing supernatant in small aliquots at −80°C.

2. For diluting lipofectin and for preparing the DNA/Lipofectin mixture use a polystyrene tube, not the polypropylene Eppendorf tube.

3. Hydroxylamine sometimes disintegrates if the pH is adjusted too quickly. Dissolve hydroxylamine in ice-cold distilled H_2O and put the solution in an ice-bath. Add solid NaOH lentils one by one and under permanent stirring until the desired pH is reached. Rapid pH changes and the appearance of a brown colour in the normally colorless hydroxylamine solution are indications for a disintegration of hydroxylamine. In this case the solution has to be discarded.

4. 100 µCi–1 mCi ^{3}H-palmitic acid/mL cell culture medium are usually used for the labeling of acylated proteins. Tritiated fatty acids are supplied as solutions in ethanol or toluene at concentrations too low to add directly to the medium. Because of its cytotoxicity, the final concentration of ethanol in the labeling medium should not exceed 0.5%. Concentration of ^{3}H-palmitic acid can be done in advance and the concentrated stock should be stored at −20°C in tightly sealed polystyrene tubes. Concentration and storage in polypropylene (e.g., Eppendorf tubes) should be avoided because this can result in irreversible loss of much of the label on the tube.

5. Use standard tissue culture medium for the cell line to be labeled. Possible addition of serum to the medium requires some consideration. Serum contains albumin, a fatty acid binding protein, which may delay ^{3}H-palmitate uptake by the cells. Thus, for short labeling periods (up to four hours) we usually use medium without serum. However, serum does not prevent ^{3}H-palmitate labeling

of proteins and can be added, if required. For long labeling periods the presence of serum may even be beneficial, because reversible binding of fatty acids to albumin may help to distribute the ^{3}H-palmitate uptake of the cells more evenly. Serum contains several poorly characterized factors with biological activities and deacylation of particular proteins upon serum treatment of cells has been reported *(3)*. Obviously, in these cases serum addition is detrimental.

6. The necessary time for optimal labeling of palmitoylated proteins is variable and has to be determined empirically for each protein. Proteins with a low but steady rate of synthesis and no turn-over of their fatty acids should be labeled as long as possible, e.g., at least 4 h up to 24 h. The amount of ^{3}H-palmitate incorporation into these proteins increases with time until it reaches saturation. Reversibly palmitoylated proteins show an increase in their labeling intensity in the beginning until a peak is reached. Because of deacylation their ^{3}H-palmitic acid labeling then decreases with time. Proteins expressed from a viral expression vector should be labeled as long as the peak period of their synthesis prevails. This is 1–2 h for the described Vaccinia virus /T7 polymerase system.

7. Sedimentation of insoluble material at higher g-values (e.g., 30 min, 100,000g) sometimes causes a cleaner immunoprecipitation. Make sure that your protein does not precipitate under these conditions. Another possibility to improve the specificity of the immunoprecipitation would be to preincubate the cell lysate with sepharose-beads, but without antibodies (1 h, 4°C). Pellet the beads and transfer the supernatant to a fresh Eppendorf tube and proceed with immunoprecipitation as described.

8. The amount of a monoclonal antibody or antiserum necessary to precipitate a protein quantitatively from a cell lysate depends on its affinity for the antigen and must be determined empirically, but 1–5 μL of an high-affinity antiserum is usually sufficient for the conditions here described. To analyze a complete precipitation of a protein, transfer the first supernatant from step 8 (**Section 3.1.1.**) to a fresh Eppendorf tube. Add antibody and protein-A-sepharose and proceed with immunoprecipitation. Antibodies of particulate subtypes do not bind to protein-A. This has to be considered when monoclonal antibodies are used. However, most of these antibodies bind to protein-G sepharose.

9. The thioester-type linkage of fatty acids to cysteine residues is labile upon treatment with reducing agents. Cleavage of the fatty acids by these compounds is concentration, time and temperature dependent. Therefore, mercaptoethanol and DTT should be omitted from the sample buffer. If the protein requires reducing agents for solubilization, heating should be done as short as possible (95°C, 2 min) or the temperature should be decreased (e.g., 15 min at 50°C). Ester-linked fatty acids are highly susceptible to basic pH values above 12. Under these conditions the fatty acids are cleaved quantitatively and rapidly (< 1 min) even at low temperatures. Therefore, basic pH values should be avoided under all circumstances. In contrast, treatment with acid pH (e.g., pH 1) for time periods up to 1 h is tolerated by the fatty acid bond. Acetic acid containing gel-fixing solutions do not lead to an obvious loss of ^{3}H-palmitate labeling.

10. Tritium derived radioactivity has a very short range. Therefore, dry the gel as thin as possible and make sure that the X-ray film is in close contact with the gel. The times to detect a signal on the film are highly variable. Endogenous cellular proteins with low rates of synthesis require exposure times from several days up to 3 mo. For highly overexpressed proteins with multiple palmitoylation sites, they can be as short as several hours.

11. This TLC-solvent system separates fatty acids according to their hydrophobicity, i.e., the number of carbon atoms and double bonds. Myristic acid (C14) runs faster than palmitic acid (C16), which runs faster than stearic acid (C18). Unsaturated fatty acids are not separated from saturated ones. A fatty acid with one double bond runs to the same position as a saturated fatty acid with two methylene groups less, i. e., oleic acid (18:1) co-migrates with palmitic acid (C16:0). ^{3}H-Arachidonic acid (20:4) is not to be expected as protein-bound fatty acid after labeling with ^{3}H-palmitate, because cells can not metabolize palmitate into arachidonate.

12. The primer/template ration strongly influences the specificity of the PCR and should be optimized empirically. Usually, a very wide range of template concentrations up to about microgram works. However, too much template may lead to an increase in mispriming events.

13. It is the sequence and the concentration of primers that determines the success of the assay. Primers should be between 25 and 45 bases in length. For site directed mutagenesis the mismatched portions should be in the middle of the primer. For creation of chimeric molecules, the priming region (3′-end of primer which act as primer on its template) and the overlap region (5′-end of primer which overlap with the sequence to be joined) of the oligonucleotides should have similar length. The primers should have a G/C content between 40% and 60% but should not have any complementarity between 3′ ends.

14. 10 × PCR buffer: 15mM MgCl2, 500 mM KCl, 100 mM Tris-HCl, pH 8.3, and 0.01% (w/v) gelatin. This buffer is available from "Perkin Elmer" which is one component of GeneAmp PCR Reagents Kit.

15. 10 × dNTPs = 2.5 mM of each dNTP. Unbalanced dNTP mixture will reduce *Taq* fidelity. DNTPs reduce free Mg^{++}, thus interfering with polymerase activity and decrease primer annealing.

16. PCR fragments purified from gel by different DNA extraction kits are often contaminated with silica beads, which strongly inhibit the subsequent PCR reaction. Therefore, it is advisable to purify DNA fragments again by phenol/chloroform extraction and ethanol precipitation.

References

1. Schlesinger, M. J., Veit, M., and Schmidt, M. F. G. (1993) Palmitoylation of cellular and viral proteins, in: *Lipid Modifications of Proteins* (M. J., Schlesinger., ed.), pp. 2–19, CRC Press, Boca Raton.
2. Resh, M. D. (1999) Fatty acylation of proteins: new insights into membrane targeting of myristoylated and palmitoylated proteins. *Biochem. Biophys. Acta* **1451**, 1–16.

3. el-Husseini Ael, D. and Bredt, D. S. (2002) Protein palmitoylation: a regulator of neuronal development and function *Nat. Rev. Neurosci.* **3**, 791–802.

4. Bijlmakers, M. J. and Marsh, M. (2003) The on-off story of protein palmitoylation. *Trends Cell Biol.* **13**, 32–42.

5. Smotrys, J. E. and Linder, M. E. (2004) Palmitoylation of intracellular signaling proteins: regulation and function. *Annu Rev Biochem* **73**, 559–587.

6. Veit, M., Reverey, H., and Schmidt, M. F. (1996) Cytoplasmic tail length influences fatty acid selection for acylation of viral glycoproteins. *Biochem J* **318**, 163–172.

7. Veit, M., Sachs, K., Heckelmann, M., Maretzki, D., Hofmann, K. P., and Schmidt, M. F. (1998) Palmitoylation of rhodopsin with S-protein acyltransferase: enzyme catalyzed reaction versus autocatalytic acylation. *Biochim Biophys Acta* **1394**, 90–98.

8. Qanbar, R. and Bouvier, M. (2003) Role of palmitoylation/depalmitoylation reactions in G-protein-coupled receptor function *Pharmacol Ther* **97**, 1–33.

9. Duncan, J. A. and Gilman, A. G. (1998) A cytoplasmic acyl-protein thioesterase that removes palmitate from G protein alpha subunits and p21 (ras). *J. Biol. Chem.* **273**, 15830–15837.

10. Veit, M. and Schmidt, M. F. (2001) Enzymatic depalmitoylation of viral glycoproteins with acyl-protein thioesterase 1 in vitro. *Virology* **288**, 89–95.

11. Veit, M., Kretzschmar, E., Kuroda, K., Garten, W., Schmidt, M. F., Klenk, H. D., and Rott, R. (1991) Site-specific mutagenesis identifies three cysteine residues in the cytoplasmic tail as acylation sites of influenza virus hemagglutinin. *J Virol* **65**, 2491–2500.

12. Parenti M., Vigano M. A, Newman C. M., Milligan G., and Magee A. I. (1993) A novel N-terminal motif for palmitoylation of G-protein alpha subunits. *Biochem J.* **291**, 349–353.

13. Veit, M., Söllner, T. and Rothman, J. E. (1996) Multiple palmitoylation of Synaptotagmin and the t-SNARE SNAP-25. *FEBS-Letters* **385**, 119–123.

14. Hancock J. F., Magee A. I., Childs J. E., and Marshall C. J. (1989) All ras proteins are polyisoprenylated but only some are palmitoylated. *Cell.* **57**, 1167–1177.

15. Berger, M., and Schmidt, M. F. (1984) Cell-free fatty acid acylation of Semliki Forest viral polypeptides with microsomal membranes from eukaryotic cells. *J Biol Chem* **259**, 7245–7252.

16. Berthiaume, L. and Resh, M. D. (1995) Biochemical characterization of a palmitoyl acyltransferase activity that palmitoylates myristoylated proteins. *J Biol Chem* **270**, 22399–22405.

17. Dunphy, J. T., Greentree, W. K., Manahan, C. L., and Linder, M. E. (1996) G-protein palmitoyltransferase activity is enriched in plasma membranes. *J Biol Chem* **271**, 7154–7159.

18. Linder M. E. and Deschenes R. J. (2004) Model organisms lead the way to protein palmitoyltransferases. *J Cell Sci.* **117**, 117521–11526.

19. Duncan, J. A. and Gilman, A. G. (1996) Autoacylation of G protein alpha subunits. *J Biol Chem* **271**, 23594–23600.

20. Bano M. C., Jackson C. S., and Magee A. I. (1998) Pseudo-enzymatic S-acylation of a myristoylated yes protein tyrosine kinase peptide in vitro may reflect non-enzymatic S-acylation in vivo. *Biochem J.* **330**, 723–731.

21. Veit, M. (2000) Palmitoylation of the 25-kDa synaptosomal protein (SNAP-25) in vitro occurs in the absence of an enzyme, but is stimulated by binding to syntaxin. *Biochem J.* **345**, 45–51.

22. Dietrich, L., Gurezka, R., Veit, M., and Ungermann, C. (2004) The SNARE Ykt6 mediates protein palmitoylation during an early stage of homotypic vacuole fusion *EMBO-Journal* **23**, 45–53.

23. Veit, M. (2004) The human SNARE protein Ykt6 mediates its own palmitoylation at C-terminal cysteine residues. *Biochem J.* **384**, 233–237.

24. Dietrich, L.E. and Ungermann, C. (2004) On the mechanism of protein palmitoylation. *EMBO Rep.* **5**,1053–1057.

25. Qanbar, R. and Bouvier, M. (2004) Determination of protein-bound palmitate turnover rates using a three-compartment model that formally incorporates [3H]palmitate recycling. *Biochemistry.* **43**, 12275–12288.

26. Kim, Y. G., Sohn, E. J., Seo, J., Lee, K. J., Lee, H. S., Hwang, I., Whiteway, M., Sacher, M., and Oh, B. H. (2005) Crystal structure of bet3 reveals a novel mechanism for Golgi localization of tethering factor TRAPP. *Nat Struct Mol Biol.* **12**, 38–45.

27. Bizzozero, O. A., Bixler, H. A., and Pastuszyn, A (2001) Structural determinants influencing the reaction of cysteine-containing peptides with palmitoyl-coenzyme A and other thioesters. *Biochim Biophys Acta.* **1545**, 278–288.

13

Metabolic Labeling and Structural Analysis of Glycosylphosphatidylinositols from Parasitic Protozoa

Nahid Azzouz, Peter Gerold, and Ralph T. Schwarz

Summary Glycosylphosphatidylinositol (GPI) is a complex glycolipid structure that acts as a membrane anchor for many cell-surface proteins of eukaryotes. GPI-anchored proteins are particularly abundant in protozoa and represent the major carbohydrate modification of many cell-surface parasite proteins. A minimal GPI-anchor precursor consists of core glycan (ethanolamine-P-Manα1-2Manα1-6Manα1-4GlcNH$_2$) linked to the 6-position of the D-*myo*-inositol ring of phosphatidylinositol. Although the GPI core glycan is conserved in all organisms, many differences in additional modifications to GPI structures and biosynthetic pathways have been reported. The preassembled GPI-anchor precursor is post-translationally transferred to a variety of membrane proteins in the lumen of the endoplasmic reticulum in a transamidase-like reaction during which a C-terminal GPI attachment signal is released. Increasing evidence show that a significant proportion of the synthesized GPIs are not used for protein anchoring, particularly in protozoa in which a large amount of free GPIs are being displayed at the cell surface. The characteristics of GPI biosynthesis are currently being explored for the development of parasite-specific inhibitors. Especially as this pathway, at least for *Trypanosoma brucei*, has been validated as a drug target.

Keywords Glycosylphosphatidylinositol; GPI; GPIs-labeling; GPI structural elucidation; parasites.

1 Introduction

Glycosylphosphatidylinositols (GPIs) have first been identified as membrane anchors of the variant surface glycoprotein of the parasitic protozoan parasite *Trypanosoma b. brucei (1,2)*. Since then GPIs have been described as protein- and glycoconjugate-membrane anchors and as free GPIs on the surface of a variety of organisms ranging from ancient eukaryotes (e.g., flagellates) to mammalian cells *(3–10)*. The basic GPI structural motif attached to protein compromises ethanolamine-PO$_4$-mannoseα1-2-mannoseα1-6mannoseα1-4glucosamineα1-6*myo*inositol-1-PO$_4$-lipid.

From: *Post-translational Modifications of Proteins.*
Methods in Molecular Biology, Vol. 446.
Edited by: C. Kannicht © Humana Press, Totowa, NJ

GPI-biosynthesis consists of a sequence of single-molecule transfers from activated donors (e.g., nucleotide sugars and dolichol-phosphate-mannose) onto inositol-phosphate containing lipids of various structures (Fig. 13.1) *(11–17)*. A prerequisite for the function of GPIs as membrane anchors of proteins is the presence of an ethanolamine-phosphate linked to the terminal mannose of the highly conserved trimannosyl core-glycan. The preassembled GPI-anchor precursor is post-translationally transferred to a variety of membrane proteins in the lumen of the endoplasmic reticulum in a transamidase-like reaction *(18–19)*. The transamidase complex leads to the replacement of the C-terminal hydrophobic GPI-attachment sequence by the preformed GPI-anchor precursor (Fig. 13.1). Besides their function as membrane anchors of proteins or glycoconjugates, additional functions of GPIs have been described in playing roles as signal for protein sorting in epithelial cells, for signal transduction, for immune responses, and for pathology of infectious

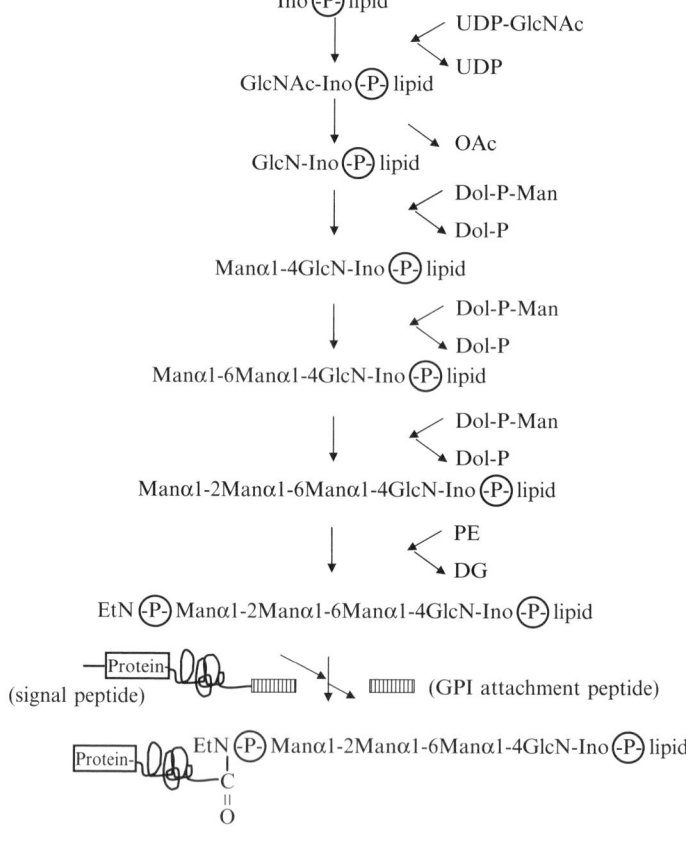

Fig. 13.1 Schematic representation of the general features of GPI-biosynthesis. UDP-GlcNAc, uridine-diphosphate-*N*-acetyl-glucosamine; OAc, O-acetyl group; Dol-P-Man, dolichol-phosphate-mannose; PE, phosphatidylethanolamine; DG, diacylglycerol

diseases *(20–23)*. GPIs of parasitic protozoa have been described as parasite-derived factors affecting host cell signalling and immunity *(24–26)*. GPIs of the human malaria parasite *Plasmodium falciparum* have been described as a novel type of toxin involved in the development of severe malaria pathology *(27–31)*. Although none of the proteins or enzymes involved in GPI-biosynthesis has been purified to homogeneity, many genes involved in different steps of GPI biosynthesis have been cloned from mammalian cells, yeast, and parasitic protozoa *(32–33)*.

In this chapter we will describe protocols to identify and analyze GPI-biosynthesis intermediates and GPI-anchor precursors in the malaria parasite *P. falciparum*, although they are identical for the analysis of GPIs of other eukaryotes apart from cell culture and labeling medium conditions.

The protocols described for identification and characterization depend upon metabolic labeling techniques using radioactive GPI-precursor molecules, organic solvent extraction procedures, and the use of specific chemical and enzymatic treatments (Fig. 13.2). GPIs are identified by their sensitivity towards nitrous acid deamination (HNO$_2$), and specific enzymatic treatments with phosphatidylinositol-specific phospholipase C (PI-PLC) and glycosylphosphatidylinositol-specific phospholipase D (GPI-PLD). Structural characterization can be achieved by analyzing hydrophilic fragments and neutral core-glycans. Hydrophilic fragments

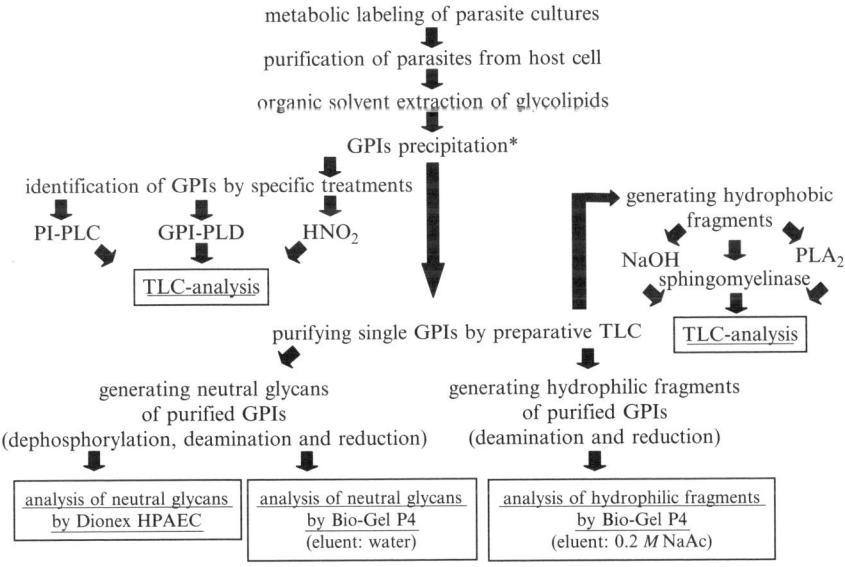

Fig. 13.2 Schematic representation of the protocol to identify and analyse non-protein-bound glycosylphosphatidylinositol. PI-PLC, phosphatidylinositol-specific phospholipase C; GPI-PLD, glycosylphosphatidylinositol-specific phospholipase D, HNO$_2$, nitrous acid deamination and reduction; NaOH, alkaline saponification; PLA$_2$, phospholipase A$_2$. *GPIs precipitation is needed for biological investigations of purified GPIs, but can be omitted if only structural studies are investigated using labeled GPIs

are generated by nitrous acid deamination. This procedure leads to the cleavage of the linkage between the inositol and the nonacetylated glucosamine, present exclusively in GPI structures, converting glucosamine to anhydromannose. Anhydromannose generated at the reducing terminus of the GPI-glycan will be reduced by sodium borohydride and converted to anhydromannitol. This reaction prevents unspecific destruction of the glycan. After purification, the hydrophilic fragments can be analyzed by size-exclusion chromatography (Bio-Gel P4). Neutral core-glycans are prepared by dephosphorylation, deamination, and reduction. The released core-glycans are intensively desalted before the analysis by high-pH anion-exchange chromatography (Dionex) along with an internal standard of β-glucan oligomers. Exoglycosidase treatments are used to confirm the predicted structures of the GPI-glycans. Hydrophobic fragments of GPIs metabolicaly labeled with fatty acids are generated by treatments such as alkaline saponification, phospholipase A_2 (PLA$_2$), and sphingomyelinase treatment and are investigated by TLC-analysis in comparison to standard lipids.

2   Materials

2.1   Metabolic labeling, Extraction and Analysis of P. falciparum Glycolipids

1. Deionized and filtered water.
2. Culture medium: RPMI-1640 medium (GIBCO-BRL) supplemented with gluthation, neomycin, Albumax I, and NaHCO$_3$.
3. Labeling medium: RPMI-1640 medium without glucose (Abimed) supplemented with fructose, gluthation, neomycin, Albumax I, and NaHCO$_3$.
4. Radiolabels were purchased from Amersham, NEN-DuPont and Hartmann Analytic.
5. Save-lock or screw-top Eppendorf tubes.
6. Analytical grade chloroform and methanol mixed in the ratio 1:1 (by vol)
7. Analytical grade chloroform, methanol, and water mixed in the ratio 10:10:3 (by vol) or 4:4:1 (by vol).
8. Butan-1-ol shaken with an equal volume of water in a clean bottle.
9. Access to a liquid scintillation counter.
10. Glass-backed silica gel 60 and silica 60 HPTLC plates (Merck), and a TLC-chamber (Desaga).
11. Access to a TLC-scanner (e.g., Berthold TLC Linear Analyzer) or to BAS-1000 BioImager (Fuji), BioImager plates (Tritum, Fuji Film), and film cassettes.

2.2   Enzymatic Characterization of GPIs

1. *Bacillus cereus* phosphatidylinositol-specific phospholipase C (Sigma). Store enzyme stock at 4°C.
2. Phosphatidylinositol-specific phospholipase C incubation buffer: 0.1% Triton X-100, 50 mM Tris-HCl, pH 7.4.
3. Glycosylphosphatidylinositol-specific phospholipase D incubation buffer: 50 mM Tris-HCl, pH 7.4, 2 mM $CaCl_2$.
4. Access to a Speed-Vac concentrator.

2.3   Chemical Characterization of GPIs

1. 0.2 M Sodium acetate (NaAc), pH 4.0, with (for lipidated glycolipids) or without (for nonlipidated glycolipids) 0.1% sodium dodecyl sulphate (SDS). Made by titrating sodium acetate solution (0.2 M final) to pH 4.0 with glacial acetic acid. Stable at room temperature for several months.
2. 1 M Sodium nitrite ($NaNO_2$). Prepare just before use.
3. Access to a water-bath sonicator.

2.4   Generating of GPIs Hydrophilic Fragments
by Deamination and Reduction

1. 0.8 M Boric acid. Stable at room temperature for several months.
2. 2 M Sodium hydroxide.
3. 1 M Sodium borohydride. $NaBH_4$ is dissolved in 0.1 M NaOH before use.
4. 5% and 50% glacial acetic acid.
5. Toluene: analytical-grade (Merck).
6. Dextran from *Leuconostoc spp.* (Serva).
7. Access to a Bio-Gel P4 system, using 0.2 M ammonium acetate as eluent.
8. 96-Well microtiter plates.
9. H_2SO_4/0.2% orcinol.

2.5   Generating of GPIs Neutral Core-Glycans
by Dephosphorylation, Deamination, and Reduction

1. 48% Aqueous HF (Sigma).
2. Dowex AG50-X12, 200–400 mesh (Bio-Rad) converted to the H⁺ form by washing 5× with 10 vol 1 M HCL and 5–7× with 10 vol water. Store at 4°C in water containing 0.02% azide.
3. Chelex 100 (Na⁺) (Bio Rad). Store at 4°C in water containing 0.02% azide.

4. Dowex AG3-X4, 200–400 mesh (Bio Rad) converted to the OH⁻ form by washing with 5 x with 10 vol 1 *M* NaOH and 5–7 × with 10 vol water. Store at 4°C in water containing 0.02% azide.

5. QAE-Sephadex-A25 (Pharmacia Biotech Inc). Swollen in water and washed with 5 × 10 vol water. Store at 4°C in water containing 0.02% azide.

2.6   Sequencing of Neutral Glycans by Exoglycosidase treatments

1. Jack bean α-mannosidase (JBAM) (Prozyme). Store enzyme stock at −20°C.
2. *Aspergillus saitoi* α-mannosidase (ASAM) (Prozyme). Store enzyme stock at −20°C.

2.7   Bio-Gel P4 Analysis of Neutral Glycans

1. Access to a Bio-Gel P4 system, using water as eluent.

2.8   Dionex HPAEC Analysis of Neutral Glycans

1. Access to a Dionex HPAEC system (Dionex Corp.).
2. 0.2 µm HPLC-filter (Schleicher and Schüll).
3. 0.1 *M* NaOH (carbonate-free).
4. 0.1 *M* NaOH (carbonate-free)/0.25 *M* NaAc.

2.9   Identification of GPI Hydrophobic Fragments

1. Phospholipase A$_2$ from bee venom (Sigma). Store enzyme stock at −20°C.
2. Reactivials (Pierce).
3. *Staphylococcus aureus* Sphingomyelinase (Sigma). Store enzyme stock at 4°C.
4. 40 µL 1 *M* Tris-acetate pH 7.6/20 µL 0.4 *M* MgCl$_2$/300 µL diethylether.

2.10   Characterization of Parasite GPI Hydrophobic Fragments

1. Soybean lyso-phosphatidylinositol (Sigma), bovine-brain phosphatidylinositol (Sigma), egg yolk phosphatidic acid (Sigma) and bovine brain ceramide (Sigma) dissolved in chloroform/methanol (1:1, by vol) at a concentration of 1 µg/µL. Store at −20°C.
2. Analytical grade chloroform/methanol/water mixed in the ratio 25:75:5 (by vol).

3   Methods

3.1   Metabolic Labeling, Extraction, and Analysis of P. falciparum Glycolipids

1. Remove medium from cultures of intraerythrocytic stages of the malaria parasite *P. falciparum* by washing two times with phosphate-buffered saline (PBS) (*see* **Note 1**).

2. Resuspend the parasites (5×10^8 parasite infected erythrocytes) in 5 mL glucose-deficient RPMI-1640 medium containing 20 mM fructose, 0.5% Albumax I, and 100–200 μCi of the tritiated precursors glucosamine, mannose, ethanolamine, or fatty acids (*see* **Note 2**) and incubate the parasite cultures for 3–4 h with the labeling medium (*see* **Note 3**) at 37°C under reduced (5%) oxygen pressure.

3. Release the parasites from their host cells by saponin lysis (*see* **Note 4**) and wash 2 times with PBS. Transfer the washed parasites to an Eppendorf tube.

4. Extract the glycolipids from the parasite pellet by addition of 7 vol chloroform/methanol (1:1, by vol) to 1 vol cells (*see* **Note 5**) and incubate at room temperature for 30 min. Pellet the insoluble material by centrifugation (5 min, 10,000*g*) and remove the supernatant to a fresh tube. Add 500 μL chloroform/methanol/water (10:10:3, by vol) to the pellet and sonicate for 5 min. Spin down the insoluble material (5 min, 10,000*g*), and combine the first and second extract.

5. Dry the combined extracts under a stream of nitrogen and resuspend the extracted glycolipids in 1 mL water-saturated butan-1-ol. Wash the organic phase with 1 mL butan 1-ol saturated water to remove hydrophilic contaminants (*see* **Note 6**). The recovered butanol phase contains GPIs but also other hydrophobic components such as sphingolipids and phospholipids. These contaminants can be removed thought GPIs precipitation *(36)* (*see* **Note 7**).

6. Count an aliquot of the glycolipid extracts in a liquid scintillation counter (*see* **Note 8**).

7. Dry about 20,000 cpm of glycolipid extracts labeled with tritiated glucosamine or mannose and about 100,000 cpm of glycolipid extracts labeled with tritiated ethanolamine or fatty acids (*see* **Note 9**) to the bottom of Eppendorf tubes (*see* **Note 10**) and resuspend in 20 μL chloroform/methanol (1:1, by volume). 5 μL at a time of the glycolipid extracts are spotted as a band 0.5 cm wide and 2 cm from the bottom onto silica 60 plates. Develop the plate in organic solvents (e.g., chloroform/methanol/water (4:4:1, by vol) or (10:10:3, by vol)) up to a line 2 cm from the top of the plate (*see* **Note 11**). Allow the plate to dry in a fume hood. Scan the plate for radioactivity with a Berthold LB 2842 automatic scanner or place it in a film cassette against a Bio-Imager Tritium plate and leave it for 10–14 days before developing.

3.2   Enzymatic Characterization of GPIs

1. About 20,000 cpm of glycolipid extracts labeled with tritiated glucosamine or mannose are dried to the bottom of Eppendorf tubes and are resuspended in 5 μL 1% Triton X-100 by mixing and sonication. Subsequently, 45 μL Tris-HCl, pH 7.4 (final concentration 50 mM), is added. Incubate the samples with 0.5 U PI-PLC for 16 h at 37°C. For GPI-PLD (rabbit serum; *see* **Note 12**) add 2 mM CaCl$_2$ (final concentration) to the PI-PLC buffer and incubate with 10% rabbit serum (final concentration) for 16 h at 37°C. Terminate the incubation by heating at 100°C for 5 min (*see* **Note 13**).
2. Add 50 μL butan-1-ol saturated water and 100 μL water-saturated butan-1-ol to do the butanol/water-phase partition (*see* **Note 6**). Pool the butanol phases in an Eppendorf tube and dry them in a Speed-Vac concentrator.
3. Analyse the organic phases on silica 60 TLC-plates (*see* **Section 3.1.**, step 7).

3.3   Chemical Characterization of GPIs

1. About 20,000 cpm of glycolipid extracts labeled with tritiated glucosamine or mannose are dried to the bottom of an Eppendorf tube and are resuspended in 50 μL 0.2 M NaAc pH 4, 0.1% SDS by mixing and sonication.
2. Add 50 μL freshly prepared 1 M NaNO$_2$ and incubate for 16 h at room temperature.
3. Add 100 μL water to the incubation mixture and 200 μL water-saturated butan-1-ol to do the butanol/water-phase partition (*see* **Note 6**). Pool the organic phases in an Eppendorf tube and dry them in a Speed-Vac concentrator.
4. Analyze the organic phases on silica 60 TLC-plates (*see* **Section 3.1.**, step 7).

3.4   Generating Hydrophilic Fragments by Deamination and Reduction

1. Single glycolipids are purified by scraping out the relevant areas from the TLC plates. Transfer the silica in a glass tube. GPIs are eluted from the silica by adding 5 vol of chloroform/methanol/water (10:10:3, by vol), sonicating for 5 min, and incubation for 30 min at room temperature. Spin down the samples at 2,800g for 5 min, and remove the supernatant in a clean tube. Repeat the elution twice and combine the supernatants. The combined extracts are dried in Eppendorf tubes (*see* **Note 10**) and the residual silica is removed by butanol/water-phase partition (*see* **Note 14**). Resuspend in a 100 μL butan-1-ol and count an aliquot in the liquid-scintillation counter. Dry about 2,000 cpm of the organic phases in a Speed-Vac concentrator.
2. Hydrophilic fragments of TLC purified [^{3}H]-mannose or [^{3}H]-glucosamine labeled GPIs were generated by nitrous acid deamination (*see* **Section 3.3.**, steps 1 and 2).

Table 13.1 Chromatographic Properties of some GPI Neutral Glycans[a]

Structure	DU values	GU values
AHM	0.9	1.7
Manα1-4AHM	1.1	2.3
Manα1–6Manα1–4AHM	2.1	3.2
Manα1–2Manα1–6Manα1–4AHM	2.5	4.2
Manα1–2Manα1–2Manα1–6Manα1–4AHM	3.0	5.2
Manα1–2Manα1–6Man(Galα1-3)α1–4AHM	3.6	5.2
Manα1–2Manα1–6Man(Galα1-3Galα01-6)α1–4AHM	3.8	6.1
Manα1–2Manα1–6Man(GalNAcß1-4)α1–4AHM	3.0	6.5
Manα1–2Manα1–6Man(GalNAcß1-4Glcα1-6)α1–4AHM	3.0	7.5
Manα1–2Manα1–6Man(GalNAcß1-4Galß1-3)α1–4AHM	3.0	7.5
Manα1–2Manα1–6Man(GlcNAcß1-4)α1–4AHM	3.3	6.2
Manα1–2Manα1–6Man(GlcNAcß1-4Galß1-6)α1–4AHM	3.8	7.2

[a]DU, Dionex units; GU, glucose units; Man, mannose; Gal, galactose; GalNAc, N-acetyl-galactosamine; GlcNAc, N-acetyl-glucosamine; Glc, glucose; AHM, anhydromannitol.

3. Add 16 μL 0.8 M boric acid followed by quickly 16 μL 2 M NaOH (*see* **Note 15**) and 30 μL 1 M NaBH$_4$. Incubate for 3 h at room temperature without closing the lid of the Eppendorf tubes.
4. Destroy the excess of NaBH$_4$ by adding 5 μL 50% glacial acetic acid until effervescence ceases.
5. Dry down the samples and flash-evaporate 2 x with 100 μL methanol, 2 x 100 μL methanol/5% acetic acid and 2×20 μL toluene.
6. Resuspend the samples in 100 μL water, add 20 mg β-glucan oligomers (*see* **Note 16**) as internal standards and spin down the sample in a microfuge at maximum speed for 1 min. Analyze the material by size-exclusion chromatography on 140×1 cm Bio-Gel P4 (mesh >400) columns using 0.2 M ammonium acetate as solvent (*see* **Note 17**). Collect fractions of about 1 mL. Take 25 μL aliquots from each fraction and transfer them in a microtiter plate. Visualize the elution of the glycans by adding 100 μL conc. H$_2$SO$_4$/0.2% orcinol (caution: highly corrosive) and heating at 100°C for 10 min. The maxima can be determined by eye. Determine the radioactivity in the samples by liquid-scintillation counting. Elution of the radioactivity is given in "glucose units" (GU).
7. Look up the GU values in Table 13.1 to see if the chromatographic properties of the unknown hydrophilic fragments correspond to the chromatographic properties of known structures.

3.5 Generating Neutral Core-Glycans by Dephosphorylation, Deamination and Reduction

1. 30,000 cpm of single TLC-purified GPIs (*see* **Section 3.4.**, step 1.) are dried in an Eppendorf tube. Add 50 μL 48% ice-cold HF (caution: highly corrosive) and incubate for 60 h at 0°C (*see* **Note 18**).

2. Dry the HF under a stream of nitrogen (*see* **Note 19**).
3. Deaminate as described (*see* **Section 3.3.**, steps 1 and 2) in the absence of SDS.
4. Reduce the sample as described (*see* **Section 3.4.**, steps 3 and 4).
5. Apply the sample to a column of 0.4 mL Dowex AG50-X12 (H⁺) and eluate with 3 vol of water. Dry in a Speed-Vac concentrator.
6. Remove volatile contaminants as described (*see* **Section 3.4.** step 5) and dry the sample.
7. Redissolve in 100 μL water, pass through a column of 0.1 mL Chelex 100 (Na⁺), over 0.3 mL Dowex AG50-X12 (H⁺), over 0.3 mL Dowex AG3-X4 (OH⁻),over 0.1 mL QAE-Sephadex-A25 (OH⁻) and elute with 3 vol water.
8. Dry the eluate and redissolve in 100 μL water. Store the neutral glycans at −20°C. Analyze the products by Bio-Gel P4 (*see* **Section 3.7.**) or Dionex HPAEC (*see* **Section 3.8.**).

3.6   Sequencing of Neutral Glycans by Exoglycosidase Treatments (see Note 20)

1. Dissolve dried purified GPI neutral glycans (5,000 cpm) in 100 μL enzyme buffers and treat for 16 h at 37°C.
2. Inactivate the enzymes by heating to 100°C for 5 min and desalt by passage through a column of 0.2 mL Dowex AG50-X12 (H⁺), over 0.2 mL Dowex AG3-X4 (OH⁻). Elute with 1 mL water.
3. For JBAM use 100 U/mL in enzyme buffer (0.1 M NaAc, pH 5.0, containing 2 mM Zn$_2$).
4. For ASAM use 2 mU/mL in enzyme buffer (0.1 M NaAc, pH 5.0).

3.7   Bio-Gel P4 Analysis of Neutral Glycans

1. Dry the desalted samples (3,000 cpm) and redissolve in 50 μL water.
2. Mix the core-glycans with 20 mg β-glucan oligomers as internal standards and spin down the sample in a microfuge at maximum speed for 1 min.
3. Apply each sample to a Bio-Gel P4 column (1 x 140 cm) and analyze as described (*see* **Section 3.4.**, steps 6 and 7).

3.8   Dionex HPEAC Analysis of Neutral Glycans

1. Dissolve the desalted neutral core glycans (1,500 cpm) (*see* **Section 3.5.**) in 100 μL water and filter through a 0.2 μm HPLC-syringe filter. Dry the filtered sample and resuspend it in 15 μL water.

2. Add 5 µL β-glucan oligomer standards (2 µg), which will be detected by pulsed amperometric detection.
3. The elution program for Dionex HPAEC analysis using a Carbopak PA1 (4 × 250 mm) and the corresponding guard column: 100% buffer A (0.1 M NaOH), 0% buffer B (0.1 M NaOH, 0.25 M NaAc) up to 6 min after injection, then increase of buffer B to 30% at 36 min, at a flow rate of 1 mL/min. Wash the column for 20 min at 100% buffer B and reequilibrate for 15 min at 100% buffer A before starting the next run. Elution of the radioactivity is given in "Dionex units" (DU).
4. Look up the DU-values in Table 13.1 to see if the chromatographic properties of the hydrophilic fragment(s) correspond to the chromatographic properties of known structures.

3.9   Identification of GPIs Hydrophobic Fragments (see Note 21)

1. About 20,000 cpm of glycolipid extracts labeled with tritiated glucosamine or mannose are dried to the bottom of Eppendorf tubes. Add 100 µL methanol and 100 µL 0.2 M NaOH. Mix and incubate the tubes at 37°C for 2 h. Dry the samples in a Speed-Vac concentrator and do a butanol/water-phase partition (see Note 6).
2. About 20,000 cpm of glycolipid extracts labeled with tritiated glucosamine or mannose are dried to the bottom of Eppendorf tubes and are resuspended in 5 µL 1% Triton X-100 by mixing for 5 min. Subsequently, 45 µL Tris-HCl, pH 7.4 (final concentration 50 mM) and 2 mM CaCl, (final concentration) is added Incubate the samples with 50 U PLA₂ for 16 h at 37°C. Terminate the incubation by heating at 100°C for 5 min. Subject the samples to a butane/water-phase partition (see Note 6).
3. About 20,000 cpm of glycolipid extracts labeled with tritiated glucosamine or mannose, and 10 µL Triton X-100 are dried to the bottom of Reacti-vials. Redissolve the sample in 40 µL 1 M Tris-acetate, pH 7.6, 20 µL 0.4 M MgCl₂, 300 µL diethylether, and 20 µL chloroform in the presence of 1 U sphingomyelinase (*Staphylococcus aureus*). Samples are incubated for 16 h at 37°C under constant stirring. Subsequently the organic solvent is evaporated under a stream of nitrogen. Subject the samples to butanol/water-phase partition (see Note 6).
4. Analyze the organic phases on TLC compared to untreated controls.

3.10   Characterisation of Hydrophobic Fragments of Parasite GPIs

1. The hydrophobic fragment is released from 10,000 cpm of TLC purified (see **Section 3.4.**, step 1) fatty acid labeled GPIs by PI-PLC or GPI-PLD (see **Section 3.2.**). The organic phase is analyzed on TLC using chloroform/methanol/water (4:4:1, by vol).

Use 10 µg phosphatidylinositol, lyso- phosphatidylinositol and phosphatidic acid as standards that can be visualized by exposing the TLC plate to iodine vapour.

2. The hydrophobic fragment is released from 10,000 cpm of TLC purified (*see* **Section 3.4.**, step 1) fatty acid-labeled GPIs by sphingomyelinase (*see* **Section 3.9.**, step 3). The organic phase is analyzed on HPTLC-plates using chloroform/methanol (9:1, by vol) (*see* **Note 22**). Use 10 µg ceramide as a standard that can be visualized by exposing the TLC plate to iodine vapour.

3. The ester-linked fatty acids are released from 10,000 cpm of TLC purified (*see* **Section 3.4.** step 1) fatty acid-labeled GPIs by alkaline treatment (*see* **Section 3.9.** step 1). Released fatty acids can be analyzed on reversed phase (RP-18) HPTLC plates using chloroform/methanol/water (25:75:5, by vol) as solvent system.

4   Notes

1. Only minor amounts of *N*-glycans have been described for *P. falciparum* proteins. Dolichol-cycle intermediates (except dolichol-phosphate-mannose), the lipid-linked precursors for protein *N*-glycosylation, have not been demonstrated in asexual stages of *P. falciparum*. Therefore, they do not interfere with the synthesis of GPIs. For efficient labeling of GPIs in other organisms, exhibiting substantial *N*-glycosylation, it is necessary to preincubate before labeling, the cells with 1–10 µg/mL, tunicamycin for 1 h. Having this antibiotic present in the medium will inhibit incorporation of radioactive sugars (glucosamine and mannose) into dolichol-cycle intermediates.

2. Labeling with radioactive ethanolamine will lead to a massive incorporation of labeled ethanolamine into phospholipids, especially phosphatidylethanolamine, whereas the labeling efficiency of GPI-anchor precursors is relatively low. Metabolic labeling with fatty acids is more efficient if they are coupled to defatted bovine serum albumin V (Sigma). Dry 500 µCi fatty acids in Eppendorf tubes and resuspend them in 10 µL ethanol and add 484 µL water and 6 µL defatted bovine serum albumin V (100 mg/mL). Mix for 1 h at room temperature immediately before use. In addition, for efficient fatty acid-labeling reduce the amount of Albumax I or serum present in the labeling medium to 1/10 of the original amount present in the culture medium.

3. To get maximum incorporation of radioactive precursors it is necessary to establish the time for steady-state labeling for each specific cell type. Usually labeling periods of 1–4 h give efficient incorporation of radioactive precursors when investigating parasitic protozoa.

4. Saponin-lysis is a *Plasmodium*-specific treatment that releases intraerythrocytic parasites from their host cells by solubilizing cholesterol from erythrocyte membranes. Parasite membranes are low in cholesterol and therefore are not affected by this treatment.

5. The extraction protocol described will extract phospholipids, neutral lipids, dolichol-cycle intermediates, sphingolipids and GPIs efficiently. For some systems a sequential extraction using chloroform/methanol (2:1, by vol) before chloroform/methanol/water (10:10:3, by vol) has the advantage to separate phospholipids,

neutral lipids, some sphingolipids and some dolichol-cycle intermediates (found in the chloroform/methanol extract) from the ethanolamine-phosphate containing GPIs, more hydrophilic sphingolipids and dolichol-cycle intermediates (found in the chloroform/methanol/water extract). Chloroform/methanol (2:1, by vol) extraction can be omitted if GPIs precipitation is performed *(34)* (*see* **Note 7**).

6. For butanol/water-phase partition mix water and butan-1-ol (1:1, by vol) and wait until the two phases are clearly separated. Resuspend glycolipids in 1 vol of water-saturated butan-1-ol (upper phase) and mix with 1 vol butan-1-ol saturated water (lower phase) by intensive mixing. Separate the two phases by centrifugation at 13,000*g* for 2 min. Remove the lower, aqueous phase into a new Eppendorf tube and add 1 vol of fresh butan-1-ol saturated water to the remaining organic phase. Mix both phases and spin for phase separation. Remove the lower, aqueous phase and combine both aqueous phases. Add 1 vol of fresh water-saturated butan-1-ol to the combined aqueous phases. Mix both phases and spin for phase separation. Remove the upper, organic phase and combine both organic phases. Add 1 vol of fresh butan-1-ol saturated water to the remaining organic phase and mix. Separate the two phases by centrifugation and remove the lower, aqueous phase. The organic phase will be almost completely free of aqueous soluble contaminants and salt.

7. The recovered glycolipids in butanol phase (organic phase) (2 mL) are gently submitted to a nitrogen stream until the abstention of a white precipitate of GPIs after 3/4 of the organic phase are evaporated. The tube is then centrifuged and the upper phase transferred to another new glass tube. The operation is repeated again after the resuspension of the white precipitate in 2 mL water-saturated butan-1-ol. The GPIs white precipitate is visualized only when enough parasites (10^9 parasites) as starting material are used. Therefore purified non labeled parasites can be added to labeled parasites before glycolipid extraction.

8. To determine the incorporation of radioactive precursors into GPIs, count an aliquot of 1/50 of the sample.

9. Labeling with radioactive ethanolamine or fatty acids will lead to a massive incorporation of radioactivity into phospholipids whereas GPIs are usually underrepresented. Therefore, to get detectable signals of labeled GPIs it is necessary to remove phospholipid contaminants through GPI precipitation (*see* **Note 7**) and to use large aliquots of the GPI extracts for TLC analysis.

10. To reduce losses of material at the wall of tubes it is necessary to dry the samples in steps of one-third of the previous sample volume using an appropriate solvent.

11. To perform reproducible TLC analyses, it is of importance to mix the solvent systems very carefully. Fill the solvent into a TLC-chamber and wait for gas-phase saturation in the chamber before you run the TLC.

12. As GPI-PLD is not commercially available, rabbit serum is used as a source for GPI-PLD. The enzyme activity present in serum varies. Therefore, a new batch should be tested before use. GPI-PLD is an unstable enzyme. It is recommended to store the serum in small aliquots at −80°C. After thawing the serum can be stored at +4°C for up to one week.

13. PI-PLC will only cleave GPIs having an unsubstituted C-2 atom at the inositol ring. Fatty acids present at this position will hinder the formation of the cyclic phosphate at the inositol ring, which is an obligate intermediate in the cleavage

of GPIs by PI-PLC. GPI-PLD is not affected by the presence of a substitution at C-2 atom of the inositol ring.

14. The glycolipids will adsorb to the residual silica. Therefore, do not dry the sample completely. Redissolve the purified glycolipids in 200 μL water-saturated butan-1-ol by sonication. Add 200 μL butan-1-ol saturated water, mix the two phases and separate the phases by centrifugation. Remove the silica from the bottom of the tube together with the aqueous phase.

15. The amount of 2 M NaOH will vary. Titrate the NaOH to give pH 10.0–11.0. If you have overtitrated the sample, quickly add acetic acid until a pH of 10.0–11.0 is reached. Be careful as pH-values above pH >11 lead to fragmentation of the glycans.

16. The β-glucan oligomers (Glc1 – Glc20) are prepared by partial hydrolysis of 100 mg of dextran (Sigma) in 1 mL 0.1 M HCl, 2 h, 100°C. The acid is removed by flash-evaporating with methanol (5×) and by passage through a column of 1 mL of Serdolit MB3 (Serva). Elute with 3 mL water and filtrate through a 0.2-μm filter. The resulting sets of β-glucan oligomers are stored at −20°C.

17. The hydrophilic fragments generated by nitrous acid deamination contain charged groups like ethanolamine-phosphate, mannose-phosphate, or sialic acid. These groups interfere with the Bio-Gel P4 size-exclusion chromatography matrix, which will result in an increase in the effective size of the hydrophilic fragment. Using 0.2 M ammonium acetate as eluent minimizes the effects of charged groups on the elution behaviour and reduces losses of material on the Bio-Gel P4 column. 0.2 M ammonium carbonate is preferable if the Bio-Gel P4 chromatography will be used for preparative purposes. This eluent can easily be removed by several rounds of lyophilisation.

18. The incubation time of 60–65 h will lead to cleavage of the phosphodiester linkages found in GPIs. Ethanolamine-phosphate linked to the mannosyl-coreglycan is completely cleaved by 48% HF within 36 h at 0°C, whereas the inositol-phosphate linkage will only be cleaved after longer incubation periods.

19. Using a stream of nitrogen is an easy and convenient way to remove the HF. This method will not lead to destruction of the neutral glycans of most GPI structures. However, few substitutions of GPIs might be destroyed using this condition. Therefore, it might be useful to neutralise the HF by adding frozen lithium hydroxide and sodium hydrogenocarbonate (35).

20. Besides the two α-mannosidases described, configuration, linkage-type and attachment-side of substitutions attached to the conserved tri-mannosyl-coreglycan can be investigated by using various exo-glycosidases (35).

21. Hydrophobic fragments of GPIs can be investigated using different specific chemical and enzymatic treatments. Alkaline saponification cleaves ester-linked fatty acids from glycerol-based GPIs. PLA$_2$ specifically releases the C-2 fatty acid ester-linked to the glycerol. Sphingomyelinase releases ceramide-based hydrophobic fragments from GPIs.

22. Radioactivity detection on RP-18 TLC-plates by a scanner or Bio-Imager is poor. Therefore, use more radioactivity than for the analysis of glycolipids on silica-60 plates.

Acknowledgements This work was supported by the Deutsche Forschungsgemeinschaft, Hessisches Ministerium für Kultur und Wissenschaft, Stiftung P.E. Kempkes and Fonds der Chemischen Industrie.

References

1. Ferguson, M. A., Low, M. G., and Cross, G. A. (1985) Glycosyl-sn-1,2-dimyristylphosphatidylinositol is covalently linked to *Trypanosoma brucei* variant surface glycoprotein. *J. Biol. Chem.* **260**, 14,547–14,555.
2. Ferguson, M. A., Homans, S. W., Dwek, R.A., and Rademacher, T. W. (1988) Glycosylphosphatidylinositol moiety that anchors *Trypanosoma brucei* variant surface glycoprotein to the membrane. *Science* **239**, 753–759.
3. Englund, P. T. (1993) The structure and biosynthesis of glycosyl phosphatidylinositol protein anchors. *Annu. Rev. Biochem.* **62**, 121–138.
4. McConville, M. J. and Ferguson M. A. (1993) The structure, biosynthesis and function of glycosylated phosphatidylinositols in the parasitic protozoa and higher eukaryotes. *Biochem. J.* **294**, 305–324.
5. Nosjean, O., Briolay, A., and Roux, B. (1997) Mammalian GPI proteins: sorting, membrane residence and functions. *Biochim. Biophys. Acta.* **1331**, 153–186.
6. Ferguson, M. A. (1999) The structure, biosynthesis and functions of glycosylphosphatidylinositol anchors, and the contributions of trypanosome research. *J. Cell Sci.* **112**, 2799–2809.
7. Tiede, A., Bastisch, I., Schubert, J., Orlean, P., and Schmidt, R. E. (1999) Biosynthesis of glycosylphosphatidylinositols in mammals and unicellular microbes. *Biol. Chem.* **380**, 503–523.
8. McConville, M. J. and Menon, A. K. (2000) Recent developments in the cell biology and biochemistry of glycosylphosphatidylinositol lipids. Mol. *Membr. Biol.* **17**, 1–16.
9. Hwa, K. Y. (2001) Glycosyl phosphatidylinositol-linked glycoconjugates: structure, biosynthesis and function. *Adv. Exp. Med. Biol.* **491**, 207–214.
10. Eisenhaber, B., Maurer-Stroh, S., Novatchkova, M., Schneider, G., and Eisenhaber, F. (2003) Enzymes and auxiliary factors for GPI lipid anchor biosynthesis and post-translational transfer to proteins. *Bioessays.* 2003 **4**, 367–385.
11. Menon, A. K., Mayor, S., Ferguson, M. A., and Cross, G. A. M. (1988) Candidate glycophospholipid precursor for the glycosylphosphatidylinositol membrane anchor of *Trypanosoma brucei* variant surface glycoproteins. *J. Biol. Chem.* **263**, 1970–1977.
12. Masterson, W. J., Doering, T. L., Hart, G. W., and Englund, P. T. (1989) A novel pathway for glycan assembly: biosynthesis of the glycosyl-phosphatidylinositol anchor of the trypanosome variant surface glycoprotein. *Cell* **56**, 793–800.
13. Doering, T. L., Masterson, W. J., Englund, P. T., and Hart, G. W. (1989) Biosynthesis of the glycosyl-phosphatidylinositol membrane anchor of the trypanosoma variant surface glycoprotein. Origin of the non-acetylated glucosamine. *J. Biol. Chem.* **264**, 11,168–11,173.
14. Masterson, W. J., Raper, J., Doering, T. L., Englund, P. T., and Hart, G. W (1990) Fatty acid remodeling: a novel reaction sequence in the biosynthesis of trypanosome glycosyl phosphatidylinositol membrane anchors. *Cell* **62**, 73–80.
15. Menon, A. K., Schwarz, R. T., Mayor, S., and Cross, G. A. M. (1990) Cell-free synthesis of glycosyl-phosphatidylinositol precursors for the glycolipid membrane anchor of *Trypanosoma brucei* variant surface glycoproteins. Structural characterization of putative biosynthetic intermediates. *J. Biol. Chem.* **265**, 9033–9042.
16. Menon, A. K., Mayor, S., and Schwarz, R. T. (1990) Biosynthesis of glycosyl-phosphatidylinositol lipids in *Trypanosoma brucei*: Involvement of mannosyl-phosphoryldolichol as the mannose donor. *EMBO J.* **9**, 4249–4258.
17. Menon, A. K., Eppinger, M., Mayor, S., and Schwarz, R. T. (1993) Phosphatidylethanolamine is the donor of the terminal phosphoethanolamine group in trypanosome glycosylphosphatidylinositols. *EMBO J.* **12**, 1907–1914.

18. Udenfriend, S., and Kodukula, K. (1995) How glycosylphosphatidylinositol-anchored membrane proteins are made. *Annu. Rev. Biochem.* **64**, 563–591.
19. Takeda, J., and Kinoshita, T. (1995) GPI-anchor biosynthesis. *Trends Biochem. Sci.* **20**, 367–371.
20. Robinson, P. J. (1991) Signal transduction by GPI-anchored membrane proteins. *Cell. Biol. Intern. Rep.* **15**, 761–767.
21. Magez, S., Stijlemans, B., Radwanska, M., Pays, E., Ferguson, M. A. J., and De Baetselier, P. (1998) The Glycosyl-inositol-phosphate and dimyristoylglycerol moieties of the glycosyl-phosphatidylinositol anchor of the trypanosome variant-specific surface glycoprotein are distinct macrophage-activating factors. *J. Immunol.* **160**, 1949–1956.
22. Tachado, S. D., Mazhari-Tabrizi, R., and Schofield, L. (1999) Specificity in signal transduction among glycosylphosphatidylinositols of *Plasmodium falciparum, Trypanosoma brucei, Trypanosoma cruzi* and *Leishmania spp. Parasite Immunol.* **12**, 609–617.
23. Ikezawa, H. (2002) Glycosylphosphatidylinositol (GPI)-anchored proteins. *Biol. Pharm. Bull.* **4**, 409–417.
24. Tachado, S. D. and Schofield, L. (1994) Glycosylphosphatidylinositol toxin of *Trypanosoma brucei* regulates IL-1 alpha and TNF-alpha expression in macrophages by protein tyrosine kinase mediated signal transduction. *Biochem. Biophys. Res. Commun.* **205**, 984–99125.
25. Schofield, L., and Tachado, S.D. (1996) Regulation of host cell function by glycosylphosphatidylinositols of parasitic protozoa. *Immunol. Cell Biol.*, **74**, 555.
26. Tachado, S. D., Gerold, P., Schwarz, R. T., Novakovic, S., McConville, M. J., and Schofield, L. (1997) Signal transduction in macrophages by glycosylphosphatidylinositols of *Plasmodium, Trypanosoma,* and *Leishmania*: activation of protein tyrosine kinases and protein kinase C by inositolglycan and diacylglycerol moieties. *Proc. Natl. Acad. Sci. USA.* **94**, 4022–4027.
27. Schofield, L., and Hackett, F. (1993) Signal transduction in host cells by a glycosylphosphatidylinositol toxin of malaria parasites. *J. Exp. Med.* **177**, 145–153.
28. Schofield, L., Novakovic, S., Gerold, P., Schwarz, R.T., McConville, M. J., and Tachado, S. D. (1996) Glycosylphosphatidylinositol toxin of *Plasmodium falciparum* up-regulates intercellular adhesion molecule-1, vascular cell adhesion molecule-1, and E-selectin expression in vascular endothelial cells and increases leukocyte kinase-dependent signal transduction. *J. Immunol.* **156**, 1886–1896.
29. Tachado, S.D., Gerold, P., McConville, M. J., Baldwin, M. J., Quilici, D., Schwarz, R. T., and Schofield, L. (1996) Glycosylphosphatidylinositol toxin of *Plasmodium falciparum* induces nitric oxide synthase expression in macrophages and vascular endothelial cells by a protein tyrosine kinase-dependent and protein kinase C-dependent signaling pathway. *J. Immunol.* **156**, 1897–1907.
30. Lim, J., Gowda, D. C., Krishnegowda, G., and Luckhart, S. (2005) Induction of nitric oxide synthase in Anopheles stephensi by *Plasmodium falciparum*: mechanism of signaling and the role of parasite glycosylphosphatidylinositols. *Infect. Immun.* **73**, 2778–2789.
31. Schofield, L., and Grau, G. E. (2005) Immunological processes in malaria pathogenesis. *Nat. Rev. Immunol.* **9**, 722–735.
32. Kinoshita, T., Inoue, N. (2000) Dissecting and manipulating the pathway for glycosylphosphatidylinositol-anchor biosynthesis. *Curr. Opin. Chem. Biol.* **6**, 632–638.
33. Delorenzi, M., Sexton, A., Shams-Eldin, H., Schwarz, R. T., Speed, T., and Schofield, L. (2002) Genes for glycosylphosphatidylinositol toxin biosynthesis in *Plasmodium falciparum. Infect. Immun.* **8**, 4510–4522.
34. Azzouz, N., Shams-Eldin H., and Schwarz, R. T. (2005) Removal of phospholipid contaminants through precipitation of glycosylphosphatidylinositols. *Anal. Biochem.* **343**, 152–158.
35. Treumann, A., Güther, M. L. S., Schneider, P., and Ferguson, M. A. J. (1996) Analysis of carbohydrate and lipid components of glycosylphosphatidylinositol structures, in *Methods in Molecular Biology*, Vol. 76: Glycoanalysis Protocols (Hounsell, E.F., ed.), Humana Press Totowa, NJ.

14

2-Dimensional Electrophoresis: *Detection of Glycosylation and Influence on Spot Pattern*

Klemens Löster and Christoph Kannicht

Summary The detailed characterization of complex protein mixtures as in samples from biological sources cannot be sufficiently performed by separation of polypeptides according to their molecular weight as is done by conventional sodium dodecyl sulfate-polyacrylamide gel electrophoresis (SDS-PAGE) (1DE). For analysis of such samples, 2-dimensional gel electrophoresis (2DE) is the preferable methodological approach because it combines separation of polypeptides according to isoelectric properties and molecular weight as well. The resulting pattern of protein spots does not only provide information on composition of samples because of the complexity of a mixture of polypeptides. It delivers also a picture on the microheterogenity of polypeptides caused by post-translational modifications. These might be of natural or artificial type and occur during biosynthetic processing of a polypeptide or within industrial scale production. The presented method describes an experimental approach to investigate the influence of glycosylation in general and sialylation exclusively on spot pattern of proteins separated by 2DE.

Key Words 2D-Gel electrophoresis; pI-shift; glycosylation; sialylation; lectin.

1 Introduction

The detailed characterization of complex protein mixtures as in samples from biological sources cannot be sufficiently performed by separation of polypeptides according to their molecular weight as is done by conventional SDS-PAGE (1DE). For analysis of such samples, 2-dimensional gel electrophoresis (2DE) is the preferable methodological approach, because it combines separation of polypeptides according to isoelectric properties and molecular weight as well. The resulting pattern of protein spots provides information on composition of samples. Because of the complexity of a mixture of polypeptides, proteins might appear as single spots or as typical spot chains on 2D-gels, depending on the microheterogenity of the polypeptides caused by posttranslational modifications.

From: *Post-translational Modifications of Proteins.*
Methods in Molecular Biology, Vol. 446
Edited by: C. Kannicht © Humana Press, Totowa, NJ

These might be of natural or artificial type and occur during biosynthetic processing of a polypeptide or within industrial scale production.

Modifications that cause protein pI shifts include e.g., protein truncation and deletions, acetylation, phosphorylation, or glycosylation. The extent of pI shifts upon post-translational modifications and the resulting difference from the theoretical pI of a protein can be calculated using online tools like the ProMoST database (http://proteomics.mcw.edu/promost/). The observed pI shift largely depends on the pI of the unmodified protein. Proteins with basic pI tend to shift to a much higher extend than acidic proteins upon phosphorylation (*1*).

The presented method describes an experimental approach to investigate the influence of glycosylation on spot pattern of proteins: (i) A standard mixture of human plasma proteins is subjected to desialylation and deglycosylation followed by 1DE and 2DE analysis. (ii) Glycosylated polypeptides are detected by lectin binding on blotted proteins using *Sambucus nigra* (SNA) and *Datura stramonium* lectin (DSA) and assigned to spots obtained by protein staining. (iii) Spot patterns obtained by protein staining of an untreated, desialylated, and deglycosylated protein mixture are compared by matching gel images.

Evaluation of electrophoretic separations shows that desialylation of glycoproteins results in shifting of spot patterns for about 0.4 pH units and a weak decline of the relative molecular weight Mr of individual polypeptides. Deglycosylation similarly impairs the isoelectric properties of polypeptides as desialylation does. Additionally, deglycosylated proteins are characterized by Mr reduced by 5–8 kDa. Most notably, typical trains of spots observed in 2DE are not affected neither by desialylation nor by deglycosylation, which indicates the noncarbohydrate related character of these polypeptide isoforms.

2   Materials

2.1   Sample Preparation

2.1.1   Protein Standard Mixture

1. Fibronectin, human (F-2006, Sigma, St. Louis, MO).
2. Ceruloplasmin, human (C-4519, Sigma, St. Louis, MO).
3. Plasminogen, human (P-7397, Sigma, St. Louis, MO).
4. Transferrin (Siderophilin), human (T-6549, Sigma, St. Louis, MO).
5. Alpha1-Antitrypsin, human (A-9024, Sigma, St. Louis, MO).
6. Alpha2-HS-Glycoprotein (Fetuin), human (G-0516, Sigma, St. Louis, MO).
7. Alpha1-acid-Glycoprotein (Orosomucoid), human (G-9885, Sigma, St. Louis, MO).
8. Fibrinogen, human (F-4883, Sigma, St. Louis, MO).

2.1.2 Desialylation

1. Thermomixer (Eppendorf, Hamburg, Germany).
2. Sialidase from *Arthrobacter ureafaciens* (Glyco, Inc., Novato, CA).
3. Sialidase incubation buffer: 500 mM sodium acetate, pH 5.5.
4. Centrifugal filter devices, Ultrafree 0.5, Biomax-5 (Millipore, Bedford, MA).

2.1.3 Deglycosylation

1. GlycoFree Deglycosylation Kit (PROzyme, San Leandro, CA) or equivalent, e.g., GlycoProfile IV Chemical Deglycosylation Kit (Sigma, St. Louis, MO).
2. Centrifugal filter devices, Ultrafree 0.5, Biomax-5 (Millipore, Bedford, MA).
3. Dialysis tube.
4. Glass syringe, 50–200 µL.
5. Freeze dryer.
6. 0.1% (v/v) Trifluoroacetic acid.
7. 0.5% (w/v) Ammonium bicarbonate.
8. Ethanol.
9. Acetone.
10. Dry ice.

2.2 Electrophoretic Separations

2.2.1 SDS-PAGE

1. 1DE sample buffer (5-fold): 300 mM Tris-HCl pH 6.8, 50 mM DTT, 0.015% (w/v) bromophenol blue, 50% (v/v) glycerol, 12.5% (w/v) SDS, dissolved in double distilled water. Store aliquots of sample buffer at −20°C.
2. Running buffer (10-fold): 250 mM Tris, 1.92M glycine, 1% (w/v) SDS dissolved in double distilled water. The pH has to be adjusted to pH 8.8 before addition of SDS. Dilute 10-fold running buffer with double distilled water to yield 1-fold running buffer.
3. Gels: Pre-Cast 8–16% Tris-glycine gels 1.0 mm × 2D well or 1.0 mm × 10 wells (Invitrogen, Carlsbad, CA) (*see* **Note 1**)
4. Protein marker: ProSieve Protein Markers (Cambrex BioScience, Rockland, ME) consisting of a set of marker proteins of 5, 10, 15, 25, 35, 50, 75, 100, 150, and 225 kDa.
5. Electrophoresis system: XCELL II Mini SureLock Cell (Invitrogen, Carlsbad, CA), Power supply PowerPac 200 (BioRad, Hercules, CA)

2.2.2 2D-PAGE

2.2.2.1   1st Dimension, Isoelectric Focusing

1. Rehydration buffer (1.2-fold): 2.4M thiourea, 8.4M urea, 4.8% (w/v) Chaps, 2.4% (v/v) Immobiline solution (GE Healthcare, Chalfont St. Giles, United Kingdom) pH 4–7 or pH 3–10 (depending on the desired pH-range of focusing), 0.45% (w/v) DTT, 0.015% (w/v) Bromophenol blue, dissolved in double distilled water *(2,3)*. Store aliquots of rehydration buffer at −20°C. (*see* **Note 2**).
2. Paraffin Oil: Immobiline Dry Strip Cover Fluid (GE Healthcare, Chalfont St. Giles, United Kingdom) (*see* **Note 3**).
3. IPG-strips: Immobiline Dry Strip, pH 4–7 or pH 3–10, 70 mm length, 3 mm width, 0.5 mm thickness (GE Healthcare, Chalfont St. Giles, United Kingdom).
4. Electrophoresis system: Multiphor II, supplied with Dry Strip Kit, electrode wicks and Reswelling tray, Power supply EPS 3500XL / 3501 (GE Healthcare, Chalfont St. Giles, United Kingdom).

2.2.2.2   2nd Dimension, SDS-PAGE

1. Basis solution for equilibration buffer: 6M urea, 30% (w/v) glycerol, 2% (w/v) SDS, 50 mM Tris-HCl, pH 8.8. Store aliquots of equilibration buffer at −20°C.
2. Equilibration buffer I: Basis solution for equilibration buffer supplemented with 0.15% (w/v) DTT. This solution has to be prepared fresh.
3. Equilibration buffer II: Basis solution for equilibration buffer supplemented with 0.24% (w/v) Iodoacetamide. This solution has to be prepared fresh.
4. Equilibration tubes: Immuno Tubes 10 mL (Nunc, Wiesbaden, Germany).
5. Agarose sealing solution: 0.5% (w/v) Agarose IEF (GE Healthcare, Chalfont St. Giles, United Kingdom) in 1-fold Running buffer supplemented with a few grains of Bromophenol blue. (*see* **Note 4**).
6. Running buffer (10-fold): 250 mM Tris, 1.92M glycine, 1% (w/v) SDS dissolved in double distilled water. The pH has to be adjusted to pH 8.8 before addition of SDS.
7. Precast gels: 8–16% Tris-glycine gel 1.0 mm × 2D well (EC6045, Invitrogen, Carlsbad, CA).
8. Electrophoresis system: XCELL II Mini Cell (Invitrogen, Carlsbad, CA), Power supply PowerPac 200 (BioRad, Hercules, CA).

2.3 Blot Transfer

1. (a) Transfer buffer for 2DE gels: 20 mM Tris, 150 mM glycine, 0.01% (w/v) SDS, dissolved in double distilled water.

(b) Transfer buffer for 1DE gels: 20 mM Tris, 150 mM glycine, 10% (v/v) ethanol in double distilled water. (*see* **Note 5**).
2. Transfer membrane: BA85 Protran Nitrocellulose 0.45 μm (Whatman, Dassel, Germany).
3. Blotting paper: Gel blotting paper 0.37 mm, 190 g/m^2 (Roth, Karlsruhe, Germany).
4. Ponceau-staining solution (1×): Made from 10-fold Ponceau S concentrate (Sigma, St. Louis, MO) by dilution with double distilled water.
5. Ponceau-destaining solution: 1% (v/v) acetic acid in double distilled water.
6. Blotting chamber: Hoefer Mighty Small Transphor Tank Unit (GE Healthcare, Chalfont St. Giles, United Kingdom).

2.4 Lectin Staining

1. Dishes for incubation of blotting membranes: Petri dish square 120×120 mm (Greiner, Frickenhausen, Germany).
2. Blocking solution: SuperBlock reagent (Pierce, Rockford, IL).
3. Reagents for glycan detection: SNA (*Sambucus nigra* agglutinin, specificity towards α(2–6) sialic acid bound to galactose), or DSA (*Datura stramonium* agglutinin, specificity for desialylated glycans), Digoxigenin-labeled (Roche Diagnostics, Mannheim, Germany); Sheep anti Digoxigenin antibodies, Fab fragments, Peroxidase (Horseradish)-labeled (Roche Diagnostics, Mannheim, Germany).
4. Lectin-dilution buffer: 50 mM Tris-HCl, 150 mM NaCl, 1 mM MgCl$_2$, 1 mM CaCl, pH 7.5.
5. Washing-buffer: 10 mM Tris-HCl, 150 mM NaCl, pH 7.4.
6. Detection system: SuperSignal West Pico Chemiluminescent Substrate (Pierce, Rockford, IL).
7. Autoradiographic film: BiomaxMR-1 or X-Omat AR (Perkinelmer, Shelton, CT).
8. Processing chemicals for manual film development: GBX developer and fixer (PerkinElmer, Shelton, CT).

2.5 Gel Staining

1. Gel staining box, Nalgene.
2. Fixing solution: 40% (v/v) ethanol, 10% (v/v) acetic acid, 0.018% formaldehyde in double distilled water.
3. Ethanol solution: 50% (v/v) Ethanol in double distilled water.
4. Reducing reagent: 0.02% (w/v) sodium thiosulfate-penta-hydrate dissolved in double distilled water.
5. Silver nitrate reagent: 0.016% (w/v) silver nitrate, 0.027% (w/v) formaldehyde dissolved in double distilled water.

6. Developer: 5% (w/v) sodium carbonate, 0.025% (w/v) sodium thiosulfate-penta-hydrate, 0.015% (w/v) formaldehyde dissolved in double distilled water.
7. Stopping reagent: 1% (w/v) glycine dissolved in double distilled water.
8. Gel preserving solution: 10% (v/v) ethanol, 5% (v/v) glycerol in double distilled water.
9. Gel drying solution: GelDry Drying Solution (Invitrogen, Carlsbad, CA).
10. Cellophane sheets (*see* **Note 6**).
11. Gel drying frames: GelAir Drying Frames (BioRad, Hercules, CA).

2.6   Gel Evaluation

1. Scanning system with transillumination unit (*see* **Note 7**).
2. Software for image processing (*see* **Note 8**).

3   Methods

3.1   Sample Preparation

3.1.1   Preparation of Standard Protein Mixture

1. Prepare protein standard mixture by dissolving proteins (see **Section 2.1.1.**) in double distilled water as follows: fibronectin, 100μg/mL; ceruloplasmin, 100μg/mL; plasminogen, 30μg/mL; transferrin, 100μg/mL; α-antitrypsin, 50μg/mL; α-HS-glycoprotein, 50μg/mL; α-acid-glycoprotein, 33μg/mL; fibrinogen, 400μg/mL.
2. Store aliquots of protein standard mixture at −20°C. Avoid repeated freezing/thawing.

3.1.2   Desialylation

This method uses enzymatic release of sialic acids in order to obtain intact desialylated proteins for further analysis. We use sialidase from *Arthrobacter ureafaciens* for its capability to split off α 2-3, 6 and 8-linked sialic acids *(4)*. The cleavage protocol consists of three steps: Transfer of sample in cleavage buffer, sialidase treatment and buffer exchange. The last step is required before submission of proteins for biochemical analysis like 2DE. Success of desialylation should be checked by 1DE/protein staining and 1DE/lectin blotting before performing a 2DE experiment (Fig. 14.1).

Terminal sialylation of glycoproteins detected by SNA-binding reveals almost equal staining of the glycoprotein standard mixture (*see* Fig. 14.1, lane 4) *(5)*. This is reasonable because plasma glycoproteins are typically sialylated.

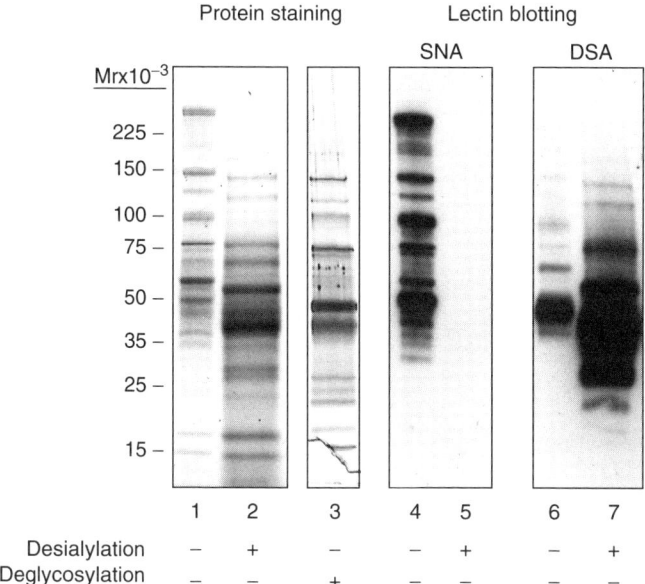

Fig. 14.1 1DE analysis of the protein standard mixture following desialylation or deglycosylation monitored by protein staining (silver staining) or blotting with SNA and DSA. Lectin blotting of deglycosylated proteins is not shown since these do not contain lectin binding sites. Proteins, rel Mr: fibronectin, 280 kDa; ceruloplasmin, 150 kDa/126 kDa; plasminogen, 98 kDa; transferrin, 79 kDa; fibrinogen α chain, 75 kDa; fibrinogen β chain, 60 kDa; α1 antitrypsin, 58 kDa; fibrinogen γ chain, 55 kDa; plasminogen HCh, 55 kDa; α$_2$-HS-glycoprotein, 49 kDa; α$_1$-acid-glycoprotein, 44 kDa; plasminogen HCh fragment, 39 kDa

Probing with DSA is used to detect terminal *N*-Acetyl-galactosamine residues *(6)*. Fully sialylated glycoproteins are not detected by DSA. DSA-reactive bands in the lectin blot of plasma proteins show incomplete sialylation. In particular, this shows the DSA reactive band at about 50 kDa (see Fig. 14.1, lane 6). Following removal of sialic acid residues by sialidase treatment, glycoproteins have lost SNA binding property (see Fig. 14.1, lane 5 vs lane 4) but strongly enhanced DSA reactivity (see Fig. 14.1, lane 7 vs lane 5).

Besides change in lectin binding, desialylation reduces the apparent Mr of proteins as can be seen in 1DE. For example, the Mr of ceruloplasmin (150/126 kDa) and the Mr of α$_1$ antitrypsin (58 kDa) shift by approximately 4–5 kDa to 145/121 kDa and 54 kDa, respectively (see Fig. 14.1, lane 2 vs lane 1). Most notably, the strong signal of DSA-reactive bands ranging between 25 and 75 kDa indicates successful removal of sialic acids, uncovering sub-terminal *N*-Acetyl-glucosamine.

1. Dissolve salt free samples in sialidase incubation buffer to a final concentration of 10-20 mg/mL.
2. If samples are not salt free, exchange buffer of protein solutions to sialidase incubation buffer by performing 3 wash cycles with centrifugal filter devices (5 kDa nominal molecular weight limit, 12,000*g*).

3. Concentrate glycoproteins by centrifugal filtration to approximately 10–20 mg/ mL in incubation buffer.
4. Resolve 0.2 U sialidase in 100 μL incubation buffer (*see* **Note 9**).
5. Add 50 μL sialidase solution to 50 μL protein solution.
6. Incubate for 18 h at 37°C under gentle shaking.
7. Exchange buffer to 1 m*M* Tris-HCl, 15 m*M* NaCl pH 7.4 by performing 3 wash cycles with centrifugal filter devices (5 kDa nominal molecular weight limit) before applying samples to 2DE.

3.1.3 Deglycosylation

Carbohydrates may be removed from proteins by enzymatic or chemical deglycosylation (for review see *(7)*). Because enzymes available for deglycosylation exhibit substrate specificity, it is difficult to achieve complete removal of carbohydrate moiety by enzymatic cleavage *(8,9,10)*. However, recently developed enzymatic deglycosylation kits may be applied as well (Enzymatic CarboReleaseit, QA-Bio, Palm Desert, CA). Here we describe chemical deglycosylation of proteins by anhydrous trifluoro-methansulfonic acid using a commercially available kit *(11)*. This method does not need special laboratory equipment.

Following a final buffer exchange step, success of deglycosylation can be simply monitored by 1DE/protein staining and 1DE/lectin blotting, as described for the desialylation procedure (see Fig. 14.1). Completely deglycosylated proteins have lost capacity for lectin binding. In addition, the decline in Mr indicates to which extend the glycan moiety contributes to the Mr of a given protein (see Fig. 14.1, lane 3 vs lane 1).

1. Dialyze samples thoroughly against water or 0.1% trifluoroacetic acid (*see* **Note 10**).
2. Transfer up to 1 mg protein to the reaction vial and dry sample by lyophilization (*see* **Note 11**).
3. Prepare a dry ice/ethanol cold bath.
4. Perform the deglycosylation process as described by the manufacturer (*see* **Note 12**). During the last step, the addition of ammonium bicarbonate, proteins may precipitate.
5. If proteins do not precipitate, remove the reaction mixture from protein by standard methods like dialysis or gel filtration.
6. In case of precipitation transfer reaction mixture to 2 mL conical reaction vials.
7. Add 1 mL cold acetone and chill samples to −20°C.
8. Centrifuge samples in a cooled centrifuge at 2,000*g*.
9. Carefully remove supernatant and dry sample using vacuum centrifugation or lyophilization.

3.2 Electrophoretic Separations

3.2.1 SDS-PAGE

1. Mix 4 volumes of sample with 1 volume of 1DE sample buffer (5-fold) and heat the mixture at 95–100°C for 5 min. If necessary, dilute the sample to appropriate protein concentration with a diluted Tris-buffer, e.g., with 1 mM Tris-HCl, 15 mM NaCl, pH 7.4. Approximately 1 µg or 2–5 µg of total protein mixture per lane are appropriate if silver staining or lectin blotting has to be performed, respectively.
2. Prepare the electrophoresis chamber by assembling Pre-Cast gels into the XCELL SureLock II Mini Cell system and place it on a magnetic stirrer.
3. Transfer a small magnetic stir bar into the chamber and fill the chamber with 1-fold running buffer prepared from 10-fold stock solution by dilution with double distilled water.
4. Load the samples (15–20 µL of sample per well) and the molecular weight marker (diluted 10-fold with 1fold 1DE sample buffer) onto the gel. Be careful to avoid swirling the running buffer.
5. Connect the Mini Cell system with the power supply. Switch the magnetic stirrer on. Start running the gels for 30 min at a constant voltage of 120 V, and continue running at 200 V until the dye front starts moving out of the gel.
6. Disconnect the Mini Cell system from the power supply.
7. Remove the gel unit out of the Mini Cell system, open the gel unit and remove the gel. Cut off the remaining parts of the stacking gel.
8. Transfer (*see* **Section 3.3**) or stain (*see* **Section 3.5**) the gels.

3.2.2 2D-PAGE

Influence of glycosylation and sialylation on 2DE spot pattern of our standard protein mixture is shown in Figure 14.2. Minigel 2DE analysis following desialylation and deglycosylation is performed applying protein staining and lectin staining. Individual gel imaging shows that desialylation and deglycosylation is accompanied by changes in the Mr and isoelectric point (pI) as well. Individual gel imaging clearly demonstrates contribution of the carbohydrate moiety to the physicochemical properties of polypeptides. In case of ceruloplasmin (150 kDa/126 kDa, see boxes within Fig. 14.2) desialylation shifts the pI by approximately 0.4-0.5 pH units towards basic pH and declines the Mr by approximately 1–2 kDa. Deglycosylation itself does not result in further increase of the pI but impairs the Mr of both polypeptide chains by approximately 4–5 kDa. Most notably, the heterogeneity of the polypeptides as displayed by trains of protein spots is not affected indicating that the carbohydrate residues do not cause these.

Gel Overlay
A: Protein staining   B: Lectin staining

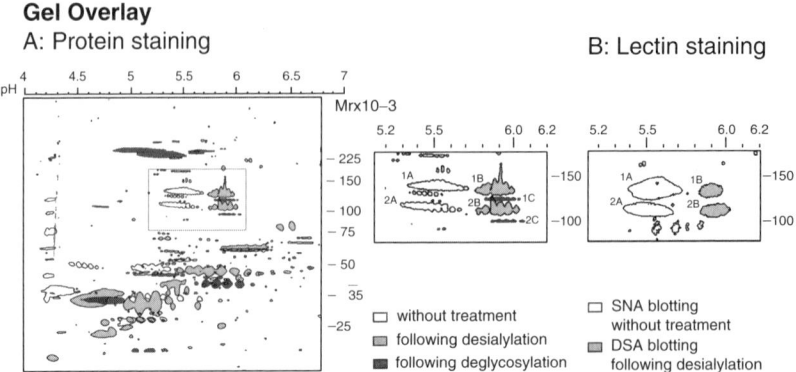

Fig. 14.2 2DE analysis of the protein standard mixture following sialidase treatment or deglycosylation monitored by protein staining (silver staining) or lectin blotting. Lectin blotting of deglycosylated proteins is not shown since these do not contain lectin binding sites. The boxes within the figure encircle ceruloplasmin (1A/2A) following desialylation (1B/2B) and deglycosylation (1C/2C)

We describe a 2DE-method using small, ready made mini-gels. Because the introduction of IPG strips with short length (7 cm), ready made mini-1DE gels can be used for the second dimension. This format has several practical advantages in comparison to large format gels. It is easy to handle, it requires only one day for doing a complete 2D analysis (first dimension, second dimension, blot transfer), and it safes material and working time. Furthermore, the mini gel system shows comparable separation efficiency of polypeptides compared to that of large format gels. We use this system for examination of protein mixtures of defined composition, for first determinations of pI/Mr of proteins after chemical/enzymatic modifications or for screening the reactivities of antibodies of until unknown specificity. The minigel 2DE is therefore the method of choice before starting to work with a gel system of large format (i.e., ≥18 × 20 cm or even larger). Finally, application of gel imaging followed by comparison of individual images may be used for evaluation of changes in spot pattern between gels related to carbohydrate specific modifications.

3.2.2.1 1st Dimension, Isoelectric Focusing

1. Change the buffer of samples to 1 mM Tris-HCl, 15 mM NaCl pH 7.4. (*see* **Note 13**)
2. Mix sample (1 volume) and 1.2-fold rehydration buffer (5 volumes), and incubate the mixture at 25°C for 1 h by gentle shaking. Approximately 5 µg or 5–10 µg of total protein mixture per lane are recommended if silver staining or lectin blotting has to be performed, respectively.
3. Add 165 µL of sample/rehydration buffer mixture (for 70-mm long IPG-strips) into each channel of the rehydration tray. Place IPG strips with the gel site towards the sample/rehydration buffer mixture (*see* **Note 14** and **15**). Overlay the

Table 14.1 Voltage Program for Isoelectric Focusing

Phase	[V]	[mA]	[W]	[h]	Vh
1	200	2	5	0:01	
2	3500	2	5	1:30	
3a (for IPG-strips pH 4–7L)	3500	2	5	1:30	5250
3b (for IPG-strips pH 3–10L)	3500	2	5	1:00	3500

channels with paraffin oil until strips are completely covered. Proceed with IPG strip rehydration overnight at room temperature.

4. Drain off the paraffin oil. Transfer strips onto a sheet of thick filter paper (0.5 –1 mm) and allow the oil to drain off.
5. Place and align rehydrated IPG strips into the focusing tray with the acidic and the basic site towards the anode and cathode, respectively. (*see* **Note 16**)
6. Cut two electrode wicks into appropriate length depending on the number of IPG strips. Wet the wicks with double distilled water (60 µL double distilled water per cm). Carefully place moistened electrode wicks on top of the strips at each end (*see* **Note 17**).
7. Place the electrodes on top of the electrode wicks by gently pressing.
8. Poor approximately 80–100 mL paraffin oil into the tray until all strips have been equally covered.
9. Set the temperature of the cooling bath to 20°C.
10. Start isoelectric focusing by applying the voltage program summarized in Table 14.1 (for 70 mm IPG strips) (*see* **Note 18**).
11. Remove IPG strips from focusing tray. Place the strips on a sheet of 1-mm thick filter paper and allow the paraffin oil to drain off. Subsequently, transfer strips into equilibration tubes (one tube per strip). Close tubes with stoppers. At this step, strips can be stored at −20°C until performing the second dimension (SDS-Page).

3.2.2.2   2nd Dimension, SDS-PAGE

1. Prepare equilibration buffer I and II (*see* **Section 2.2.2.2**)
2. Add 5 mL of equilibration buffer I into each tube, and gently rotate the tubes for 15 min.
3. Displace equilibration buffer I by 5 mL equilibration buffer II, and continue gently rotation of tubes for 15 min. Meanwhile, heat the Agarose sealing solution (water bath or microwave oven).
4. Transfer a small magnetic stir bar into the chamber and fill the chamber with 1-fold running buffer prepared from 10-fold stock solution by dilution with double distilled water.
5. Remove equilibration buffer II, and wash the strips twice with 10 mL Running buffer.
6. Cut the IPG strip by 5 mm at the cathodic and anodic site because the sample cup of the Precast Mini 2D gel only fits to a 60-mm long strip. Carefully place the

strips within the sample cup of the Pre-Cast Mini 2D gel. Avoid contact of the gel layer with the minigel plate and formation of any bubbles between the strip and the gel. Gently press the strip on top of the gel.

7. Fill the sample cup with the heated Agarose sealing solution. Be careful that the agarose sealing solution does not enter the reference well.

8. Prepare the electrophoresis chamber by assembling Pre-Cast gels completed with IPG strips into the XCELL SureLock II Mini Cell system. Transfer a small magnetic stir bar into the chamber and place the chamber on a magnetic stirrer.

9. Fill the electrophoresis chamber with 1-fold running buffer.

10. Load the molecular weight marker into the reference well of the gel. (*see* **Note 19**)

11. Connect the Mini Cell system with the power supply. Switch the magnetic stirrer on. Start running the gels for 30 min at a constant voltage of 120 V, and continue running at 200 V until the dye front starts moving out of the gel.

12. Disconnect the Mini Cell system from the power supply.

13. Remove the gel unit out of the Mini Cell system and open the gel unit. Cut off the remaining parts of the stacking gel and remove IPG strips and agarose.

14. Transfer (*see* **Section 3.3**) or stain (*see* **Section 3.5**) the gels.

3.3 Blot Transfer

1. Cut the nitrocellulose membrane and the gel blotting paper into gel size.

2. Fill the blotting chamber with cooled transfer buffer (Transfer buffer for 2DE gels or Transfer buffer for 1DE gels; *see* **Section 2.3**). Transfer a small magnetic stir bar into the blotting chamber and place it on a magnetic stirrer. Connect the blotting chamber to an external cooling device. (*see* **Note 20**)

3. Soak the blotting paper and the blotting membrane in Transfer buffer for 2DE gels or Transfer buffer for 1DE gels (*see* **Section 2.3**).

4. Prepare the blotting unit as follows: blot sponge, two sheets of blotting paper, nitrocellulose membrane, gel, two sheets of blotting paper, blot sponge. Avoid formation of air bubbles between the gel and the blotting membrane and between the stacks of blotting paper.

5. Transfer the blotting unit into the blotting chamber. The gel site has to be oriented towards the cathode.

6. Run the blot transfer at a constant voltage of 100 V for 1 h at 10°C.

7. Disassemble the transfer unit. Discard gel and blotting paper.

8. Drain the blotting membrane with double distilled water.

9. Control quality of blot transfer by staining the membrane with Ponceau Red: Place the nitrocellulose membrane into a dish with Ponceau-staining solution. Incubate the membrane for approximately 5–10 min by gently shaking, and remove non-specific background staining by several washes of the membrane in 1% (v/v) acetic acid in double distilled water until the membrane turns to white. At this time, the molecular weight marker can be labeled with a pencil.

10. Destain the membrane by several incubations in washing buffer (*see* **Section 2.4**).

3.4   Lectin Staining

Gently shaking of the blot membranes is recommended. All incubations are performed at room temperature.

1. Incubate the blotting membrane for 10 min in lectin-dilution buffer followed by a 30-min incubation in Blocking solution.
2. Prepare working dilutions of the lectins: Dilute digoxigenin-labeled SNA and DSA 1:2500 in lectin-dilution buffer. (*see* **Note 21**)
3. Incubate the blotting membrane in the lectin solution for 2h at room temperature.
4. Wash membrane 3 times with lectin-dilution buffer for 10 min.
5. Prepare working dilutions of anti-Digoxigenin-antibody: Dilute antibody 1:5,000 in lectin-dilution buffer.
6. Incubate the blotting membrane in the anti-Digoxigenin-antibody solution for at least 1 h at room temperature.
7. Wash membrane 5 times with Washing buffer for 10 min. (*see* **Note 22**)

The procedures of steps 8–12 have to be continued in a dark room.

8. Prepare the film processing solutions (developer, fixer). Poor the solutions into trays of appropriate size. Prepare a third tray with double distilled water. (*see* **Note 23**)
9. Transfer the membrane into a separate dish, and add 4 mL of chemiluminescent substrate onto the membrane. Incubate the membrane for 5 min.
10. Drain off the chemiluminescent substrate. Immediately place the membrane between two sheets of plastic wrap. (*see* **Note 24**)
11. Expose autoradiographic film on the blot surface for several periods of time ranging between 20 s and 5 min.
12. Develop the film by incubation in developer (approximately 3 min) followed by a short wash in double distilled water and subsequent incubation in fixer (approximately 10 min). All incubations should be performed by gently shaking.
13. Wash the film with deionized water and dry it on air.

3.5   Gel Staining

This recipe follows the protocol originally described by Blum et al. (1987) *(12)*. Gently shaking of gels is recommended.

1. Incubate gels for 1h 30 min in fixation reagent.
2. Wash gels 3 times with ethanol solution for 30 min.
3. Incubate gels for 1 min in reducing reagent.
4. Wash gels 3 times with double distilled water for 1min.
5. Incubate gels for 1 h 30min in silver nitrate reagent.

6. Wash gels 3 times with double distilled water for 1 min.
7. Place gels in Developer and incubate for several minutes depending on progress of staining.
8. Transfer gels immediately into stopping reagent and incubate for 30 min.
9. Wash gels twice with double distilled water for 30 min to remove the stopping reagent.

   (a) If gels shall be stored for some days in wetted form, transfer gels into gel preserving solution.
   (b) If gels shall be dried, incubate gels for 30 min in gel drying solution before placing between two sheets of cellophane that have been soaked for 1 min in gel drying solution. The cellophane/gel sandwich is placed within a gel drying frame and dried at room temperature for approximately 1–2 d.

3.6   Gel Evaluation

A range of diverse image processing software may be used for comparison of gel and immunoblot images. Among them are professional gel matching programs that allow automatic processing, statistical evaluation and graphic visualization of gels and matching results. However, we believe that the acquisition of this rather expensive software makes only sense if multiple numbers of gels have to be evaluated routinely. Alternatively, we recommend application of common office software for desktop publishing (e.g., AdobePhotoshop™), which allow comfortable processing of several pictures at once. In the following, a short description is given on how we evaluate gel images by use of AdobePhotoshop5.5.
Scanning procedure:

1. Transfer the wet gel or film onto the surface of a desktop scanner.
2. Scan gels/films in transmission mode: Select gray scale (8 bit/256 gray levels), and set input to 300 ppi, scale to 100%, descreen to 250 lpi, at high sharpness.
3. Normalize the image by auto color correction.
4. Save image as 8 bit gray level image.

Image overlay (described for overlay of two different colored images) (*see* **Note 25**):

1. Open images (I1 and I2) in AdobePhotoshop 5.5.
2. Make a copy of image I1.
3. Create a new document (D1). This document has the size of image I1.
4. Paste image I1 into document D1.
5. Make a copy of image I2.
6. Paste image I2 into document D1. This results in a picture consisting of 2 layers. Only 1 layer can be seen now.
7. Switch the RGB mode on (image>mode>RGB color). Do not reduce the whole picture (layers) to background layer.
8. Color the images/layers by working with the layer tool as follows: (a) Hide display of layer 2. Adjust the color of layer 1 (choose image>adjust>hue/saturation) by

switching colorize on; hue: 0; saturation: 100; lightness: 20. Image I1 has now red coloration. (b) Hide display of layer 1. Switch display of layer 2 on. Adjust the color of layer 2 (image>adjust>hue/saturation) by switching colorize on; hue: 120–140; saturation: 100; lightness: 20. Image I2 has now green coloration.

9. Combine both layers: Activate both layers and choose the multiplication mode (layer>layer options>multiply). Now, both images are seen at once.

The combined images show the whole set of proteins present in both images (red stained - image I1; green stained - image I2). In case of overlapping spots/bands, color addition results in brown stained spots/bands. If necessary, the image sizes can fit each other by application of the transformation mode (edit>transform/free transformation) (*see* **Note 26**).

4 Notes

1. We generally prefer using mini gels with acrylamide gradients because these give sharper bands and a better overall picture of proteins with molecular masses differing between 50–100 kDa.

2. This buffer has a rather high viscosity. Preparation of this buffer is performed at 37°C to insure complete dissolving and mixing of all reagents. Before use of rehydration buffer aliquots, the buffer cup is prewarmed to 25°C.

3. The paraffin oil used for covering the IPG strips during isoelectric focusing (*see* **Section 3.2.2.1**) may be used for several times. We only use fresh paraffin oil when IPG strips are rehydrated.

4. We prepare 10 and 20 mL aliquots of agarose sealing solution that are stored at 4°C, and used once.

5. A better cooling efficiency during blot transfer can be achieved if transfer buffers are stored at 4°C.

6. Several companies offer cellophane sheets for gel drying. We usually apply cellophane sheets used for conservation of food.

7. An office desktop scanner (300–1,200 dpi; 8,10, or 12 bit color) with transillumination unit gives sufficient optical resolution.

8. We are using AdobePhotoshop for routine image processing.

9. Sialidase preparations may contain BSA as stabilizer. This may impair analysis of proteins with similar molecular weight, depending on ratio of sample and BSA concentration.

10. This step is critical for successful performance of the method. Samples have to be essentially free of salts, metal ions and detergents. Low SDS concentrations are tolerable, if SDS concentration is kept below 2 mg per sample.

11. Samples have to be thoroughly dry for proper deglycosylation. Lyophilization should be performed <0.5 milliTorr for >24h. Do not allow samples to be exposed to moisture after lyophilization.

12. Please pay attention to the safety data sheet given by the manufacturer. Trifluoromethanesulfonic acid causes burns on skin and eyes.

13. Because high concentrations of salts disturb separation of proteins according their isoelectric properties, it remains necessary to transfer the sample into a buffer of low ionic strength (e.g., 1 mM Tris-HCl, 15 mM NaCl pH 7.4) or water. This should be preferably performed by gel filtration (e.g. by use of a PD10 column). In our laboratory, this method gives the best efficiency in buffer exchange. In general, avoid use of phosphate-containing solutions because these ions interfere with isoelectric focusing.

14. The cover foil of IPG strips has to be carefully removed by gently peeling up. Otherwise, parts of the acrylamid layer might be detached.

15. Prevention of air bubble formation between the IPG strip and sample/rehydration buffer mixture is absolutely necessary because the success of the isoelectric focusing strongly depends on even rehydration of the acrylamide layer.

16. It is necessary to process at least two IPG strips for isoelectric focusing. Because of the low conductivity of one IPG strip, the circuit protection of the power supply switches off the current.

17. The usage of excess of double distilled water impairs the resolution power of the first dimension resulting in smeared protein separation.

18. In case of samples with higher concentration of salts, we recommend application of a prefocusing step of 2 h at 200 V before starting the voltage gradient. This prefocusing step should remove salt ions and other undesired charged low molecular components of the sample.

19. For silver staining, 10 μL of 1:20 diluted (dilution made with 1-fold 1DE sample buffer) molecular weight marker are applied. If blot transfer is intended, 5 μL of undiluted marker are used.

20. The external cooling device may consist of an icebox filled with water/ice and a simple aquarium pump for water circulation connected to the blotting chamber by silicone tubes.

21. At least 5 mL of lectin or antibody solution are needed for soaking a membrane of 8 × 6 cm. We prefer to soak a 8 × 6 cm nitrocellulose membrane in 10 mL of solution.

22. The efficiency of washing (reduction of background staining) depends on the applied volume of washing buffer but not on the time of incubation.

23. Preparation of film processing chemicals follows the provided instructions. Storage of these reagents should be done in brown bottles in the dark at 4°C. Before starting with film processing, pre-warm agents to room temperature.

24. Avoid partial drying of the membrane. This may result in signal reduction and/ or elevated background staining.

25. For evaluation of silver stained gels, we recommend application of the positive mode (colored spots/bands and transparent background). This is because application of the inversion mode results in a rather high color signal of the background even in case of faint background staining of silver gels. In opposite, the homogenous background of films allows application of the inversion mode (colored spots/bands and black background).

26. Changes in gel size by image/layer transformation only results in linear changes of its dimensions.

15
Carbohydrate Composition Analysis of Glycoproteins by HPLC Using Highly Fluorescent Anthranilic Acid (AA) Tag

George N. Saddic, Shirish T. Dhume, and Kalyan R. Anumula

Summary Oligosaccharides in glycoproteins by their very nature influence many aspects of protein function, e.g., half-life and activity/potency. Recombinant IgGs constitute a major portion of therapeutic proteins. Though the glycans in IgGs account for about 2% of the total weight, they influence biologic activity apart from antigen binding. Characterization of the carbohydrates is not only a regulatory requirement but it may allow understanding of structure-function of proteins. Current advances in analytical techniques permit structural elucidation of small quantities of glycoproteins. At a first glance monosaccharide analysis may provide insight into the types of glycosylation similar to information afforded by amino acid composition. It is the only stand-alone technique by which individual sugar residues can be identified and quantitated (mol/mol). Fluorescent anthranilic acid (AA) has been extensively used as a high sensitivity detection tag for carbohydrates. HPLC methods with fluorescence detection described in this chapter are suitable for the analysis of monosaccharides (including sialic acids) on a routine basis. AA is used for the determination of hexoses and hexosamines, and o-phenylenediamine for sialic acids. These methods were validated and found to be highly reproducible compared to HPAEC-PAD and CE methods.

Key Words Monoclonal; antibodies; MAbs; recombinant; IgG; *N*-linked; monosaccharides; glycans; HPLC; fluorescence; anthranilic acid; composition.

1 Introduction

Proteins with covalently bound sugars are known as glycoproteins and are widely distributed in nature, e.g., in plants, animals, bacteria, and viruses. Antibodies, hormones, enzymes, and toxins are examples of proteins that are both biologically active and glycosylated. In addition, a number of proteins on the cell's surface, in cytosol and nucleus are glycosylated. During the last few years, enormous advances have been made in the understanding of glycoproteins, specifically their structure and biochemistry. This increased understanding is primarily because of the advent

From: *Post-translational Modifications of Proteins.*
Methods in Molecular Biology, Vol. 446.
Edited by: C. Kannicht © Humana Press, Totowa, NJ

of new tools for the study of complex carbohydrates. These new analytical methods have allowed for the reporting of a large number of well-characterized glycoproteins. The complete analysis of a glycoprotein provides information on the primary structure of the oligosaccharides as well as their variation at individual glycosylation sites. Such analysis requires a multipronged approach involving mapping and characterization of oligosaccharides, which is described in accompanying articles of this book, and determination of carbohydrate composition.

Carbohydrate composition analysis of glycoproteins is similar to amino acid analysis of proteins. Just as an accurate amino acid composition is critical to protein structure determination and identification by database searching, the accurate determination of the carbohydrate composition of a glycoprotein provides information as to the type and extent of glycosylation. To determine and understand the structure of a glycoprotein, the individual monosaccharides present must be identified and quantitated. This is critical since as a class, hexoses and hexosamines have the same molecular weights, and therefore, cannot be distinguished by mass spectrometry.

Another aspect to carbohydrate analysis became apparent with the advent of glycoprotein biopharmaceuticals. With respect to biopharmaceuticals, there is a need to demonstrate consistency of glycosylation in production lots that are intended for human therapy. In addition, there is an increasing demand to provide a well-characterized product description for regulatory submissions. One aspect, which can be used to determine the consistency of glycoprotein production lots, is the amount of carbohydrate it contains (expressed as% carbohydrate). The other aspect is monitoring of the sialic acid content in glycoprotein drugs which may be critical for biologic function *(1,2)*.

For the last decade, high performance anion exchange chromatography with pulsed amperometric detection method (HPAEC-PAD) was used for determining carbohydrate composition *(3,4)*. Today, newer highly sensitive and reproducible methods using reversed phase high performance liquid chromatography (HPLC) with fluorescence detection are available and described in this chapter. The HPLC methods for carbohydrate composition namely, monosaccharides and sialic acids are based on pre-column derivatization with fluorescent tags.

Carbohydrate composition is determined by acid hydrolysis of a glycoprotein sample to release the individual monosaccharides. After hydrolysis, the monosaccharides (neutral and amino sugars) are derivatized with anthranilic acid (AA, 2-aminobenzoic acid, 2AA), and then separated from each other and from excess reagent using reversed phase HPLC *(5,6)*. The resulting peak areas are compared to those of concomitantly derivatized and analyzed monosaccharide standards to determine the amount of each monosaccharide in the sample. A recent evaluation of monosaccharide methods suggests that, indeed, analysis by HPLC using AA label is far superior compared to capillary electrophoresis *(7)*.

For sialic acid determination, the sialic acids are initially released from the glycoprotein by mild acid hydrolysis followed by derivatization with *o*-phenylenediamine (OPD) to yield a fluorescent quinoxaline derivative. The derivative is separated from excess reagent by reverse phase HPLC for quantitation using fluorescence detection *(6,8)*. The resulting peak areas are compared to those of concomitantly derivatized sialic acid standards to determine the amount of the sialic acids (*N*-acetyl and *N*-glycolylneuraminic acids). In addition, OPD method offers more advantages than the 1,2-diamino-4,5-methylenedioxybenzene (DMB) method (e.g., it is sensitive, has linear reaction kinetics and does not require internal standard) *(9)*.

The methods described here are based on the use of highly fluorescent tags. They offer the most sensitive approach to analyze glycoproteins at this time, and therefore, these methods are suitable for analyzing samples available in limited amounts *(10)*.

2 Materials (*see* Notes 1 and 2)

2.1 Excipient Removal and Protein Concentration Determination

1. 2% (w/v) Ammonium bicarbonate.
2. Ethanol: ethyl acetate mixture 1: 1 (v/v) with 0.5% acetic acid (*see* **Note 3**).
3. 5% (v/v) Acetic acid solution (used with basic glycoproteins only).
4. 50 m*M* sodium hydroxide (used with acidic glycoproteins only). Store in a plastic container.
5. Glass tubes 13 × 100 mm with Teflon-lined screw caps (Sigma, Z281174).
6. Vacuum centrifuge.

2.2 Monosaccharide Analysis

1. Neat trifluoroacetic acid (TFA).
2. 1% (w/v) aqueous sodium acetate solution.
3. Polypropylene vials (1.6 mL) with O-ring seal screw caps (Fisher, 118448 or National Scientific, BC16NA-BP) (*see* **Note 4**).
4. 4% (w/v) Sodium acetate (trihydrate)-2% (w/v) boric acid (granular) in methanol (*see* **Note 3**).
5. Anthranilic acid (AA) solution: Weigh approximately 45 mg of anthranilic acid (2-amino benzoic acid) into a polypropylene vial. Add approximately 45 mg of sodium cyanoborohydride. Dissolve the solids in 1.5 mL of the sodium acetate-boric acid- methanol solution. (Note: Sodium cyanoborohydride is a poison and tends to absorb moisture readily from the air, which may affect the derivatization reaction. Limit the exposure of this chemical to air when weighing. If desired, 50 μL of 1.0M sodium cyanoborohydride in THF (Aldrich, cat. no. 296813) may be substituted in the reaction mixture).
6. Monosaccharide standard solution (1.0 mM): Weigh exactly 43.0 mg each of glucosamine hydrochloride and galactosamine hydrochloride and 36.0 mg each

of galactose, mannose, and glucose into a 200-mL volumetric flask. Add 32.8 mg of fucose into the same volumetric flask and bring to volume with Milli-Q water. Mix well and aliquot small volumes for storage. All the monosaccharides were from either Sigma or Pfansteihl Labs. Expiration: 1 yr at $-20°C$, if used more than once (*see* **Note 5**). Monosaccharide working standard solution: Dilute the 1.0 m*M* monosaccharide standard solution 1: 100 with Milli-Q water (*see* **Note 6**).

7. Chromatographic solvents: see **Section 2.4.**
8. Temperature controlled oven and/or heating block.

2.3   Sialic Acid Analysis

1. 0.5*M* Sodium bisulfate (NaHSO4).
2. 0.25*M* NaHSO4: dilute the 0.5*M* sodium bisulfate solution with an equal volume of Milli-Q water.
3. OPD derivatization solution: prepare a 20 mg/mL OPD solution in 0.25*M* sodium bisulfate.
4. 0.05% (v/v) Acetic acid-water solution for preparing the sialic acid stock standard.
5. 1.0 m*M* Sialic acid standard stock solution made in 0.05% acetic acid-water (*see* **Note 7**). Sialic acid working standard solution: dilute the sialic acid standard solution 1:100 to 500 with 0.25M NaHSO4 (*see* **Note 8**).
6. Chromatographic solvents: see **Section 2.4.**
7. Temperature controlled oven and/or heating block.

2.4   Chromatography System

1. HPLC with a fluorescence detector (highly sensitive fluorescence detectors such as Jasco FP 920, Waters 474, and HP (now Agilent) 1100 were used in these studies).
2. Thermostatted column compartment.
3. C18 Column: Waters Symmetry (3.9 × 150 mm, 5 μm, WAT045905) for monosaccharide analysis.
4. C18 Column: Ultrasphere ODS (4.6 × 150 mm, 5 μm, Beckman, cat no. 235330) for sialic acid analysis.
5. Column prefilter (Upchurch Scientific, A-315) and 0.2 μm insert (Upchurch Scientific, A-1O1X).
6. The following solvents were used for the analysis of both monosaccharides and sialic acids: Solvent A is composed of 0.2% (v/v) 1 -butylamine (Aldrich), 0.5% (v/v) phosphoric acid, and 1.0% (v/v) tetrahydrofuran (inhibited, Aldrich) in Milli-Q water (*see* **Notes 9 and 10**). Solvent B: Dilute solvent A with an equal volume of acetonitrile (HPLC grade) (*see* **Note 10**).

3   Methods

3.1   *Excipient Removal and Protein Concentration Determination*

Sugars (sucrose, mannitol), detergents, buffers, and so on, may also be present as excipients in the glycoprotein or biopharmaceutical samples. These excipients are likely to interfere in the analysis, and therefore, these substances must be removed prior to analysis of the sample. If the sample quantity is limited, a variety of desalting methods including dialysis and desalting using non-Sephadex resins, new types of centrifuge filters and mini-columns, and so on, could be used to remove the excipients. Samples can also be desalted by drying directly onto polyvinylidene difluoride (PVDF) membranes. In addition, glycoproteins that have been electroblotted onto a PVDF membrane following gel electrophoresis can be analyzed. This line of approach is useful when small amounts of the sample are available or several proteins exist as a mixture. Any gel technique may be used to perform this step. Once the stained band of interest has been identified (preferably on PVDF), it is excised and cut into small pieces. The pieces are placed in the bottom of the vial before proceeding with the hydrolysis step in **Section 4.3.1.** (*see* **Note 11**).

The following procedure is routinely used for removal of excipients from formulated biopharmaceuticals. If the amount of glycoprotein sample is limited, the procedure may be followed by reducing the reagents proportionally or by selecting another method for desalting as mentioned previously.

3.1.1   Excipient (Formulation Ingredients) Removal (see Note 12)

1. Place about 1.0–2.0 mg of each sample into each of 2 or 4 separate glass tubes (13 × 100 mm) with Teflon-lined screw caps.
2. Add 0.5 mL of 2% ammonium bicarbonate solution to each and vortex briefly (*see* **Note 13**).
3. Add 2 mL of the ethyl acetate/alcohol mixture to each vessel. Cap the vials and mix on a Vortex for about 20 s (*see* **Note 14**).
4. Allow the samples to stand for at least 10 min after vortexing.
5. Centrifuge the samples for about 5 min at the maximum setting in a suitable centrifuge.
6. Remove the samples from the centrifuge without disturbing the pellet.
7. Decant the supernatant from the pellet into a suitable waste container.
8. Invert the vial on a clean tissue paper (Kimwipe) placed at the bottom of a test tube rack for a minute to allow the residual solvent to drain (*see* **Note 15**).
9. Repeat steps 2–8 for a total of 4 precipitations.
10. If the glycoprotein is basic: after the fourth extraction, invert the vial on a tissue paper (Kimwipe) and allow the protein pellet to dry to insure that all the organic solvent has been removed (approximately 5 min). Add about 300 μL of 5% acetic

acid solution to each vial and mix well. Approximate protein concentration should be 5 mg/mL. Skip to step 14. If the glycoprotein is acidic: after draining the supernatant from the fourth extraction, cover the vials with several layers of parafilm and puncture the parafilm 5–7 times with a clean needle.

11. Dry the pellets in a vacuum centrifuge for 5–10 min (*see* **Note 16**).
12. Add about 300 μL of 50 m*M* sodium hydroxide to each vial and mix well. Approximate protein concentration should be 5 mg/mL.
13. Vortex the samples for about 30 s and allow them to stand for at least 48 h to clarify. Intermittent vortexing is recommended throughout the 48 h period (*see* **Note 17**).

3.1.2   Protein Concentration Determination

Each sample should be clear at this point (*see* **Note 18**). Pool the contents of the clarified samples if additional amount of protein is required (*see* **Note 19**).

1. Prepare a 10-fold dilution of the pooled material using Milli-Q water (acidic samples) or 5% acetic acid solution (basic samples). This diluted purified sample will be used in the monosaccharide analysis.
2. Read the absorbance of the diluted material at 280 nm against a blank, which has been diluted in similar manner to the sample (*see* **Note 20**).
3. Divide the observed absorbance by the extinction coefficient of the protein to obtain the protein concentration (*see* **Note 21**).

3.2   Monosaccharides Analysis

This section describes a method for the determination of carbohydrate composition of glycoproteins (**5,8**). Carbohydrate composition is determined by acid hydrolysis of a glycoprotein sample to release monosaccharides. After hydrolysis, the monosaccharides are derivatized with anthranilic acid (a highly sensitive fluorescent tag) and separated from excess reagent using RP-HPLC. Monosaccharide standards are concomitantly derivatized with the samples and analyzed. The resulting peak areas are compared to determine the amount of each monosaccharide present.

3.2.1   Hydrolysis

1. Place 0.1-mL aliquots of the sample into three separately labeled sample vials or an excised and minced PDVF band into a vial for monosaccharide analysis. Place 0.1 mL of the appropriate blank into the fourth vial (*see* **Note 22**).
1. Add 0.3 mL of Milli-Q water and mix gently.
2. Add 0.1 mL of neat TFA to each vial using a positive-displacement pipet, cap tightly, seal with 4–6 layers of Teflon tape and vortex gently on a low setting.
3. Hydrolyze the samples by placing them in a temperature-controlled oven at 100 ± 2°C for 6 h ± 5 min (*see* **Note 23** and Fig. 15.1).

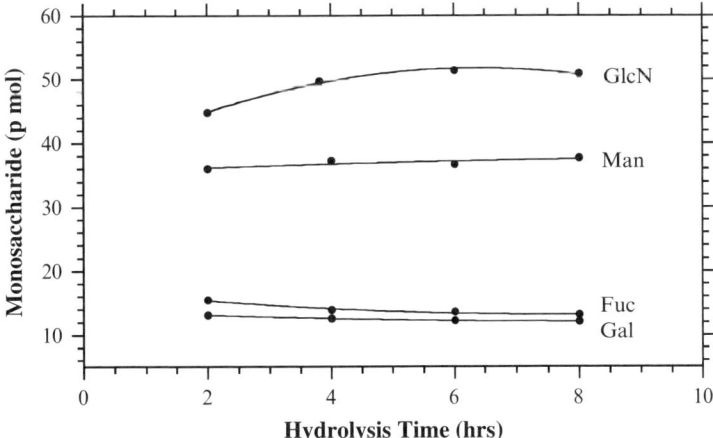

Fig. 15.1 Effect of hydrolysis time on the recovery of monosaccharides

4. Remove the samples from the oven and allow them to cool to room temperature. Centrifuge these samples briefly to collect the solution at the bottom of the tube.
5. Remove the Teflon tape and open the caps half way. Dry the samples overnight without heat in a vacuum centrifuge.
6. Similarly, excised and minced PDVF bands containing glycoproteins and appropriate blanks are prepared for monosaccharide analysis. However, 0.3 mL glass crimp-top micro vials are used with a 75 µL hydrolysis volume (8).

3.2.2 Derivatization of Samples and Monosaccharide Standards

1. Dissolve the dried hydrolysis sample replicates in 100 µL each of 1% sodium acetate solution and vortex vigorously for at least 2 min at the highest possible setting (*see* **Note 24**). Allow the sample to sit at room temperature for at least 30 min, vortex intermittently (approximately every 10 min) to ensure the pellet is completely dissolved (*see* **Note 25**).
2. Centrifuge the tubes briefly to spin the solution to the bottom of the tube and transfer 50 µL of each hydrolyzed sample to separately labeled 1.6-mL polypropylene vials (*see* **Note 26**). Appropriately label additional 1.6 mL polypropylene vials as the standards and aliquot 50 µL of the monosaccharide working solution into each vial.
3. Add 100 µL of anthranilic acid reagent to each sample and monosaccharide standard, cap the vials tightly, vortex, and centrifuge briefly to collect the solution at the bottom of the tube.
4. Heat the tubes for 45 min at 80 ± 2°C in a thermal heating block (*see* **Note 27**).
5. After the incubation, remove all tubes from the heating block and allow them to cool to room temperature.
6. Add 850 µL of solvent A to each tube, cap, and vortex vigorously.

7. Centrifuge all tubes for 5 min at maximum speed in the centrifuge to obtain a solution that is free of particulates.
8. Transfer each solution to an autosampler vial (*see* **Note 28**). Install the vials on the HPLC system autosampler and analyze using appropriate injection volumes (typically 50 μL).

3.2.3 Chromatography

1. Equilibrate a Waters Symmetry column (C18, 3.9 × 150 mm) in 5% Solvent B at a flow rate of 1 mL/min.
2. Make duplicate injections of 50 μL from samples and standards.

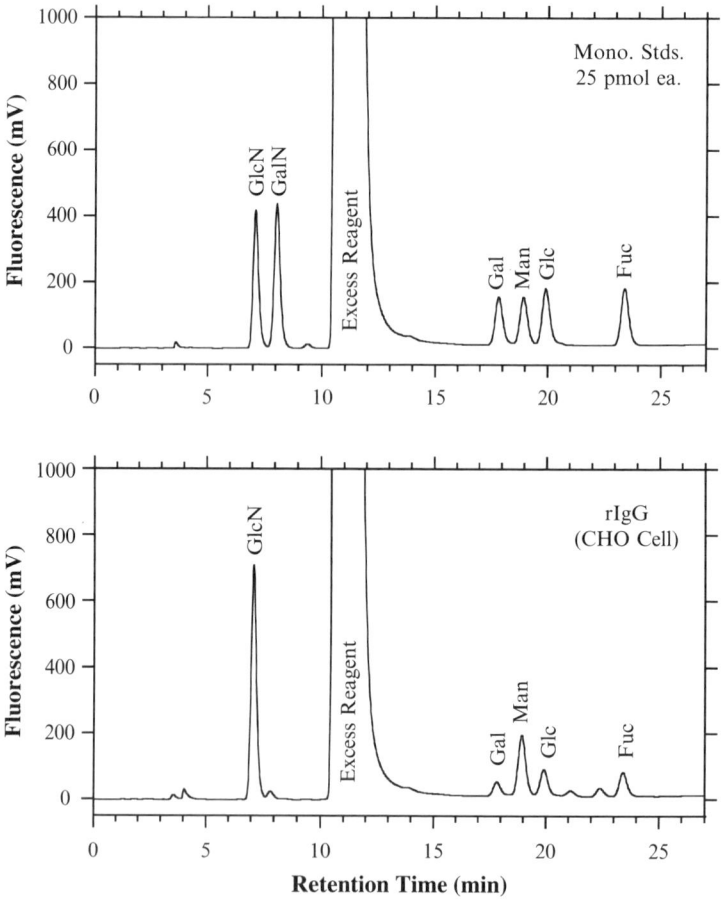

Fig. 15.2 Representative monosaccharide chromatograms obtained with a standard (top) and a rIgG produced in Chinese hamster ovary cells (bottom)

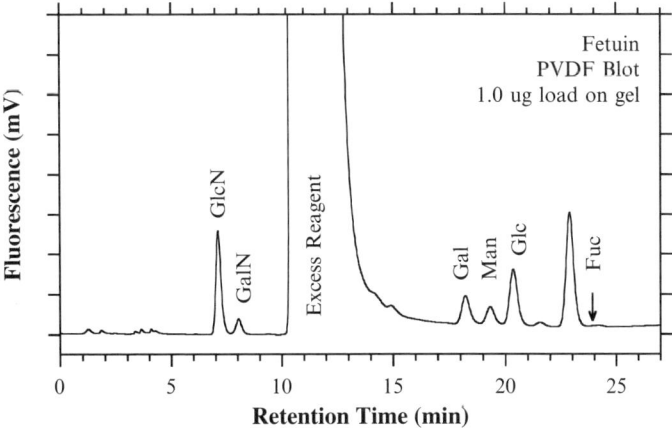

Fig. 15.3 An example of monosaccharides obtained from one of the three bands of fetuin electroblotted onto PVDF following SDS-PAGE. Glucose is a contaminant in this analysis *(10)*

3. Separate the monosaccharides as follows: 5% solvent B isocratic for 7 min followed by a linear gradient from 5 to 8% B over 18 min. This is followed by a 5 min wash using 100% B and an 8 minute equilibration at initial condition prior to next injection. The column temperature is maintained at 17°C (*see* **Note 29**). Total run time is 40 min with 28 min of data collection. The fluorescence detector settings are excitation λ = 360 nm and emission = 425 nm (*see* **Note 30**). See Fig. 15.2 for representative standard and sample chromatograms (*see* **Note 31**). See Fig. 15.3 for a chromatogram of fetuin monosaccharides using a PVDF membrane.

3.2.4 Calculations

1. Program the computer data system to calculate the peak area response for each sample (in pmol) from the average peak area response of the monosaccharide standard replicates (total six data sets, i.e., duplicate injections from 3 vials).
2. Calculate the amount of each monosaccharide in pmol per mg of protein taking into account dilution factors, sample volumes, and injection volumes used.
3. Calculate the moles of monosaccharide per mole of protein by the following formula:

$$\text{Moles of each monosaccharide per mole of protein} = \frac{(\text{pmol/mg of protein x MW})}{10^9}$$

where:

10^9 is the conversion factor from pmol to moles, and mg to g

MW is the molecular weight as determined from the amino acid sequence of the protein

4. Calculate the percent carbohydrate (w/w):

Percent carbohydrate (w/w) =

$$\frac{\text{all monosaccharides per mole of protein x mol. wt. of that monosaccharide x 100}}{\text{all monosaccharides per mole of protein x mol. wt. of that monosaccharide} + \text{MW}}$$

Use the following molecular weights for the monosaccharides: Glucosamine 221.2; Galactose; 180.2; Mannose; 180.2; Fucose; 164.2 and Sialic Acid; 309.3 (*see* **Note 32**).

3.2.5 Acceptance Criteria for Test Results

The relative standard deviation between six standard injections is less than 3.0% for each monosaccharide (usually 1.0%). The relative standard deviation between sample injections is less than 5.0% for each monosaccharide (usually 2.0%) (*see* **Note 33**).

3.3 Sialic Acid Analysis

This section describes the method of quantitation of sialic acid (N-acetyl and N-glycolylneuraminic acids) in glycoproteins (6). The sialic acids are released from the glycoproteins by mild acid hydrolysis followed by derivatization with OPD to yield a fluorescent quinoxaline derivative. The derivative is separated from excess reagent by RP-HPLC for quantitation using fluorescence detection.

3.3.1 Mild Acid Hydrolysis

1. Place 50 μL of the excipient free undiluted sample into labeled sample vials.
2. Prepare a vial with the appropriate blank (*see* **Note 34**).
3. Add 50 μL of 0.5 M sodium bisulfate to the blank and each sample, cap each vial tightly, and vortex each slowly for several seconds.
4. Hydrolyze the samples by placing them in a temperature controlled oven or heating block at 80 ± 2°C for 20 min.
5. Remove the vials from the heating block and allow them to cool to room temperature.

3.3.2 Derivatization of the Samples and Sialic Acid Standards

1. Aliquot 100 μL of the sialic acid working solution to labeled vials.
2. Add 100 μL of OPD solution to the hydrolyzed samples and standard vials. Cap the tubes tightly and vortex.

3. Heat the tubes for 40 min at 80 ± 2°C in a thermal heating block (*see* **Note 27**).
4. Remove all vials from the heating block and allow them to come to room temperature.
5. Add 800 µL of solvent A to each sample and standard. Cap the tubes and vortex them vigorously.
6. Centrifuge all the vials for 5 min at the maximum setting to obtain a particulate-free solution.
7. Transfer each solution to an autosampler vial (*see* **Note 26**). Install the vials on the autosampler and analyze.

3.3.3 Chromatography

1. Equilibrate a Beckman ODS column (4.6 × 150 mm, 5 µm) in 8–12% solvent B at a flow rate of 1 mL/min (*see* **Note 35**).
2. Inject 100 µL of each sample or standard in duplicate.
3. Separate the sialic acids as follows: isocratic at initial solvent B for 15 min followed by a 10 min wash at 95% solvent B and 10-min equilibration at initial conditions. The column temperature is maintained at 17°C (*see* **Note 29**). Total run time is 35 min with 20 min of data collection. The fluorescence detector settings are excitation λ = 230 nm and emission λ = 425 nm. See Fig. 15.4 for representative standard and sample chromatograms (*see* **Note 36**).

3.3.4 Calculations

1. Calculate sialic acid amount (in pmol) per injection in samples using the average peak area response of the 100 pmol sialic acid standard.
2. Calculate the quantity of sialic acid (in pmol) per mg of protein taking into account dilution factors, sample volumes, and injection volumes used.
3. The moles of sialic acid per mole of protein is determined by the following formula:

$$\text{moles of sialic acid per mole of protein} = \frac{(\text{pmol/mg of protein x MW})}{10^9}$$

where:

10^9 is the conversion factor from pmol to moles, and mg to g
MW is the molecular weight of the protein as determined from the amino acid sequence.

3.3.5 Acceptance Criteria for Test Results

The relative standard deviation for six sialic acid standard injections is less than 3.0% as determined by peak area. The relative standard deviation between sample injections is less than 5.0% for sialic acid as determined by peak areas (usually 2.0%) (*see* **Note 37**).

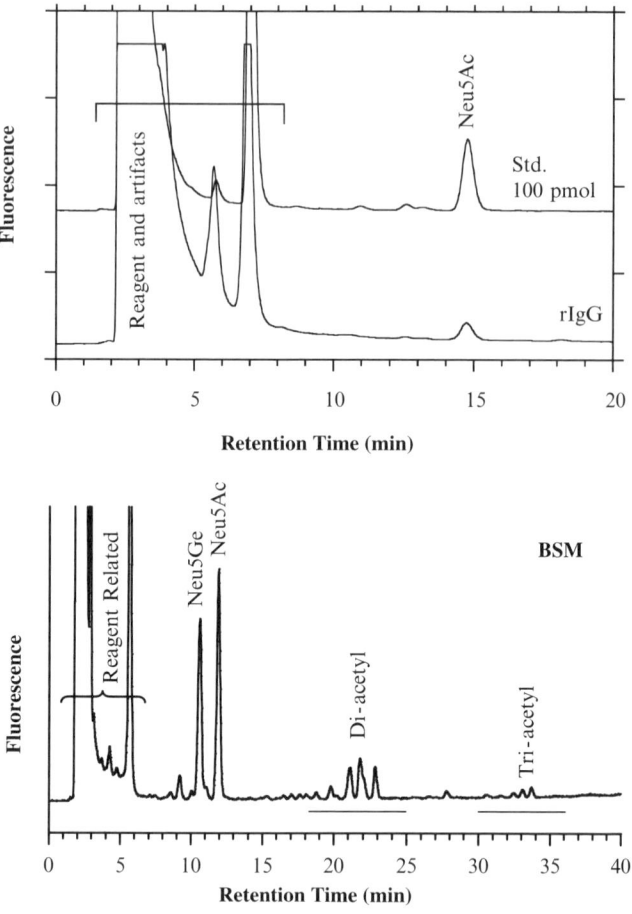

Fig. 15.4 Top panel shows representative sialic acid chromatograms obtained with a standard (top trace) and a rIgG produced in Chinese hamster ovary cells (bottom trace). The rIgG typically contains about 2% (w/w) carbohydrate and less than 5% of the oligosaccharides are sialylated. Bottom panel shows a sialic acid profile of bovine submaxillary mucin and separation of these into groups based on degree of acetylation *(11)*

4   Notes

1. CAUTION: Sodium cyanoborohydride, trifluoroacetic acid, 1-butylamine, THF, acetonitrile and OPD are toxic and/or flammable. Avoid contact with skin or inhalation. Wear gloves when preparing all solutions. When acetonitrile, phosphoric acid, THF, TFA, 1-butylamine, and cyanoborohydride are not in use they should be stored in an appropriate place. Dispose off these chemicals appropriately.

2. The manufacturer's part numbers for the equipment and reagents are meant as a guide. Equivalent substitutions may be made.
3. Expiration: 6 mo at room temperature.
4. This type of vial and cap are critical and should not be substituted.
5. An unthawed stock standard stored at −20°C is accurate for 5 yr.
6. The working standard contains 10 nmol of each monosaccharide/mL.
7. Aliquot the solution into 500 μL portions. Expiration: 1 yr at −15 to −25°C.
8. Typical working standard contains 2–10 nmoL/mL of sialic acid.
9. Make sure the butylamine has completely dissolved before proceeding to the next step. For convenience, use graduated glass pipet to add the THF.
10. Expiration: 1 mo at room temperature if containers are sealed.
11. Make sure to prepare simultaneously a blank piece of PVDF as a control.
12. If the glycoprotein is basic then follow the procedure using acetic acid as the diluent. If the glycoprotein is acidic then follow the procedure using sodium hydroxide as the diluent.
13. It is important to mix the sample vigorously to disperse the pellet before proceeding to the next step.
14. The solution should become cloudy. If the protein settles at the bottom then intermittent vortexing may be appropriate during this time.
15. Be careful: the pellet may slide out of the inverted vial. Before proceeding, it is helpful to gently tap the opening of the inverted vial on a clean tissue paper (Kimwipe) to remove any residual solvent remaining in the treads of the vial.
16. Do not over-dry. The pellet may not go into solution.
17. Most proteins will clarify in less time but the actual time may vary depending on the protein.
18. Additional dilution and/or clarification time may be necessary if the samples are not completely clear. Inspect the bottom of the tube for undissolved protein.
19. It is recommended that the pooled samples be allowed to stand overnight, however this is not a necessity.
20. This diluted blank would be either a 1:10 dilution of the 50 mM sodium hydroxide solution in water or the 5% acetic acid solution depending upon which diluent was used in **Section 3.2.**
21. Multiply the calculated protein concentration by the dilution factor (i.e., 10×). It is best if the protein concentration is between 3–5 mg/mL after correcting for the dilutions (e.g., rIgGs with approximately 2% (w/w) carbohydrate). Lower concentrations may make the accurate quantitation of residual sialic acid difficult for proteins with very low sialic acid content (e.g., rIgGs). Higher concentrations may produce off-scale peaks for glucosamine and galactosamine, which would require an additional dilution prior to analysis. The recommended ranges are dependent on the degree of glycosylation of the protein.
22. For the blank, use either the 5% acetic acid solution or a 10-fold dilution of the 50 mM sodium hydroxide solution prepared in **Section 3.2.**
23. It may be necessary to determine the optimum hydrolysis time through the use of a time-course study over 4–8 h.

24. It is crucial to completely dissolve the pellet before proceeding. It is recommended that the end of a pipet tip be used to physically crush/scrape the pellet from the bottom of the tube. Rinse the tip with the sample's sodium acetate diluent several times to ensure that any residual pellet is cleared from the tip.
25. If a flat-top Vortex mixer is available, samples can be bound together by a rubber band or placed in a small box and placed on the vortex for 15 min.
26. Transfer only the supernatant and none of the pellet.
27. Cover the top of the heating block with insulating material (e.g., foam, paper towels, and/or wool pads, etc.) to maintain the temperature.
28. Some samples may contain particulates at the bottom of the tube. These particulates should not be transferred and injected into the HPLC. It is best to use amber autosampler vials to minimize light exposure to the derivatized samples.
29. The actual temperature is not critical, however, the column should be maintained at a constant temperature in order to obtain reproducible retention times. A column cooler (Cool Pocket, Keystone Scientific) can be used for this purpose.
30. For less sensitive fluorescence detectors, an excitation of 230 nm may be used.
31. Sample chromatograms may have some additional peaks but the monosaccharides peaks should be resolved from any artifact peaks. It is common to observe varying levels of glucose in the samples and blank. Due to the abundance of glucose- containing polymers (e.g., lint) in the environment, complete elimination is nearly impossible. Glucose levels, therefore, cannot be accurately determined.
32. The value for the number of moles of sialic acid is determined by the procedure described in the sialic acid method.
33. This value may be greater with very low amount of a particular monosaccharide in the glycoprotein. The tailing factor for each of the peaks in the standard chromatogram must be less than 1.2. The resolution of the mannose in the standard chromatogram peak must be greater than 1.2.
34. The blank would either be a similarly diluted blank of the 50 mM sodium hydroxide solution or the 5% acetic acid solution, depending on which solution was used in **Section 3.2.**
35. It is usually necessary to optimize the chromatography on a particular lot of columns by adjusting the initial percentage of solvent B to between 8 and 12%. Sialic acid peak must separate from the artifacts.
36. Sialic acid should elute within 15 min.
37. This value may be greater with very low abundance of sialic acid in the glycoprotein. The tailing factor for the sialic acid peaks in the standard chromatogram must be less than 1.4. The standard theoretical plate count should be greater than 4,000.

Acknowledgement We thank Ms. Ping Du and Ms. Mary Beth Ebert for their contributions to validation of methods.

References

1. Varki, A. (1993) Biological roles of oligosaccharides: all of the theories are correct. *Glycobiology.* **3**, 97–130.
2. Dwek, R. A. (1996) Towards understanding the function of sugars. *Chem. Rev.,* **96**, 683–720.
3. Townsend, R. R. (1995) Analysis of glycoconjugates using high-pH anion-exchange chromatography, in *Carbohydrate Analysis: High-Performance Liquid Chromatography and Capillary Electrophoresis* (El Rassi, Z., ed.), Elsevier, New York, NY, 181–209.
4. Davies, M. J. and Hounsell, E. F. (1998) HPLC and HPAEC of oligosaccharides and glycopeptides, in *Methods in Molecular Biology, Glycoanalysis Protocols* (Hounsell, E. F., ed.), Humana Press, Totowa, NJ.
5. Anumula, K. R. (1994) Quantitative determination of monosaccharides in glycoproteins by high performance liquid chromatography with highly sensitive fluorescence detection. *Anal. Biochem.* **220**, 275–283.
6. Anumula, K. R. (1995) Novel fluorescent methods for quantitative determination of monosaccharides and sialic acids in glycoproteins by reversed phase high performance liquid chromatography, in *Methods in Protein Structure Analysis* (Atassi, M. Z. and Apella, E., ed.), Plenum Press, New York, NY, 195–206.
7. Račaitytė, K., Kiessig, S., and F. Kálmán, (2005) Application of capillary zone electrophoresis and reversed-phase high-performance liquid chromatography in the biopharmaceutical industry for the quantitative analysis of the monosaccharides released from a highly glycosylated therapeutic protein. *J Chromatog A.* **1079**, 354–365.
8. Anumula, K. R. (1997) Highly sensitive pre-column derivatization procedures for quantitative determination of monosaccharides, sialic acids and amino sugar alcohols of glycoproteins by reversed phase HPLC, in *Techniques in Glycobiology* (Townsend, R. R. and Hotchkiss, A. T., eds.), Marcel Dekker, New York, NY, 349–357.
9. Anumula, K. R. (2006) Advances in fluorescence derivatization methods for high-performance liquid chromatographic analysis of glycoprotein carbohydrates. *Anal. Biochem.* **350**, 1–23.
10. Anumula, K. R. and P. Du (1999) Characterization of carbohydrates using highly fluorescent 2-aminobenzoic acid tag following gel electrophoresis of glycoproteins. *Anal. Biochem.* **275**, 236–242.
11. Anumula, K. R. (1995) Rapid quantitative determination of sialic acids in glycoproteins by high-performance liquid chromatography with sensitive fluorescence detection. *Anal. Biochem.* **230**, 24–30.

16
Enzymatical Hydrolysis of *N*-Glycans from Glycoproteins and Fluorescent Labeling by 2-Aminobenzamide (2-AB)

Rolf Nuck

Summary When performing a structural analysis of *N*-glycans, a number of aspects should be considered. *N*-Glycans may be hydrolyzed from purified glycoproteins, serum glycoprotein mixtures, or delipidated membrane fractions by chemical hydrolysis using hydrazine or enzymatic hydrolysis using PNGase F. Chemical deglycosylation may be an economical alternative to produce *N*-and *O*-glycans in a preparative scale, but it is less suitable for analytical purposes. By chemical hydrazinolysis the protein core is destroyed completely and all acyl groups are cleaved from neuraminic acid residues as well as from *N*-acetylhexosamine residues. If not only a structure analysis of *N*-glycans is intended but a sequencing of the protein core, an analysis of the different types of neuraminic acids or an elucidation of the carbohydrate structures in distinct glycosylation sites has to be performed in addition, enzymatical deglycosylation using PNGase F is the most suitable way to hydrolyze *N*-glycans from the protein backbone.

Key Words *N*-Glycan; enzymatic deglycosylation; PNGase F; fluorescent labeling; 2-aminobenzamide.

1 Introduction

Performing a structural analysis of *N*-glycans a number of aspects should be considered before. *N*-Glycans may be hydrolyzed from purified glycoproteins, serum glycoprotein mixtures or delipidated membrane fractions by chemical hydrolysis using hydrazine or enzymatic hydrolysis using PNGase F.

Chemical deglycosylation may be an economical alternative to produce *N*-and *O*-glycans in a preparative scale *(1)* but it is less suitable for analytical purposes. By chemical hydrazinolysis the protein core is destroyed completely and all acyl groups are cleaved from neuraminic acid residues as well as from *N*-acetylhexosamine residues. Re-*N*-acetylation is therefore necessary before the carbohydrate analysis procedure *(2,3)*.

From: *Post-translational Modifications of Proteins.*
Methods in Molecular Biology, Vol. 446.
Edited by: C. Kannicht © Humana Press, Totowa, NJ

If not only a structural analysis of N-glycans is intended but a sequencing of the protein core, an analysis of the different types of neuraminic acids or an elucidation of the carbohydrate structures in distinct glycosylation sites has to be performed in addition, enzymatical deglycosylation using PNGase F is the most suitable way to hydrolyze N-glycans from the protein backbone.

Enzymatical hydrolysis may be performed from the intact glycoprotein or from glycopeptide fractions as obtained by previous proteolytic digestion of the glyco-protein by different proteases of sequencing grade such as trypsin or proteases from *Pseudomonas fragi* (Asp-N), *Staphylococcus aureus* V8 (Glu-C), and *Lysobacter enzymogenes* (Lys-C). Because in many cases complete enzymatic deglycosylation of the intact glycoprotein requires the use of detergents and enzyme concentrations of up to 100 mU/ml of PNGase F *(4)*, a prior proteolytic digestion *(5)* is advanta-geous, minimizing the concentration of the enzyme required to 2–10 mU/mL, and does not necessitate a purification of the *N*-glycans from detergent *(6)*.

Note that some glycoproteins of plant and insect origin may be resistant to PNGase F from *Flavobacterium meningosepticum*, if they contain a fucose residue, α1–3 linked to the core *N*-acetylglucosamine, which inhibits a deglycosylation by PNGase F. In this case the use of PNGase A from almonds *(7)* is recommendable.

Note that for enzymatical hydrolysis of glycoproteins and/or glycopeptides for an analytical purpose, a mixture of endoglycosidase F and PNGase F, which is offered by some companies should not be used. Because endoglycosidase F is hydrolyzing the oligosaccharides in the chitobiose core between the two *N*-acetlyglucosamine residues, but PNGase F is hydrolyzing the intact *N*-glycans directly from asparagine, a mixture of both enzymes is producing 2 oligosaccharide chains differing in size by an *N*-acetylglucosamine residue from a defined homoge-neous asparagine linked glycan *(8)*.

Moreover some critical aspects should be taken into account with respect to the analysis of *N*-glycans. Oligosaccharides in small amount (picomol to nanomol range) are difficult to detect, because they do not contain a chromophore. A sensitive detection of sugars without previous derivatization may be performed by HPAEC combined with pulsed amperometric detection. However, in most cases this separation method is not sufficient to obtain homogeneous *N*-glycans. A multi-dimensional separation of complex oligosaccharide mixtures has been proven to be superior to come to homogeneous *N*-glycans, consisting of HPLC on anion exchange (Mono Q, GlycoSep C), hydrophilic interaction (APS 2-Hypersil) and hydrophobic interaction columns (ODS-Hypersil) *(6,9)*. However, the multidimen-sional separation methods described require a sensitive labeling of the carbohydrate chains for their detection, such as reductive amination using 2-aminobenzamide *(9)* or radioactive labeling using sodium borotritide *(3)*.

To perform an optimized labeling reaction, a purification of the *N*-glycans from detergent, protein/peptides, and salt is essentially necessary. If the *N*-glycans to be analyzed are exclusively of neutral character such as in the case high mannose oligosaccharides, *N*-glycans may be very effectively purified by passing it through mixed bed resins. However it should be kept in mind that most *N*-glycans are negatively charged containing sialic acid, phosphate, or sulfate residues. In principal in those cases gel permeation chromatography using Bio-Gel P-6 or Sephadex G-25 is usually applied for desalting. However this approach has been found to cause substantial losses of material if only small amounts of *N*-glycans have to be analyzed.

Keeping in mind the considerations mentioned above, a protocol was developed that allows an economical, short and simple procedure for the enzyme hydrolysis and the fluorescent labeling of *N*-glycans, which may be applied in each laboratory and does not require expensive special instrumentation.

2   Materials

2.1   *Trypsin Digestion of Glycoproteins*

1. Trypsin (13500 U/mg) TPCK treated from bovine pancreas (Sigma-Aldrich, St. Louis, USA).
2. 1% (w/v) acetic acid.
3. Deglycosylation buffer: 50 mM *N*-Methyl-2,2'-iminodiethanol-trifluoroacetat, pH 8.0. 0.575 mL (0.595 g) *N*-Methyl-2,2'-iminodiethanol (VWR International, Bruchsal) are dissolved in 100 mL Milli Q water and adjusted to pH 8.0 using trifluoroacetic acid (Uvasol, VWR International).

2.2   *Enzymatical Deglycosylation with PNGase F*

1. PNGase F from the culture medium of Flavobacterium meningosepticum or the recombinant enzyme expressed in *Escherichia coli* (Roche Diagnostics GmbH, Mannheim)
2. Volatile deglycosylation buffer: 50 mM *N*-Methyl-2,2'-iminodiethanol-trifluoro-acetat, pH 8.0 (*see* **Section 2.1.3**).
3. 1N acetic acid and 1% (v/v) acetic acid.
4. 50% aqueous ethanol.
5. Nanosep centrifugal devices with 3 K Omega Membrane (Pall GmbH, Dreieich, Germany)
6. Vacuum centrifuge.
7. Thermosensitive centrifuge.
8. Blue Sepharose Fast Flow (GE Healthcare Life Sciences, Germany)
9. Dowex 50 W-X12 (H+) - cation exchanger (Bio-Rad Laboratories, Richmond) or equivalent.
10. Disposable Polystyrene Columns, 2 mL (Pierce, Rockford, USA).
11. Acilit pH-indicator-strips (VWR, Darmstadt).
12. Thermosensitive shaker.
13. Falcon Conical Centrifuge Tubes (15 mL) (BD Biosciences Europe).

2.3 *Fluorescent Labeling of N-Glycans with 2-Aminobenzamide*

1. Ludger Tag 2-AB Glycan labeling kit (Ludger Ltd, Abingdon, UK).
2. Screw capped microvials (500 µL or less) with conical bottom.
3. Whatman 3 MM Paper (3 × 10 cm).
4. Heating oven.
5. Chromatography chamber (5 × 7 × 14.5 cm).
6. Solvent for paper chromatography: *n*-Butanol/ethanol/distilled water (4/1/1, each by vol.).
7. Clothes pegs (Clothespins) and cord.
8. Fume cupboard.
9. Millex-HV Filter Unit, 0,45 µm (Millipore GmbH, Schwalbach, Germany).
10. Luer-Lock Syringes (B-D hypodermic syringe, 3 mL, Sigma-Aldrich Corporation).

3   Methods

3.1 *Trypsin Digestion of Glycoproteins*

1. Dissolve 400 µg of glycoprotein to be analyzed, in 400 µL of 50 mM volatile deglycosylation buffer.
2. Dissolve 100 µg of trypsin in 100 µL of 50 mM volatile deglycosylation buffer immediately before use.
3. Add 8 µL (8 µg) of trypsin in volatile deglycosylation buffer (ratio of trypsin/protein 1/50 (w/w) to the dissolved glycoprotein (step 2.1.1).
4. Incubate the solution for 4 h at 37°C.
5. Prepare a fresh solution of trypsin again and add again 16 µL (16 µg) of trypsin and incubate the solution overnight.
6. Inactivate the enzyme by incubation at 95°C in a water bath and cool the digest for 30 min on ice.

   7.1 If sequencing of the peptides should be performed, trypsin sequencing grade from porcine (1 µg/100 µg glycoprotein) (Promega, Madison, USA) should be used. Evaporate the obtained glycopeptide mixture to dryness in a vacuum centrifuge, dissolve it in 500 µL of water and apply it to reversed phase HPLC (5).

   7.2 If deglycosylation of the complete glycoprotein should be performed only, apply the tryptic digest directly to enzymatic deglycosylation.

3.2 *Enzymatic Deglycosylation with PNGase F*

1. Redissolve 250 U of commercially available *N*-Glycosidase F (*see* **Note 1**) in 250 µL of distilled water and apply it to dialysis by threefold ultrafiltration (*see* **Note 2**) in Nanosep centrifugal devices (3 K) by centrifugation at 5,000*g* for

about 30 min at 4°C. For a complete exchange of buffer dilute the concentrated enzyme in the upper layer (final volume is about 50 µL) each time with volatile deglycosylation buffer to volume of 500 µL. After 3 circles of centrifugation concentrated enzyme is diluted with volatile deglycosylation buffer to an enzyme concentration of 250 U in 250 µL of water.

2. Let the trypsinized glycoprotein come to room temperature (step 3.1.6).

3. Take an aliquot corresponding to about 0.1–0.5 nmol of glycans for monosaccharide analysis (Analysis 1) (11).

4. Add 10 µL (10 U) of PNGase F to the proteolyzed glycoprotein solution (about 400 µL) and incubate for 4 h at 37°C in a thermosensitive shaker.

5. Add another 10 µL of PNGase F and continue incubating overnight at a final enzyme concentration of 50 U/mL.

6. Remove glycerol and sodium azide from the Nanosep centrifugal devices by twice addition of 500 µL distilled water to the top and centrifugation at room temperature for 15 min at 5,000g. Discard the flow through.

7. Transfer the digest from step 3.2.5 onto the top of the purified Nanosep centrifugal devices and apply it to ultrafiltration by centrifugation at 5,000g at 4°C for 30 min. Collect the filtrate, containing the *N*-glycans, and separate the remaining amount of hydrolyzed *N*-glycans, peptides and salt in the top of the concentrator tubes from protein (trypsin and PNGase F) by twice addition of 500 µL distilled water to the top of the tubes and centrifugation as described above.

8. Collect the combined filtrates containing the *N*-glycans in 15 mL Falcon Centrifuge Tubes and evaporate it to dryness in a vacuum centrifuge tempered to 30°C at 0.05 mbar.

9. Remove main part of volatile buffer by twice adding of 1 mL 50% (v/v) aqueous ethanol and evaporation in a vacuum centrifuge.

10. Remove residual volatile buffer by twice adding of 1 mL methanol and evaporation to dryness.

11. Redissolve the dried sample in 0.5 mL distilled water, adjust the pH to 3.0 with 1N acetic acid and improve the pH by dipping an aliquot of 0.5–1 µL to Acilit pH-indicator strips (*see* **Note 3**).

12. Prepare cation exchange/affinity-columns by filling 0,5 mL Affi-Blue Gel into a polystyrene column. Wash the column with 3 mL distilled water and allow the gel to settle down completely. Add 0.5 mL of Dowex 50 WX-H+ to the top of the columns and wash the resulting cation exchange/affinity column with the 5-fold column volume (5 mL) of distilled water.

13. Apply the acidified digest from step 3.2.11 to the cation exchange/affinity column (*see* **Note 12**) and wash the column with 5 mL (5-fold bed volume) of distilled water.

14. Collect the column eluate to a 15 mL centrifugation tube and evaporate it to dryness using a vacuum centrifuge at 30°C (*see* **Note 3**).

15. Redissolve the dried sample in 500 µL of distilled water.

16. Take an aliquot corresponding to about 0.1 to 0.5 nmol of glycan for monosaccharide analysis (Analysis 2) (*see* **Note 3**).

17. Transfer residual sample to a 500 µL conical screw capped microvial and evaporate the sample in a vacuum centrifuge to dryness. Remove acid and residual volatile buffer by two times redissolving the samples in 200 µL of 50% (v/v) aqueous ethanol and then in 200 µL of methanol followed by evaporation in vacuo as described above.
18. Apply the desalted and completely dried sample to fluorescence labeling.

3.3   Fluorescent Labeling of N-Glycans with 2-Aminobenzamide

1. Concentrate the released N-glycans (5–10 nmol) carefully from 10 µL of water to the bottom of a screw capped microvial by evaporation at 30°C in a vacuum centrifuge (*see* **Note 6**).
2. Let the commercially available Ludger Tag™ 2-AB Glycan labeling kit, which should be held at 4°C, come to room temperature to thaw dimethylsulfoxide and acetic acid. Prepare the 2-AB labeling reagent as recommended by the manufacturer. The labeling kit contains two batches of kit reagents, which are sufficient to label up to 10 samples of N-glycans each. Each sample can contain up to 50 nmols glycans.
3. Label the sample with 10 µL of the reducing amination reagent, containing 2-AB (250 µg) and sodium cyanoborohydride (500 µg), dissolved in dimethyl-sulfoxide/acetic acid (7/4, by vol.) for 2 h at 65°C in an heating oven in the dark.
4. Allow the reaction mixture to come to room temperature, and spot the samples in two separate portions of 5 µL onto a paper strip each (3 × 8 cm).
5. Dry the paper strips upright in a fume cupboard overnight (*see* **Note 7**).
6. Separate the labeled glycans from unreacted 2-AB by paper chromatography for about 1 h using n-butanol/ethanol/water (4/1/1, by vol.) as chromatography solvent.
7. Dry the papers for 2 h in a fume cupboard.
8. Improve the positions of N-glycans near to the origin and of unreacted dye, which moves to about 6–8 cm from the origin under an UV-lamp operating at 254 nm or 366 nm (*see* **Note 8**).
9. Label the positions of the N-glycans using a pencil and cut out the paper where the N-glycans are located.
10. Put the cut out and folded papers to the top of prewashed Luer-locked syringes equipped with 0.45 mm HV-filters. Put the syringes in 15-mL Falcon Centrifuge Tubes and dissolve the N-glycans by incubating the papers with 1 mL of distilled water for 5 min, so that they are completely wetted. Centrifuge the tubes together with the syringes for 5 min at 750g. Complete the elution by twice incubation with 1 mL water followed by centrifugation.
11. Combine the filtrates of each sample, containing the labeled oligosaccharides and evaporate to dryness in a vacuum centrifuge.
12. Redissolve samples in 1 mL of water and keep at −20°C in the dark until further use (*see* **Note 9**).

4 Notes

1. PNGase F is commercially available as solution in buffer, containing 50% glycerol, pH 7.2 or as a lyophilisate without glycerol, which when reconstituted as recommended by the manufacturer, results in a concentration of 100 mM sodium phosphate buffer, 25 mM EDTA, pH 7.2. Note that, 1 Unit is defined as the enzyme activity which hydrolyzes 1 nmol of dabsyl fibrin glycopeptide or 0.2 nmol dansyl fetuin glycopeptide within 1 min at 37°C at pH 7.8, if the enzyme used is obtained from Roche. Keep in mind that the recombinant enzyme, expressed in *E. coli*, is thought to have an optimal enzyme activity at pH 7.8. In contrast, the pH optimum of the enzyme purified from the culture medium of *Flavobacterium meningosepticum* is reported to be 8.6, also the enzyme is as least 80% active over the range 7.5–9.5.

2. PNGase F preparations should not contain glycerol, which may be present as stabilizing reagent in some commercially obtainable preparations, since this can interfere with subsequent fluorescence labeling reaction efficiencies. The enzyme is therefore dialyzed to a volatile buffer, which may be removed in vacuo without the necessity to desalt the solution by gel filtration in addition.

3. At pH 3.0 most peptides are cationic and are removed by passing through a cation-exchange resin (capacity: 2.1 meq/mL resin bed). Uncharged peptides may be removed by passing through Affi-Gel Blue (capacity: 11.7 mg albumin per milliliter gel).

4. By passing of the *N*-glycan mixture through cation-exchange/affinity-columns all salts are converted to their corresponding acids. The presence and enrichment of phosphoric acid during the following evaporation has to be avoided by prior dialysis of the commercially obtained enzyme solution to volatile deglycosylation buffer (*see* **Section 3.2.1**). It is highly important not to increase the temperature above 30°C, because acetic acid and traces of trifluoroacetic acid which are in the flow through may hydrolyze fucose and sialic acids residues from the *N*-glycan at higher temperatures.

5. From the comparison of the monosaccharide analyses after PNGase F digestion, before and after the passage through a cation exchange/affinity-column, it can be concluded to a successful deglycosylation reaction as well as to the presence of *O*-glycans which do not pass the column.

6. To avoid an incomplete derivatization of glycans, a careful concentration of the sample to the bottom of the microvial, as well as a twice volume of 10 μL labeling reagent is essentially necessary. By MALDI-TOF mass spectrometry an incomplete derivatization may be recognized by the occurrence of additional peaks showing a lowered mass of m/e minus 120 Da.

7. Careful drying of the papers overnight is essentially necessary to remove DMSO and acetic acid from the paper strips, which would cause an inefficient separation of excess label from *N*-glycans during paper chromatography.

8. Note, that low molecular oligosaccharides from mono- to penta-saccharides may also migrate from the origin near to the position of unreacted dye (6–8 cm from origin).

9. Working under illumination by fluorescent light may cause a decrease of fluorescence in 2-AB labeled glycans even at 4°C. Therefore, if possible, all steps during the purification procedure should be performed at daylight and the labeled oligosaccharides are stored at −20°C in the dark.

References

1. Patel, T., Bruce, J., Merry, A., Bigge, C., Wormald, M., Jaques, A., and Parekh, R. (1993) Use of Hydrazine to release an intact and unreduced form both from N- and O-linked oligosaccharides from glycoproteins. *Biochemistry* **32**, 679–693.
2. Patel, T. P. and Parekh, R. B. (1994) Release of oligosaccharides from glycoproteins by hydrazinolysis. *Methods Enzymol.* **230**, 57–66.
3. Takasaki, S., Mizuochi, T., and Kobata, A., (1982) Hydrazinolysis of asparagine-linked sugar chains to produce free oligosaccharides. *Methods Enzymol.* **83**, 263–268.
4. Nuck, R., Zimmermann, M., Sauvageot, D., Josic, D., and Reutter, W., (1990) Optimized deglycosylation of glycoproteins by peptide-N4-(N-acetyl-beta-glucosaminyl)-asparagine amidase from *Flavobacterium meningosepticum. Glycoconj.* J., **7**, 279–286.
5. Gohlke, M., Baude, G., Nuck, R., Grunow, D., Kannicht, C., Bringmann, P., Donner, P., and Reutter, W., (1996) O-linked L-fucose is present in Desmodus rotundus salivary plasminogen activator. *J. Biol. Chem.* **271**, 7381–7386.
6. Nuck, R. and Gohlke, M. (**1997**) Characterization of subnanomol amounts of N-glycans by 2-amnobenzamide labeling, matrix-assisted laser desorption ionization time-of-flight mass spectrometry and computer assisted sequence analysis, in *Techniques in Glycobiology,* Townsend, R., and Hotchkiss, A. eds., Marcel Dekker, pp. 491–508.
7. Fan, J. Q. and Lee, Y. C. (**1997**) Detailed studies on substrate structure requirements of glycoamidases A and F. *J. Biol. Chem.* **272**, 27058–27064.
8. Tarentino. A. L. and Plummer Jr, T. H. (1994) Enzymic deglycosylation of asparagine-linked glycans: purification, properties, and specificity of oligosaccharide-cleaving enzymes from *Flavobacterium meningosepticum. Methods Enzymol.* **230**, 44–57.
9. Townsend, R. R., Lipniunas, P. H., Bigge, C., Ventom, A., and Parekh R. (1996) Multimode high-performance liquid chromatography of fluorescently labeled oligosaccharides from glycoproteins. *Anal. Biochem.* **239**, 200–207.
10. Bigge, J. C., Patel, T. P., Bruce, J. A., Goulding, P. N., Charles, S. M. and Parekh, R. B. (1995) Nonselective and efficient fluorescent labeling of glycans using 2-amino benzamide and anthranilic acid. *Anal. Biochem.* **230**, 229–238.
11. Hardy, M. H. (1989) Monosaccharide analysis of glycoconjugates by high-performance anion-exchange chromatography with pulsed amperometric detection. *Methods Enzymol.* **179**, 76–82.

17
Separation of N-Glycans by HPLC

Martin Gohlke and Véronique Blanchard

Summary Most glycoproteins carry a very heterogeneous mixture of oligosaccharides and even a single glycosylation site of a pure glycoprotein is often heterogeneously glycosylated. The structural diversity of oligosaccharides arises from linkage variants, from differences in the size and number of charges of glycans, and from differences in the monosaccharide composition of glycans. Fortunately, the biosynthetic pathway is subject to certain restrictions, so that structural diversity is limited and amenable to laboratory investigation. Different approaches have been developed to the structural characterization of oligosaccharides, including nuclear magnetic resonance (NMR), mass-spectrometry, linkage analysis by gas chromatography-mass spectrometry (GC-MS), sequence analysis using specific exoglycosidases and others, but a crucial part of these strategies is the separation of the glycan mixture into homogeneous glycan fractions. In this chapter some high-performance liquid chromatography (HPLC)-techniques are described for the isolation of oligosaccharides, in particular N-linked glycans.

Key Words N-Glycan; anion-exchange-HPLC; normal phase-HPLC and RP-HPLC; fluorescence labeling.

1 Introduction

Most glycoproteins carry a very heterogeneous mixture of oligosaccharides and even a single glycosylation site of a pure glycoprotein is often heterogeneously glycosylated. The structural diversity of oligosaccharides arises from linkage variants, from differences in the size and number of charges of glycans, and from differences in the monosaccharide composition of glycans. Fortunately, the biosynthetic pathway is subject to certain restrictions, so that structural diversity is limited and amenable to laboratory investigation. Different approaches have been developed to achieve the structural characterization of oligosaccharides, including nuclear magnetic resonance, mass-spectrometry, and linkage analysis by gas chromatography-mass

From: *Post-translational Modifications of Proteins.*
Methods in Molecular Biology, Vol. 446.
Edited by: C. Kannicht © Humana Press, Totowa, NJ

spectrometry, sequence analysis using specific exoglycosidases and others, but a crucial part of these strategies is the separation of the glycan mixture into homogeneous glycan fractions. In this chapter some high-performance liquid chromatography (HPLC)-techniques are described for the isolation of oligosaccharides, in particular N-linked glycans.

The complex heterogeneity of N-glycans demands refined systems for the separation of oligosaccharide mixtures, both for 'mapping' analysis and for further structural characterization (1,2). Normal phase columns with primary, secondary or quaternary amines, as well as reversed phase (RP) columns have been used for the separation of glycoprotein-derived oligosaccharides (3–5). In a highly organic phase, neutral glycans interact with the stationary phase via hydrogen bonding and are eluted by increasing the polarity of the eluents (hydrophilic interaction chromatography). Normal phase HPLC gives a high resolution and a detailed profile of fluorescence-labeled N-glycans from glycoproteins (6). In addition to hydrophilic interaction chromatography, the basic properties of amine-bonded resins can be used for anion-exchange chromatography for the separation of negatively charged oligosaccharides. High-pH anion-exchange chromatography with pulsed amperometric detection (HPAEC-PAD) has proved a useful tool for the separation of sialylated glycans, as well as uncharged oligosaccharides. Since the introduction of HPAEC-PAD in 1988 by Hardy and Townsend (7) for the separation of N-linked oligosaccharides, this method has become standard for the separation of N-glycans; it produces a high resolution and allows the sensitive detection of underivatized oligosaccharides. Depending on the chromatographic conditions applied, negatively charged glycans or neutral oligosaccharides can be separated to provide a 'fingerprint' map for the glycans of a given protein. Sugars are separated using strong basic eluents (pH 13). Under these conditions the oligosaccharides are present as oxyanions and can bind to the amino groups of the stationary phase of the column. The selectivity of the chromatographic behavior may be determined by the particular hydroxyl groups of the oligosaccharides that become deprotonated and undergo interaction with the quaternary ammonium groups of the column resin. The separation depends on charge, molecular size, sugar composition and linkage of the monosaccharides, but prediction of the elution order is largely empirical. At this high base concentration an epimerization of the reducing GlcNAc to ManNAc has been observed. Thus a single oligosaccharide species may elute as two peaks differing only in the monosaccharide at the reducing end (8,9,10). Further characterization of the oligosaccharides requires neutralization and removal of eluent salts. Automated systems for on-line desalting have been introduced (11). While HPAEC is usually coupled with PAD, the combination of HPAEC and fluorescence detection of labeled glycans has also been reported (12).

Despite rapid progress in the development of separation techniques in the last decade and availability of high resolution methods, a complete separation of all structures (even after release of the sialic acid residues) present in a mixture is rarely achieved in a single chromatographic step. Thus rechromatography using a complementary HPLC method is often needed. It is important to note that for both one- and multidimensional HPLC-based methods, the use of retention times is not sufficient for the structural identification of sugars.

In addition to the problems presented by the structural diversity of N-glycans, sensitive on-line detection is difficult because oligosaccharides do not contain a chromophore. The requirement for sensitive detection of carbohydrates has led to the development of 2 approaches, based on electrochemical detection and fluorimetric detection. Underivatized glycans can be monitored by pulsed amperometric detection (PAD) at low picomole concentrations, but this method is nonspecific for carbohydrates *(13)*. Pulsed amperometric detection is usually combined with HPAEC to provide a sensitive and selective detection system for carbohydrates and other oxidizable species. In principle, a repeating waveform potential is applied in a flow-through cell. The standard waveform is a triple potential waveform, which has worked well in our hands (the triple pulse voltage sequence is given in **Section 3.3.1**). The potential E1 is used for oxidation of the oligosaccharide; this generates the signal that is detected for the analyte. The subsequent two steps at potentials of more positive or negative potential are necessary for cleaning the gold electrode. Detection is based on the measurement of a current, which is proportional to the oxidation rate of the analyte, which in turn depends on various factors. Changes in the cell, like a decrease in the reaction area of the gold electrode, therefore impair the response. The requirement for highly sensitive detection in carbohydrate analysis has promoted increased interest in the fluorescence labeling of glycans. Generally derivatization of the oligosaccharides with a fluorescent label is nonselective and allows the detection of glycans in subpicomolar concentrations in their correct molar proportions. Different labels such as 2-aminopyridine (2-AP) *(14)*, 2-aminobenzoic acid (2-AA) *(15)*, 2-aminobenzamide (2-AB) *(16)* and others have been introduced for specific detection at very high sensitivity. The chromatographic behavior of 2-AP-*(10,18)*, 2-AA-*(19)* and 2-AB-tagged oligosaccharides *(6,20)* has been extensively investigated.

In this chapter techniques for separating sialylated and/or neutral oligosaccharides using anion-exchange-HPLC, normal phase-HPLC and RP-HPLC are introduced.

2   Materials

All salts used for the preparation of the eluents should be of analytical quality.

1. Double-distilled water was used for HPLC. Double-distilled water and aqueous buffers should be filtered through a 0.45 μm membrane (Millipore, Durapore Membrane Filter, 0.45 μm HV, Bedford, USA). NaOH (50%) is available from J. T. Baker (Deventer, Netherlands).
2. A common HPLC-system capable of delivering accurate gradients at flow rates of 0.5–1.5 mL/min is needed. A low dead volume of the HPLC-system and a decreased injection volume increase the resolution. The loop size depends on the amount of the sample and the separation mode. For analytical purposes a 10- to

20-µL loop and for preparative separation a 50- 100-µL loop is adequate. For the detection of 2-AB-labeled glycans a fluorescence detector is required (excitation wavelength: 330 nm, excitation wavelength: 420 nm).
3. For HPAEC, a DIONEX system, e.g., ICS-3000 is preferred, equipped with a pulsed electrochemical detector (e.g., DIONEX ED), and a Helium degassing system. If a different HPLC-system is used it must be capable of resisting the very basic conditions (pH 13) and must consist of titan or peek parts.
4. Samples should be stored at –20°C. Before injection, samples should be sonicated for 3 min to ensure a homogeneous solution.

3 Methods

In "charge profiling," negatively charged N-linked oligosaccharides are separated by various HPLC-methods to provide a characteristic elution pattern. Depending on the columns used and the chromatographic conditions applied, sialylated glycans can be separated at low resolution or at a high resolution.

3.1 Charge Profiling of N-Glycans (Low Resolution)

Oligosaccharides can be separated strictly according to their number of negative charges. Under appropriate chromatographic conditions, other structural features of the oligosaccharides do not significantly influence the elution profile (*see* **Note 1**). Charge profiling at a low resolution is often used as a preparative method to fractionate oligosaccharides according to their charge state, before subjecting the glycans to further structural characterization. Two methods using a strong anion exchanger (Mono Q-column) or a weak anion exchanger (GlycoSepC-column), are described. Both columns can be used for the separation of 2-AB-labeled glycans with fluorimetric detection (*see* **Notes 2–4**). Mono Q was also used for the separation of underivatized glycans detected by PAD (21). However the post-column addition of 0.5*M* NaOH is required for high sensitivity.

3.1.1 Charge Profiling on a Mono Q-Column (Fig. 17.1)

Column: TRICORN Mono Q 5/50GL (GE Healthcare)
Eluent 1: H_2O
Eluent 2:   0.6*M* NH_4OAc, pH 7
Flow: 1 mL/min

1. Wash the column by either running the complete gradient or using a shortened gradient (in 15 min from 0 to 100% eluent 2, wash the column 10 min with 100% eluent 2, equilibrate the column 20 min in eluent 1) (*see* **Notes 1 and 5**).

Fig. 17.1 "Charge profiling" on a Mono Q-column (low resolution). 2-AB labeled N-linked oligosaccharides from α1 acid glycoprotein were separated according to the number of negative charges using a Mono Q-column. The elution positions of neutral, mono-, bi-, tri- and tetracharged glycans are marked A0, A1, A2, A3, and A4, respectively

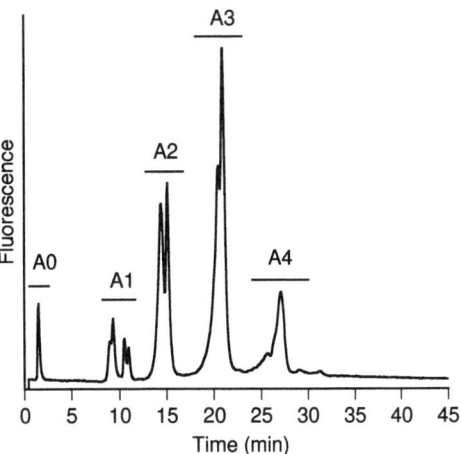

2. Inject the sample in water while the sample loop is filled with eluent 1 and elute the analyte with the following gradient. For detection of 2-AB-labeled glycans use fluorimetric detection (excitation 330 nm, excitation: 420 nm).

Gradient for Mono Q-HPLC

Time	%1	%2	Event
0	100	0	Injection
5	100	0	
45	75	25	
47	0	100	
57	0	100	
59	100	0	
80	100	0	Restart

3. Store the column in 80% H_2O and 20% ethanol.

3.1.2 Charge Profiling on a GlycoSepC-Column (Fig. 17.2)

The GlycoSepC-column can be used for multiple applications in glycan analysis including the separation of charged glycans, neutral glycans and a mixture of both, depending on the eluents and gradients applied (22). This column has been described for the separation of 2-AB oligosaccharides, but might be generally applicable for fluorescent-tagged oligosaccharides.

Column: GlycoSepC (4.6 × 100 mm, PPC134, ProZyme, San Leandro, CA USA, *see* **Note 6**)
Eluent 1: 20% acetonitrile, 80% H_2O
Eluent 2: 20% acetonitrile, 30% H_2O, 50% 0.5M NH_4OAc, pH 4.5
Flow: 0.5 mL/min

1. Wash the column by either running the complete gradient or using a shortened gradient (in 15 min from 0 to 100% eluent 2, wash the column 10 min with 100% eluent 2, equilibrate the column 20 min in eluent 1).

Fig. 17.2 Charge profiling on a GlycoSepC-column (low resolution). 2-AB labeled N-linked oligosaccharides from α1 acid glycoprotein were separated according to the number of negative charges using a GlycoSepC-column. The elution positions of neutral, mono-, bi-, tri- and tetracharged glycans are marked A0, A1, A2, A3, and A4, respectively

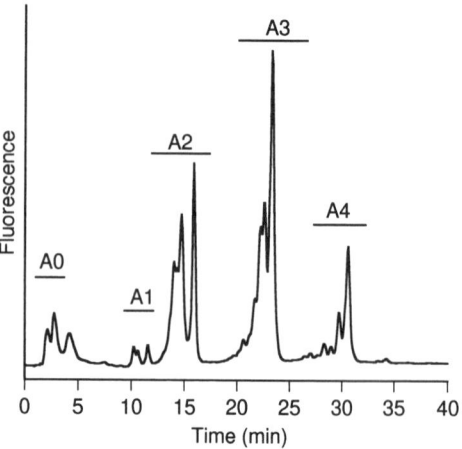

2. Inject the sample in a small volume of water while the sample loop is filled with eluent 1 or inject the sample in 20% acetonitrile, 80% H_2O. Elute the analyte with the gradient. For detection of 2-AB-labeled glycans use fluorimetric detection (excitation: 330 nm, excitation: 420 nm).

Gradient for GlycoSepC-HPLC

Time	%1	%2	Event
0	100	0	Injection
2	100	0	
45	25	75	
50	0	100	
60	0	100	
61	100	0	
81	100	0	Restart

3. Store the column in 80% H_2O and 20% acetonitrile.

3.2 Charge Profiling of N-Glycans (High Resolution) (Fig. 17.3)

A separation technology providing a high resolution is required to give a detailed profile of the overall glycosylation of a protein, including neutral and sialylated glycans. HPAEC-PAD has proved to be a high resolution system for mapping sialylated N-glycans (23).

Column: Carbopak PA-100 and guard PA -100
Eluent 1:   0.1M NaOH (5.75 mL 50% NaOH / 1 L)
Eluent 2:   0.1M NaOH, 0.6M NaOAc (5.75 mL 50% NaOH / 1 L + 49.2 g/1 L NaOAc anhydrous, crystalline)

1. Filter 1 L double-distilled water for each eluent through a 0.45 μm membrane (Millipore, Durapore Membrane Filter, 0.45 μm HV) and sparge with helium for 15 min (*see* **Note 7**).

Fig. 17.3 Charge profiling of N-glycans from α1 acid glycoprotein on HPAEC-PAD (high resolution). Underivatized N-linked oligosaccharides from α1 acid glycoprotein were separated using the gradient for sialylated N-glycans on HPAEC-PAD. The elution pattern of the total moiety is shown in the upper panel. The traces AO, A1, A2, A3, and A4 show the rechromatography of corresponding glycan fraction obtained by Mono Q-HPLC and defines the elution times for neutral, mono-, bi-, tri-, and tetrasialylated N-glycans. Because α1 acid glycoprotein carries only traces of neutral and monosialylated glycans, no signal was detected for these fractions. Neutral N-glycans elute at about 15–25 min and monosialylated at about 25–35 min

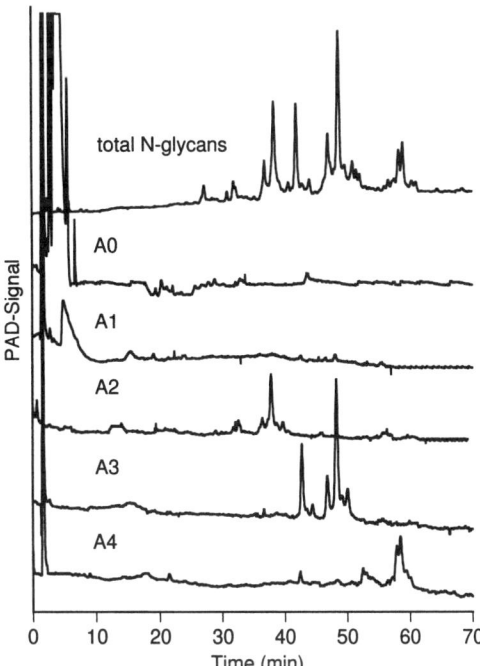

2. For preparation of eluent 1 add 5.75 mL of 50% NaOH (Baker, Deventer, Netherlands) to the 1 L double-distilled water and mixing by gently shaking the flask (*see* **Notes 8** and **9**).

3. Dissolve 49.2 g NaOAc (anhydrous, crystalline) in 1 L double-distilled water and gently shake. After 5 min of sparging with helium, add 5.75 mL of 50% NaOH and mix gently (eluent 2). Continue sparging for 5 min for both eluents.

4. Wash the column by either running the complete gradient or using a shortened gradient (in 15 min from 0 to 100% eluent 2, wash the column 10 min with 100% eluent 2, equilibrate the column 25 min in eluent 1, *see* **Note 10**).

5. Inject the sample in water and elute the analyte with the gradient (*see* **Notes 11–13**).

Gradient for HPAEC-PAD of sialo-N-glycans

Time	%1	%2	Event
0	100	0	Injection
1	100	0	
2	100	0	
82	60	40	
85	0	100	
93	0	100	
94	100	0	
130	100	0	Restart

The detector settings for PAD-detection are (*see* **14**):

Integration   0.20–0.40 s
E1 = 0.05 Volt   0.00–0.40 s
E2 = 0.75 Volt   0.41–0.60 s
E3 = −0.15 Volt   0.61–1.00 s

3.3   Separation of Neutral N-glycans

In this section two different methods are described: (i) for the separation of underivatized sugars using HPAEC-PAD, and (ii) for the separation of glycans fluorescence-labeled with 2-aminobenzamide, using two dimensional NH$_2$-HPLC and RP-HPLC.

3.3.1   HPAEC-PAD (Fig. 17.4)

HPAEC-PAD is a high resolution, sensitive method for the separation of native oligosaccharides. The relationship between N-glycan structures and their retention times is summarized in Table 17.I. In general larger structures of the same charge elute later but separation is greatly influenced by structural features like branching, type of linkage and monosaccharide composition. The gradient is useful for the separation of neutral oligosaccharides but not for sialylated N-glycans *(21)*. The main difference between the gradients for separation of neutral or sialylated sugar chains is the higher salt concentration needed for the elution of negatively charged structures. A gradient build-up from a low salt concentration is favored for the separation of neutral oligosac-

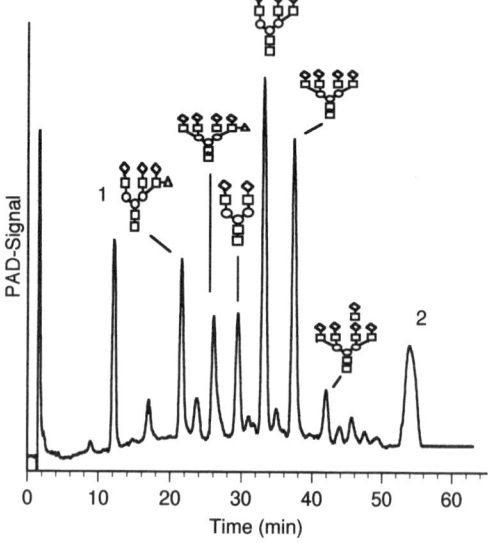

Fig. 17.4 Separation of desialylated N-glycans from α1 acid glycoprotein using HPAEC-PAD. Underivatized N-linked oligosaccharides from α1 acid glycoprotein were separated using the gradient for neutral N-glycans on HPAEC-PAD. Peak 1 and 2 are internal standards used for calibration of the chromatographic system. The potential structures of the oligosaccharides are given in symbols. For structural details see *(23)*

Table 17.1 Relationship between N-linked glycan structure and retention times on HPAEC-PAD and NH$_2$-HPLC. The influence of the carbohydrate structure on the retention time given in this table was empirically determined (see *ref. 25* for HPAEC-PAD). Changes in retention times depend on the gradients used. The time values in brackets correspond to differences in retention times estimated in our laboratory applying the methods described in this chapter

Structural characteristic	Change in retention time on NH$_2$-HPLC	Change in retention time on HPAEC-PAD
Increasing antennarity, bi-, tri- and tetraantennary glycans	Increasing retention times (about 5 min per antenna)	Increasing retention times (about 4 min per antenna)
Fucose α1-6 linked to the core	Increasing retention times (about 2 min)	Decreasing retention times (about 3 min)
Fucose linked to the antennae (Fucα1-3GlcNAc)	Increasing retention times (about 2 min)	Decreasing retention times (about 12 min)
Galß1-4GlcNAc (type II chain)/ Galß1-3GlcNAc (type I chain)	Slighly different	Galß1-3GlcNAc elutes later (about 6 min)
Addition of an N-acetyllactosamine-repeat	Increasing retention times (about 4 min)	Increasing retention times (about 4 min)
2,4-branched triantennary/ 2,6-branched triantennary glycans	2,4-branched triantennary elutes earlier (about 2 min)	2,4-branched triantennary elutes earlier (about 4 min)
Reduction or labeling with 2-AB	Based on 2-AB-labeling	Decreasing retention time
Addition of a bisecting GlcNAc	Not measured	Increasing retention time
Glycans with a complete core/ glycan with only one GlcNAc in the core	Decreasing retention time	Increasing retention time (about 2 min)

charides because even subtle differences in the glycan structure influence the retention time. However, as mentioned in the introduction, a complete separation of all structures present in a mixture is rarely achieved in a single chromatographic step.

1. Wash the column by either running the complete gradient or using a shortened gradient (in 15 min from 0 to 100% eluent 2, wash the column 10 min with 100% eluent 2, elute 5 min with eluent 4 and equilibrate the column 20 min in eluent 1).
2. Inject the sample in water and elute the analyte with the gradient.

Column: Carbopak PA-100 and guard PA -100
Eluent 1: 0.1*M* NaOH (5.75 mL 50% NaOH / 1 l)
Eluent 2: 0.1*M* NaOH, 0.6 M NaOAc (5.75 mL 50% NaOH / 1 l + 49.2 g / 1 l
        NaOAc anhydrous, crystalline)
Eluent 3: 0.1*M* NaOH, 120 m*M* NaOAc (5.75 mL 50% NaOH / 1 l + 9.84 g / 1 l
        NaOAc anhydrous, crystalline)
Eluent 4: 0.2*M* NaOH (11.5 mL 50% NaOH / 1 l)

Gradient for HPAEC-PAD of neutral N-glycans

Time	%1	%2	%3	%4	Event
0	100	0	0	0	Injection
1	100	0	0	0	
2	100	0	0	0	
72	50	0	50	0	
82	0	0	100	0	
87	0	100	0	0	
97	0	100	0	0	
98	0	0	0	100	
103	0	0	0	100	
104	100	0	0	0	
125	100	0	0	0	Restart

The detector settings for PAD-detection are:

Integration   0.20–0.40 s
E1 = 0.05 Volt   0.00–0.40 s
E2 = 0.75 Volt   0.41–0.60 s
E3 = −0.15 Volt 0.61–1.00 s

3.3.2   Separation of Neutral N-glycans Using two-Dimensional-HPLC

Two and three-dimensional techniques have been developed either for sugar mapping *(17,18)* e. g., in combination with exoglycosidase digestion, or for fractionation to obtain homogeneous glycan fractions for further analysis *(20,24)*. Multidimensional HPLC-techniques have been preferentially applied for the separation of fluorescently-labeled oligosaccharides.

These techniques often use amine-bonded columns, because these columns can perform as both hydrophilic interaction media and as an anion-exchange phase. RP-HPLC can be applied as a complementary system with a different chromatographic characteristic. Different combinations of amine-bonded columns, working either in the hydrophilic mode or used as an anion-exchanger, and RP-HPLC have been reported for the separation of oligosaccharides. In this section, a complimentary technique for 2-AB-labeled glycans using NH$_2$-bonded HPLC in the first chromatographic step and RP-18-HPLC in the second chromatographic dimension is described (*see* **Note 15**).

3.3.2.1   Separation of Neutral N-glycans Using NH$_2$-HPLC (Fig. 17.5)

1. Wash the column by either running the complete gradient or using a shortened gradient (in 15 min from 0 to 100% eluent 2, wash the column 10 min with 100% eluent 2, equilibrate the column 20 min in eluent 1).
2. Inject the sample in a small volume of water while the sample loop is filled with acetonitrile (*see* **Notes 16** and **17**). Elute the analyte with the following gradient. For detection of 2-AB-labeled glycans use fluorimetric detection (excitation: 330 nm, excitation: 420 nm).

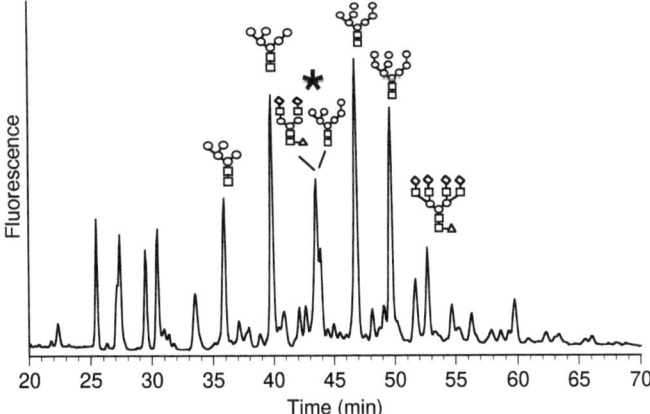

Fig. 17.5 Separation of desialylated 2-AB-labeled N-glycans. N-linked glycans released from K562-cells were desialylated and fluorescently labeled with 2-AB. Neutral oligosaccharides were separated mainly according to their size on NH_2-HPLC. Fractions were subjected to mass determination and the potential structures of the main fractions are given in symbols. The glycan fraction marked with an asterics was rechromatographed using RP-18-HPLC (Fig. 17.6)

Column: Luna 3μ NH_2 100A (4.6 × 150 mm, 3 μm, Phenomenex,
                 Aschaffenburg, Deutschland) (*see* **Notes 18** and **19**)
Eluent 1:   acetonitrile
Eluent 2:   15 mM Ammonium acetate pH 5.2
Flow: 1.5 mL/min

Time	%1	%2	Event
0	100	0	Injection
10	80	20	
80	50	50	
82	30	70	
92	30	70	
95	100	0	
115	100	0	Restart

3. Store the column in isopropanol.

3.3.2.2   Separation of Neutral N-glycans Using RP-18 HPLC (Fig. 17.6)

1. Wash the column by either running the complete gradient or using a shortened gradient (in 15 min from 0 to 100% eluent 2, wash the column 10 min with 100% eluent 2, equilibrate the column 20 min in eluent 1).
2. Inject the sample in water while the sample loop is filled with eluent 1 (*see* **Notes 20** and **21**). Elute the analyte with the gradient. For detection of 2-AB-labeled glycans use fluorimetric detection (excitation: 330 nm, excitation: 420 nm) (*see* **Note 22**).

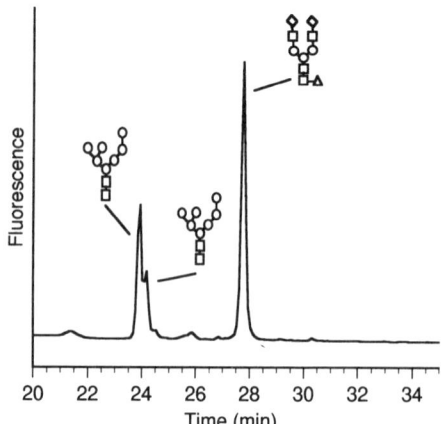

Fig. 17.6 Rechromatography of N-glycans using RP18-HPLC. The resulted glycan fraction from NH$_2$-HPLC (Fig. 17.5, fraction marked with an asterics) was rechromatographed on RP-18-HPLC. Mass spectrometry revealed the presence of a biantennatry and two isomers of high mannose-sugars (Man7)

Column: ODS-Hypersil column (4 × 250 mm, 3 μm Knauer (Berlin, Germany))
Eluent 1:   H$_2$O
Eluent 2:   acetonitrile
Flow: 1 mL/min

Time	%1	%2	Event
0	100	0	Injection
6	100	0	
46	80	20	
47	0	100	
57	0	100	
58	100	0	
80	100	0	Restart

3. Store the column in acetonitrile.

3.4   Simultaneous Separation of Neutral and Charged N-Glycans (Fig. 17.7)

A high resolution and extremely sensitive method for the separation of 2-AB-labeled N-glycans, that combines features of hydrophilic and anion-exchange chromatography using an NH$_2$-bonded column was originally published by Anumula et al. (19, see also this volume). This accounts for a one step characterization of samples carrying reasonable amounts of neutral besides of multiply sialylated glycans, as e.g., immunoglobulins or cell membrane extracts. Furthermore, this column has the potential to separate N-glycans differing in the type and linkage of terminal sialic acid.

The separation is adaptable to all types of fluorescent labeled glycans with minor modifications. Thus it can be integrated in analytical strategies based on multidimensional HPLC separation.

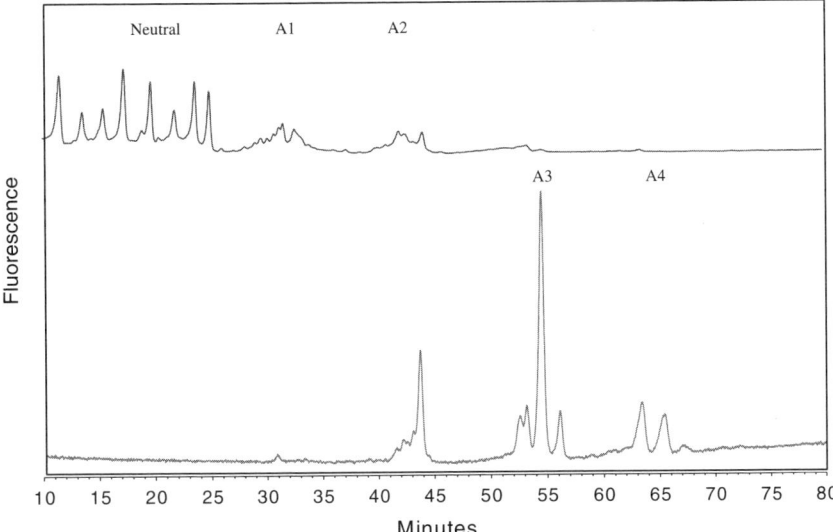

Fig. 17.7 Profiling 2-AB-labeled N-glycans using Asahipak-NH$_2$-column. N-linked glycans released from K562-cell membrane preparation (upper part) and from α1 acid glycoprotein (lower part) were fluorescently labeled with 2-AB and separated on a Asahipak-NH$_2$-column. The cellular N-glycans, which are of the high-mannose and hybrid types (indicated 'neutral'), are not resolved by other anion-exchange chromatographies. The sialylated complex-type N-glycans are separated according to their charges (A1, A2, A3, and A4)

1. Wash the column by either running the complete gradient or using a shortened gradient (in 15 min from 0 to 100% eluent 2, wash the column 10 min with 100% eluent 2, equilibrate the column 20 min in eluent 1).
2. Inject the sample in less than 10 μL of water or dissolved in the starting buffer (*see* **Notes 16** and **17**). Elute the analyte with the following gradient. For detection of 2-AB-labeled glycans use fluorimetric detection (excitation: 330 nm, excitation: 420 nm).

Column:	Asahipak NH2P-50 4E (4,6 × 100 mm. 5 μm, Shodex, Showa Denko, Kawasaki, Japan)
Eluent 1:	Acetonitrile containing 2% acetic acid and 1% tetrahydrofuran
Eluent 2:	Water containing 5% acetic acid, 3% triethylamine and 1% tetrahydrofuran
Flow:	0,8 mL/min
Temperature:	50°C

Gradient for Asahipak NH$_2$-HPLC

Time	%1	%2	Event
0	70	30	Injection
2	70	30	
82	5	95	
97	5	95	
98	30	70	
113	30	70	Restart

3. Store column in isopropanol.

4 Notes

1. It is essential for samples to be free of salts, because anions present in the sample behave as eluents.
2. A proper re-equilibration of the column is important for reproducible results.
3. The quantification of the neutral sugars can be difficult because they elute in the void volume, close to the signal of residual 2-AB-label. The ratio of the neutral and charged oligosaccharides can be calculated after rechromatography of the resulting fractions. We use amino-phase HPLC for rechromatography of glycan fractions after digestion with neuraminidase (described in 4.2). If no high mannose or hybrid-type structures are present, a monosaccharide-composition analysis is helpful for quantification of all resulting fractions (A0–A4).
4. Both Mono Q and GlycoSepC columns are eluted with NH_4OAc, which can be removed in a vacuum centrifuge. For glycan fractions containing higher salt concentrations, the total removal of salts can be difficult even after repeated evaporation of the samples. Alternatively glycan fractions can be desalted by gel-chromatography on a Sephadex G-25 superfine column (0.5 × 20 cm, Pharmacia, Uppsala, Sweden) at a flow rate of 0.5 mL/min or on Carbograph extract-clean columns (Alltech, Deerfeld, IL).
5. Adjust the limit for maximum back pressure to 50 bar.
7. The careful preparation of the eluents is extremely important for HPAEC-PAD to ensure reproducibility of retention times and elution profiles.
8. NaOH-containing eluents must be prepared so as to minimize the content of Na_2CO_3. Do not use NaOH pellets, because these are coated with a Na_2CO_3 film. We recommend 50% NaOH solution (J. T. Baker, Deventer, Netherlands).
9. Store the 50% NaOH-solution under argon to minimize absorption of CO_2 from the air.
10. Re-equilibration of the system influences the retention time. Thus if several analyses are performed, use the same intervals for injection or use an autosampler with titan or peek equipment.
11. Avoid high salt concentrations in the sample, or the analyte may elute earlier or with poor resolution. Anions present in the sample act as eluents.
12. Generally, detergents, Tris and hydroxylated compounds should be avoided (e.g., glycerol, other alcohols).
13. Samples should be dissolved in water before injection, no organic solvent (e.g., acetonitrile) is allowed.
14. The gold electrode of the detector can be cleaned by rubbing. The reference electrode should be regenerated in KCl-solution depending on the time used.
15. The eluent salts are usually no obstacle for mass determination by MALDI-TOF-MS. DHB (2,5-dihydroxybenzoic acid) is a suitable matrix, which is relatively tolerant of salt contamination.
16. Injection of more than 30 μL of the sample might result in doubling of the peaks.
17. Samples should be free of salts.

18. Columns filled with 3 µm instead of 5 µm material give a slightly enhanced resolution. Back pressure can be high (about 300 bar), but this is not a problem for conventional HPLC-systems.
20. Salts present as contamination of the analyte do not affect the separation in this system.
21. The injection volume is not critical (10 µL –100 µL are adequate).
22. Volatile mobile phases can be easily removed in a vacuum centrifuge. Thus eluents used in this system are compatible with mass spectrometry and other techniques for analyzing the resulting glycan fractions.
23. GlycoSepC is supplied with new matrix and column dimensions since December. 2005 (PPC149). Slight modifications of the gradient program may be necessary to achieve equivalent separations.
24. Stability of amino-covered matrices is critical. Performance of many aminophase-HPLC columns decreases significantly even after a few runs, a non-proportional shift of retention times is then observed. In our hands, besides the Luna-column, the APS 2-Hypersil column (4 × 250 mm, 3 µm, Knauer, Berlin, Germany, run with 1,5 mL/min) also gave satisfactory results.

References

1. Davies, M. J., and Hounsell, E. F. (1998) HPLC and HPAEC of oligosaccharides and glyco-peptides. *Methods Mol. Biol.* **76**, 79–100.
2. Davies, M. J., and Hounsell, E. F. (1996) Carbohydrate chromatography: towards yoctomole sensitivity *Biomed. Chrom.* **10**, 285–289.
3. Churms, S. C. (1995) in *Carbohydrate Analysis: High performance Liquid Chromatography and Capillary Electrophoresis* (El Rassi, Z., Ed.), pp. 103–146, Elsevier, New York.
4. Huber, C. G., and Bonn, G. K. (1995) in *Carbohydrate Analysis: High performance Liquid Chromatography and Capillary Electrophoresis* (El Rassi, Z., Ed.), pp. 147–180, Elsevier, New York.
5. Townsend, R. R. (1995) in *Carbohydrate Analysis: High performance Liquid Chromatography and Capillary Electrophoresis* (El Rassi, Z., Ed.), pp. 181–209, Elsevier, New York.
6. Guile, G. R., Rudd, P. M., Wing, D. R., Prime, S. B., and Dwek, R. A. (1996) A rapid high-resolution high-performance liquid chromatographic method for separating glycan mixtures and analyzing oligosaccharide profiles. *Anal. Biochem.* **240**, 210–226.
7. Hardy, M. R., and Townsend, R. R. (1988) Separation of positional isomers of oligosaccharides and glycopeptides by high-performance anion-exchange chromatography with pulsed amperometric detection. *Proc. Natl. Acad. Sci. USA* **85**, 3289–3293.
8. Hardy, M. R., and Townsend, R. R. (1994) High-pH anion-exchange chromatography of glyc-oprotein-derived carbohydrates. *Methods Enzymol.* **230**, 208–225.
9. Hoffman, R. C., Andersen, H., Walker, K., Krakover, J. D., Patel, S., Stamm, M. R., and Osborn, S. G. (1996) Peptide, disulfide, and glycosylation mapping of recombinant human thrombopoietin from Ser1 to Arg246. *Biochemistry* **35**, 14849–14861.
10. Stroop. C.J.M., Weber, W., Gerwig, G.J., Nimtz, M., Kamerling, J.P., and Vliegenthart, J.F.G. (2000) Characterization of the carbohydrate chains of the secreted form of the human epidermal growth factor receptor. *Glycobiology* **10**, 901–917.

11. Thayer, J. R., Rohrer, J. S., Avdalovic, N., and Gearing, R. P. (1998) Improvements to in-line desalting of oligosaccharides separated by high-pH anion exchange chromatography with pulsed amperometric detection. *Anal. Biochem.* **256**, 207–216.

12. Kotani, N., and Takasaki, S. (1998) Analysis of 2-aminobenzamide-labeled oligosaccharides by high-pH anion-exchange chromatography with fluorometric detection. *Anal. Biochem.* **264**, 66–73.

13. Weitzhandler, M., Pohl, C., Rohrer, J., Narayanan, L., Slingsby, R., and Avdalovic, N. (1996) Eliminating amino acid and peptide interference in high-performance anion-exchange pulsed amperometric detection glycoprotein monosaccharide analysis. *Anal. Biochem.* **241**, 128–134.

14. Hase, S., Ikenaka, T., and Matsushima, Y. (1978) Structure analyses of oligosaccharides by tagging of the reducing end sugars with a fluorescent compound. Biochem. *Biophys. Res. Commun.* **85**, 257–263.

15. Anumula, K. R., (1994) Quantitative determination of monosaccharides in glycoproteins by high-performance liquid chromatography with highly sensitive fluorescence detection. *Anal. Biochem.* **220**, 275–283.

16. Bigge, J. C., Patel, T. P., Bruce, J. A., Goulding, P. N., Charles, S. M., and Parekh, R. B. (1995) Non selective labeling of glycans using 2-aminobenzamide and anthranilic acid. *Anal. Biochem.* **230**, 229–238.

17. Hase, S. (1994) High-performance liquid chromatography of pyridylaminated saccharides. *Methods Enzymol.* **230**, 225–237.

18. Takahashi, N. J. (1996) Three-dimensional mapping of N-linked oligosaccharides using anion-exchange, hydrophobic and hydrophilic interaction modes of high-performance liquid chromatography. *J. Chromatogr. A.* **720**, 217–25.

19. Anumula, K. R., and Dhume, S. T. (1998) High resolution and high sensitivity methods for oligosaccharide mapping and characterization by normal phase high performance liquid chromatography following derivatization with highly fluorescent anthranilic acid. *Glycobiology* **7**, 685–94.

20. Nuck, R., and Gohlke, M. (1997) in *Techniques in Glycobiology: Characterization of subnanomolar amounts of N-glycans by 2-aminobenzamide labelling, matrix-assisted laser desorption ionization time-of-flight mass spectrometry, and computer-assisted sequence analysis* (Townsend, R. R., Ed.), pp. 491–507, Marcel Dekker, New York.

21. Gohlke, M., Nuck, R., Kannicht, C., Grunow, D., Baude, G., Donner, P., and Reutter, W. (1997) Analysis of site-specific N-glycosylation of recombinant Desmodus rotundus salivary plasminogen activator rDSPAα1 expressed in Chinese hamster ovary-cells. *Glycobiology* **7**, 67–77.

22. Townsend, R. R., Lipniunas, P. H,. Bigge, C., Ventom, A., and Parekh, R. B. (1996) Multimode high-performance liquid chromatography of fluorescently labeled oligosaccharides from glycoproteins. *Anal. Biochem.* **239**, 200–207.

23. Hermentin, P., Witzel, R., Doenges, R., Bauer, R., Haupt, H., Patel, T., Parekh, R. B., and Brazel, D. (1992), The mapping by high-pH anion-exchange chromatography with pulsed amperometric detection and capillary electrophoresis of the carbohydrate moieties of human plasma α1-acid glycoprotein. *Anal. Biochem.* **206**, 419–429.

24. Gohlke, M., Mach, U., Nuck, R., Zimmermann-Kordmann, M., Volz, B., Fieger, C., Grunow, D., Tauber, R., and Werner Reutter (2000) Carbohydrate structures of soluble human L-selectin recombinantly expressed in baby hamster kidney cells. *Biotechnol. Appl. Biochem.* **32**, 41–51.

25. Rohrer, J. S. (1995), Separation of asparagine-linked oligosaccharides by high-pH anion-exchange chromatography with pulsed amperometric detection: empirical relationships between oligosaccharide structure and chromatographic retention. *Glycobiology* **5**, 359–360.

18

Enzymatic Sequence Analysis of *N*-Glycans by Exoglycosidase Cleavage and Mass Spectrometry — detection of Lewis X Structures

Christoph Kannicht, Detlef Grunow, and Lothar Lucka

Summary Enzymatic sequencing of oligosaccharides gives structural information on sequence of monosaccharides and type of linkage within the oligosaccharide chain. This data can be obtained by stepwise enzymatic digestion of a single, isolated oligosaccharide using individual or mixtures of specific exoglycosidases. *N*-glycans have to be fractionated from mixtures prior to sequence analysis to assign this type of structural information to a specific glycan. Enzymatic sequencing can be applied to oligosaccharide mixtures as well to evaluate the occurrence of distinct oligosaccharide motives of functional and/or structural interest.

Here we describe the application of enzymatic sequence analysis to a mixture of *N*-glycans released from α1-acid glycoprotein. The experimental conditions are optimized for detection of possible Lewis X structures after stepwise exoglycosidase digestion by MALDI-TOF mass spectrometry. However, the described method is generally applicable to analyze other structural properties of N-glycans by use of the respective specific exoglycosidases.

Key Words Enzymatic sequencing; *N*-glycan; exoglycosidase; MALDI-TOF; Lewis x; oligosaccharide structure.

1   Introduction

Enzymatic sequencing of oligosaccharides gives structural information on sequence of monosaccharides and type of linkage within the oligosaccharide chain. This data can be obtained by stepwise enzymatic digestion of a single, isolated oligosaccharide using individual or mixtures of specific exoglycosidases. *N*-glycans have to be fractionated from mixtures prior to sequence analysis to assign this type of structural information to a specific glycan. However, enzymatic sequencing can be applied to oligosaccharide mixtures as well to evaluate the occurrence of distinct oligosaccharide motives of functional and/or structural interest.

Exoglycosidases cleave the glycan stepwise from the nonreducing terminal linkage releasing monosaccharides. They are highly specific for monosaccharide type,

From: *Post-translational Modifications of Proteins.*
Methods in Molecular Biology, Vol. 446.
Edited by: C. Kannicht © Humana Press, Totowa, NJ

anomeric configuration, linkage, and branching. The digestion will stop at the monosaccharide, which cannot be cleaved off with the enzyme(s) available in the respective digestion step. Enzymatic digestion results in truncated forms of the analyzed *N*-glycan.

For evaluation of the assay, the size of the analyzed oligosaccharide and the respective fragments resulting from exoglycosidase-treatment have to be determined. From this data and the known specificity of the exoglycosidases, type, order and linkage of monosaccharide within the *N*-glycan chain can be deduced. Size of oligosaccharide fragments can be determined by gel permeation chromatography with internal standard *(1)*, a special gel-electrophoresis system *(2)* or mass-spectrometry *(3)*. Here we describe a method using MALDI-TOF mass spectrometry for size determination of the exoglycosidase digestion products. Using this method, oligosaccharides without fluorescence label can be measured as well (*see* **Note 1**). Moreover, one does not need special instrumentation used solely for oligosaccharide analysis.

Occurrence and action of the glycosyltransferases involved in synthesis of *N*-glycans is limited and only a number of monosaccharides are accepted as substrates in *N*-glycan-synthesis (*see* Table 18.1). As a result, the number of possible oligosaccharide chains found in mammalian cells is limited. Moreover, *N*-glycans share a common core structure and sugar chains bound to the core follow some structural rules. *N*-linked glycans are divided into 3 subgroups: high mannose type, complex type, and hybrid type *(4,5)*. For this reason, it is possible to perform sequence analysis of most *N*-glycans from mammalian proteins using only few exoglycosidases.

Here we describe the application of enzymatic sequence analysis to a mixture of *N*-glycans released from α1-acid glycoprotein. The experimental conditions are optimized for detection of possible Lewis X structures after stepwise exoglycosidase digestion by MALDI-TOF mass spectrometry *(6,7)*. Other structural properties of *N*-glycans can be covered by use of the respective specific exoglycosidases.

Table 18.1 Monosaccharides and Possible Linkage Commonly Found in Mammalian Glycoproteins

Monosaccharide	Anomer	Bound to C#
N-Acetylneuraminic acid	α	3, 6, 8
N-Glycolylneuraminic acid	α	3, 6
D-Galactose	α	3
	β	3, 4, 6
N-Acetyl D-glucosamine	β	2, 3, 4, 6
N-Acetyl D-galactosamine	α	3
	β	4
D-Mannose	α	2, 3, 6
	β	4
L-Fucose	α	2, 3, 4, 6
D-Xylose	β	2

2 Materials

2.1 Desialylation

1. Thermomixer (Eppendorf, Hamburg, Germany).
2. Glyko Sialidase III, recombinant from *Arthrobacter ureafaciens*, expressed in *Escherichia coli* (PROzyme, San Leandro, CA).
3. Sialidase incubation buffer: 100 mM ammonium acetate, pH 5.0.
4. Mixed bed column for desalting of oligosaccharides: add one after the other 0.5–1 mL 50% aqueous suspensions of (I) anion exchange resin (Dowex AG3-X4, OH⁻ form) and (II) cation exchange resin (Dowex AG 50 W-X12, H⁺ form) or equivalent into a small column (5 mL) (BioRad, Hercules USA).
5. Centrifugal evaporator.

2.2 Enzymatic Sequencing

1. Centrifugal evaporator.
2. Exoglycosidases according to the desired structural information. For example we use following enzymes for detection of Lewis X structures on *N*-glycans: β-*N*-acetylhexosaminidase from jack bean (EC 3.2.1.52), β-galactosidase from bovine testes (EC 3.2.1.23), (1–4)-specific β-galactosidase from *Streptococcus pneumoniae* (EC 3.2.1.23) and α-fucosidase III (1–3,4)- from *Xanthomonas manihotis* (EC 3.2.1.51) (PROzyme, San Leandro, CA).
3. Enzyme reaction buffers: β-galactosidase / β-*N*-acetylhexosaminidase: 1:1 (v/v) mixture of 0.1 M ammonium acetate pH 6.0 and 0.1 M sodium citrate/phosphate pH 5.0; α-fucosidase III: 50 mM sodium phosphate pH 5.0; ß-1-4-specific galactosidase: 0.1 M sodium acetate pH 6.0. For other enzymes please refer to the instructions of the manufacturer.
4. Thermomixer (Eppendorf, Hamburg, Germany).
5. TopTip carbon spin column, type P2-Carbon (Glygen Corp., Columbia, USA) for sample purification: Wash spin column two times with 10 μL 80% (v/v) acetonitrile / 0.1% (v/v) trifluoroacetic acid (TFA, for safety instructions and disposal please refer to the Safety Data Sheet of the manufacturer) by centrifugation for 10 s at 1,400 g and subsequently 3 times with 10 μL 0.1% (v/v) TFA by centrifugation at 1,400 g for 10 s. Alternatively, a mixed bed column can be used for sample purification: add one after the other 0.5 mL 50% aqueous suspensions of (I) protein-binding resin (Mimetic Blue AX6LSA), (II) anion exchange resin (Dowex AG3-OH⁻ form or equivalent), (III) cation exchange resin (Dowex AG50-H⁺ form or equivalent) and (IV) anion exchange resin (Dowex AG1-OH⁻ form or equivalent) into a 5 mL plastic column.
6. Pyridine (for safety instructions and disposal please refer to the Safety Data Sheet of the manufacturer).

2.3 Mass Determination

1. MALDI-TOF Mass Spectrometer equipped with a 337-nm nitrogen laser, e.g., Biflex (Bruker, Germany).
2. Matrix solution: 5 mg/mL of D-arabinosazone in 80% (v/v) ethanol. Alternatively, a saturated solution of 2,5-Dihydroxybenzoic acid (DHB) in 60% ethanol can be used (*see* **Note 2**).
3. Calibration standard for mass spectrometry. For example dextran hydrolysate (Glucoseα1–6)*n*, (Glucose Homopolymer Standard, PROzyme, San Leandro, CA).

3 Methods

3.1 Desialylation

This section describes the enzymatic release of sialic acids from oligosaccharides to obtain desialylated neutral *N*-glycans for use in positive ion mass spectrometry or enzymatic sequencing as described in **Section 3.2.** We recommend the use of sialidase from *A. ureafaciens* (EC 3.2.1.18) for its ability to split off α2–3, −6, −8, and −9 linked sialic acids (*8*).

In many cases, desialylation has already been performed before separation of *N*-glycans by HPLC. Neutral oligosaccharides obtained from chromatographic fractionation can be subjected to mass spectrometry or enzymatic sequencing without further sialidase treatment.

N-glycan mixtures obtained from enzymatic or chemical cleavage of oligosaccharides from glycoproteins or glycopeptides can be subjected to MALDI-TOF mass spectrometry using negative ion mode for measurement of sialylated glycans (*9*) or have to be desialylated prior to positive ion MALDI-TOF mass spectrometry (*see* **Section 3.3.** and **Note 3**).

1. Dissolve 0.1 U sialidase from *A. ureafaciens* in 50–100 μL incubation buffer (*see* **Note 4**).
2. Dissolve salt free oligosaccharides in 50–100 μL sialidase solution.
3. Check pH of the incubation buffer.
4. Incubate 24 h at 37°C under gentle shaking.
5. Load the sample onto the pre-washed TopTip P2-Carbon column. Wash three times with 5 μL 0.1% (v/v) TFA each by centrifugation for 10 s at 1,400*g*. Elute the carbohydrate sample three times with 5 μL 25% (v/v) acetonitrile/ 0.1% (v/v) TFA by centrifugation at 1,400*g* for 10 sec. Alternatively the mixed bed column can be used for sample desalting. Elute the sample from the mixed bed column with 5-fold bed volume water.
6. Dry the combined eluates by centrifugal evaporation.

3.2   *Enzymatic Sequencing of N-Glycans*

The principle of enzymatic sequencing is to stepwise release the terminal mono-saccharides until digestion stops at a fragment that the enzyme(s) available in the respective sample can not cleave. The sequence can then be deduced from (I) substrate specificity of the applied enzyme(s), and (II) size of *N*-glycan fragment(s) resulting from enzymatic cleavage. The specificity of the endoglycosidases applied for sequencing is known and the fragment sizes can be determined by MALDI-TOF mass spectrometry (*see* **Section 3.3.**).

Enzymatic sequencing of *N*-glycans can be performed either with single, fractionated oligosaccharides as well as with oligosaccharide mixtures, depending on the desired structural information.

If linkage and sequence analysis of single oligosaccharides is intended, *N*-glycans from mixtures have to be fractionated first. This can be performed for example by aminophase- and/or reversed phase-chromatography *(9)* or use of columns specifically developed for separation of oligosaccharides (Prozyme, San Leandro, USA). Depending on the chromatographic methods employed, oligosaccharides have to be desialylated before chromatography as described in **Section 3.1.** Please note, that fractionated samples may contain organic eluent like acetonitrile, which has to be removed before exoglycosidase treatment.

Here we describe the application of enzymatic sequence analysis to a mixture of *N*-glycans released from α1-acid glycoprotein. The experimental conditions are optimized for detection of possible Lewis X structures after stepwise exoglycosidase digestion by MALDI-TOF mass spectrometry. Other structural properties of *N*-glycans can be covered by use of the respective specific exoglycosidases.

1. Remove sialic acids as described in **Section 3.1.**
2. Transfer the neutral *N*-glycan sample to a 0.5 or 1.5 mL tube (*see* **Note 5**).
3. Remove possible organic solvents and dry the sample in a centrifugal evaporator.
4. Add 0.15 U ß-*N*-acetylhexosaminidase and 1.5 U ß-galactosidase to the oligosaccharide-mixture and add reaction buffer to a final volume of 100 µL.
5. Incubate at 37°C for 18 h (*see* **Note 6**).
6. Take an aliquot of about 10–20% of the reaction mixture and apply it to the solid phase extraction column. Wash and elute the sample as described in **Section 2.2.** This aliquot is further analyzed by MALDI-TOF mass spectrometry as described below.
7. Dry the remaining sample by centrifugal evaporation.
8. Add 1.5 mU α-fucosidase III and add reaction buffer to a final volume of 30 µL.
9. Incubate at 37°C for 3 h.
10. Repeat steps 6 and 7.
11. Add 8 mU ß-1-4-specific galactosidase and add reaction buffer to a final volume of 100 µL.

12. Incubate at 37°C for 16 h.
13. Repeat step 6.

3.3   Mass Determination

In principle, relative masses of sialylated and desialylated neutral N-glycans can be determined by MALDI-TOF mass spectrometry. Sialylated N-glycans can be analyzed using negative ion mode. For measurement of desialylated neutral N-glycans, positive ion mode is applied. The required concentration of N-Glycans necessary for MALDI-TOF mass spectrometry depends on sample purity, residual salt concentration, type of instrument and typically ranges between 1 and 10 pmol/μL oligosaccharide (*see* **Note 7**).

Please note that N-glycans may be labeled with fluorophores like 2-aminobenzamide (2-AB) *(8)*. 2-AB labeled N-glycans can be subjected to MALDI-TOF mass spectrometry without any problems, but other fluorophores may not (*see* **Note 8**).

3.3.1   Sample Preparation and Measurement

1. Freshly prepare arabinosazone matrix solution.
2. Dissolve N-glycan sample from enzymatic sequencing with water to a final concentration of typically 2–50 pmol/μL of each digestion product.
3. Place 0.5–1 μL of the sample or standard solution onto the target.
4. Subsequently add an equal volume of matrix solution.
5. Mix the sample/ matrix solution by drawing into the pipet and dispensing onto the target (*see* **Note 9**).
6. Let the sample dry at room temperature (*see* **Note 10**).
7. Perform measurements. Refer to the appropriate parameter setting recommended by the manufacturer of your instrument.
8. Calibrate mass spectrometer using an oligosaccharide standard sample.
9. For mass determination of digestion products from enzymatic sequencing use the positive ion mode and the reflector if available (*see* **Note 11**).

3.3.2   Interpretation of Results

Figure 18.1 shows the mass spectrograms of N-glycans released from α1-acid glycoprotein after sequential treatment with sialidase (sia) (A), a mixture of β-N-acetylhexosaminidase and β-galactosidase (hex/gal) (B), α(1-3,4)-specific fucosidase (fuc III) (C) and a β(1–4)-specific galactosidase (gal) (D). For specificities of the applied enzymes please refer to Table 18.2. The interpretation of mass shifts is exemplarily summarized in Table 18.3 for a complex, triantennary, monofucosylated N-glycan with a measured relative mass of 2174.3 (see Fig. 18.1).

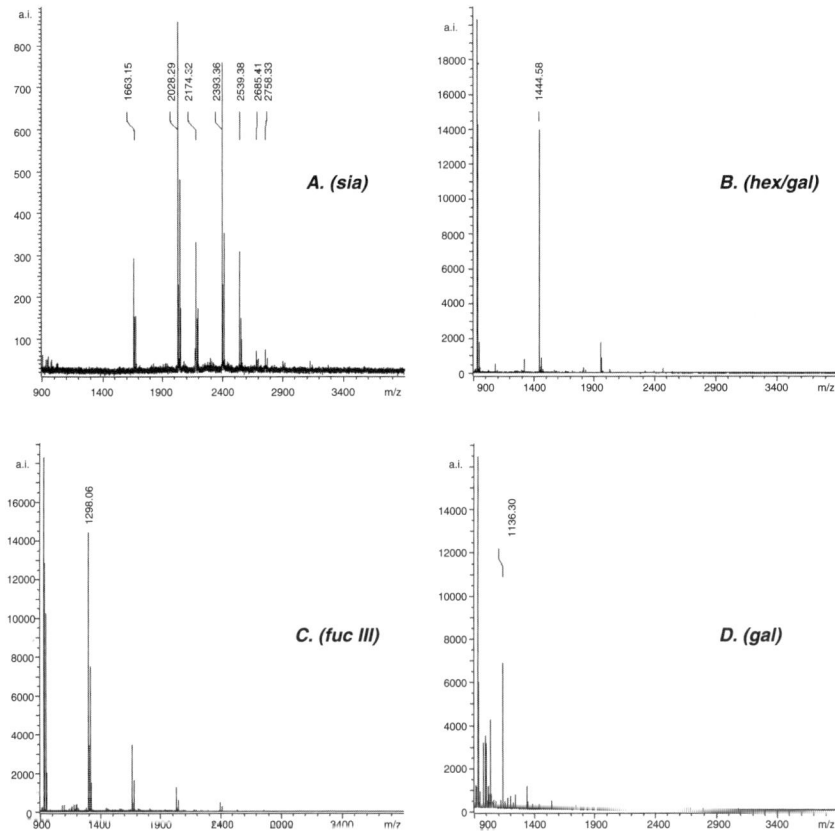

Fig. 18.1 MALDI-TOF mass spectra of *N*-glycans released from α1-acid glycoprotein following treatment with different glycosidases. *N*-glycans were stepwise digested by sialidase (**A**), α(1–3,4)-specific fucosidase (**B**), a mixture of β-galactosidase and β-*N*-hexosaminidase (**C**) and a ß(1–4)-specific galactosidase (**D**) and measured by MALDI-TOF mass spectrometry. Detected ions could be assigned to Na-adducts (M+Na)⁺ of the glycans. Mass values correspond to the data presented in Table 18.3. Mass peaks corresponding to the K⁺-adduct are not indicated. Reprinted from *(6)* by permission of Oxford University Press

Table 18.2 Specificity of Exoglycosidases used for Detection of Lewis X Structures on *N*-Glycans from α1-Acid Glycoprotein

Enzyme name	Specificity
Sialidase (*Arthrobacter ureafaciens*)	(α2-3,6,8)-linked *N*-acetylneuraminic acid
β-Galactosidase (Bovine Testes)	(β1-3,4,6)-linked galactose
β-*N*-Acetylhexosaminidase III (Jack bean)	(β1-2,3,4,6)-linked *N*-acetylglucosamine
α-Fucosidase III (*X. manihotis*)	(α1–3,4)-linked, terminal fucose
β-Galactosidase III (*S. pneumoniae*)	(β1-4)-linked galactose

Table 18.3 Enzymatic Sequencing Procedure for the Identification of Lewis Structures, Demonstrated for an α1-Acid Glycoprotein-Linked N-Glycan[a]

Enzyme	Enzyme Specificity	Cleaved structures (putative examples)		Resulting structure $(M_{calc}.+Na)^+ / (M+Na)^+$
Sialidase (*Arthrobacter ureafaciens*)	Release of α2-3,6,8,9 linked N-acetylneuraminic acid			(Hex)3 (HexNAc)3 (Deoxyhexose)1 + (Man)3(GlcNAc)2
				2174.772 / **2174.3**
β-Galactosidase (Bovine testes) plus β-N-Acetylhexosaminidase (*Jack bean*)	Release of β1-3,4,6 linked galactose and β1-2,3,4,6 linked N-acetylglucosamine, if fucose is not bound		Oligosaccharide analysis (MALDI-MS)	(Hex)1 (HexNAc)1 (Deoxyhexose)1 + (Man)3(GlcNAc)2
				1444.507 / **1444.6**
Fucosidase III (*Xanthomonus manihotis*)	Release of terminal α1-3,4 linked terminal fucose; <u>does not release core fucose</u>			(Hex)1 (HexNAc)1 + (Man)3(GlcNAc)2
				1298.449 / **1298.1**
β-Galactosidase (*Streptococcus pneumoniae*)	Release of β1-4 linked galactose			(HexNAc)1 + (Man)3(GlcNAc)2
				1136.397 / **1136.3**

◇ : N-acetylneuraminic acid; ◆ : galactose; □ : N-acetylglucosamine; ● : mannose; △ : fucose

Dashed line: cleaved linkage; ○: Le^X-structure.

[a]Used enzymes, enzyme specificity, putative structures cleaved by these enzymes, resulting structures with calculated masses and measured m/z from MALDI-MS (see Fig. 18.1 for corresponding mass spectra). Mcalc refers to the calculated average molecular mass (Na$^+$-adduct, in daltons) summed from the likely carbohydrate composition and verified by the GlycoMod tool of the Expasy Molecular Biology Server.

Reprinted from Lucka et al. 'Identification of Lewis x structures of the cell adhesion molecule CEACAM1 from human granulocytes.' *Glycobiology* (2005) **15**, 87–100 by permission of Oxford University Press.

The application of β-galactosidase with β(1–3,4,6) cleavage specificity and β-N-acetylhexosamidase with the cleavage specificity β(1–2,3,4,6) results in a mass shift of 729.7 to m/z 1444.6 reflecting the removal of 2 galactose and 2 GlcNAc residues from the antennae that does not carry a fucose residue (see Figure 18.1B and Table 18.3). The mass shift of 146.5 to m/z 1298.1 after α-fucosidase III treatment with cleavage specificity for α(1–3,4)-linked fucose indicates the removal of 1 terminal linked fucose (Figure 18,1C). Finally, the mass shift of 161.8 to m/z 1136.3 after incubation with β(1–4)-specific galactosidase reflects the release of β(1,4)-linked galactose, which carried a α(1–3)-linked fucose residue prior to the fucosidase treatment (Figure 18,1D). The last sequencing step demonstrates that the previously released fucose residue was part of Lewis X and not Lewis A structures.

For general interpretation of mass spectrometry data obtained from enzymatic sequencing, calculated masses [M+H]⁺ and [M+Na]⁺ of some widely found complex *N*-glycans are given in Table 18.4. Neutral *N*-glycans may form [M+H]⁺-, [M+Na]⁺-, or [M+K]⁺-ions depending on the extent of salts in the sample. Mainly Na⁺-adducts and small amounts of K⁺-adducts of *N*-glycans are found using MALDI-TOF mass spectrometry. Corresponding data for high-mannose type *N*-glycans is given in Table 18.5. For calculation of further theoretical masses of *N*-glycans the GlycanMass software available on the Expert Protein Analysis System (ExPASy) proteomics server from the Swiss Institute of Bioinformatics (SIB) web site is very helpful *(10)* <www.expasy.org>. Alterations of molecular weight of *N*-glycans due to labeling with fluorophores such as 2-aminobenzamide (2-AB) have to be taken into account for interpretation of mass spectrometry data as well. Calculated mass differences caused by cleavage of different monosaccharides are summarized in Table 18.6.

1. Check the mass spectrum of standard glycans for [M+H]⁺-, [M+Na]⁺-, or [M+K]⁺-ions.
2. Consider mass difference caused by possible fluorophore label of the *N*-glycan.

Table 18.4 Theoretical Monoisotopic Masses of Neutral *N*-Glycans of the Complex and Hybrid Type

| Structure (core + monosaccharides) | | | Calculated mass | |
GlcNAc	Gal	Fuc	[M+H]⁺	[M+Na]⁺
biantennary				
2			1316.5	1338.5
2	2		1640.6	1662.6
2		1	1462.5	1484.5
2	2	1	1786.6	1808.6
triantennary				
3			1519.6	1541.6
3	3		2005.7	2027.7
3		1	1665.6	1687.6
3	3	1	2151.8	2173.8
tetraantennary				
4			1722.6	1744.6
4	4		2370.9	2392.9
4		1	1868.7	1890.7
4	4	1	2516.9	2538.9
Bisecting GlcNAc e.g.: biantennary, 1 bis. GlcNAc			1843.7	1865.7
Hybrid Type e.g.: 1 GlcNAc, 2 Man			1437.5	1459.5

Table 18.5 Calculated Monoisotopic Masses [M+H]⁺ and [M+Na]⁺ of High Mannose Type N-Glycans

Core + no. of mannoses	[M+H]⁺	[M+Na]⁺
2	1234.4	1256.4
3	1396.5	1418.5
4	1558.5	1580.5
5	1720.6	1742.6
6	1882.6	1904.6

Table 18.6 Mass Differences Caused by Addition or Loss of Monosaccharides, Fluorophore Labeling or Ionization of N-Glycans. Calculated Single Monoisotopic Mass Values

	Example	calculated mass difference
Hexose	Gal, Man, Glc	162.0
HexNAc	GlcNAc, GalNAc	203.1
Deoxyhexose	Fuc	146.1
Pentose	Xyl	132.0
NeuAc	Sialic Acid	291.1
Core	GlcNAc₂-Man₃	910.3
Fluorophore	2-AB-label	122.1
H+		1.0
Na+		23.0

3. Compare the mass found in the mass spectrum of the sample obtained before enzymatic sequencing with corresponding calculated masses listed in Tables 18.4 and 18.5.

4. Try to classify the N-glycan as complex, hybrid or high-mannose structure.

5. Calculate mass differences between digestion product(s) found in the mass spectra after each exoglycosidase digestion step and the mass of the glycan before exoglycosidase treatment.

6. Compare the measured mass differences with values given in Table 18.6. For example mass differences of 162, 324, or 486 indicate loss of 1, 2, or 3 hexoses, respectively.

7. Check for the specificity of the applied exoglycosidases. For example mass difference of 324 following incubation with β-galactosidase III (S. pneumoniae) indicates loss of two β(1–4)-linked galactoses (see Table 18.2). In general, refer to the specifications given in the data sheets of the applied exoglycosidase.

8. Deduce the suggested N-glycan structure from the relative mass of the uncleaved N-glycan, the mass differences of cleavage products found in the mass spectra, the enzyme mixtures used for sequencing and the specificity of the applied enzymes (see **Note 12**).

4 Notes

1. For electrophoretic separation *N*-glycans have to be labeled with ANTS fluophore to introduce negative charges. ANTS labeled *N*-glycans can be analyzed using a special 'FACE™' electrophoresis system.
2. Though D-arabinosazone works better for mass spectrometry of oligosaccharides in general by facilitating to obtain mass spectra, DHB matrix supports the evaluation of relative peak intensities by generation of more consistent and reproducible pattern than a D-arabinosazone matrix.
3. We recommend performing mass determination of the complete mixture of oligosaccharides prior to chromatographic fractionation. Perform positive ion spectrum for neutral oligosaccharides or negative ion spectrum for sialylated or charged oligosaccharides (9).
4. Alternatively 10 mU sialidase from Newcastle disease virus can be used instead of 0.1 U sialidase from Arthrobacter ureafaciens.
5. The amount of *N*-glycan needed for mass determination by MALDI-TOF depends on purity of the sample, i.e. lack of salts and detergents and on the instrumentation. Typically 10 to 100 pmol oligosaccharides are needed for the measurement. The initial sample amount has to be adapted to the number of enzymatic cleavage steps needed to gain the targeted structural information accordingly.
6. The exact incubation time is not critical for these enzyme mixtures and may vary between 16 to 20 h. Please note that the incubation time might be critical for some enzymes in order to achieve specific cleavage.
7. In general, measurement of sialylated *N*-glycans in negative ion mode is less sensitive compared to the measurement of neutral *N*-glycans using positive ion mode.
8. The fluorophore 8-aminonaphtalene-1,3,6-trisulphonate (ANTS) used for labeling oligosaccharides for analysis by the FACE system contains sulfonic acid groups. ANTS imparts negative charge to drive the electrophoretic separation. These may cause problems for measurement of ANTS labeled *N*-glycans. However, we did not perform mass spectrometry with these derivatives so far.
9. Mixing of the sample and matrix solution on the target helps to use as little sample as possible for the analysis. Alternatively, mix sample and matrix solution in an Eppendorf cup upfront and place 0.5-1 µL sample/matrix solution onto the target.
10. The target can be stored at room temperature in the dark for several days.
11. Appropriate parameter setting of the mass spectrometer depends on the instrument used for measurements. Please refer to the recommendation of the manufacturer and use these settings as starting point. We use a Bruker Biflex instrument. Possible parameter settings in positive ion mode with reflector for measurements of neutral glycans are (I) high voltage IS/1 19 kV, IS/2 12.8 kV, Refl. 20 kV, Lens 7 kV, (II) cut-off mass 1200, deflection HV on, (III) detector neutrals refl. 1.55 kV, lin. 1.6 kV. A typical parameter set for the measurement of glycans in negative ion mode with reflector is (I) high voltage IS/1 19 kV, IS/2 12.8 kV to 13.3 kV, Refl. 20 kV,

Lens 6.8 to 7 kV, (II) cut-off mass 400, deflection HV on, (III) detector neutrals refl. 1.55 kV, lin. 1.6 kV.

12. The specificity of some exoglycosidases depends on the applied incubation conditions, i.e. incubation buffer, concentration, temperature and incubation time. For these enzymes, it is important to follow the manufacturer's instructions closely. For instance non-ideal incubation conditions may lead to incomplete or unspecific cleavage to a certain extent. This has to be taken into consideration during interpretation of the results.

References

1. Edge, C. J., Rademacher, T. W., Wormald, M. R., Parekh, R. B., Butters, T. D., Wing, D. R., and Dwek, R.A. (1992) Fast sequencing of oligosaccharides: the reagent-array analysis method. *Proc. Natl. Acad. Sci. U.S.A.* **89**, 6338–6342.
2. Frado, L. Y. and Strickler, J.E. (2000) Structural characterization of oligosaccharides in recombinant soluble interferon receptor 2 using fluorophore-assisted carbohydrate electrophoresis. *Electrophoresis* **21**, 2296–2308.
3. Gohlke, M., Mach, U., Nuck, R., Zimmermann-Kordmann, M., Grunow, D., Flieger, C., Volz, B., Tauber, R., Petri, T., Debus, N., and Reutter, W. (2000) Carbohydrate structures of soluble human L-selectin recombinantly expressed in baby-hamster kidney cells. *Biotechnol. Appl. Biochem.* **32**, 41–51.
4. Kornfeld, R. and Kornfeld, S. (1985) Assembly of asparagine-linked oligosaccharides. *Ann. Rev. Biochem.* **54**, 631–664.
5. Snider, M. D. (1984) Biosynthesis of glycoproteins: formation of N-linked oligosaccharides, in Ginsburg, V. and Robbins, P.W. (eds.), *Biology of Carbohydrates. 2.* John Wiley and Sons, New York, N.Y., pp. 163–198.
6. Lucka L, Fernando M, Grunow D, Kannicht C, Horst AK, Nollau P,, and Wagener C. (2005) Identification of Lewis x structures of the cell adhesion molecule CEACAM1 from human granulocytes. *Glycobiology* **15**, 87–100.
7. Bogoevska V., Nollau P., Lucka L., Grunow D., Klampe B., Uotila L.M., Samsen A., Gahmberg C.G., and Wagener C. (2006) DC-SIGN binds ICAM-3 isolated from peripheral human leukocytes through Lewis x residues. *Glycobiology* Epub ahead of print.
8. Uchida, Y., Tsukada, Y. and Sugimori, T. (1979) Enzymatic properties of neuraminidases from Arthrobacter ureafaciens. *J. Biochem* **86**, 1573–1585.
9. Kannicht, C., Lucka, L., Nuck, R., Reutter, W., and Gohlke, M. (1999) *N*-Glycosylation of the carcinoembryonic antigen related cell adhesion molecule, C-CAM, from rat liver: detection of oversialylated bi- and triantennary structures. *Glycobiology* **9**, 897–906.
10. Appel, R. D., Bairoch A., and Hochstrasser D. F. (1994) A new generation of information retrieval tools for biologists: the example of the ExPASy WWW server. *Trends Biochem. Sci.* **19**, 258–260.

19

Immunochemical Methods for the Rapid Screening of the *O*-Glycosidically Linked *N*-Acetylglucosamine Modification of Proteins

Michael Ahrend, Angela Käberich, Marie-Therese Fergen, and Brigitte Schmitz

Summary For the rapid screening of specific post-translational modifications antibody-based methods are very well suited and applicable without demanding expenditure. Here we describe the immunochemical detection of the *O*-glycosidically linked cytosolic *N*-acetylglucosamine modification of proteins, which has attracted increasing interest in the last years. Two different monoclonal antibodies were used in enzyme-linked immunosorbent assays (ELISA), Western blots of 1- and 2- dimension (1D and 2D) separated proteins and immunohistochemical analysis of tissue sections. Slight differences in the recognition of this post-translational epitope by the 2 antibodies are observed and will be discussed.

Key Words O-GlcNAc modification of proteins; posttranslational modification; immunochemical screening methods; ELISA; 1D and 2D Western blots; immuno-histochemical analysis.

1   Introduction

Glycosylation of serine and threonine residues of proteins with the *N*-acetylglucosamine monosaccharide (*O*-GlcNAc) is a ubiquitous post-translational modification of meta-zoan proteins of the nucleus, cytoskeleton, and cytoplasm and has also been detected on cytosolic tails of membrane proteins (for review see *(1–3)*). Since its first discovery in 1984 by Torres and Hart *(4)* the increasing interest in this post-translational protein modification has led to discoveries supporting the idea that proteins may be endowed with different functional activities or binding capacities in either the phosphorylated, *O*-GlcNAc modified or nonmodified form.

The observation that *O*-GlcNAc is dynamically turned over by *O*-GlcNAc transferase (OGT) and *O*-GlcNAc hydrolase (OGH) in an analogous manner to phosphate by kinases and phosphatases led to the assumption that *O*-GlcNAc might be involved in signaling *(5,6)*. It has been shown that *O*-GlcNAc levels can respond reciprocally to inhibition or activation of certain kinases or phosphatases *(7–9)*. In addition, OGT is tightly associated with protein phosphatase 1β and 1γ,

From: *Post-translational Modifications of Proteins.*
Methods in Molecular Biology, Vol. 446.
Edited by: C. Kannicht © Humana Press, Totowa, NJ

is tyrosine phosphorylated and is itself modified by *O*-GlcNAc *(10)*. These and further observations, as for example, that caseinkinase II and Yes tyrosine kinase are *O*-GlcNAc modified and that PKC translocates from the membrane to the cytosol at high cellular *O*-GlcNAc concentration, add another dimension of complexity to signaling networks of cells *(11–13)*.

The regulatory function of *O*-GlcNAc has been shown to play a role in many physiological processes of multicellular organisms, e.g., in transcriptional regulation, cell cycle regulation, proteasomal degradation, apoptosis, and insulin signaling *(14,15)*. Alterations of *O*-GlcNAc turnover have been observed under pathological conditions such as stress, ischemia, hyperglycemia, and insulin resistance so that it is not surprising that *O*-GlcNAc has been associated with human diseases like diabetes and possibly also cancer and neurodegenerative diseases *(16–19)*.

In recent years much effort has been invested in developing techniques for the identification of *O*-GlcNAc modified proteins and particularly of the sites of *O*-GlcNAc addition within a protein. As for the identification of phosphopeptides one problem is that the stoichiometry of *O*-GlcNAc modification is often low. Chemical substitution or chemoenzymatic derivatization of *O*-GlcNAc allowed the enrichment of originally *O*-GlcNAc-modified peptides through affinity methods and subsequent identification of sites of *O*-GlcNAc glycosylation by LC-MS/MS methods *(3,20–23)*. Using such techniques it was recently possible to identify about 200 *O*-GlcNAc-modified peptides in a single high throughput analysis *(24)*. A further improvement of *O*-GlcNAc proteomic analyses is the application of the electron capture dissociation MS technique that, combined with wheat germ agglutinin affinity chromatography resulted in the identification of *O*-GlcNAc peptides from protein mixtures without chemical derivatization *(25)*.

However, these techniques require particular chemical expertise and highly sophisticated, expensive analytical and mass spectrometric devices not available in most biological laboratories. For the rapid screening of *O*-GlcNAc expression of a particular protein, cells or tissue immunologic techniques are convenient and have already been used frequently in the past. We here describe immunological detection methods using *O*-GlcNAc specific antibodies, which have advantages in time, cost, and technical expenditure over those techniques previously described, but which can of course not provide detailed information on e.g., sites of *O*-GlcNAc modification. Detailed protocols for the detection and other analytical aspects of *O*-GlcNAc have been published by Zachara et al. *(26)*.

We previously reported the *O*-GlcNAc reactivity of the monoclonal antibodies HGAC 85, Mud 50, and RL2 *(27)*. The latter is still frequently being used together with the more recently generated CTD 110.6 *(28)*. We here compare the application of these 2 antibodies in ELISA, 1D and 2D Western blots, and immunohistochemical analysis.

The RL2 monoclonal antibody is a mouse IgG1 and was raised against a nuclear pore complex-lamina fraction isolated from rat liver nuclear envelopes, which are multiply *O*-GlcNAc modified (29). One drawback of this antibody is that it appears to have a significant requirement for a peptide epitope of nuclear pore proteins. Nevertheless, the antibody reacts with many other *O*-GlcNAc modified proteins as shown by several groups *(26,28–30)*.

The CTD 110.6 monoclonal antibody was raised against the synthetic peptide YSPTS(*O*-GlcNAc)PSK derived from RNA polymerase II carboxyterminal domain peptide *(28)*. It is a mouse antibody of the IgM type and available as ascites. The advantage of this antibody is that it is much more specific for the Ser/Thr-*O*-GlcNAc motif independent of the peptide to which GlcNAc is attached than the RL2 antibody. This has been shown, for example, by inhibition of the antibody binding to *O*-GlcNAc modified proteins from HeLA cells in the presence of 10 m*M* GlcNAc *(28)*.

2   Materials

2.1   *Antibodies*

1. The RL2 antibody is commercially available from Affinity Bioreagents Inc. (catalogue no. MA1-072), Abcam (ab 2739) or from Alexis Corporation (ALX-804-111). The original description of the antibody is given in Snow et. al *(29)*. It was kept at −20°C in glycerol/PBS (1:1) at a concentration of 0.5 mg/mL.
2. The CTD 110.6 antibody is available from Covance Research Products (MMS-248R), from Abcam (ab24687) or from Pierce (included in Western blot detection kit No 24565). The antibody was originally described by Comer et al. *(28)*. It is an ascites containing 3–5 mg protein/mL and was stored at −20°C in glycerol / PBS (1:1).

2.2   *ELISA*

1. GlcNAc modified bovine serum albumin (GlcNAc-BSA): bovine *p*-aminophenyl-*N*-acetyl-β-D-glucosaminide albumin (Sigma No. A 1034; Sigma, Deisenhofen, Germany).
2. Horse serum (Sigma, Deisenhofen, Germany).
3. Bovine serum albumin (BSA Fraction V, Sigma No A7906).
4. Peroxidase (POD) conjugated goat anti-mouse antibodies (Jackson Immunoresearch, No. 115 035 068 from Dianova, Hamburg, Germany).
5. ABTS solution: 0.5 mL 2,2~-azino-di-3-ethylbenzthiazolinesulfonate (ABTS) stock solution (2 g ABTS in 100 mL aqua dem.; Novartis, Mannheim, Germany), 9.5 mL acetate buffer (0.1*M* sodium acetate and 0.05*M* sodium dihydrogenphosphate adjusted to pH 4.2 with glacial acetic acid) and 3 μL hydrogen peroxide. The solution should be freshly prepared directly before use.

2.3 Isoelectric Focussing (for 2-Dimensional Electrophoresis)

1. Urea (Sigma, Taufkirchen, Germany).
2. Dithiotreitol (DTT; Fluka, Seelze, Germany).
3. Pharmalyte 3–10, (GE Healthcare, Munich, Germany).
4. 3-3′-(Cholamidopropyl)-3,3-dimethylammonium-propylsulfat (CHAPS; Roth, Karlsruhe, Germany).

2.4 Western Blots

1. Ponceau reagent: 0.2% Ponceau S, 3% glacial acetic acid.
2. Peroxidase (POD) conjugated goat anti-mouse antibodies (Jackson Immunoresearch, No. 115 035 068 from Dianova, Hamburg, Germany).
3. Roti®-Block (Roth, Karlsruhe, Germany).
4. Fount India (Pelikan, Germany).

2.5 Detection of Total Proteins with "Fount India"

1. PBS-Tween: 48.8 g of NaCl, 14.5 g Na_2HPO_4, 1.17 g NaH_2PO_4, 0.5 g NaN_3, 0.5 mL Tween 20, make up to 5 L with distilled water.
2. Staining solution: 250 mL of PBS-Tween + 2.5 mL of acetic acid + 250 μL of Fount India, filter the staining solution.

2.6 Immunohistochemistry

1. Bovine serum albumin (BSA Fraction V, Sigma No A7906).
2. Blocking solution: 2% BSA/PBS, 0.2% Tween 20.
3. Cy-2 conjugated goat anti-mouse IgG and IgM Jackson Immunoresearch; No 115-225-068 from Dianova, Hamburg, Germany).
4. Permafluor (Beckmann-Coulter, Krefeld, Germany).

3 Methods

3.1 Enzyme Linked Immunosorbent Assay (ELISA)

1a. Coat Nunc Maxisorp 96-well flat bottom plates overnight at 4°C with 50 μL GlcNAc-BSA in 0.1M $NaHCO_3$ at a concentration of 12.5 μg per well. Continue with step 3, 4 and then from step 6 on as for the samples described in 1b.

1b. To analyze detergent soluble and detergent insoluble fractions from mouse brain (dsf and dicf, respectively, *see* **Note 1**), coat microtiter plates 30 min with 0.5% glutaraldehyde solution, wash after the removal of glutaraldehyde 3 times with PBS.

2. Add 50 µL of the protein fractions at a concentration of 500 µg/mL for 1 h at room temperature.

3. Remove protein solutions and wash 15 min with 150 mM Tris buffered saline (TBS, pH 7.2, *see* **Note 2**).

4. Block 30 min with 350 µL/well 1% horse serum.

5. Add 4 µL 100% formic acid (*see* **Note 3**), mix well and incubate for 2 additional hours at room temperature.

6. Wash coated plates 3 times with TBS.

7. Add the primary antibody in appropriate dilutions (1:4,000 for the RL2 antibody and 1:30,000 for the CTD 110.6 antibody in 50 µL/well TBS containing 0.1% BSA (w/v) and 0.1% Tween 20 (v/v) and incubate at 4°C for 16 h. Wells for 2nd antibody control are kept during this time in 1% horse serum.

8. Perform 3 washing steps with TBS, then incubate wells with 50 µL POD-conjugated anti-mouse antibodies diluted 1:7,500 in BSA/TBS-Tween 20 for 1 h at room temperature and wash as described previously.

9. For the development incubate each well with 50 µL ABTS solution for up to 30 min at room temperature and terminate the reaction by adding 50 µL 0.6% aqueous SDS solution per well.

10. Determine antibody reactivity by measuring the optical density of the reaction products of the POD conjugated to the secondary antibodies at 405 nm wavelength (OD 405) in an ELISA reader.

Figure 19.1 shows that in ELISA the CTD 110.6 antibody has a weaker reactivity toward mouse brain proteins than the RL2 antibody whereas it is the opposite for the GlcNAc-BSA. But see Figure 19.2 for the reactivity in Western blots (*see* also **Note 9**).

3.2 Gel Electrophoresis

Determine protein concentration using the D$_c$ Protein Assay Kit from Biorad (Munich, Germany). For 1D gel electrophoresis use 10% (Fig. 19.2) or 8% (Fig. 19.3) polyacrylamide gels for SDS-PAGE according to the method of Laemmli *(31)*. Load each lane with approximately 25 µg protein (*see* **Note 4**).

For 2D gel electrophoresis first isoelectric focussing is performed.

1. Rehydrate IPG-strips (linear pH 5–9; 10 cm) prepared according to Westermeier *(32)* for 12 h in the presence of 8M urea, 6.5 mM DTT, 2% Pharmalyte 3–10 and 4% CHAPS.

2. Remove excess rehydration solution.

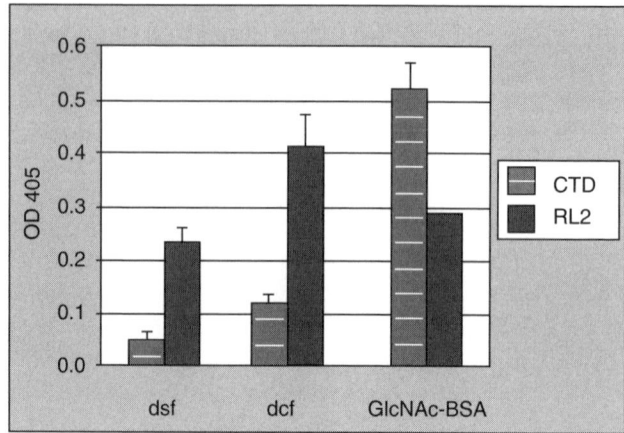

Fig. 19.1 Comparison of the reactivity of *O*-GlcNAc specific monoclonal antibodies RL2 and CTD 110.6 with GlcNAc-BSA and with detergent soluble and detergent insoluble cytoskeletal protein fractions from mouse brain (dsf and dicf, respectively). The optical density at 405 nm (OD 405) of reaction products with POD conjugated secondary antibodies was measured, mean values ± standard deviations of 1 representative out of 3 experiments carried out in triplicate are shown. Note that RL2 exhibit stronger reactivity toward dsf and dicf proteins than the CTD 110.6 antibody whereas the opposite reactivity is detected toward *O*-GlcNAc-BSA

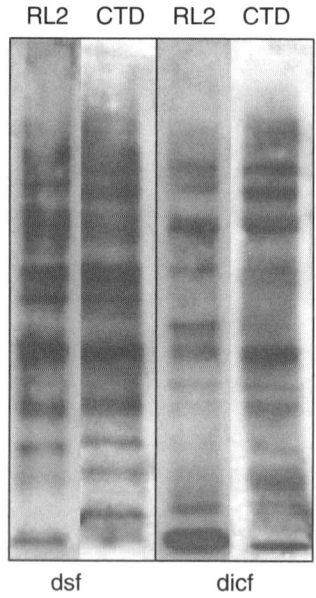

Fig. 19.2 Western blot of detergent soluble and detergent insoluble cytoskeletal protein fractions from mouse brain (dsf and dicf, respectively) with the *O*-GlcNAc specific monoclonal antibodies RL2 and CTD 110.6 antibody. Note that the antibodies recognize slightly different protein bands

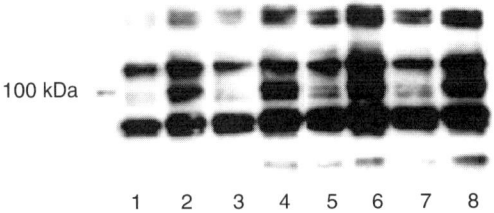

Fig. 19.3 Western blot of *O*-GlcNAc modified proteins from HEK cells cultivated in the presence of the *O*-GlcNAc hydrolase inhibitors PUGNAc and bha-4-37A. Shown are proteins from cell lysates after 4 h (lane 1–4) and after 24 h incubation (lane 5–8) with 40 μ*M* PUGNAc (lane 2 and 6) or 40 μ*M* bha-4-37A (lane 4 and 8). Control cells were incubated in the presence of vehicle H$_2$O (lane 1 and 5) and ethanol (lane 3 and 7). Note that both inhibitors produce a similar time dependent increase of the reactivity with the CTD 110.6 antibody

3. Place the sample application pieces (2 × 5 mm) onto the middle of each strip (3 mm × 10 cm) and 5 mm in front of the anode.
4. Apply 32.5 μg of protein to each sample application piece (*see* **Note 5**).
5. For isoelectric focussing a Multiphor II (GE Healthcare) operating in the linear mode was used.
6. Perform the focussing at 500 V, 2 mA, 5 W for 1 min; 3,500 V, 2 mA, 5 W for 90 min and 3,500 V, 2 mA, 5 W for 10 h, respectively.
7. After focussing store the IPG- strips either at −80°C or immediately equilibrate:

   a. 5 min in 2 mL equilibration stock solution (ESS; 6*M* urea 0,1 m*M* EDTA, 0.01% bromphenol blue, 50 m*M* Tris-HCl pH 6.8, 30% glycerol, v/v),
   b. 15 min in 2 mL ESS I (10 mL ESS containing 200 mg SDS, 100 mg DTT)
   c. 15 min in ESS II (10 mL ESS containing 200 mg SDS, 480 mg Iodoacetamide).

8. Perform protein separation in the 2nd dimension by electrophoresis on 12.5% SDS polyacrylamide according to Laemmli *(31)* with the following program: 1 h at 12.5 mA and approximately 180 min at 30 mA (1000 V and 50 W limits).

3.3   *Western Blot Analysis*

Perform Western blotting according to the method of Towbin et al. *(33)*.

1. Control success of the electrophoretic transfer of proteins after semi-dry blotting onto nitrocellulose sheets by incubating the blot with Ponceau solution for 1–2 min.
2. Destain blot by washing several times with water. Transferred proteins are visible as red bands.
3. Keep blot either for 1 h at 70°C in PBS or for 5 min in boiling PBS (*see* **Note 6**).
4. Block unspecific binding sites with Roti-Block for 1 h at room temperature.
5. Incubate blots with the appropriate dilution of the respective antibody in Roti-Block overnight at 4°C (1:1,000 for RL2 and 1:3,000 up to 1:10,000 for CTD 110.6, *see* **Note 7**).

6. Perform 5 washing steps with Roti-Block for 10 min each.
7. Use POD-conjugated anti-mouse antibodies in a 1:5,000 dilution in Roti-Block as secondary antibodies. Incubate for 1 h at room temperature.
8. Repeat 5 washing steps as described in step 6.
9. Detect peroxidase activity by enhanced chemiluminescence using the Super Signal West Dura Kit from Pierce (Perbio, St. Augustin, Germany) according to the description of the manufacturer.

3.4 Detection of Total Proteins on the Membrane with "Fount India" (see Note 8).

1. Remove antibodies from the nitrocellulose by stripping the membrane with 200 mM glycine pH 2.8 for 1 h at RT.
2. Wash the membrane 4 × 10 min with PBS-Tween (250 mL per wash).
3. Stain 2 h or overnight with staining solution.
4. Remove the dye by washing 3 × 10 min with PBS-Tween.
5. Conserve the membrane by washing 2 × 2 min with distilled water.
6. Dry the membrane.
7. Document the membrane by scanning.

In Figure 19.2, the reactivity of the RL2 and CTD 110.6 antibodies with O-GlcNAc modified proteins from mouse brain detergent soluble and detergent insoluble cytoskeletal fractions (dsf and dicf) is compared (see **Note 9**). Figure 19.3 shows the detectability of the time-dependent increase of O-GlcNAc on proteins from human embryonic kidney (HEK) cells in the presence of the novel OGH inhibitor bha-4-37A in comparison to the known OGH inhibitor O-(2-amido-2-deoxy-D-glucopyranosylidene)amino-N-phenylcarbamate (PUGNAc; see **Note 10**).

In Figure 19.4 Western blot analysis of the O-GlcNAc modification of crystallins from pig lens is shown after 2D separation(see **Note 11**).

3.5 Immunohistochemistry

1. Deparaffinize paraffin-embedded sections according to standard protocols and keep sections in PBS.
2. Block 60 min at room temperature in blocking solution.
3. Incubate sections with CTD 110.6 antibody diluted 1:200 in 1% BSA/PBS, 0.2% Tween 20 overnight at 4°C.
4. Wash 3 times with PBS.
5. Incubate sections with Cy-2 conjugated goat anti mouse IgG + M diluted 1:50 in 1% BSA/PBS, 0.2% Tween 1 h in the dark at RT.

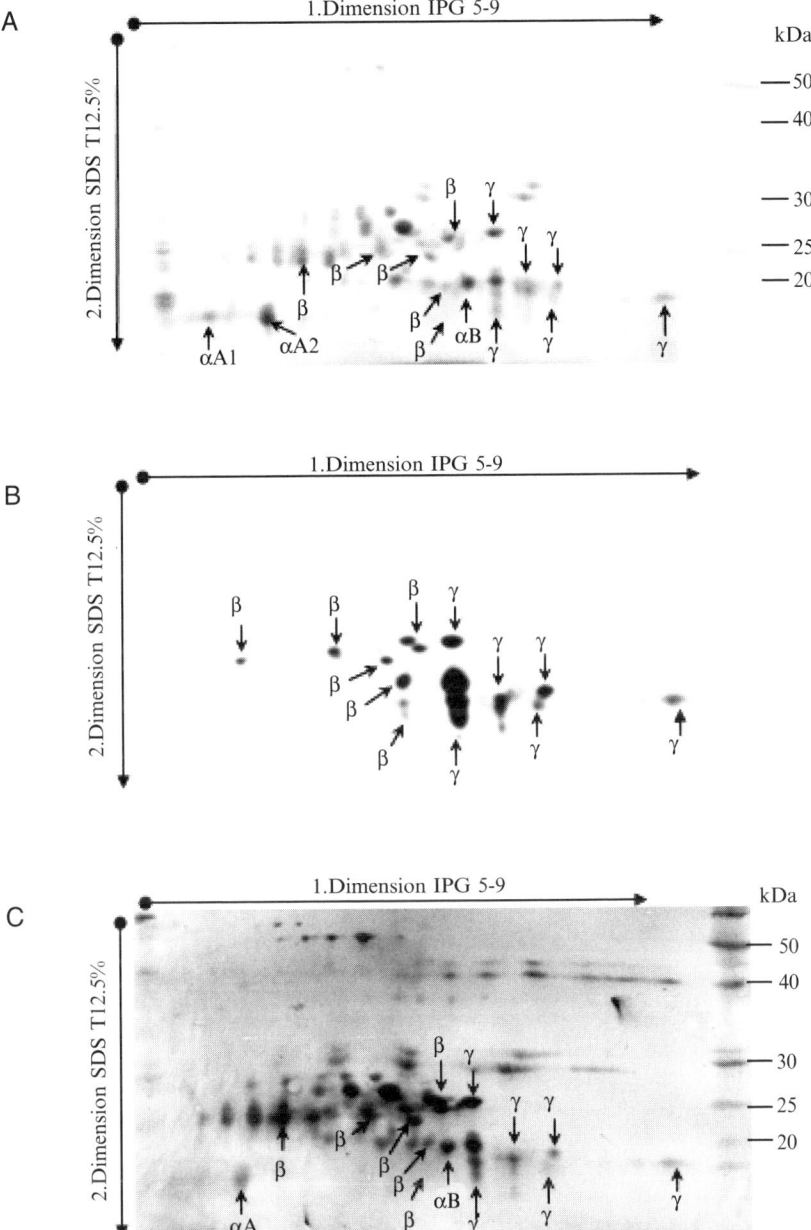

Fig. 19.4 Detection of *O*-GlcNAc on 2 D separated crystallins from pig lenses. Coomassie staining is shown in (**A**), Western blot with the CTD 110.6 antibody in (**B**) and in (**C**) Fount India staining after stripping of the antibodies with 200 m*M* glycine from the blot shown in (**B**)

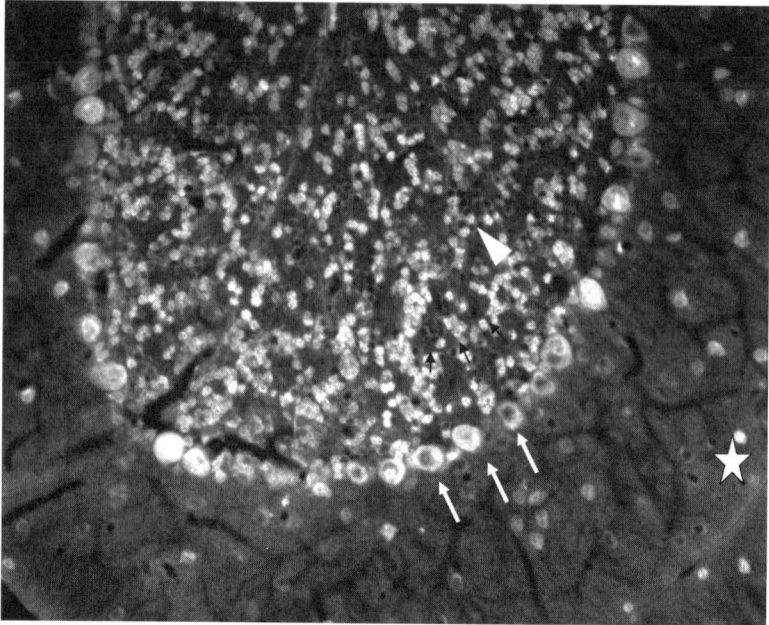

Fig. 19.5 Detection of *O*-GlcNAc in cerebellum of 30-day-old mice using CTD 110.6 antibody by indirect immunofluorescence analysis. The Cy2-conjugated secondary antibodies shows a marked staining of the cytoplasm of Purkinje cells (arrows) and cells in the interior granule cell layer (arrowhead). The surrounding molecular layer is poor of cells that are also *O*-GlcNAc positive (*). Preincubation of the antibody with 10 m*M* GlcNAc completely abolishes the staining (*see* also **Note 12**). Control sections without first antibody showed also no staining

6. Wash 3 times with PBS, rinse once with distilled water.
7. Embed coverslips in permafluor and keep at 4°C in the dark.

Figure 19.5 shows a representative section of 30 day old mouse cerebellum. Note the strong cytoplasmic staining of Purkinje cells (*see* **Note 12** for comparison of immunohistochemical staining with other *O*-GlcNAc specific antibodies).

4   Notes

1. Because we were interested in the detection of an altered expression of *O*-GlcNAc on proteins in specific subcellular fractions we prepared detergent soluble and detergent insoluble cytoskeletal fractions (dsf and dicf respectively). The preparation of these fractions is described in Griffith and Schmitz (*7*). The modified coating protocol using glutaraldehyde is particularly useful for the effective coating of proteins from solutions containing high amounts of detergents (*7,34,35*).

2. This step is necessary to block free glutaraldehyde groups.
3. The treatment with % formic acid serves as an additional denaturation step to release possibly masked epitopes.
4. For preparation of dsf and dicf proteins *see* **Note 1**. Proteins from human embryonic kidney cells were obtained by lysis of cells for 20 min at 4°C in RIPA lysis buffer (CC Pro, Neustadt/W, Germany) containing 1 m*M* phenylmethylsulfonylfluoride followed by centrifugation at 20,000*g*.
5. Freeze-dried pig eye lens crystalline material was prepared according to Ahrend et al *(36)* and solved in 200 μL 8*M* urea, 1% CHAPS, 0.1% DTT, 2% Pharmalyte 3–10 and 40 μ*M* O-(2-amido-2-deoxy-D-glucopyranosylidene)amino-*N*-phenylcarbamate (PUGNAc; *O*-GlcNAc hydrolase inhibitor).
6. As described by Snow et al. *(29)* an additional denaturation step was carried out by either keeping the blot for 1 h at 70°C in PBS (phosphate buffered saline) or for 5 min in boiling PBS. The heat treatment is recommended as a necessary step to obtain reproducibly high levels of labelling of antigen bands on immunoblots with the monoclonal antibodies used.
7. The dilution used for the CTD 110.6 antibody varies from lot to lot and although a dilution is recommended by the manufacturers, the optimal working concentration should be determined before use.
8. Fount India is the manufacturer denotation. Indian Ink is the dye.
9. The antibodies RL2 and CTD 110.6 give similar, but not identical staining patterns in Western blots. This is probably because of differences in the epitopes recognized by the antibodies with the CTD 110.6 being more specific for the GlcNAc than the RL2. This is also supported by the ELISA results and by the observation that 10 m*M* GlcNAc abolishes binding of CTD110.6, but inhibits only weakly RL2 antibody binding *(27,28)*. A comparison of the reactivity of these two antibodies is also described in the paper by Liu et al *(35)*.
10. Figure 19.3 shows that it is easily possible to manipulate *O*-GlcNAc expression in cultured cells by adding inhibitors of *O*-GlcNAc hydrolase. PUGNAc has been used frequently in the past for this purpose *(37–39)*. The other inhibitor bha-4-37A was synthesized by Shanmugasundaram and Vasella (MS in preparation). and is more stabile in aqueous solution than PUGNAc. Both inhibitors have been added to the culture medium at a concentration of 40 μ*M*.
11. The characterization of lens crystallins is according to *(40–42)*. Note that γ-crystallins exhibit a strong staining with the CTD antibody whereas α-crystallins give only very weak staining that is not visible on the blot shown. In contrast, mass spectrometric analysis clearly showed that alpha crystallins carry *O*-GlcNAc *(22,39,43–46)*. The Fountain Ink staining of the crystallins after blot stripping (Fig. 19.4C) is almost identical to the Coomassie staining of the gel (Fig. 19.4A) and represents, therefore, an important tool to identify protein spots recognized by specific antibodies directly on the same blot. In addition some unknown proteins of 30 kDa and higher are visible with Fountain Ink but not with Coomassie.
12. The indirect immunofluorescence staining with the CTD antibody is different to that described by Rex-Mathes et al *(27)* for the RL2, MUD 50 and HGAC 85 antibodies because these antibodies appear to stain nuclei stronger than the

cytoplasm whereas the CTD110.6 stains the cytoplasm much more strongly than the nuclei. Binding of the CTD antibody could be inhibited completely in the presence of 10 m*M* GlcNAc (not shown), whereas HGAC 85 and MUD 50 antibody staining could be reduced almost completely only in the presence of 100 m*M* GlcNAc and staining by the RL2 antibody was only partially inhibitable even with 0.5*M* GlcNAc *(27)*. Another difference is that the above mentioned antibodies did only function with cryosections and not with paraffin embedded sections as shown here for the CTD antibody. Indirect immunofluorescence detection of *O*-GlcNAc in mouse brain neurons in culture or of neuroblastoma cells was also carried out after fixation with 4% PFA 15 min at room temperature and permeabilization in cold ethanol for 5 min.

Acknowledgements We would like to thank A. Vasella and B. Shanmugasundaram, ETH Zurich, Switzerland, for the kind gift of the O-GlcNAc hydrolase inhibitor bha-4-37A *(5R,6R,7S,8S)-8-acetamido-5,6,7,8-tetrahydro-6,7-dihydroxy-5-(hydroxy-methyl)-N-phenylimidazo [1,2-a]pyridine-2-carboxamide)* and U-Munzel for expert secretarial assistance.

References

1. Wells, L., Vosseller, K., and Hart, G. W. (2001) Glycosylation of nucleocytoplasmic proteins: signal transduction and O-GlcNAc. *Science* **291**, 2376–2378.
2. Slawson, C., Housley, M. P., and Hart, G. W. (2006) O-GlcNAc cycling: how a single sugar post-translational modification is changing the way we think about signaling networks. *J. Cell Biochem.* **97**, 71–83.
3. Whelan, S. A. and Hart, G. W. (2003) Proteomic approaches to analyze the dynamic relationships between nucleocytoplasmic protein glycosylation and phosphorylation. *Circ. Res.* **93**, 1047–1058.
4. Torres, C. R. and Hart, G. W. (1984) Topography and polypeptide distribution of terminal N-acetylglucosamine residues on the surfaces of intact lymphocytes. Evidence for O-linked GlcNAc. *J. Biol. Chem.* **259**, 3308–3317.
5. Comer, F. I. and Hart, G. W. (2001) Reciprocity between O-GlcNAc and O-phosphate on the carboxyl terminal domain of RNA polymerase II. *Biochemistry* **40**, 7845–7852.
6. Wells, L., Whelan, S. A., and Hart, G. W. (2003) O-GlcNAc: a regulatory post-translational modification. *Biochem. Biophys. Res. Commun.* **302**, 435–441.
7. Griffith, L. S. and Schmitz, B. (1999) O-linked N-acetylglucosamine levels in cerebellar neurons respond reciprocally to pertubations of phosphorylation. *Eur. J. Biochem.* **262**, 824–831.
8. Lefebvre, T., Alonso, C., Mahboub, S., Dupire, M.-J., Zanetta, J.-P., Caillet-Boudin, M.-L., and Michalski, J.-C. (1999) Effect of okadaic acid on O-linked N-acetylglucosamine levels in a neuroblastoma cell line. *Biochim. Biophys. Acta* **1472**, 71–81.
9. Kamemura, K., Hayes, B. K., Comer, F. I., and Hart, G. W. (2002) Dynamic interplay between O-glycosylation and O-phosphorylation of nucleocytoplasmic proteins: alternative glycosylation/phosphorylation of THR-58, a known mutational hot spot of c-Myc in lymphomas, is regulated by mitogens. *J. Biol. Chem.* **277**, 19229–19235.
10. Wells, L., Kreppel, L. K., Comer, F. I., Wadzinski, B. E., and Hart, G. W. (2004) O-GlcNAc transferase is in a functional complex with protein phosphatase 1 catalytic subunits. *J. Biol. Chem.* **279**, 38466–38470.
11. Kreppel, L. K. and Hart, G. W. (1999) Regulation of a cytosolic and nuclear O-GlcNAc transferase. *J. Biol. Chem.* **274**, 32015–32022.

12. Lazarus, B. D., Love, D. C., and Hanover, J. A. (2006) Recombinant O-GlcNAc transferase isoforms: identification of O-GlcNAcase, yes tyrosine kinase, and tau as isoform-specific substrates. *Glycobiology* **16**, 415–421.

13. Matthews, J. A., Acevedo-Duncan, M., and Potter, R. L. (2005) Selective decrease of membrane-associated PKC-alpha and PKC-epsilon in response to elevated intracellular O-GlcNAc levels in transformed human glial cells. *Biochim. Biophys. Acta* **1743**, 305–315.

14. Slawson, C., Zachara, N. E., Vosseller, K., Cheung, W. D., Lane, M. D., and Hart, G. W. (2005) Perturbations in O-linked beta-N-acetylglucosamine protein modification cause severe defects in mitotic progression and cytokinesis. *J. Biol. Chem.* **280**, 32944–32956.

15. Wells, L. and Hart, G. W. (2003) O-GlcNAc turns twenty: functional implications for post-translational modification of nuclear and cytosolic proteins with a sugar. *FEBS Lett.* **546**, 154–158.

16. Zachara, N. E., O'Donnell, N., Cheung, W. D., Mercer, J. J., Marth, J. D., and Hart, G. W. (2004) Dynamic O-GlcNAc modification of nucleocytoplasmic proteins in response to stress. A survival response of mammalian cells. *J. Biol. Chem.* **279**, 30133–30142.

17. Lefebvre, T., Caillet-Boudin, M. L., Buee, L., Delacourte, A., and Michalski, J. C. (2003) O-GlcNAc glycosylation and neurological disorders. *Adv. Exp. Med. Biol.* **535**, 189–202.

18. McClain, D. A., Lubas, W. A., Cooksey, R. C., Hazel, M., Parker, G. J., Love, D. C., and Hanover, J. A. (2002) Altered glycan-dependent signaling induces insulin resistance and hyperleptinemia. *Proc. Natl. Acad. Sci. U S A* **99**, 10695–10699.

19. Griffith, L. S. and Schmitz, B. (1995) O-linked N-acetylglucosamine is upregulated in Alzheimer brains. *Biochem. Biophys. Res. Commun.* **213**, 424–431.

20. Vosseller, K., Hansen, K. C., Chalkley, R. J., Trinidad, J. C., Wells, L., Hart, G. W., and Burlingame, A. L. (2005) Quantitative analysis of both protein expression and serine / threonine post-translational modifications through stable isotope labeling with dithiothreitol. *Proteomics* **5**, 388–398.

21. Vocadlo, D. J., Hang, H. C., Kim, E. J., Hanover, J. A., and Bertozzi, C. R. (2003) A chemical approach for identifying O-GlcNAc-modified proteins in cells. *Proc. Natl. Acad. Sci. U S A* **100**, 9116–9121.

22. Khidekel, N., Ficarro, S. B., Peters, E. C., and Hsieh-Wilson, L. C. (2004) Exploring the O-GlcNAc proteome: direct identification of O-GlcNAc-modified proteins from the brain. *Proc. Natl. Acad. Sci. U S A* **101**, 13132–13137.

23. Nandi, A., Sprung, R., Barma, D. K., Zhao, Y., Kim, S. C., and Falck, J. R. (2006) Global identification of O-GlcNAc-modified proteins. *Anal. Chem.* **78**, 452–458.

24. Zachara, N. E., Cheung, W. D., and Hart, G. W. (2004) Nucleocytoplasmic glycosylation, O-GlcNAc: identification and site mapping. *Methods Mol. Biol.* **284**, 175–194.

25. Vosseller, K., Trinidad, J. C., Chalkley, R. J., Specht, C. G., Thalhammer, A., Lynn, A. J., Snedecor, J. H., Guan, S., Medzihradszky, K. F., Maltby, D. A., Schoepfer, R., and Burlingame, A. L. (2006) O-GlcNAc proteomics of postsynaptic density preparations using lectin weak affinity chromatography (LWAC) and mass spectrometry. *Mol. Cell. Proteomics* epup ahead of print, manuscript Nr. TS00040–MCP200.

26. Zachara, N. E. (2002) Detection and Analysis of Proteins Modified by O-linked N-Acetylglucosamine. Current Protocols in Molecular Biology, Wiley & Sons.

27. Rex-Mathes, M., Koch, J., Werner, S., Griffith, L. S., and Schmitz, B. (2002) Immunological detection of O-GlcNAc. *Methods Mol. Biol.* **194**, 73–87.

28. Comer, F. I., Vosseller, K., Wells, L., Accavitti, M. A., and Hart, G. W. (2001) Characterization of a mouse monoclonal antibody specific for O-linked N-acetylglucosamine. *Anal. Biochem.* **293**, 169–177.

29. Snow, C. M., Senior, A., and Gerace, L. (1987) Monoclonal antibodies identify a group of nuclear pore complex glycoproteins. *J. Cell. Biol.* **104**, 1143–1156.

30. Han, I. and Kudlow, J. E. (1997) Reduced O-glycosylation of Sp1 is associated with increased proteasome susceptibility. *Mol. Cell. Biol.* **17**, 2550–2558.

31. Laemmli, U. K. (1970) Cleavage of structural proteins during the assembly of the head of bacterophage T 4. *Nature* **227**, 680–685.

32. Westermeier, R. (1990) *Electrophoresis in Practice*. VCH Verlagsgesellschaft mbH.
33. Towbin, H., Staehelin, T., and Gordon, J. (1979) Electrophoretic transfer of proteins from polyacrylamide gels to nitrocellulose sheets: procedure and some applications. *Proc. Natl. Acad. Sci. USA* **76**, 4350–4354.
34. Evan, G. I. (1984) A simple and rapid solid phase enzyme linked immunoadsorbance assay for screening monoclonal antibodies to poorly soluble proteins. *J. Immunol. Methods* **73**, 427–435.
35. Liu, F., Iqbal, K., Grundke-Iqbal, I., Hart, G. W., and Gong, C. X. (2004) O-GlcNAcylation regulates phosphorylation of tau: a mechanism involved in Alzheimer's disease. *Proc. Natl. Acad. Sci. U S A* **101**, 10804–10809.
36. Ahrend, M. H. J., Bours, J., and Födisch, H. J. (1987) Water-soluble and insoluble crystallins of the developing human fetal lens, analyzed by agarose/polyacrylamide thin-layer isoelectric focusing. *Ophtalmic Res.* **19**, 150–156.
37. Dong, D.-Y. and Hart, G. W. (1994) Purification and characterization of an O-GlcNAc selective N-Acetyl-beta-D-Glucosaminidase from rat spleen cytosol. *J. Biol. Chem* **269**, 19321–19330.
38. Haltiwanger, R. S., Grove, K., and Philipsberg, G. A. (1998) Modulation of O-linked N-acetylglucosamine levels on nuclear and cytoplasmic proteins in vivo using the peptide O-GlcNAc-beta-N-acetylglucosaminidase inhibitor O-(2-acetamido-2-deoxy-D-glucopyranosylidene) amino-N-phenylcarbamate. *J. Biol. Chem.* **273**, 3611–3617.
39. Park, S. Y., Ryu, J., and Lee, W. (2005) O-GlcNAc modification on IRS-1 and Akt2 by PUGNAc inhibits their phosphorylation and induces insulin resistance in rat primary adipocytes. *Exp Mol Med* **37**, 220–229.
40. Bloemendal, H., Van de gaer, K., Benedetti, E. L., Dunia, I., and Steely, H. T. (1997) Towards a human crystallin map. Two-dimensional gel electrophoresis and computer analysis of water-soluble crystallins from normal and cataractous human lenses. *Ophthalmic Res.* **29**, 177–190.
41. Bours, J. (1971) Isoelectric focusing of lens crystallins in thin-layer polyacrylamide gels. A method for detection of soluble proteins in eye lens extracts. *J. Chromatogr.* **60**, 225–233.
42. Datiles, M. B., Schumer, D. J., Zigler, J. S., Jr., Russell, P., Anderson, L., and Garland, D. (1992) Two-dimensional gel electrophoretic analysis of human lens proteins. *Curr. Eye Res.* **11**, 669–677.
43. Chalkley, R. J. and Burlingame, A. L. (2003) Identification of novel sites of O-N-acetylglucosamine modification of serum response factor using quadrupole time-of-flight mass spectrometry. *Mol. Cell. Proteomics* **2**, 182–190.
44. Haynes, P. A. and Aebersold, R. (2000) Simultaneous detection and identification of O-GlcNAc-modified glycoproteins using liquid chromatography-tandem mass spectrometry. *Anal. Chem.* **72**, 5402–5410.
45. Roquemore, E. P., Chevrier, M. R., Cotter, R. J., and Hart, G. W. (1996) Dynamic O-GlcNAcylation of the small heat shock protein alpha B-crystallin. *Biochemistry* **35**, 3578–3586.
46. Roquemore, E. P., Dell, A., Morris, H. R., Panico, M., Reason, A. J., Savoy, L. A., Wistow, G. J., Zigler, J. S., Jr., Earles, B. J., and Hart, G. W. (1992) Vertebrate lens alpha-crystallins are modified by O-linked N-acetylglucosamine. *J. Biol. Chem.* **267**, 555–563.

20
Analysis of *O*-Glycosylation

Juan J. Calvete and Libia Sanz

Summary Secreted as well as membrane-associated eukaryotic proteins are most commonly glycosylated. Saccharides are attached to proteins mainly through *N*- and *O*-glycosydic bonds or as part of the glycosylphosphatidyinositol-membrane anchor. In contrast to *N*-glycosylation, which involves the co-translational transfer in the endoplasmic reticulum (ER) of the glycan portion of Glc3Man9GlcNAc2-PP-dolichol to suitable Asn residues on nascent polypeptides, *O*-glycosylation begins with the addition of a single monosaccharide. Contrary to *N*-glycosylation, which involves an asparagine residue in the sequon Asn-Xaa-Thr/Ser (Xaa can be any amino acid except Pro, and it is rarely Cys), no particular sequence motif has been described for *O*-glycosylation. This may reflect the fact that: (1) the specificity of the UDP-GalNAc:polypeptide *N*-acetylgalactosaminyltransferase is presently unknown; and (2) seems to be modulated by sequence context, secondary structure, and surface accessibility *(1)*. An internet server, accessible at http://www.cbs.dtu. dk/netOglyc/cbsnetOglyc.html, produces neural network predictions of mucin-type GalNAc *O*-glycosylation sites in mammalian proteins based on 299 known and verified mucin-type *O*-glycosylation sites. The sequence context of glycosylated threonines was found to differ from that of serine, and the sites were found to cluster. Nonclustered sites had a sequence context different from that of clustered sites, and charged residues were disfavored at position −1 and +3.

Key Words O-glycosylation; GalNAc O-glycosylation; mucin-type O-glycosylation.

1 Introduction

Secreted as well as membrane-associated eukaryotic proteins are most commonly gly-cosylated. Saccharides are attached to proteins mainly through *N*- and *O*-glycosydic bonds or as part of the glycosylphosphatidyinositol membrane anchor. In contrast to *N*-glycosylation, which involves the cotranslational transfer in the endoplasmic reticulum (ER) of the glycan portion of $Glc_3Man_9GlcNAc_2$-PP-dolichol to suitable Asn residues on nascent polypeptides, *O*-glycosylation begins with the addition of a single monosac-charide. Contrary to *N*-glycosylation which involves an asparagine residue in the

From: *Post-translational Modifications of Proteins.*
Methods in Molecular Biology, Vol. 446.
Edited by: C. Kannicht © Humana Press, Totowa, NJ

sequon Asn-Xaa-Thr/Ser (Xaa can be any amino acid except Pro, and it is rarely Cys), no particular sequence motif has been described for O-glycosylation. This may reflect the fact that i) the specificity of the UDP-GalNAc:polypeptide N-acetylgalactosaminyl-transferase is presently unknown and ii) seems to be modulated by sequence context, secondary structure and surface accessibility (1). An internet server, accessible at http://www.cbs.dtu.dk/netOglyc/cbsnetOglyc.html, produces neural network predictions of mucin type GalNAc O-glycosylation sites in mammalian proteins based on 299 known and verified mucin-type O-glycosylation sites. The sequence context of glycosylated threonines was found to differ from that of serine, and the sites were found to cluster. Nonclustered sites had a sequence context different from that of clustered sites, and charged residues were disfavoured at position −1 and +3.

In animal cells, O-glycosylation is normally initiated in the Golgi apparatus by enzymatic transfer of a N-acetylgalactosamine (GalNAc) residue to the hydroxyl side chain of a serine or a threonine residue. This reaction is catalyzed by GalNAc transferase (GalNAc-T, EC 2.4.1.41) using UDP-GalNAc as the sugar donor. O-glycosylation of hydroxylysine and hydroxyproline (in collagen and in plant proteins) and O-linked GlcNAc (often on cytoplasmic and nuclear proteins as a reversible regulatory modification) have also been characterized. In addition, O-linked fucose occurs together with O-linked glucose on specific sequons in the EGF domains of several proteins (2).

In yeast, unlike in animal cells, O-glycosylation begins in the endoplasmic reticulum by addition of a single mannose from Man-P-dolichol to selected Ser/Thr residues; once transported to the Golgi, sugar transferases add one or more $\alpha 1,2$-linked mannoses that may be capped with one or more $\alpha 1,3$-linked mannose residues (3,4). In animal cells, however, stepwise enzymatic elongation by specific transferases yields several core structures, which can be further elongated or modified by sialylation, sulphation, acetylation, fucosylation, and polylactosamine extension. The 8 core structures identified to date (2,5) are shown in Table 20.1.

The diversity of O-linked oligosaccharides linked to eukaryotic glycoprotein is comparable to that described for N-linked glycans, and site microheterogeneity and variable site occupancy are both common phenomena. The consequence of this is that many isolated glycoproteins are actually a set of glycoforms exhibiting small differences in their physicochemical properties. Thus, in the practice, glycosylation analysis is performed on a mixture of glycoforms.

Most analytical strategies involve several stages that seek to address questions like "is the protein glycosylated?"; "does it contain N- and/or O-glycans?"; "which amino acid residues are glycosylated?"; "which are the structures of the sugar chains?". The methods outlined below are simple protocols that can be carried out in many noncarbohydrate specialized biochemical laboratories.

1.1   Occurrence and Extent of O-Glycosylation

Diverse strategies may aid in determining the occurrence of O-glycosylation in a protein. As a first approach monosaccharide compositional analysis provides a good estimate of the presence of N- and/or O-glycosylation sites. Thus, GalNAc is

Table 20.1 The Ser/Thr O-Linked Core
Sequences of Animal Glycoproteins

1	Galβ1,3-GalNAcol
2	GlcNAcβ1,6
	|
	GalNAcol
	|
	Galβ1,3
3	GlcNAcβ1,3-GalNAcol
4	GlcNAcβ1,6
	|
	GalNAcol
	|
	GlcNAcβ1,3
5	GalNAcα1,3-GalNAcol
6	GlcNAcβ1,6-GalNAcol
7	GalNAcα1,6-GalNAcol
8	Galα1,3-GalNAcol

present in all *O*-linked oligosaccharides, while this amino sugar is uncommon in *N*-linked glycans. On the other hand, all N-linked carbohydrate chains contain mannose residues, but it is not present in most *O*-linked oligosaccharide chains. Thus, detection of GalNAc and absence of Man strongly indicate that the protein may be *O*-glycosylated. The situation is normally not so simple and a given glycoprotein may exhibit both *N*- and *O*-glycosylation sites.

2 Materials

2.1 Determination of Amino Sugars using an Amino Acid Analyzer

1. D-(+/−) Glucosamine hydrochloride (2-amino-2-deoxy-D-glucopyranoside) and D(+/−). galactosamine (2-amino-2-deoxy-D-galactopyranoside), crystalline, minimum 99%.
2. Amino acid standard solution (Sigma) for protein hydrolysates.
3. L-norleucine (D-α-amino *n*-caproic acid).
4. Hydrochloric acid (HCl) 37% (12*M*) for analysis.
5. Deionized water (MilliQ (18 MΩ) or equivalent).
6. Glass ampules or Pasteur pipets.
7. Blowlamp.
8. Vacuum pump.
9. Thermobloc (temperature range up to 130°C).
10. Amino acid analyzer with postcolumn (ninhidrin) derivatization.

2.2 Determination of Dansyl Hydrazone Derivatives of Neutral Sugars by Reversed-Phase HPLC

1. 2N Hydrochloric acid (HCl, for analysis).
2. Standard neutral sugars (maltose, lactose, galactose, glucose, mannose, fucose, xylose; minimum 99%).
3. Dowex 50W X4 (200–400 mesh, H$^+$ form) resin.
4. 2N NaOH.
5. Deionized water (MilliQ (18 MΩ) or equivalent).
6. 10% (v/v) Trichloroacetic acid (TCA).
7. 5% solution of dansyl hydrazine (dansyl-(5-dimethylaminonaphtalene-1-sulphonyl)-hydrazine) in acetonitrile.
8. Rotary evaporator (SpeedVac).
9. Water bath.
10. C18 Sep-Pak cartridges (Waters Ass.).
11. 5%, 18%, and 20% (v/v) acetonitrile in water.
12. HPLC equipment.
13. Reversed-phase C18 analytical column (25 × 0.4 cm, 5 μm particle size, 100 Å pore size).

2.3 Molecular Mass Determination of Intact Glycoproteins

1. 0.1% trifluoroacetic acid.
2. Acetonitrile.
3. Matrix solution: 22 mg of 3,5-dimethoxy-4-hydroxycinnamic acid in 1 mL of 40:60 (v/v) acetonitrile: 0.1% trifluoroacetic acid in water.
4. MALDI mass spectrometer.

2.4 Release of O-Linked Oligosaccharides

2.4.1 Mass Spectrometric Analysis of O-Deglycosylated Protein

1. 0.1M NaOH.
2. Thermobloc (temperature range up to 60°C).
3. Gel filtration column (Bio-Gel P10 or P30 (BioRad); Superdex HR10/30 (Pharmacia) or equivalent).
4. Lyophilisator.
5. Deionized water (MilliQ (18 MΩ) or equivalent).
6. MALDI mass spectrometer.

2.4.2 Recovery and Initial Characterization of Released Oligosaccharides

1. 0.1M NaOH.
2. 1M NaBH$_4$.

3. Thermobloc (temperature range up to 60°C).
4. 4*M* Acetic acid.
5. Rotary evaporator (SpeedVac).
6. Methanol:acetic acid (99:1, v/v).
7. Sephadex G25 Superfine Fast-Desalting HR 10/10 column.
8. FPLC system.
9. Matrix solution: mixture of 2,5-dihydroxybenzoic acid (DHB) and α-cyano-4-hydroxycinnamic acid, each at a concentration of 10 mg/mL in 30% (v/v) acetonitrile/water.

2.5 Fractionation of Oligosaccharides: Anion-Exchange Chromatography

1. FPLC system.
2. MonoQ HR 5/5 column (Pharmacia).
3. Deionized water (MilliQ (18 MΩ) or equivalent).
4. 0.5*M* NaCl

2.6 Fractionation of Oligosaccharides: Amino-Bonded Phase Chromatography

1. Rotary evaporator (Speed Vac)
2. 85% (v/v) acetonitrile: 15% 5–15 m*M* potassium phosphate, pH 7.0
3. 5–15 m*M* potassium phosphate, pH 7.0
4. Lichrosorb-NH2 (Merck) 25 × 0.4 cm, 5 μm particle size)
5. HPLC system

2.7 Location of O-Glycosylation Sites

1. 100 m*M* ammonium bicarbonate, pH 8.3
2. Proteolytic enzymes (TPCK-trypsin, α-chymotrypsin, endoproteinase Lys-C or endoproteinase Asp-N).
3. HPLC system
4. (25 × 0.4 cm) C18 (5 μm particle size, 100 Å pore size) column
5. 0.1% (v/v) trifluoroacetic acid in water
6. 0.1% (v/v) trifluoroacetic acid in acetonitrile
7. Amino acid analyzer with postcolumn (ninhidrin) derivatisation
8. N-terminal sequencer
9. MALDI mass spectrometer

3   Methods

3.1   Determination of Amino Sugars Using an Amino Acid Analyzer

Amino sugars (GlcNAc and GalNAc) can be easily identified and quantitated using an automated amino acid analyzer with postcolumn (ninhydrin) derivatization *(6)*.

1. Make a stock solution of glycoprotein in water or 0.1% TFA.
2. Put aliquots of 0.5–1 nmol glycoprotein in two glass ampules (*see* **Note 1**).
3. Add HCl to final concentrations of 6*M* (for total protein hydrolysis) and 4*M* (for amino sugar analysis).
4. To avoid decomposition of hydroxy amino acids, the ampules should be evacuated to below 50 microns for 1–2 min.
5. Seal the ampules under vacuum using a blowlamp.
6. Hydrolyze samples in a thermobloc at 110°C for 24 h (for protein hydrolysis) and 4 h (for amino sugar analysis).
7. Open the ampules, add a known amount (i.e., 5–10 nmoles) of norleucine as internal standard, and remove HCl in a rotary evaporator (SpeedVac).
8. Run the samples in an automated amino acid analyzer calibrated with a mixture of standard amino acids containing known amounts of galactosamine and glucosamine (*see* **Note 2**).

3.2   Determination of Dansyl Hydrazone Derivatives of Neutral Sugars by Reversed-Phase HPLC

The method outlined below for labeling of the reducing end of sugars with a fluorescent tag followed by HPLC separation and quantitation of the resulting dansyl hydrazones of neutral sugars has been employed in the author's laboratory for compositional analysis of human platelet $\alpha_{IIb}\beta_3$ integrin subunits *(8,9)*. It is simple, reproducible, and does not need a dedicated carbohydrate analyzer.

1. The optimal conditions for the release of neutral sugars may vary from glycoprotein to glycoprotein. Therefore optimal conditions should be determined in a case-to-case manner. However, hydrolysis with 2*N* HCl for 2 h at 110°C in sealed tubes or ampules (*see* steps 1–6 of **Section 3.1.**) are good starting conditions.
2. Treat a mixture of neutral sugars (10 nmoles of each maltose, lactose, galactose, glucose, mannose, fucose, and xylose) in a separate tube the same way as the glycoprotein (*see* **Note 3**).
3. After acid hydrolysis, dry standards and samples in their hydrolysis tubes under reduced pressure, at 40–45°C, using a rotavapor or SpeedVac evaporator.
4. Dissolve dried samples in 0.5 mL of deionized water (MilliQ) and add 10 nmoles of an internal standard (maltose or lactose) to each hydrolysate.

5. Prepare 0.5 mL columns of Dowex 50W X4 (200–400 mesh, H⁺ form) (Fluka) using a Pasteur pipets containing a piece of glass-wool as filter. Wash the resin exhaustively with 2*N* NaOH, water, 2 N HCl, and water.

6. Apply the samples and wash the columns with 5 mL water. Neutral sugars are not retained in the Dowex columns and will be, thus, recovered in the flow-through fraction.

7. Dry the samples in a rotavapor or SpeedVac evaporator and dissolve the residues containing the neutral sugars in 100 µL water.

8. Add 10 µL of TCA 10% and 50 µL of a 5% solution of dansyl hydrazine (dansyl-(5-dimethylaminonaphtalene-1-sulphonyl)-hydrazine) (Sigma) in acetonitrile.

9. Cap the tubes and incubate for 20 min at 65°C in a water bath in the dark.

10. Place the reaction mixtures in an ice-cold water bath and dilute with 2 mL water.

11. Activate a Sep-Pak C18 cartridge (Waters Associated) by rinsing with 3 mL acetonitrile followed by 5 mL water.

12. Load slowly the derivatized samples (point 10.) into the activated Sep-Pak cartridge and rinsed it slowly with 5 mL of 5% (v/v) acetonitrile. (*see* **Note 4**).

13. Recover the dansyl hydrazones by elution with 6 mL of 20% (v/v) acetonitrile (*see* **Notes 5** and **6**).

14. Dry the dansyl hydrazone fraction in a SpeedVac.

15. Dissolve in 0.3 mL of 18% (v/v) acetonitrile and store in the dark and at 4°C until use (*see* **Note 7**).

16. Inject aliquots of 25–100 µL of the dansyl hydrazone solution into an analytical (25 × 0.4 cm) HPLC C18 reversed-phase column (100 Å pore size, 5 µm particle size) eluted at a flow rate of 1 mL/min isocratically with 18% (w/w) acetonitrile at room temperature (around 22°C) (Fig. 20.1) (*see* **Note 8**).

17. Wash the column with 100% acetonitrile until baseline indicates no further elution of interfering material, and equilibrate with 18% (w/w) acetonitrile for subsequent analyzes.

3.3   Molecular Mass Determination of Intact Glycoproteins

Mass spectrometric analysis of the glycoprotein before and after release of *O*-glycans provide an estimate of the relative amount of total *O*-linked carbohydrate. The molecular mass of intact glycoproteins may be accurately determined by matrix-assisted laser-desorption ionization (MALDI) mass spectrometry:

1. Dissolve the glycoprotein (approximately 10–50 pmol/µL) in 0.1% trifluoroacetic acid (TFA). Add acetonitrile in 10%-steps if necessary for solubilization.

2. Prepare the matrix solution: dissolve 22 mg of 3,5-dimethoxy-4 hydroxycinnamic acid in a 1 mL of 40:60 (v/v) acetonitrile: 0.1% TFA in water.

3. Mix equal volumes (0.5–1 µL) of protein and matrix solution onto the stainless steel MALDI target and dry at room temperature.

4. Set an acceleration voltage of 30 kV and operate in the linear mode.

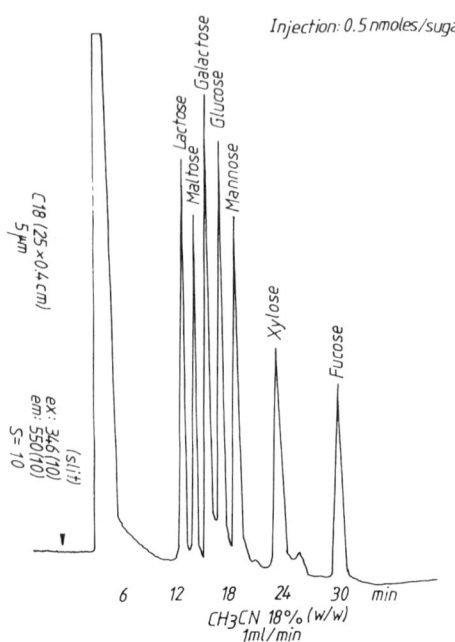

Fig. 20.1 Separation of the dansyl hydrazones of standard sugars by reverse-phase HPLC using a 5 μm C18 analytical column (25 × 0.4cm) and isocratic elution with 18%(w/w) acetonitrile, at room temperature and a flow-rate of 1 mL/min 0.5 nmol of each sugar were injected. Fluorescence detection was done in a Perkin-Elmer MPF 3 spectrofluorimeter equipped with a 15μL flow cell. Excitation wavelength: 346 nm (slit 10); emission wavelength: 550nm (slit 10); sensitivity: 10

3.4   Release of O-linked Oligosaccharides

3.4.1   Mass Spectrometric Analysis of O-Deglycosylated Protein

Although the enzyme O-glycosidase from *Streptococcus pneumoniae* (endo-α-N-acetyl-galactosaminidase, EC 3.2.1.97 or 3.2.1.110) can be used for cleavage of O-linked oligosaccharides from glycoproteins, it has a narrow substrate specificity, releasing only unsubstituted galactose β-3-N-acetylgalactosamine. Thus, a mixture of exoglycosidases (sialidase, α-fucosidase, α-N-acetylgalactosaminidase) is often required to be used with or before digestion with this enzyme. For mass spectrometric analysis of deglycosylated glycoproteins, O-linked oligosaccharides can be optimally released by mild alkali-catalyzed β-elimination:

1. Dissolve the glycoprotein (approximately. 10 mg/mL) in 0.05–0.1M NaOH.
2. Incubate for 16–24 h at 40–50°C.
3. Desalt on a gel-filtration column (approximately 30×0.5 cm) Bio-Gel P10 or P30 (BioRad) or Superdex HR10/30 (Pharmacia).

4. Dialyze against deionized water and lyophilize.
5. Determine the molecular mass by MALDI mass spectrometry as in **Section 3.3.**

3.4.2 Recovery and Initial Characterization of Released Oligosaccharides

1. For release the *O*-linked glycans by β-elimination dissolve the glycoprotein as point 1. of **Section 3.4.1.** However, to prevent breakdown of the oligosaccharide chains, 0.5–1*M* NaBH4 must be included in the reaction mixture. The released oligosaccharides are recovered as alditols. Because of the inclusion of the borohydride, release of *N*-glycans may also occur.
2. Incubate the reaction mixture for 16–24 h at 40–50°C.
3. Cool on ice and neutralize to pH 6 with 4*M* acetic acid.
4. Remove boric acid by repetitive co-evaporation (using a rotavapor) with methanol containing 1% acetic acid.
5. Resuspend in 1 mL deionized water and desalt on a Sephadex G25 Superfine Fast-Desalting HR 10/10 column using an FPLC system (Pharmacia). Monitor the elution of oligosaccharides at 206 nm.
6. The degree of compositional heterogeneity and the relative abundance of oligosaccharides in the pool of released glycans may be conveniently analyzed by mass spectrometry. Underivatized oligosaccharides can be analyzed by MALDI-MS using a matrix such as a mixture of 2,5-dihydroxybenzoic acid (DHB) and α-cyano-4-hydroxycinnamic acid, each at a concentration of 10 mg/mL in 30% acetonitrile/water with a detection limit of about 100 fmol *(7)*. Alternatively, determination of the molecular masses of reduced and permethylated *(7)* oligosaccharides (~10 pmol/μL) can be determined by MALDI mass spectrometry (MS) (using 10 mg/mL DHB in 10% ethanol as matrix) or by electrospray ionisation (ESI) mass spectrometry. For ESI-MS, the reduced and permethylated oligosaccharides are dissolved in methanol saturated with NaCl at a concentration of ~10 pmol/μL. In this case, oligosaccharides are detected as the natrium adducts (M + 23 Da) (*see* **Note 9**).

3.5 Fractionation of Oligosaccharides: Anion-Exchange Chromatography

Released oligosaccharide chains can be fractionated according to their charge (i.e., neutral, monosialylated, and disialylated) by anion-exchange chromatography on a MonoQ HR 5/5 column using a Pharmacia FPLC system.

1. Dissolve the oligosaccharide pool in 0.5 mL of deionized (MilliQ) water.
2. Apply the sample and run the column at 1 mL/min with a mixture of water (solvent A) and 0.5*M* NaCl (solvent B). Separation of different mixtures may require optimization of the elution conditions. Tentatively starting conditions could be: isocratic (100% A) for 5 min followed by a linear gradient from 100% A–100% B for 30–45 min. Oligosaccharides are detected at 206 nm.

3.6   Fractionation of Oligosaccharides: Amino-Bonded Phase Chromatography

Oligosaccharide fractions isolated by anion-exchange chromatography might be further subfractionated by means of amino-bonded HPLC. Oligosaccharides are eluted in order of increasing molecular weight.

1. Dry the MonoQ fractions (isolated as described in **Section 3.5.**) containing 10–200 µg oligosaccharide using a Speed-Vac.
2. Dissolve each fraction in 0.5 mL of 85% (v/v) acetonitrile containing 5–15 mM potassium phosphate pH 7.0 (solvent A).
3. Apply to a Lichrosorb-NH2 (0.4 × 25 cm, 5 µm particle size) Merck HPLC column eluting at 1 mL/min with linear gradients of solvent A and 5–15 mM potassium phosphate pH 7.0 (solvent B). Monitor oligosaccharide elution by absorbance at 206 nm.

3.7   Location of O-Glycosylation Sites

The simple strategy for determination of the position of O-glycosylated residues along the amino acid sequence of a glycoprotein includes fragmentation of the protein, separation of the resulting peptides, and structural characterization of the isolated fragments by combination of N-terminal sequencing and carbohydrate analysis (*see* **Note 4**) (*12*). This approach requires that the primary structure of the protein is known.

1. Degrade the glycoprotein (1–2 mg/mL in 100 mM ammonium bicarbonate, pH 8.3 (*see* **Notes 10** and **11**)) with the chosen proteolytic enzyme (usually TPCK-trypsin, α-chymotrypsin, endoproteinase Lys-C, or endoproteinase Asp-N) at an enzyme:substrate ratio of 1:100 (w/w) overnight at 37°C.
2. Separate peptides by reversed-phase HPLC (*see* **Note 12**).
3. Determine the presence of galactosamine in the HPLC-separated fractions by amino acid analysis as in 3.1.
4. Determine the N-terminal sequence of those peptides containing galactosamine. A blank in the sequence may indicate the existence at that position of a glycosylated residue (*see* **Note 13**).
5. Determine the molecular mass of galactosamine-containing peptides. The difference between the experimentally determined mass and that calculated from the amino acid sequence indicates the size of the glycan.

3.8   Sequence Analysis

Sequencing of O-linked oligosaccharides involves two separate steps: identification of the monosaccharide units followed by the linkages between units in the sequence. Carbohydrate compositional analysis of glycans or glycopeptides isolated as in **Sections 3.4.2.** or **3.7.** involve the same protocols as described in **Sections 3.1.** and **3.2.** Several techniques are available for the determination of the anomericity of the glycosidic linkages, specific linkages between monosaccharides, and branch configuration, including

i) methylation analysis in combination with diverse mass spectrometric techniques (fast-atom-bombardment mass spectrometry; collision-induced dissociation using tandem [electrospray] mass spectrometry; postsource decay by MALDI mass spectrometry); ii) enzymatic analysis using exo- and endoglycosidases; and iii) 1H-NMR. However, a comprehensive review of these techniques, which are also used for structural characterization of *N*-linked oligosaccharides, is beyond the aim of this chapter.

4 Notes

1. Ampules can be easily made from Pasteur pipets using a small blowlamp. Because of their manufacture process, Pasteur pipets from local suppliers are free from amino acids, and need not be pretreated.
2. Although glycoproteins contain *N*-acetylgalactosamine and *N*-acetylglucosamine, these residues are hydrolyzed to their respective deacetylated amino sugars. The amount of galactosamine and glucosamine added to the amino acid standard (Sigma) should be roughly the same as the amount of any amino acid present in the mixture.
3. Mixtures of standard sugar may be adequately prepared by carefully weighting of desiccated pure sugars.
4. Because the recovery of eluting sugars is affected by the flow-rate through the Sep-Pak cartridge, loading and elution should not exceed about 2 mL/min.
5. This sample clean-up procedure removes both high- and low-polarity material, consisting primarily of dansyl sulphonic acid and excess of dansyl hydrazine, reducing thereby the presence of junk peaks in the chromatogram. Recovery of dansyl hydrazones of neutral sugars is usually greater than 90%.
6. The Sep-Pak cartridge can be reused following regeneration and activation with 3 mL of acetonitrile followed by 5 mL water.
7. The derivatized sugar fraction is 85–100% stable if stored 2 d at 4°C or 2–3 wk at −20°C in the dark.
8. Detection of dansylhydrazones can be done with a spectrophotometer monitoring at 240 nm or with a fluorimeter recording the emission at 550 nm after excitation at 346 nm. The detection limit is of the order of 25–50 picomoles, although for accurate compositional analysis, 100–500 picomoles are recommended.
9. A description of derivatization (permethylation or peracetylation) and fragmentation behavior of oligosaccharides is far beyond the scope of this chapter. The methods and analytical techniques are the same as for structural elucidation of *N*-linked carbohydrates described elsewhere in this book and in excellent previous book chapters and reviews (*10, 11*).
10. The composition and pH of the digestion buffer may vary. Ammonium bicarbonate buffer is a convenient medium for digestion with many enzymes, including the serine proteinases listed. It is a volatile buffer and therefore easy to remove by lyophilization. Other suitable buffers in the pH range 7.5–8.5, with or without 50–150 m*M* NaCl, are based on Tris/HCl, phosphate, *N*-ethylmorpholine-HCl, etc. If necessary for solubilization of the glycoprotein, guanidine hydrochloride up to a final concentration of 2*M* may be added. Most serine proteinases retain 40–100% proteolytic activity under these conditions.

11. If the polypeptide chain is strongly crosslinked by disulphide bonds, the protein must be reduced and alkylated before cleavage. To this end, dissolve the protein (5–10 mg/mL) in a buffer pH 8–9 containing $6M$ guanidine-hydrochloride, add 2-mercaptoethanol (5% (v/v) final concentration and incubate for 2 min in a boiling water bath. Thereafter, let cool at room temperature and add a $2M$ fold excess of alkylating reagent (i.e., iodoacetamide or 4-vynilpyridine) over reducing agent. Let react for 30 min, and dialyze extensively against digestion buffer (*see* **Note 10**).

12. Many different chromatographic conditions can be employed for reversed-phase peptide mapping. Most common conditions employ a C18 25 × 0.4 cm (5 μm particle size, 100 Å pore size) column eluted at 1 mL/min with a linear gradient of 0.1% trifluoroacetic acid in water (solution A) and acetonitrile (solution B). For detection of peptides, monitor the eluate at 206–220 nm.

13. If the sequence of the glycosylated fragment is not completely determined by *N*-terminal sequencing, it might be necessary to subfraction the glycopeptide by digestion with another enzyme followed by separation of the secondary peptides and sequence analysis.

References

1. Hansen, J. E., Lund, O., Tolstrup, N., Gooley, A. A., Williams, K. L., and Brunak, S. (1998) NetOglyc: Prediction of mucin type O-glycosylation sites based on sequence context and surface accessibility. *Glycoconjugate J.*, **15**, 115–130.
2. van den Steen, P., Rudd, P. M., Dwek, R. A., and Opdenakker, G. (1998) Concepts and principles of O-linked glycosylation. *Crit. Rev. Biochem. Mol. Biol.* **33**, 151–208.
3. Gemmill, T. R. and Trimble, R. B. (1999) Overview of N- and O-linked oligosaccharide structures in various yeast species. *Biochim. Biophys. Acta* **1426**, 227–237.
4. Strahl-Bolsinger, S., Gentzsch, M., and Tanner, W. (1999) Protein O-mannosylation. *Biochim. Biophys. Acta* **1426**, 297–307.
5. Hounsell, E. F., Davies, M. J., and Renouf, D. V. (1996) O-linked protein glycosylation structure and function. *Glycoconjugate J.*, **13**, 19–26.
6. Fountoulakis, M. and Lahm, H-W. (1998) Hydrolysis and amino acid composition of proteins. *J. Chromatogr. A*, **826**, 109–134.
7. Whittal, R. M., Palcic, M. M., Hindsgaul, O., and Li, L. (1995) Direct analysis of enzymatic reactions of oligosaccharides in human serum using matrix-assisted laser desorption ionization mass spectrometry. *Anal. Chem.*, **67**, 3509–3514.
8. Eirín, M. T., Calvete, J. J., and González-odríguez, J. (1986) New isolation procedure and further biochemical characterization of glycoproteins IIb and IIIa from human platelet plasma membrane. *Biochem. J.*, **240**, 147–153.
9. Calvete, J. J. and González-Rodríguez, J. (1986) Isolation and biochemical characterization of the α- and β-subunits of glycoprotein IIb of human platelet plasma membrane. *Biochem. J.* **240**, 155–161.
10. Montreuil, J., Bouquelet, S., Debray, H., Lemoine, J., Michalski, J-C., Spik, G., and Strecker, G. (1994) Glycoproteins *in Carbohydrate Analysis. A Practical Approach* (Chaplin, M. F. and Kennedy, J. F., eds.) IRL Press, Oxford, pp. 181–293.
11. Dell, A., Khoo, K-H., Panico, M., McDowell, R. A., Etienne, A. T., Reason, A. J., and Morris, H. R. (1993) FAB-MS and ES-MS of glycoproteins in *Glycobiology. A Practical Approach* (Fukuda, M. and Kobata, A., eds.) IRL Press, Oxford, pp. 187–222.
12. Alving, K., Paulsen, H., and Peter-Katalinic, J. (1999) Characterization of O-glycosylation sites in MUC2 glycopeptides by nanoelectrospray QTOF mass spectrometry. *J. Mass Spectrom.*, **34**, 395–407.

21
Characterization of Site-specific
N-Glycosylation

Katalin F. Medzihradszky

Summary Even if a consensus sequence has been identified for a post-translational modification, the presence of such a sequence motif only indicates the possibility, not the certainty that the modification actually occurs. Proteins can be glycosylated on certain amino acid side-chains, and these modifications are designated as N- and O-glycosylation. N-glycosylated species are modified at Asn residues. There is a consensus sequence for N-glycosylation: AsnXxxSer/Thr/Cys, where Xxx can be any amino acid except proline. N-linked oligosaccharides share a common core structure of GlcNAc$_2$Man$_3$. In addition, an enzyme, peptide N-glycosidase F (PNGase F), removes unaltered most of the common N-linked carbohydrates from proteins while hydrolyzing the originally glycosylated Asn residue to Asp. O-glycosylation occurs at Ser or Thr-residues, usually in sequence-stretches rich in hydroxy amino acids, but there has been no consensus sequence determined for this modification. In addition, O-glycosylation lacks a common core structure: mammalian proteins have been reported bearing O-linked N-acetylgalactosamine, fucose, glucose, and corresponding elongated structures, as well as N-acetylglucosamine. Chemical methods are used to liberate these oligosaccharides because no enzyme has been discovered that would cleave all the different O-linked carbohydrates. Characterization of both types of glycosylation is complicated by the fact that the same amino acids within a population of protein molecules may be derivatized with an array of different carbohydrate structures, or remain unmodified. This site-specific heterogeneity may vary by species, tissue, and may be affected by physiological changes, and so on.

For addressing site-specific carbohydrate heterogeneity mass spectrometry has become the method of choice. Although matrix-assisted laser desorption ionization mass spectrometry of collected HPLC-fractions has been used successfully for this purpose, reversed phase HPLC directly coupled with electrospray ionization mass spectrometry (LC/ESIMS) offers better resolution. Using a mass spectrometer as on-line detector not only assures the analysis of every component eluting (mass mapping), but at the same time diagnostic carbohydrate ions can be generated by collisional activation in the ion-source that permit the selective detection of glycopeptides.

From: *Post-translational Modifications of Proteins.*
Methods in Molecular Biology, Vol. 446.
Edited by: C. Kannicht © Humana Press, Totowa, NJ

Keywords *N*-glycosylation; *O*-glycosylation; site-specific glycosylation; electrospray ionization mass spectrometry; LC/ESIMS.

1 Introduction

Based on our present knowledge post-translational modifications cannot be predicted from genomic information, and have to be studied on the protein level. Even if a consensus sequence has been identified for any particular modification, the presence of such a sequence motif only indicates the possibility, not the certainty that the modification actually occurs.

Proteins can be glycosylated on certain amino acid side-chains, and these modifications are designated as *C, N-* and *O*-glycosylation (http://glycores.ncifcrf. gov). *C*-glycosylation features an α-mannopyranosyl group on the indol side-chain of a Trp residue. A consensus sequence for this modification, Trp-Xxx-Xxx-Trp has been reported, in which the first Trp residue will be modified *(1)*. Mass spectrometry analysis of *C*-glycosylation is relatively straightforward task: the modification is stable both chemically and in the gas-phase. *N*-glycosylated species are modified at Asn residues. There is a consensus sequence for *N*-glycosylation: Asn-Xxx-Ser/ Thr/Cys *(2,3)*, where Xxx can be any amino acid except proline. *N*-linked oligosaccharides share a common core structure of $GlcNAc_2Man_3$. In addition, an enzyme, peptide *N*-glycosidase F (PNGase F), removes unaltered most of the common *N*-linked carbohydrates from proteins while hydrolyzing the originally glycosylated Asn residue to Asp. *O*-glycosylation occurs at Ser or Thr residues, usually in sequence-stretches rich in hydroxy amino acids, but there has been no consensus sequence determined for this modification. In addition, *O*-glycosylation lacks a common core structure: mammalian proteins have been reported bearing *O*-linked *N*-acetylgalactosamine *(4)*, fucose *(5)*, glucose *(5, 6)*, and corresponding elongated structures, as well as *N*-acetylglucosamine *(7)*. Chemical methods are used to liberate these oligosaccharides because no enzyme has been discovered that would cleave all the different *O*-linked carbohydrates.

The characterization of *N-* and *O*-glycosylation is complicated by the fact that the same amino acids within a population of protein molecules may be modified with an array of different carbohydrate structures, or remain unoccupied. This site-specific heterogeneity may vary by species, tissue *(8–11)*, and may be affected by physiological changes *(12–17)*, etc.

Glycosylation of a given protein has usually been characterized from its enzymatically or chemically released carbohydrate pool, using a wide variety of methods, such as high pH anion exchange chromatography *(18)* or fluorophore-assisted carbohydrate electrophoresis (FACE) *(19)*, sequential exoglycosidase digestions *(20,21)*, mass spectrometry *(22–26)* and NMR *(26)*. This approach is perfect for carbohydrate structure elucidations.

For occupied *N*-glycosylation site assignment a novel proteomics approach has been developed recently, using selective glycopeptide enrichment followed by

PNGase F digestion and LC/MS/MS analysis: Glycopeptides are isolated either after oxidation and chemical binding or by lectin chromatography *(27,28)*. However, this proteomics approach does not provide information about the carbohydrate structure(s), and will not reveal whether a site is 100% occupied or only partially glycosylated.

For addressing site-specific carbohydrate heterogeneity mass spectrometry has become the method of choice. Although matrix-assisted laser desorption ionization mass spectrometry of collected HPLC-fractions has been used successfully for this purpose *(29,30)*, reversed phase HPLC directly coupled with electrospray ionization mass spectrometry (LC/ESIMS) offers better resolution *(9,31–39)*. Using a mass spectrometer as on-line detector not only assures the analysis of every component eluting (mass mapping), but at the same time diagnostic carbohydrate ions can be generated by collisional activation in the ion-source that permit the selective detection of glycopeptides *(33)*. These characteristic nonreducing end oxonium ions at *m/z* 204, 274 and 292, 366, and 657 indicate the presence of *N*-acetylhexosamine, neuraminic (sialic) acid, hexosyl-*N*-acetylhexosamine, and sialyl-hexosyl-*N*-acetylhexosamine, respectively *(see* **Note 1**). During LC/ESIMS experiments fragmentation can be induced and these diagnostic ions can be monitored (selective ion monitoring - SIM) in alternate scans, while every second scan will yield molecular weight information of the intact species. Glycopeptide-containing fractions should exhibit components with molecular weights that do not match any predicted tryptic peptide, and these species would be further analyzed. Before LC/ESIMS analysis *N*-linked carbohydrates can be eliminated by PNGase F leaving only the Ser- Thr-glycosylated species *(34)*. However, for the characterization of *O*-linked sites, other methods are frequently required, such as tandem mass spectrometry, additional proteolytic and/or glycosidase cleavages, Edman sequencing or the combination thereof, depending on the actual carbohydrate structure *(36,40–42)*. Thus, this chapter will focus on the characterization of *N*-glycosylated proteins, where a more general protocol can be followed because of the shared features of these modifications. From a known amino acid sequence, potential *N*-glycosylation sites can be identified, and a proteolytic enzyme can be selected that would separate these sites on individual peptides. Then the digestion mixture can be analyzed by LC/ESIMS. Glycopeptide-containing fractions will be identified by SIM, and from the molecular weights measured for the glycopeptides the site-specific heterogeneity may be addressed *(9,31,35–38)*. Studying recombinant proteins may help in overcoming some of the hurdles, for example, developing new approaches and techniques, and/or working out optimal conditions for the characterization of the native species available in much lower quantities.

2   Materials

2.1   *Reduction and Alkylation of Glycoprotein*

1. 1 nmole of the purified glycoprotein, little salt, *no* detergent *(see* **Note 2**).

In our example, thrombin activated recombinant human Factor VIII (rhFVIIIa) was separated into its four polypeptides by reversed-phase HPLC on a semipreparative C4 column *(37)*.

2. 6M guanidine hydrochloride, 200 mM NH$_4$HCO$_3$ buffer (pH~8.0).
3. Freshly prepared DTT solution in water (250 nM/µL).
4. Iodoacetic acid sodium salt solution in H$_2$O (550 nM/µL). Prepare just before use.
5. Microdialyzer, dialysis membrane of the appropriate MW cut off.
6. 20 mm NH$_4$HCO$_3$ buffer (pH~7.4).

2.2   Tryptic Digestion (see Note 3)

1. Side-chain protected porcine trypsin (Promega).
2. 1 mM TFA in H$_2$O.

2.3   LC/ESIMS Analysis of the Tryptic Digest

1. HPLC system with a microbore C18 column (1.0 × 150 mm).
2. UV detector, variable wavelength.
3. HPLC solvents: 0.1% TFA in water, 0.08% TFA in acetonitrile (*see* **Note 4**).
4. Make-up solvent: 2-methoxyethanol/isopropanol 1:1 mixture.
5. Additional pump providing stable flow at 50 µL/min flow rate, for the make-up flow.
6. Mixing tee, splitting tee, fused silica capillaries for plumbing, teflon sleeves.
7. Mass spectrometer equipped with an electrospray source and with hardware and software permitting selective ion monitoring and data acquisition with alternating scans at high and low cone-voltage.

In our example the digests were injected onto a microbore (C18, 1.0 × 150 mm, Vydac) reversed-phase column fitted to an Applied BioSystems 140B HPLC system. Fused silica capillaries 285 µm OD/75 µm ID were used for plumbing (LC Packings). The UV-detector (Applied Biosystems 759A) was equipped with a U-Z View flow cell (LC Packings). Capillaries were connected wherever it was necessary with LC Packings teflon sleeves. Make-up solvent was added after UV-detection, and to accomplish this a separate syringe pump (Isco µLC-500) delivered a 2-methoxyethanol/isopropanol 1:1 mixture to a 3.1 µL dead volume PEEK mixing tee (Upchurch Scientific) at a flow rate of 50 µL/min. After the mixing tee a zero dead volume tee (Upchurch Scientific) was included to split the column effluent at a rate of 1:4 (*see* **Note 5**). The split ratio was adjusted by cutting the outlet capillary to the right size while measuring the flow rates. Mass spectrometric detection was performed on a Micromass AutoSpec SE mass analyzer equipped with an electrospray source (*see* **Note 6**).

3   Methods

3.1   *Reduction and Alkylation of Glycoprotein*

1. Dissolve 1 nmole of the glycoprotein in 100 µL of a 6*M* guanidine hydrochloride/200 m*M* NH$_4$HCO$_3$ pH~8.0 solution.
2. Add 2 µL DTT in H$_2$O (250 n*M*/µL). Incubate the mixture at 60°C, for 1 h.
3. Add 2 µL sodium iodoacetate in H$_2$O (550 n*M*/µL). Incubate the mixture at room temperature, for 2 h, in the dark.
4. Dialyze the mixture against 20 m*M* NH$_4$HCO$_3$ buffer overnight to remove the reagent excess.
5. Concentrate the sample by vacuum centrifugation to approximately 100 µL.

In our example approximately 1 nmole of the rhFVIIIa proteins was derivatized with 1,000-fold excess of vinylpyridine in 6*M* guanidine hydrochloride, 200 m*M* NH$_4$HCO$_3$ solution, at 20°C, for 3.5 h, in the dark. The only purpose of this derivatization step is the identification of free sulfhydryls in the protein; therefore for glycosylation studies it could have been eliminated. Incubation with 2,000-fold excess with dithiothreitol followed, at 60°C, for 1 h., then new sulfhydryls were derivatized with 4,400-fold excess of iodoacetic acid sodium salt at 20°C, for 2 h, in the dark. Finally the reagent excess was eliminated by dialysis against 20–50 m*M* NH$_4$HCO$_3$ solution *(37)*.

3.2   *Tryptic Digestion*

1. Side-chain protected trypsin dissolved in 1 m*M* TFA solution (0.1–1 µg/µL) is added to the glycoprotein solution. The amount of trypsin should be approximately 4% w/w.
2. Incubation for 8 h, at 37°C (*see* **Note 7**).
3. Stop the digestion by boiling the mixture for 5 min (*see* **Note 8**).

The reduced and alkylated rhFVIIIa proteins required longer incubation time: 16–24 h.

3.3   *LC/ESIMS Analysis of the Tryptic Digest*

1. Prepare solvents.
2. Set up the microbore HPLC system with a make-up solvent added after the UV-detector and the flow-splitter. Measure flow rates. Equilibrate the column in 5% solvent B.
3. Set up an LC/ESIMS method of alternating high and low cone-voltage scans. In high cone-voltage mode monitor ions at *m/z* 204 and 292. Usually an approximately

1 Da mass window is monitored around these m/z values. In most cases these two values are sufficiently informative, because all _N_-linked oligosaccharides contain at least two N-acetylhexosamines, while the presence or absence of neuraminic (sialic) acid will provide some indication of what class of _N_-linked structure is present. In low cone-voltage scans monitor the mass range from _m/z_ 350–2,000 (*see* **Note 9**).

4. Inject approximately 100 pmoles (approximately 10 μL) of the tryptic digest. Start mass spectrometry data acquisition.
5. Keep the flow isocratic (5% B) for 5 min, then start a linear gradient increasing the percentage of solvent B to 50% over 90 min (*see* **Note 10**). Collect fractions at the splitting tee, if further analysis is planned.

Approximately 100 pmoles of a tryptic digest of a reduced and alkylated rhFVIIIa protein (73 kDa or B-domain *(37)*) was injected onto a microbore(C18, 1.0 ˣ 150 mm, Vydac) reversed-phase column that was equilibrated in 5% solvent B, at a flow rate of 50 μL/min. The gradient was started 5 min after the injection, the amount of solvent B was linearly increased to 40% or 50% over 80 min. Data shown here were collected while applying a make-up solvent after UV-detection, at a flow rate of 50 μL/min. Two identical LC/ESIMS analyses were performed using an AutoSpec SE mass spectrometer as the detector. The source voltage was 4000 V, the needle voltage was 7,800 V. During the first analysis fragmentation was induced in the ion source by increasing the sample cone voltage to 313 V above the source potential and the skimmer voltage to 325 V. During the second analysis the sampling cone and skimmer voltages were kept at a lower value, 196 and 235 V, respectively. Under these lower voltage conditions no carbohydrate fragmentation was observed. For normal scans the magnet was scanned to monitor the spectra from _m/z_ 2,000 to 350, while for the high cone voltage scans the lower mass limit was set at m/z 200 (*see* **Note 11**). Mass resolution was approximately 1,500 (*see* **Note 6**).

3.4 Data Interpretation

3.4.1 Glycopeptide Identification

1. Using Protein Prospector (http://prospector.ucsf.edu/) MS-digest (or other similar programs) print out the list of expected tryptic peptides. Do not forget to select the appropriate modification for the Cys-residues (*see* **Note 12**).
2. Print out the total ion current (TIC) and SIM chromatograms. The TIC chromatogram will show a UV-like elution profile, while the SIM-peaks will indicate the glycopeptides.
3. Print out the full scan electrospray spectra of glycopeptide-containing TIC peaks: i.e., combine the scans, beginning where the SIM trace starts to rise to the edge of the SIM peak.

4. Determine the MH⁺ values of the coeluting digest components: based on charge state determination (high resolution) or based on the presence of a set of differently charged ions (low resolution) (*see* **Note 13**, and see following detailed explanation).

5. Compare these MH⁺ values (monoisotopic for high resolution and average for low resolution) to the MS-digest list. Identify unmodified peptides.

6. Identify glycopeptide ion-series from mass differences that reflect carbohydrate heterogeneity (*see* **Note 13** and **14**, Tables 21.1 and 21.2, and see following).

7. Identify the modified peptide (see following; Table 21.3).

8. Determine/confirm the carbohydrate structure by sequential exoglycosidase digestions/LC/ESIMS analyses of the collected glycoprotein fractions (Table 21.4, and see following).

LC/ESIMS analysis of the 73 kDa protein of rhFVIIIa is presented here as an example for addressing site-specific glycosylation by SIM and mass mapping. Figure 21.1 shows the extracted ion chromatograms (the intensity changes of given m/z value ions during the analysis) of diagnostic carbohydrate fragments at *m/z* 204

Table 21.1 Some Common Components of <u>N</u>- and <u>O</u>-Linked Oligosaccharides

Sugar	M_r	In Chain: nominal mass	Exact mass	Average mass
Pentose Arabinose, Xylose	150 $C_5H_{10}O_5$	132 $C_5H_8O_4$	132.0423	132.1161
DeoxyHex Fucose	164 $C_6H_{12}O_5$	146 $C_6H_{10}O_4$	146.0579	146.1430
Hexose	180 $C_6H_{12}O_6$	162 $C_6H_{10}O_5$	162.0528	162.1424
Hexuronic acid HexA	194 $C_6H_{10}O_7$	176 $C_6H_8O_6$	176.0321	176.1259
HexNAc	221 $C_8H_{13}NO_6$	203 $C_8H_{13}NO_5$	203.0794	203.1950
NeuAc/SA *N*-acetyl-neuraminic acid/sialic acid	309 $C_{11}H_{19}NO_9$	291 $C_{11}H_{17}NO_8$	291.0954	291.2579
NeuGc *N*-glycolyl-neuraminic acid	325 $C_{11}H_{19}NO_{10}$	307 $C_{11}H_{17}NO_9$	307.0903	307.2573
H_2O		+18	18.0106	18.0153
Phosphate		+80	79.9663	79.9799
Sulphate		+80	79.9468	80.0642

Table 21.2 Carbohydrate Heterogeneity in Mass Differences

Saccharide structure	Mass difference			
	for 2+	3+	4+	5+ ions
Fucose	73	48.7	36.5	29.2
Hexose	81	54	40.5	32.4
N-Acetylhexosamine	101.5	67.7	50.8	40.6
Neuraminic acid	145.5	97	72.8	58.2
HexHexNAc	182.5	121.7	91.3	73
SAHexHexNAc	328	218.7	164	131.2

Table 21.3 Mass Addition to Peptides: - Some Common $\underline{N}$-Linked Oligosaccharides

Glycan	Composition	Mass added to peptide monoisotopic	Average
Oligomannose structures			
Man3 core	$Man_3GlcNAc_2$	892.32	892.82
Man3 core fucosylated	$Man_3GlcNAc_2Fuc$	1038.37	1038.96
Man5	$Man_5GlcNAc_2$	1216.42	1217.10
Man6	$Man_6GlcNAc_2$	1378.48	1379.24
Man7	$Man_7GlcNAc_2$	1540.53	1541.39
Man8	$Man_8GlcNAc_2$	1702.58	1703.53
Man9	$Man_9GlcNAc_2$	1864.63	1865.67
bisecting hybrid	$Man_5GlcNAc_4$	1622.58	1623.49
Bi-antennary structures			
asialo	$Gal_2Man_3GlcNAc_4$	1622.58	1623.49
asialo, fucosylated	$Gal_2Man_3GlcNAc_4Fuc$	1768.64	1769.64
bisecting asialo fucosylated	$Gal_2Man_3GlcNAc_5Fuc$	1971.72	1972.83
bisecting asialo	$Gal_2Man_3GlcNAc_5$	1825.66	1826.69
asialo agalacto	$Man_3GlcNAc_4$	1298.48	1299.21
asialo agalacto fucosylated	$Man_3GlcNAc_4Fuc$	1444.54	1445.35
bisecting asialo agalacto fucosylated	$Man_3GlcNAc_5Fuc$	1647.61	1648.55
monosialo	$SAGal_2Man_3GlcNAc_4$	1913.68	1914.75
monosialo fucosylated	$SAGal_2Man_3GlcNAc_4Fuc$	2059.74	2060.89
disialo	$SA_2Gal_2Man_3GlcNAc_4$	2204.77	2206.01
disialo fucosylated	$SA_2Gal_2Man_3GlcNAc_4Fuc$	2350.83	2352.15
Tri-antennary structures			
asialo	$Gal_3Man_3GlcNAc_5$	1987.71	1988.83
asialo fucosylated	$Gal_3Man_3GlcNAc_5Fuc$	2133.77	2134.97
asialo agalacto	$Man_3GlcNAc_5$	1501.56	1502.40
asialo agalacto fucosylated	$Man_3GlcNAc_5Fuc$	1647.61	1648.55
monosialo	$SAGal_3Man_3GlcNAc_5$	2278.81	2280.09
monosialo fucosylated	$SAGal_3Man_3GlcNAc_5Fuc$	2424.87	2426.23
disialo	$SA_2Gal_3Man_3GlcNAc_5$	2569.90	2571.35
disialo fucosylated	$SA_2Gal_3Man_3GlcNAc_5Fuc$	2715.96	2717.49
trisialo	$SA_3Gal_3Man_3GlcNAc_5$	2861.00	2862.60
trisialo fucosylated	$SA_3Gal_3Man_3GlcNAc_5Fuc$	3007.06	3008.75
tetrasialo	$SA_4Gal_3Man_3GlcNAc_5$	3152.10	3153.86
tetrasialo fucosylated	$SA_4Gal_3Man_3GlcNAc_5Fuc$	3298.15	3300.00

and 292 and the TIC chromatogram during the high cone voltage LC/ESIMS analysis of a tryptic digest of this protein. There are 4 consensus sequences in the 73 kDa protein, and from the SIM data *alone* the presence of 4 glycopeptide-containing fractions can be concluded. The first two partially resolved peaks exhibited the diagnostic ion for neuraminic acid, thus those glycopeptides should be modified by complex or hybrid oligosaccharides. From the lack of neuraminic acid oligomannose structures may be suspected for the other fractions. The next step is the analysis of the full scan data of the chromatographic peaks producing the

Table 21.4 Recommendations for Exoglycosidase Digestion

Enzymes (in sequencing order)	Enzyme source	Linkage specificity	Concentration (U/mL)	Digestion buffer[a]	Digestion time (hrs)
1. Neuraminidase	Newcastle Disease Virus	NeuAcα2-3,8R NeuGAcα2-3,8R	0.02–0.05	A	3-7
2. β-Galactosidase	*Streptococcus pneumoniae*	Galβ1-4GlcNAc Galβ1-4GalNAc	0.03–0.07	A	3-7
3. β-GlcNAse	Chicken liver	GlcNAcβ1-3,4R	1.0	A	3-7
4. α-Mannosidase	Jack bean	Manα1-2,3,6Man	3.5	A+25 mM ZnCl$_2$	3-6
5. β-Mannosidase	*Helix pomatia*	Manβ1-4GlcNAc	2.0–4.0	A+25 mM ZnCl$_2$	3-6
6. β-GlcNAse	Chicken liver	GlcNAcβ1-3,4R	2.0	A	3-7
7. α-Fucosidase	Bovine epididymis	Fucα1-6(>2,3,4)R	0.2	A	3-6

[a]Buffer A - 30mM sodium acetate, pH 5.0
Reprinted with permission from (*42*).

Factor VIII, 73 kDa, tryptic digest   **LC/ESIMS AutoSpec**

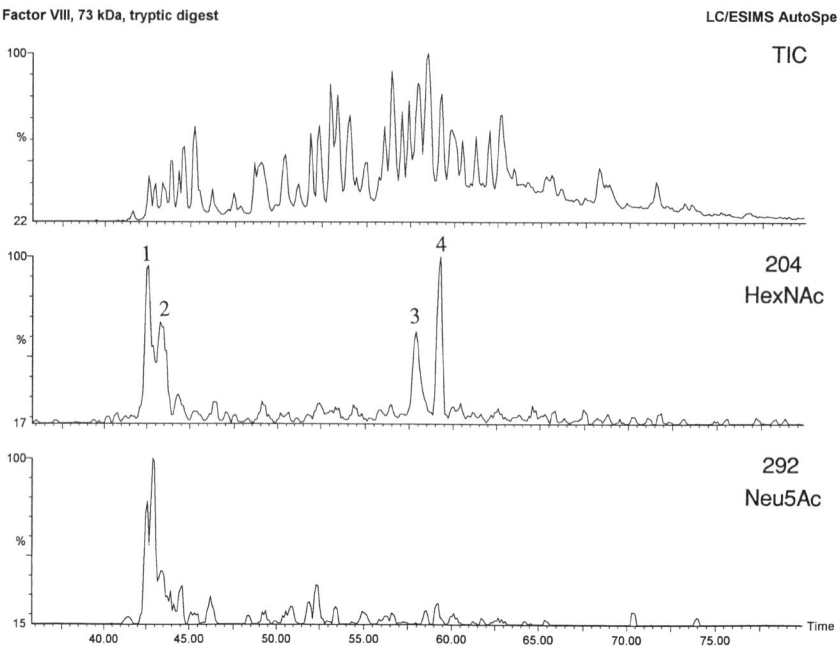

Fig. 21.1 Total ion chromatogram and selected ion chromatograms of diagnostic carbohydrate ions, m/z 292 (Neu5Ac) and 204 (HexNAc), acquired during high cone voltage LC/ESIMS analysis of a tryptic digest of the 73 kDa protein of rhFVIIIa. Reprinted with permission from ref. 37. Copyright (1997) American Chemical Society

diagnostic carbohydrate fragments. The AutoSpec SE mass spectrometer afforded a high enough mass resolution to allow the charge state and the monoisotopic mass to be determined for each ion (*see* **Note 13**). Figure 21.2 shows the electrospray spectrum of the first partially resolved glycopeptide-containing fraction. The peaks are labeled with their monoisotopic masses and charge states. Whenever a molecule was represented by ions of different charge states usually the more highly resolved ions were used for the molecular weight determination, if they were sufficiently abundant. From these values the protonated molecular mass of each component was calculated: $n \times (MH_n)^{n+} - (n-1)1.0$ (1.0 is the approximate mass of a proton). The MH⁺ value was selected over the molecular mass because the software for protein digestion, MS-digest, lists this value. MH⁺ values calculated for the fraction in Figure 21.2 were 1,204.9, 2,974.1, 3,264.4, and 3,556.0 Da, from ions at *m/z* 602.9, 991.9, 1,089.1, and 1,186.7, respectively. First the predicted tryptic peptides were identified: *m/z* 1,204.9 corresponds to peptide [115]Lys-Lys[124] (MH⁺$_{cal}$ at *m/z* 1,204.7). Surprisingly, this sequence contains a consensus sequence, obviously unmodified. The 3 higher masses did not match predicted tryptic products and differed by approximately 291 Da. Different glycoforms of the same glycosylation site usually elute very close together, generally the larger the oligosaccharide, the shorter the retention time (slightly). Heterogeneity of the carbohydrate structure is reflected in

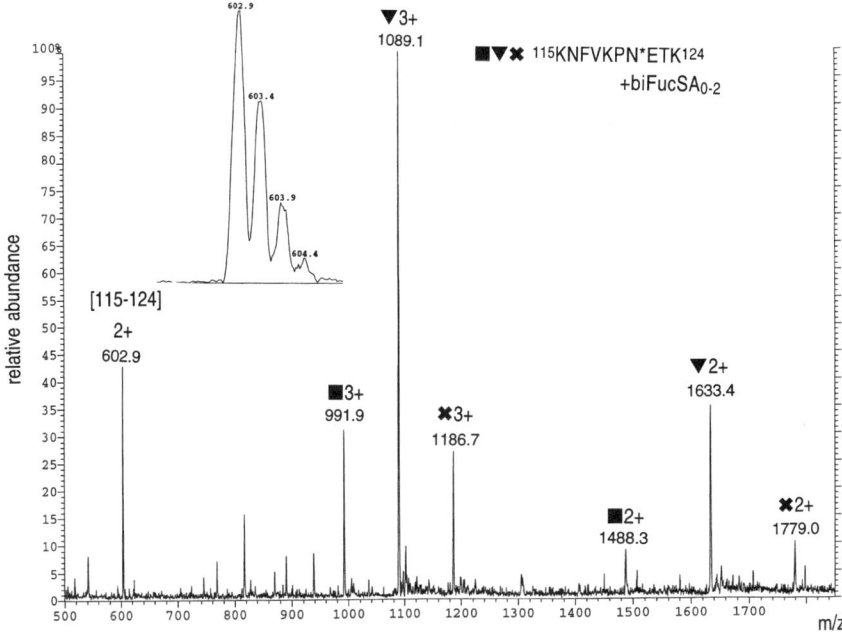

Fig. 21.2 Electrospray mass spectrum of the first partially resolved glycopeptide-containing peak from the tryptic digest of the 73 kDa protein of rhFVIIIa (Fig. 21.1). These data were acquired at normal cone voltage. Insert illustrates the resolution during this LC/ESIMS experiment. Reprinted with permission from ref. 37. Copyright (1997) American Chemical Society

closely related masses (*see* **Note 14**, Table 21.1 and 21.2). Thus, a series of ions differing in mass by 291 Da indicate different numbers of neuraminic acids in the oligosaccharides. Similarly, an ion series with 162 Da differences would indicate the presence of an array of oligomannose structures. A mass difference of 365 Da shows the presence of an additional HexHexNAc unit, that may indicate another antenna or an addition to the repetitive *N*-acetyllactosamine chain, and a 146-Da mass difference indicates fucosylation (*see* **Note 15**). The modified peptides can be identified readily if the structure of the oligosaccharides, separated from the protein have been studied prior to the LC/ESIMS analysis, which is frequently the case for recombinant protein pharmaceuticals. The masses of these structures can be calculated (Table 21.3) and subtracted from the molecular weights determined. If the correct oligosaccharide has been considered the result of this subtraction will be the MH$^+$ value of a predicted tryptic peptide containing a potential glycosylation site. For example, it had been reported that the major component of the *N*-linked carbohydrate pool of rhFVIIIa was a monosialo biantennary core-fucosylated oligosaccharide *(19)*. The addition of this carbohydrate would increase the peptide molecular mass by 2,059.7 Da (monoisotopic, Table 21.3). Thus, this number was subtracted from the masses above and the results were compared to the mass table of the predicted tryptic peptides. 3,264.4-2,059.7 = 1,204.7, is indeed a tryptic peptide with a consensus site: [115]Lys-Asn[121]-Lys[124] and interestingly the same peptide that had previously been identified unmodified in this fraction. The glycoforms represent this sequence bearing the above described complex carbohydrate with 0, 1, and 2 neuraminic acids. Thus, this consensus site occurs in 4 different forms (*see* **Notes 16**).

The full scan mass spectrum of fraction 2 was analyzed as described previously. Its glycopeptide-component was identified as [116]Asn-Asn[121]-Lys[124] bearing the same oligosaccharides as reported for fraction 1 (data not shown).

Figure 21.3 shows the electrospray mass spectrum of fraction 3. Let us focus only on the glycopeptides. A series of doubly charged ions was detected in this fraction that differed by approximately 81 Da (the corresponding triply charged ions are present too, and labeled accordingly). When multiplied by 2 this number yields the residue weight of a hexose unit (Table 21.1, and 21.2), which together with the lack of sialic acids seems to suggest the presence of an oligomannose-modified peptide. To identify the modified sequence, the predicted molecular masses of the unmodified tryptic peptides with the potential glycosylation sites ("consensus" peptides): 1,204.7 (115–124), 1,186 (116–124), 2,666.3/2,619.3 (309–331) with S-β-4-ethylpyridyl (PE) or carboxymethyl (CM) Cys, 2,031.0 (428–447) and 2,386.0/2,291.9 (475–494) with PE or CM-Cys, respectively were subtracted from the MH[+] values determined

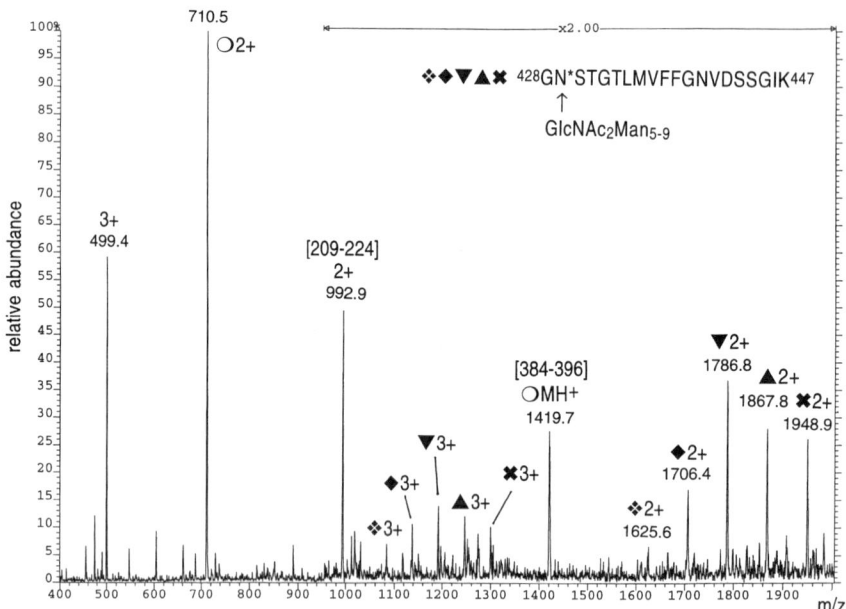

Fig. 21.3 Electrospray mass spectrum of the third glycopeptide-containing peak from the tryptic digest of the 73 kDa protein of rhFVIIIa (Fig. 21.1). These data were acquired at normal cone voltage. Reprinted with permission from (*37*). Copyright (1997) American Chemical Society

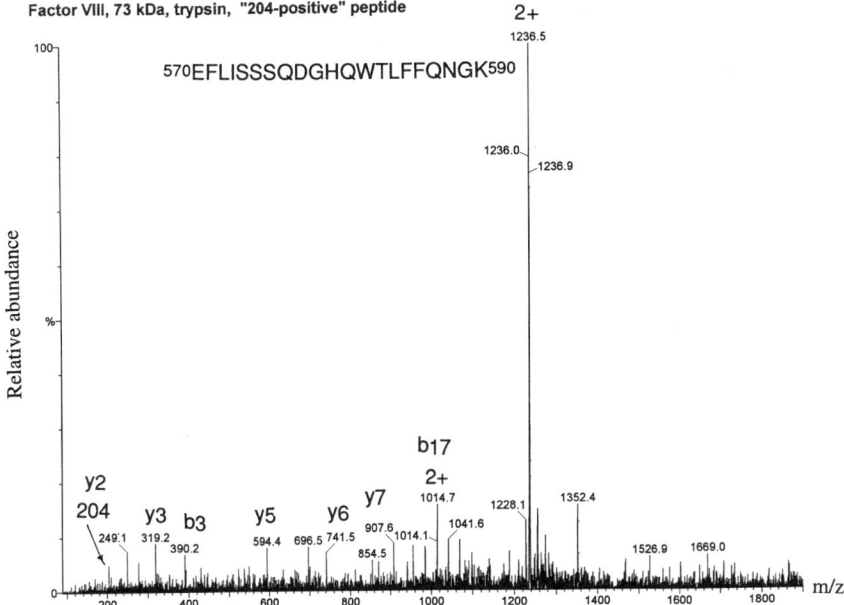

Fig. 21.4 Electrospray mass spectrum of the last 204-positive peak from the tryptic digest of the 73 kDa protein of rhFVIIIa (peak 4 in Fig. 21.1). These data were acquired at high cone voltage. Peptide fragments are labeled according to Biemann *(43)*. All peptides with Gly-Lys C-terminus may yield a y_2 ion at m/z 204

for the glycopeptides: 3,247.0, 3,410.2, 3,571.3, 3,733.9, and 3,896.5 Da. The resulting mass differences revealed the potential additional masses of the carbohydrates. This number must be bigger than 892 Da (core structure). First, let us suppose that the protein was modified with one of the known oligosaccharides (see Table 21.3, *see* **Note 17**). Results of the subtractions were compared with masses corresponding to asialo carbohydrate structures commonly described for mammalian glycoproteins (Table 21.3). From this analysis the presence of [428]Gly-Lys[447] bearing Man_5-Man_9 oligomannose structures was established. Indeed, FACE-analysis identified these carbohydrates present in the carbohydrate pool *(19)*.

The electrospray mass spectrum of chromatographic peak 4 (Fig. 21.4) exhibits no obvious glycopeptide ion series, only a doubly charged ion at m/z 1,236.0. MH[+] calculated from this ion yields 2,471.0, in good agreement with the predicted mass for peptide Glu[570]-Lys[590]. It is apparent that the ion at *m/z* 204 in this fraction did not come from carbohydrates. Under conditions that induce in-source carbohydrate fragmentation some peptides may fragment as well. Peptide Glu[570]-Lys[590] yielded a series of fragments (Fig. 21.4) *(43)*. Fragment y_2 formed *via* peptide bond cleavage at the N-terminus of the penultimate Gly-residue, with the charge retained on the newly formed dipeptide is at *m/z* 204. All peptides with GlyLys C-terminus may yield this ion, and thus are potential false carbohydrate-positives.

LC/ESIMS analysis combined with SIM of diagnostic carbohydrate ions is of only limited use for highly glycosylated proteins. Thrombin digestion of Factor VIII yields also the B-domain, a 909 amino acid protein featuring 20 consensus sequences for *N*-glycosylation. This protein also has been reported *O*-glycosylated. The comparison of SIM and TIC chromatograms of its tryptic digest rather reveals which fractions do *not* contain glycopeptides (Fig. 21.5). Unfortunately, glycopeptides that display strong SIM signals, frequently produce ions of low abundance in the full scan spectra, especially in comparison to the signals of coeluting peptides. This may be partly because of their substoichiometric quantities due to the carbohydrate heterogeneity. The electrospray spectrum of a carbohydrate-positive fraction (Fig. 21.6, peak 1 in Fig. 21.5) showed a series of weak doubly charged signals at *m/z* 1,676.3, 1,773.3, and 1,894.9. MH⁺ values were calculated as 5,026.9, 5,317.9, and 5,682.7, respectively. Subtracting the molecular masses of the predicted "consensus" tryptic peptides, peptide 513–520 was identified, doubly glycosylated at Asn-515 and -519 with 2 biantennary core-fucosylated oligosaccharides and 2 sialic acids, and a bi- and a triantennary core-fucosylated oligosaccharide and 3 sialic acids, respectively. From these data it cannot be determined which site bears the triantennary structure, or the location of the sialic acids. A tryptic digest of the

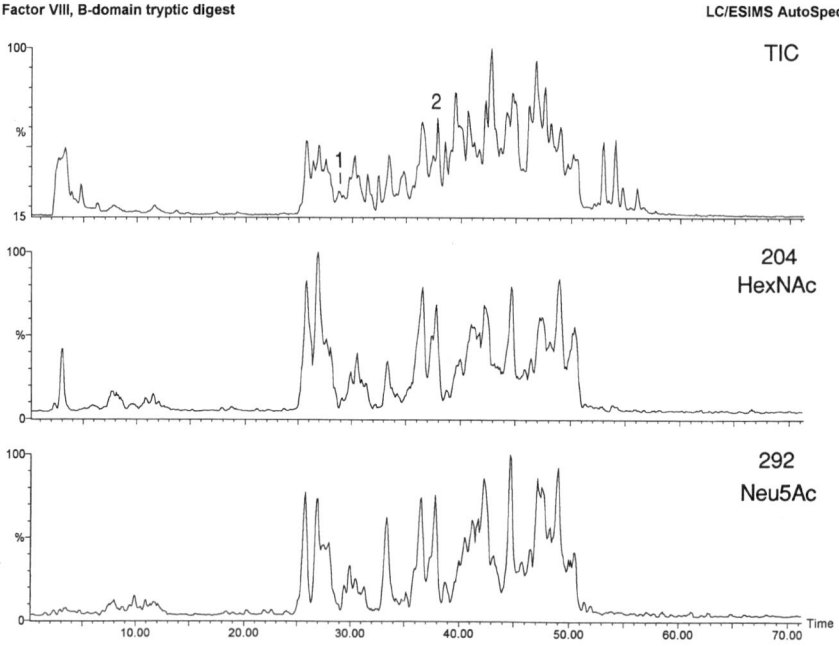

Fig. 21.5 Total ion chromatogram and selected ion chromatograms of diagnostic carbohydrate ions, *m/z* 292 (Neu5Ac) and 204 (HexNAc), acquired during high cone voltage LC/ESIMS analysis of a tryptic digest of the B-domain of rhFVIIIa. Reprinted with permission from ref. 37. Copyright (1997) American Chemical Society. Electrospray spectra of peaks labeled 1 and 2 are shown in Figs. 21.6 and 21.7

B-domain was analyzed by LC/ESIMS after PNGase digestion *(37)*. PNGase removes the *N*-linked carbohydrates while transforming the modified Asn into Asp, increasing the MH⁺ value for the unmodified peptide by 1 Da per *N*-linked site. Peptide 513–520 gave a double peak showing both 1 and 2 Da molecular weight increases. The glycoforms described above contained both sites modified. Careful inspection of the full scan data revealed the presence of 2 previously overlooked doubly charged ions at *m/z* 1,483.7 and 1,629.0, glycoforms with 1 monosialo and disialo biantennary core-fucosylated oligosaccharide, respectively (Fig. 21.6). Figure 21.7 illustrates the complexity of the digest mixture (Peak 2 in Fig. 21.5). Abundant ion series representing different glycoforms were identified from mass differences reflecting the carbohydrate heterogeneity as discussed above. Because the high mass resolution allowed the charge determination of the peaks, even minor glycoforms represented by a single ion could be identified quite convincingly.

Obviously, these experiments determined only the masses of the carbohydrate(s) at any given site, and yielded some information on the carbohydrate heterogeneity. The more accurate the mass measurement, the more precisely the composition of the oligosaccharide could be determined. However, the identity of these residues and their linkage and stereochemistry cannot be deduced from these data. For rhFVIIIa the enzymatically freed carbohydrate pool was analyzed by FACE-analysis

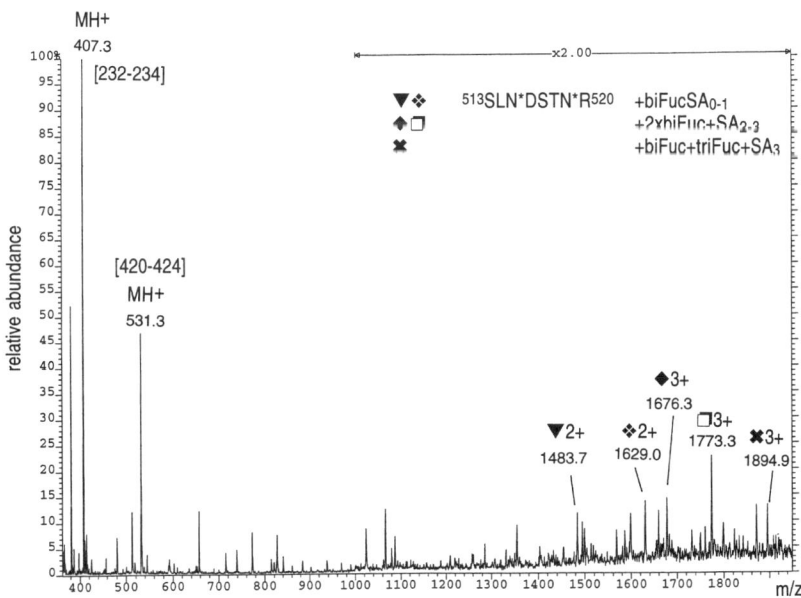

Fig. 21.6 Electrospray mass spectrum of the fraction labeled 1 in Figure 21.5. These data were acquired at normal cone voltage. For simplicity only the discussed glycopeptide ions are labeled in the high mass region with their monoisotopic mass, and charge state. The peptide is modified with bi- and triantennary core-fucosylated oligosaccharides. From these data in cannot be determined which structures are sialylated

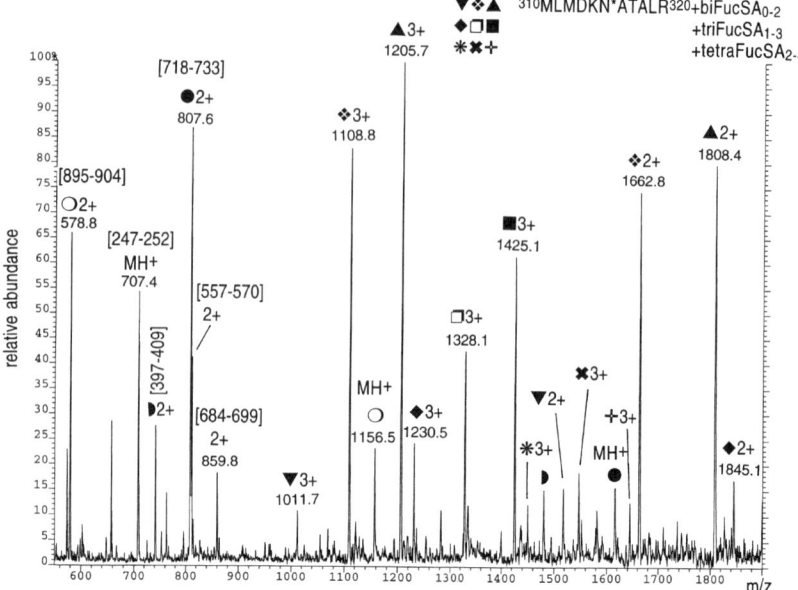

Fig. 21.7 Electrospray mass spectrum of the fraction labeled 2 in Figure 21.5. These data were acquired at normal cone voltage. Reprinted with permission from ref. 37. Copyright (1997) American Chemical Society

(19), the results of which were combined with LC/ESIMS results to draw conclusions about the carbohydrate structure at any given site. However, glycopeptide-containing fractions can be collected on-line at the splitting tee and subjected to exoglycosidase digestions and consequent mass spectrometric analyses that may reveal information about the identity and linkage of the saccharide units. Table 21.4 shows the specificity and optimal digestion conditions of some exoglycosidases recommended for these analyses. Figure 21.8 illustrates the results that can be expected from these experiments with a glycopeptide as isolated from a bovine fetuin tryptic digest (Panel A), then treated with neuraminidase (Panel B) and β-galactosidase (Panel C) *(35)*. Panel A shows the glycopeptide bearing disialo biantennary and di-, tri- and tetrasialo-triantennary oligosaccharides. As expected neuraminidase digestion yielded only 2 components, when all the sialic acids were removed, a biantennary and triantennary glycoform (Panel B). *N*-linked carbohydrates of bovine fetuin have been studied extensively. The Gal-units are mostly in β(1,4)GlcNAc linkages, but it has been reported that the tetrasialo triantennary structure also contains Galβ(1,3)GlcNAc linkages (~20%) *(35, 44, 45)*. The β-galactosidase (from *Streptococcus pneunomiae*) used for the second digestion step was expected to cleave only the Galβ(1,4)GlcNAc linkages, yielding three components: biantennary glycoform-2 $\times$ 162 Da, triantennary glycoform-3 $\times$ 162 and -2 $\times$ 162 Da (~20%). The carbohydrates were cleaved as expected. However, one additional component was detected due to some exopeptidase activity exhibited by the β-galactosidase preparation (Fig. 21.8, panel C).

A

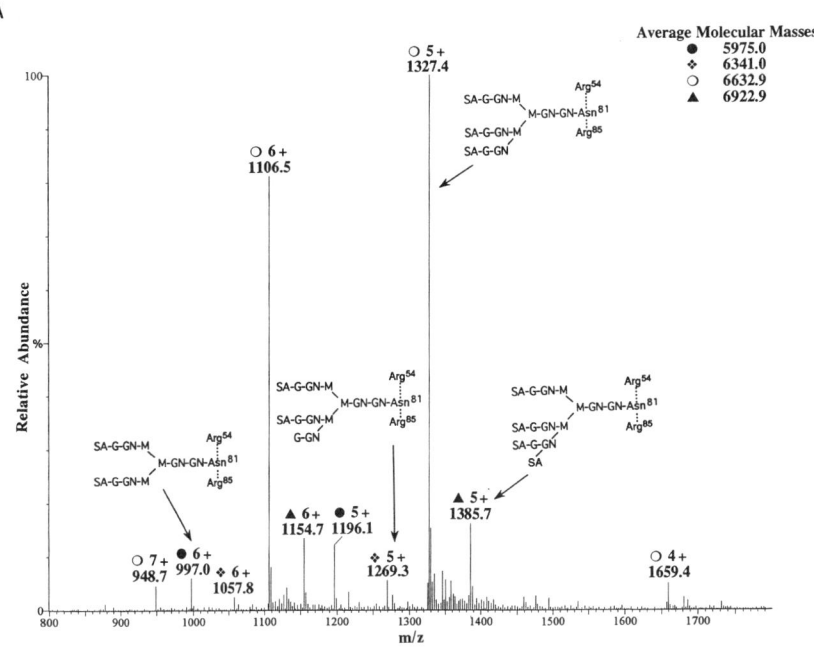

Fig. 21.8 Electrospray mass spectrum of bovine fetuin tryptic glycopeptide [54]Arg-Asn[81]-Arg[85], as isolated from a digest (Panel A); after neuraminidase digestion (Panel B) and after β-galactosidase (from *Streptococcus pneunomiae, see* Table 21.4) digestion (Panel C). The *N*-linked carbohydrate at this site contains approximately 20% Galβ1-3GlcNAc, and the enzyme is specific for Galβ1-4GlcNAc linkage. Reprinted with permission from ref. 35

3.4.2 Complete Analysis of LC/ESIMS Data

1. Consider TIC peaks as chromatographic fractions and print out their corresponding full scan mass spectra.
2. Determine the MH[+] values for every signal more abundant than approximately 5% of the base peak. Identify the peptides by comparing these values to the MS-digest mass list (*see* **Note 18**).
3. Search more thoroughly for missing peptides using selected ion extraction: use a ~2 Da window around the m/z value of interest or less depending on the mass accuracy.
4. Prepare a summary of your data: including sequence coverage, peptides still missing, relatively abundant MH[+] values unaccounted for. In this process consider missed alkylation of Cys residues, trypsin-autolysis (peptides of MH[+] at m/z 842.5, 1,045.6 and 2,211.1 are the most frequently observed), chymotryptic cleavages, and other contaminating proteins. One of the most frequent side-reactions, the oxidation of Met-residues, is included in MS-digest.

B

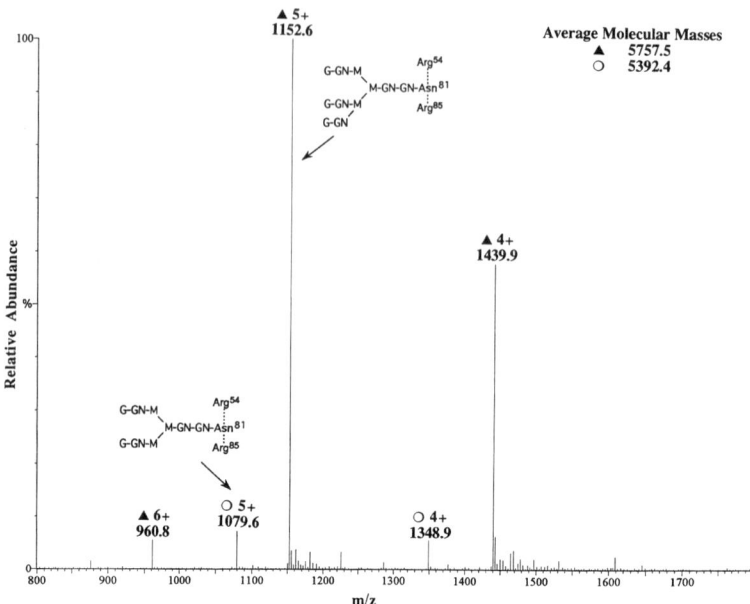

C

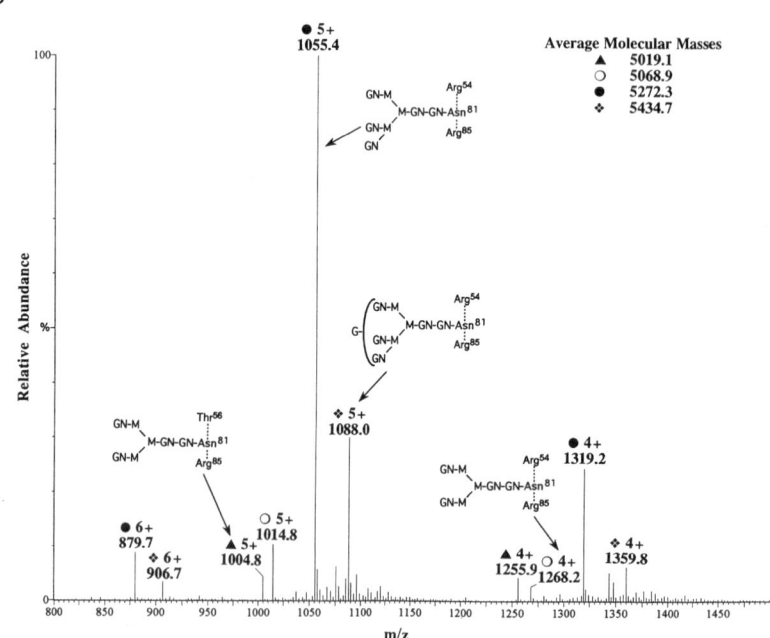

Fig. 21.8 (continued)

Discussing the complete interpretation of LC/ESIMS data for the tryptic digest of the 73 kDa protein of rhFVIIIa goes beyond this chapter. However, only two glycosylation sites were characterized above, and two other potential glycosylation sites, Asn-330 and -483 were unaccounted for when only the carbohydrate-positive peaks were analyzed. Data processing software allows creation of the ion chromatogram of any given mass (ion extraction), thus the LC/ESIMS data can be searched for any predicted peptides. Doubly, triply charged, etc., masses of peptides are also calculated for this search. If the relative abundance of ions at this m/z value was changing during the LC/ESIMS analysis, ion extraction will produce some "chromatographic" peaks. Then full scan data of these chromatographic regions should be reviewed. If the full scan data show ions of other charge states of the same species, its presence is confirmed. If the extracted ion is the only signal from a peptide it should show an abundance comparable with some other components identified unambiguously (*see* **Note 18**). Based on these criteria "consensus" peptides Val^{309}-Lys^{331} and Met^{475}-Lys^{494} were both observed unmodified: giving MH$^+$ values 2,665.9/2,619.1 and 2,386.3/2,292.4 with PE- and CM-Cys, respectively. For additional confirmation, the CM-Cys peptides eluted slightly earlier than the PE-Cys peptides, showing similar retention time differences as other Cys-containing peptides in the digest.

4 Notes

1. The numbers listed here are ions diagnostic of common oligosaccharide structures shared by a wide variety of species. However, any structure that yields a characteristic fragment of relatively unique m/z value can be monitored. For example, *m/z* 803 was monitored when tracing sialyl Lewis^x antigen on α1-acid glycoprotein *(39)*. The mass of any B-type oxonium ion can be calculated by adding together the residue weights of its components + 1 (H) (Table 21.1).

2. Since the different glycoforms are present at substoichiometric quantities, and the glycopeptide-containing fractions may have to be recollected, further digested and reanalyzed, if there is enough sample available reduce, alkylate and digest at least 1 nmole glycoprotein.

   When preparing samples for mass spectrometry experiments one has to avoid involatile salts, or they must be removed thoroughly by dialysis or chromatography. Salts interfere with mass spectrometric detection and they may clog the capillaries.

   Detergents may form micelles and thus may remain in the sample even after dialysis. Due to their exceptional mass spectrometric response and evenly spaced ion series, several detergents can be used for calibration, but they will completely suppress every other signal.

3. The proteolytic enzyme used for digestion should separate the consensus sequences into individual peptides and must show some cleavage specificity so that the

molecular masses for the cleavage products can be predicted. Trypsin, endopro-teases Lys-C and Asp-N are considered the most reliable enzymes in this regard. However, if digestion periods are too long and too much enzyme is employed, any of these may cleave at nonspecific sites, and/or undergo autolysis.

4. Using a water/acetonitrile/TFA mobile phase for the HPLC separation pro-vides the best chromatographic resolution, however TFA adversely affects mass spectrometric sensitivity. Addition of the make-up solvent, 2-methox-yethanol:isopropanol 1:1 mixture, significantly (5- to 10-fold) increases the detection sensitivity for hydrophilic components eluting at high water concentrations.

   Another HPLC mobile phase: solvent A: 0.1% formic acid in water, solvent B: 0.08% formic acid in ethanol/propanol (5:2), yields comparable chromato-graphic resolution while displaying improved ESIMS detection sensitivity without any make-up solvent addition. When formic acid is used as ion-pairing agent glycoforms may elute separated, and in reverse order, the most sialylated featuring the longest retention time. In addition, hydrophobic peptides and glycopeptides may give wide, tailing peaks.

5. The splitting ratio depends on the optimal flow rate for the electrospray source of the mass spectrometer used as detector.

6. The study presented here was finished in 1997. Since then new instrumentation has been developed with higher detection sensitivity for on-line LC/MS analysis. Orthogonal-acceleration-time-of flight instruments afford higher resolution and mass accuracy than the quadrupole mass spectrometers. Ion traps exhibit higher resolution and mass accuracy only in zoom-in mode for selected digest compo-nents. However, when data are acquired using the recently developed ion trap FT-ICR hybrid or ion trap - orbitrap analyzers all components may be detected with high resolution and mass accuracy.

7. Unfortunately, the optimal digestion period depends on the individual protein. For example, bovine fetuin is fully digested by trypsin (~1% w/w) in an hour, while other proteins require longer incubation times and even the addition of denaturing agents, such as urea or organic solvents. Trypsin tolerates both. However, long incubation with urea may result in carbamoylation of the ε-amino groups of Lys residues as well as the newly formed N-termini. Similarly, longer incubation times or larger amounts of trypsin may lead to chymotryp-tic and other nonspecific cleavages and to enzyme autolysis, despite the side-chain protection.

8. Acidification of the mixture to stop the digestion may lead to loss of sialic acid.

9. Even the best quality solvents used for the LC/ESIMS experiments produce abundant series of low mass ions. Thus, the low mass region of the mass spectra will be affected by this background. In most cases no significant background ions were observed at m/z values corresponding to diagnostic carbohydrate fragments.

10. This linear, approximately 0.5%/min gradient, started at 5% B provides a good chromatographic separation for most of the digests. If there are concerns about

the loss of hydrophilic peptides, the column should be equilibrated at 2% B. Similarly, when more hydrophobic peptides are present the final organic content of the eluent could be increased to higher level.

11. Many electrospray instruments permit alternate acquisition of high and low cone-voltage scans in a single LC/ESIMS analysis, although the AutoSpec did not. They also permit monitoring a series of diagnostic masses, hence the name selective ion monitoring, usually at higher sensitivity, rather than scanning the entire mass range.

    The voltage settings required for in-source fragmentation depend on the electrospray source and the mass spectrometer, as do all the other tuning parameters.

12. Either the sequence of the protein can be imported from a word processing program, or for Protein Prospector MS-digest its database accession number is sufficient information.

13. Remember, that mass spectrometers separate ions according to their mass-to-charge ratios, i.e., *m/z* values. Thus, stable isotope pacing can be used for charge state determination. The isotope peaks of a singly charged ion are separated by approximately 1 Da, while this mass difference is reduced to 0.5 Da for doubly charged ions: $[MH_2]^{2+}/2$ is measured, etc. The monoisotopic ion is the lowest mass member of this cluster, containing only ^{12}C, ^{1}H, ^{14}N, ^{16}O, ^{32}S etc. atoms. Monoisotopic masses are different from values calculated from the chemical atomic weights that are a weighted average for all the isotopes. Average masses are measured when the isotope peaks are not separated.

14. Table 21.1 shows the elemental composition and mass of some oligosaccharide units frequently encountered in glycoproteins.

15. For low resolution mass spectrometers these carbohydrate unit differences (Table 21.2) can aid the charge state determination. For example, complex structures almost always exhibit heterogeneity in the number of neuraminic (sialic) acids or the antennae. Thus, one can look for mass differences that would separate differently sialylated structures at different charge states. For example, a 97-Da mass difference would indicate triply charged ions, while the differently sialylated quadruply charged ions will be separated by approximately 73 Da. A similar approach can be used to trace other carbohydrate heterogeneities. Table 21.2 lists multiply charged mass differences corresponding to different carbohydrate structures.

16. In general mass spectrometry is not a quantitative method unless careful measurements are made with known quantities of internal standards. However, approximate relative amounts of different glycoforms may be estimated from the relative abundances, taking all the glycopeptide ions into consideration.

17. Unusual carbohydrate compositions can be dealt with using a simple computer program that generates the oligosaccharide compositions of any given mass. **GlycoMod** is such a tool and is available on the internet: http://www.expasy. ch/tools/glycomod/. The program can be used for free or derivatized oligosaccharides and for glycopeptides. However, elucidating the structure of unusual oligosaccharides is beyond this chapter, although the steps described here may

serve as starting point for that research. ExPasy also offers the free GlycanMass software to calculate masses from the oligosaccharide structures. A more comprehensive package, GlycoSuit is a commercially available glycan database with software tools by Proteome Systems (Sydney, Australia).

18. Deciding whether or not a peptide is really present sometimes requires judgment to be made by the analyst. If the signals are of low abundance, independent proof or repeated analysis is desirable, especially with mass spectrometers giving low resolution, lower mass accuracy data.

Acknowledgements I would like to thank Candy Stoner and Prof. Michael A. Baldwin for their assistance in the preparation of this manuscript. My work was supported by NIH grants NCRR RR01614, RR01296, RR014606 and RR015804 to the UCSF Mass Spectrometry Facility, Director: A.L. Burlingame.

References

1. Furmanek, A., Hofsteenge, J. (2000) Protein C-mannosylation: facts and questions. Review. Acta Biochim Pol. 47, 781–789.
2. Pless, D. D. and Lennarz, W. J. (1977) Enzymatic conversion of proteins to glycoproteins. *Proc. Natl. Acad. Sci. USA* **74**, 134–138.
3. Satomi, Y., Shimonishi, Y., Takao, T. (2004) N-glycosylation at Asn(491) in the Asn-Xaa-Cys motif of human transferrin. FEBS Lett. 576, 51–56.
4. Hanisch, F. G. and Peter-Katalinic, J. (1992) Structural studies on fetal mucins from human amniotic fluid. Core typing of short-chain O-linked glycans. Eur. J. Biochem. 205, 527–535.
5. Harris, R. J. and Spelmann, M. W. (1993) O-linked fucose and other post-translational modifications unique to EGF modules. Glycobiology, 3, 219–224.
6. Nishimura, H., Kawabata, S., Kisiel, W., Hase, S., Ikenaka, T., Takao, T., Shimonishi, Y., and Iwanaga, S. (1989) Identification of a disaccharide (Xyl-Glc) and a trysaccharide (Xyl2-Glc) O-glycosidically linked to a serine residue in the first epidermal growth factor-like domain of human factors VII and IX and protein Z and bovine protein Z. J. Biol. Chem. 264, 20320–20325.
7. Snow, D. M. and Hart, G. W. (1998) Nuclear and cytoplasmic glycosylation. Int. Rev. Cytol. 181, 43–74.
8. Hironaka, T., Furukawa, K., Esmon, P. C., Yokota, T., Brown, J. E., Sawada, S., Fournel, M. A., Kato, M., Minaga, T., and Kobata, A. (1993) Structural study of the sugar chains of porcine factor VIII - tissue- and species-specific glycosylation of factor VIII. Arch. Biochem. Biophys. 307, 316–330.
9. Bloom, J. W., Madanat, M. S., and Ray, M. K. (1996) Cell line and site specific comparative analysis of the N- linked oligosaccharides on human ICAM-1des454-532 by electrospray ionization mass spectrometry. Biochemistry 35, 1856–1864.
10. Zamze, S., Harvey, D. J., Chen, Y. J., Guile, G. R., Dwek, R. A., Wing, D. R. (1998) Sialylated N-glycans in adult rat brain tissue–a widespread distribution of disialylated antennae in complex and hybrid structures. Eur. J. Biochem. 258, 243–270.
11. Suzuki, N., Khoo, K. H., Chen, H. C., and Lee, Y. C. (2001) Isolation and characterization of major glycoproteins of pigeon egg white: ubiquitous presence of unique N-glycans containing Galalpha1-4Gal. J. Biol. Chem. 276, 23221–23229.
12. Yamashita, K., Koide, N., Endo, T., Iwaki, Y., and Kobata, A. (1989) Altered glycosylation of serum transferrin of patients with hepatocellular carcinoma. J. Biol. Chem. 264, 2415–2423.

13. Wada, Y., Nishikawa, A., Okamoto, N., Inui, K., Tsukamoto, H., Okada, S., and Taniguchi, N. (1992) Structure of serum transferrin in carbohydrate-deficient glycoprotein syndrome. Biochem. Biophys. Res. Com. 189, 832–836.

14. Nemansky, M., Thotakura, N. R., Lyons, C. D., Ye, S., Reinhold, B. B., Reinhold, V. N., and Blithe, D.L. (1998) Developmental Changes in the Glycosylation of Glycoprotein Hormone Free α Subunit during Pregnancy. J. Biol. Chem. 273, 12068–12076.

15. Landberg, E., Huang, Y., Stromqvist, M., Mechref, Y., Hansson, L., Lundblad, A., Novotny, M. V., and Pahlsson, P. (2000) Changes in glycosylation of human bile-salt-stimulated lipase during lactation. Arch. Biochem. Biophys. 377, 246–254.

16. Hakomori, S. (2002) Glycosylation defining cancer malignancy: new wine in an old bottle. Proc. Natl. Acad. Sci. USA 99, 10231–10233.

17. Higai, K., Aoki, Y., Azuma, Y., and Matsumoto, K. (2005) Glycosylation of site-specific glycans of alpha1-acid glycoprotein and alteration in acute and chronic inflammation, *Biochim Biophys Acta*, **1725**, 128–135.

18. Townsend, R. R. and Hardy, M.R. (1991) Analysis of glycoprotein oligosaccharides using high-pH anion exchange chromatography. *Glycobiology* 1, 139–147.

19. Kumar, H. P. M., Hague, C., Haley, T., Starr, C. M., Besman, M. J., Lundblad, R., and Baker, D. (1996) Elucidation of *N*-linked oligosaccharide structures of recombinant human factor VIII using fluorophore-assisted carbohydrate electrophoresis. *Biotechnol. Appl. Biochem.* **24**, 207–214.

20. Watzlawick, H., Walsh, M. T., Yoshioka, Y., Schmid, K., and Brossmer, R. (1992) structure of the *N*- and *O*-glycans of the A-chain of human plasma α2HS-glycoprotein as deduced from the chemical compositions of the derivatives prepared by stepwise degradation with exoglycosidases. *Biochemistry* **31**, 12198–12203.

21. Tyagarajan, K., Forte, J. G, and Townsend, R. R. (1996) Exoglycosidase purity and linkage specificity: assessment using oligosaccharide substrates and high-pH anion-exchange chromatography with pulsed amperometric detection. Glycobiology, 6, 83–93.

22. Gillece-Castro, B. L. and Burlingame, A. L. (1990) Oligosaccharide characterization with high energy collision-induced dissociation mass spectrometry. Meth. Enzymol. 193, 689–712.

23. Duffin, K. L., Welply, J. K., Huang, E., and Henion, J. D. (1992) Characterization of N-linked oligosaccharides by electrospray and tandem mass spectrometry. Anal. Chem. 64, 1440–1448.

24. Thomsson, K. A., Karlsson, N. G., Karlsson, H., and Hansson, G. C. (1997) Analysis of permethylated glycoprotein oligosaccharide fractions by gas chromatography and gas chromatography-mass spectrometry. in Techniques in Glycobiology (Townsend R. R. and Hotchkiss A. T. eds.) Marcel Decker, Inc., New York, pp. 335–347.

25. Papac, D. I., Jones, A. J. S., and Basa, L. J. (1997) Matrix-assisted laser desorption/ionization time-of-flight mass spectrometry of oligosaccharides separated by high pH anion-exchange chromatography. in Techniques in Glycobiology (Townsend R.R. and Hotchkiss A.T. eds.) Marcel Decker, Inc., New York, pp. 33–52.

26. Fu, D., Chen, L., and O'Neill, R. A. (1994) A detailed structural characterization of ribonuclease B oligosaccharides by 1H NMR spectroscopy and mass spectrometry. Carbohydr. Res. 261, 173–186.

27. Zhang, H., Li, X. J., Martin, D. B., and Aebersold, R. (2003) Identification and quantification of N-linked glycoproteins using hydrazide chemistry, stable isotope labeling and mass spectrometry. Nat. Biotechnol. 21, 660–666.

28. Bunkenborg, J., Pilch, B. J., Podtelejnikov, A. V., and Wisniewski, J. R. (2004) Screening for N-glycosylated proteins by liquid chromatography mass spectrometry. Proteomics 4, 454–465.

29. Sutton, C., O'Neill, J., and Cottrell, J. (1994) Site-specific characterization of glycoprotein carbohydrates by exoglycosidase digestion and laser desorption mass spectrometry. Anal. Biochem. 218, 34–46.

30. Ploug, M., Rahbek-Nielsen, H., Nielsen, P. F., Roepstorff, P., and Dano, K. (1998) Glycosylation Profile of a Recombinant Urokinase-type Plasminogen Activator Receptor Expressed in Chinese Hamster Ovary Cells. J. Biol. Chem. 273, 13933–13943.

31. Ling, V., Guzzetta, A. W., Canova-Davis, E., Stults, J. T., Hanckock, W. S., Covey, T. R., and Shushan, B.I. (1991) Characterization of the tryptic map of recombinant DNA derived tissue plasminogen activator by high-performance liquid chromatography-electrospray ionization mass spectrometry. Anal. Chem. 63, 2909–2915.

32. Wada, Y., Nishikawa, A., Okamoto, N., Inui, K., Tsukamoto, H., Okada, S., and Taniguchi, N. (1992) Structure of serum transferrin in carbohydrate-deficient glycoprotein syndrome. Biochem. Biophys. Res. Com. 189, 832–836.

33. Huddleston, M.J., Bean, M.F., and Carr, S.A. (1993) Collisional fragmentation of glycopeptides by electrospray ionization LC/MS and LC/MS/MS: methods for selective detection of glycopeptides in protein digests. Anal. Chem. 65, 877–884.

34. Carr, S.A., Huddleston, M.J., and Bean, M.F. (1993) Selective identification and differetiation of N- and O-linked oligosaccharides in glycoproteins by liquid chromatography-mass spectrometry. Protein Science, 2, 183–196.

35. Medzihradszky, K. F., Maltby, D. A., Hall, S. C., Settineri, C. A., and Burlingame, A. L. (1994) Characterization of protein N-glycosylation by reversed-phase microbore liquid chromatography/electrospray mass spectrometry, complementary mobile phases, and sequential exoglycosidase digestion.J. Am. Soc. Mass Spectrom. 5, 350–358.

36. Schindler, P. A., Settineri, C. A., Collet, X., Fielding, C. J., and Burlingame, A. L. (1995) Site-specific determination of the N- and O-linked carbohydrates on human plasma lecithin: cholesterol acyl transferase by HPLC/electrospray mass spectrometry and glycosidase digestion. Protein Sci. 4, 791–803.

37. Medzihradszky, K. F., Besman, M. J., and Burlingame, A. L. (1997) Structural Characterization of site-specific N-glycosylation of recombinant human factor VIII by reversed-phase high-performance liquid chromatography-electrospray ionization mass spectrometry. Anal. Chem. 69, 3986–3994.

38. Kapron, J. T., Hilliard, G. M., Lakins, J. N., Tenniswood, M. P. R., West, K. A., Carr, S. A., and Crabb, J. W. (1997) Identification and characterization of glycosylation sites in human serum clusterin. Protein Sci. 6, 2120–2133.

39. Dage, J. L, Ackermann, B. L., and Halsall, H. B. (1998) Site localization of sialyl Lewisx antigen on α1-acid glycoprotein by high performance liquid chromatography-electrospray mass spectrometry. Glycobiology 8, 755–760.

40. Medzihradszky, K. F., Gillece-Castro, B. L., Hardy, M. R., Townsend, R. R., and Burlingame, A. L. (1996) Structural elucidation of O-linked glycopeptides by high energy collision-induced dissociation. J. Am. Soc. Mass Spectrom. 7, 319–328.

41. Neumann, G. M., Marinaro, J. A., and Bach, L. A. (1998) Identification of O-glycosylation sites and partial characterization of carbohydrate structure and disulfide linkages of human insulin-like growth factor binding protein 6. Biochemistry, 37, 6572–6585.

42. Settineri, C. A. and Burlingame, A. L. (1996) Structural characterization of protein glycosylation using HPLC/electrospray ionization mass spectrometry and glycosidase digestion. Meth. Mol. Biol. 61, 255–278.

43. Biemann, K. (1990) Nomenclature for peptide fragment ions. Meth. Enzymol. 193, 886–887.

44. Townsend, R. R., Hardy, M. R., Wong, T. C., and Lee, Y. C. (1986) Binding of N-linked bovine fetuin glycopeptides to isolated rabbit hepatocytes: Gal/GalNAc hepatic lectin discrimination between Galβ(1,4)GlcNAC and Galβ(1,3)GlcNAc in a triantennary structure. Biochemistry 25, 5716–5725.

45. Cumming, D. A., Hellerqvist, C. G., Harris-Brandts, M., Michnik, S. W., Carver, J. P., and Bendiak, B. (1989) Structures of asparagine-linked oligosaccharides of the glycoprotein fetuin having sialic acid linked to N-acetylglucosamine. Biochemistry 28, 6500–6512.

22
Monitoring Glycosylation of Therapeutic Glycoproteins for Consistency by HPLC Using Highly Fluorescent Anthranilic Acid (AA) Tag

Shirish T. Dhume, George N. Saddic, and Kalyan R. Anumula

Summary Majority of protein drugs in development today are glycoproteins e.g. recombinant antibodies expressed in various cell lines. Oligosaccharides through conformational changes can modulate therapeutic value (potency) of glycoproteins e.g. complement dependent cell cytotoxicity (CDCC) and antibody-dependent cell cytotoxicity (ADCC) activities of MAbs. Carbohydrate structure analysis in detail is an integral part of protein drug characterization. This not only allows understanding of carbohydrates, but may allow deeper insight into the structure-function of the whole protein molecule. Oligosaccharide mapping by HPLC with fluorescence detection is a powerful technique that sheds considerable light into understanding of glycan structures with minimal effort. Oligosaccharide analysis using pulsed amperometric and/or chromophore detection methods lack resolution, sensitivity and ease of operations. In addition, these older methods are not highly reproducible. Simple labeling chemistry of anthranilic acid (AA) described here provide robust methods with the highest sensitivity and resolution for oligosaccharide analysis. Further, post-chromatography techniques such as mass spectrometry and NMR are amenable to this AA technology for detailed structure analysis.

Key Words Monoclonal; antibodies; MAbs; recombinant; IgG; N-linked; oligosaccharides; glycans; HPLC; fluorescence; anthranilic acid; mapping.

1 Introduction

Understanding the structure and function of proteins is a major area of interest and significance in biochemistry. Many proteins frequently reported in scientific journals are glycosylated. The oligosaccharide chains attached to the proteins may elicit important functional or structural properties *(1,2)*. The characterization of oligosaccharide chains in addition to its protein part is of importance to fully understand the structure and function of glycoproteins. There are cases where either no apparent function is obvious or not yet identified for the carbohydrate part e.g., IgGs intended for neutralizing activity only. However, world wide regulatory agencies require

From: *Post-translational Modifications of Proteins.*
Methods in Molecular Biology, Vol. 446.
Edited by: C. Kannicht © Humana Press, Totowa, NJ

demonstration of consistency in glycosylation of the manufactured lots intended for human therapy. Therefore, characterization of carbohydrate units is necessary to define the structure of therapeutic glycoproteins. This may include studying the extent of glycosylation in terms of its monosaccharide composition (including sialic acids and percent carbohydrate) and the nature of oligosaccharide chains.

Monoclonal antibodies constitute a major part of current protein drug portfolio in biopharmaceutical industry. The efficacy of recombinant/monoclonal IgGs has been demonstrated to depend on their glycoforms, specifically, if they are designed to kill the target cells. Therefore, glycosylation analysis of IgGs is of considerable interest. IgGs contain a conserved N-glycosylation site in the CH2 domain of the Fc region of the heavy chain and carbohydrate accounts for ~2% (w/w) of the IgG weight. Recombinant IgGs are characterized by heterogeneous population of fucosylated bi-antennary complex structures with Gal, GlcNAc and sialic acid as terminal sugars. The terminal sugar residues, including fucose of the 'core', have been shown to be responsible for activation of Fc effector functions (see (3) for a review). Therefore, glycosylation analysis of therapeutic mAbs is essential since the biologic activity/potency is directly related to both complement dependent cell cytotoxicity (CDCC) and antibody-dependent cell cytotoxicity (ADCC) activities. In this regard, better analytical tools offer better assurance of quality. Clearly, methods with excellent reproducibility, e.g., described here, are highly desirable for this purpose.

High performance anion exchange chromatography with pulsed amperometric detection (HPAEC-PAD) had been the method of choice for carbohydrate analysis (4). This method does not require derivatization but requires specialized dedicated equipment. However, it has many disadvantages e.g., high noise, unstable baseline, high pH, high salt, and a lack of user-friendliness. An additional problem with HPAEC-PAD is the interference of amino acids, peptides and other oxidizable compounds in the carbohydrate analysis (5). High salt present in the eluents precludes the use of HPAEC-PAD in recent techniques such as LC-MS and LC-NMR. Often, resolution of oligosaccharides is a problem with HPAEC-PAD in cases where a high level of carbohydrate microheterogeneity exists. Recently, a number of methods based on fluorescent labeling of carbohydrates to address some of these problems have been introduced. Here, facile derivatization of carbohydrates with anthranilic acid (AA, 2-aminobenzoic acid, 2AA) is found to be far superior compared to all other tags (see reference (6) for a review).

Oligosaccharide analysis with fluorescent detection provides a great improvement over HPAEC-PAD in terms of sensitivity (100–200x) and resolution (7) and AA is suitable for glycomics (8). In addition, it does not require a dedicated instrument. AA labeling is specific for reducing monosaccharides and a variety of oligosaccharides, precluding the need to purify glycans from peptides or proteins prior to derivatization (7). N-linked oligosaccharides can be conveniently released with the enzyme PNGase (A or F), and both N- and O-linked oligosaccharides can be released using hydrazinolysis (9). We have reported two chromatographic methods for high-resolution analysis of glycans released from glycoproteins (7). The mapping/profiling method offers a high resolution map in which the degree of sialylation, desialylated complex and high mannose type of oligosaccharides can be identified in a single run. The second method is mainly used for further characterization of oligosaccharides following desialylation. In this method, branch and linkage isomers are easily separated, and the method can be used for determining the sequence of sugars following treatment with an array of glycosidases.

Oligosaccharide mapping/profiling following derivatization with anthranilic acid (AA) affords very high-sensitivity and high-resolution technology for studying glycoproteins. In fact, to date, the highest resolution of sialylated as well as neutral oligosaccharides has been achieved by oligosaccharide mapping with AA technology. For example, in contrast to AA analysis, high-mannose and asialo-complex glycans from Chinese hamster ovary (CHO) cell derived rIgG coeluted in HPAEC-PAD analysis which may lead to erroneous conclusions. Such HPAEC-PAD profiles may be interpreted as a "normal" distribution of glycans in a rIgG and the presence of high-mannose structures can go undetected. However, in the AA map the high mannose and asialo-complex oligosaccharides are easily discernible since they elute as separate sharp peaks *(7)*. In addition, the fluorescent mapping method can be used for oligosaccharide analysis of the glycoproteins run on SDS-PAGE gels preferably after electroblotting onto PVDF membranes in a manner similar to monosaccharide analysis *(10)*.

The oligosaccharide methods described here are compatible with physical techniques (e.g., MS and NMR), since the AA mapping method uses only volatile organic solvents. The oligosaccharide fractions can be isolated simply by evaporation for further analysis e.g., MALDI-TOF *(7)*. The AA derivatized oligosaccharides can be easily analyzed by MALDI-TOF in either mode for neutral and acidic glycans using 2,5-dihydroxybenzoic acid (DHB) as a matrix. MALDI-TOF has also been used for more detailed structural studies in conjunction with exo-glycosidase treatments without the need to remove buffers or salts.

The fluorescent AA tag was reported to be at least twice as intense as its amide derivative, 2-aminobenzamide (AB) *(7)*. In fact, among many fluorescent tags tested to date *(11)*, AA is the most sensitive tag of choice for oligosaccharide analysis. In addition, the AA derivatized oligosaccharides are stable for over three years at −20°C. The reagents and solvents used in the method have a "shelf-life" of several months. The mapping solvents (A and B), washing solution (95% acetonitrile in water), eluting solution (80% water in acetonitrile), all stored at room temperature, and AA reagent, stored at −20°C, for more than 5 mo gave oligosaccharide maps that were comparable to those obtained with freshly made reagents and solvents.

Monitoring of production lots or in-process samples for glycosylation using the AA technology is done by comparison of their profiles with a well-defined reference standard. In order to pass the test, all peaks in the sample are expected to co-elute with the corresponding peaks in the reference standard within a set of chromatographic runs. Acceptance criteria are set such that high confidence in the results is assured. For example, ±0.25 min standard deviation for retention times is used for peaks in the experimental sample compared to corresponding peaks in the reference standard. Additionally, relative standard deviation (RSD) for percent area of peaks is set at less than 15% for the main peaks. Main peaks are defined as peaks with relative percent area greater than 10% of total area of all the glycan peaks. All main peaks should have resolution greater than 1.0 and tailing less than 1.4. In addition to retention time, resolution and tailing are important parameters for monitoring system performance and are used in all the analysis. Often, additional peaks are encountered in the map and they require further characterization, for example, by using mass spectrometry, co-elution with standards and/or exo-glycosidase studies. However, these may not necessarily present a problem for the release of clinical materials if the amounts are below previously set criteria.

For detailed characterization of CHO cell derived IgGs by HPLC, commercially available sialyl- and asialo-complex glycans with core fucose and oligosaccharides from well defined human IgG can be used as standards. For IgGs that contain high mannose structures in addition to complex glycans, RNase B glycans can be used as a reference, either alone or in combination with the standards suggested above. Some oligosaccharide standards are not easily available, such as the monogalactosyl complex glycan, and oligosaccharides with defined sialic acid linkages (α2-3 versus α2-6). In rIgGs, there are two isomers resulting in the presence of a galactose on either arm of the biantennary oligosaccharide chain. Interestingly, these isomers are easily separated by normal phase HPLC method described earlier *(7)*. Thus, each can be isolated, characterized by NMR or mass spectrometry for subsequent use as a glycan standard. For sequence and linkage determinations, specific exo-glycosidases are now available that can distinguish various linkages in an oligosaccharide chain. NMR and mass spectrometry offer a definite proof of structure but require a level of expertise that may not be available in most biopharmaceutical companies or biotechnology laboratories.

The AA oligosaccharide mapping method is now being used routinely in both academic and biotechnology laboratories, and has been tested for most validation parameters. It passed rigid acceptance criteria set for accuracy, precision, repro-ducibility, robustness, ruggedness, etc. In addition, it has been rigorously tested in different laboratories by different analysts for monitoring consistency of carbohydrates in various glycoprotein drug lots as part of regulatory submissions. Oligosaccharide mapping with AA tag is routinely used in conjunction with monosaccharide composition analysis *(12–14)* for comparison of various production lots intended for human use.

Furthermore, the technology described here is simple in terms of derivatization of saccharides with fluorescent AA and analysis by HPLC. Therefore, any analyst with an access to HPLC system with a fluorescence detector can successfully analyze the glycoprotein of interest.

2   Materials

2.1   Preparation of the Oligosaccharides using rIgG as an Example

1. Fetuin, bovine (Sigma).
2. Recombinant IgG.
3. Polypropylene vials (1.6 mL) with O-ring seals (Fisher, cat. no. 118448).

4. 1: 100 Diluted ammonium hydroxide (0.15*M*, prepare fresh for each use).
5. 10% (w/v) SDS stock solution.
6. Diluted ammonium hydroxide-0.5% SDS-1% 2-mercaptoethanol solution for PNGase F digestion (prepare fresh for each use from respective stock solutions).
7. Heating Block.
8. 5% (w/v) Nonidet P-40.
9. Peptide *N*-glycosidase F (Glyko).

2.2 Derivatization and Purification

1. Acetic acid, glacial.
2. 4% (wlv) Sodium acetate trihydrate and 2% (w/v) boric acid in methanol.
3. Anthranilic acid reagent: 30 mg/mL anthranilic acid (Aldrich) and approximately 30 mg/mL sodium cyanoborohydride in acetate-borate-methanol solution. Although it is stable for a week in the dark at ambient temperature, it is recommended that this reagent be prepared fresh. NOTE: Sodium cyanoborohydride is a poison and tends to absorb moisture readily from the air, which may affect the derivatization reaction. Transfers should be made in a chemical hood. Limit the exposure of this chemical to air when weighing. If desired, 50 μL of 1.0*M* sodium cyanoborohydride in THF (Aldrich, cat. no. 296813) may be substituted in the reaction mixture.
4. Heating block.
5. 3- or 5-mL Plastic syringe with luer lock.
6. Nylon Acrodisc syringe filter, 0.45 μm (Gelman, cat. no. 4438, Fisherbrand, 09-719-5).
7. 95% (v/v) Acetonitrile in water.
8. 20% (v/v) Acetonitrile in water.

2.3 Chromatography

1. HPLC with fluorescence detector (highly sensitive fluorescence detectors such as Jasco FP 920, Waters 474, and HP (now Agilent) 1100 were used in these studies).
2. Thermostat column compartment.
3. Polymeric-amine bonded HPLC column, 0.46 × 25 cm (Astec, cat. no. 56403 or Asahipak, NH2P-50 4E, Phenomenex, cat. no. CHO-2628) for mapping.
4. TSK Amide-80 column, 0.46 × 25 cm (Toso Haas, cat. no. 13071) for neutral oligosaccharide analysis.
5. Column prefilter with column prefilter insert (0.2 μm).
6. Autosampler vials (amber).

7. Eluent A: 2% (v/v) acetic acid and 1% (v/v) tetrahydrofuran (inhibited) in acetonitrile (all HPLC grade).
8. Eluent B: 5% (v/v) acetic acid, 1% (v/v) tetrahydrofuran (inhibited) and 3% (v/v) triethylamine (all HPLC grade) in Milli-Q filtered water.
9. Eluent C: 0.2% (v/v) acetic acid in acetonitrile (all HPLC grade): for neutral oligosaccharide chromatography.
10. Eluent D: 0.2% (v/v) acetic acid and 0.2% (v/v) triethylamine (all HPLC grade) in Milli-Q water: for neutral oligosaccharide chromatography.

3 Methods

3.1 Preparation of the Oligosaccharides Using rIgG as an Example

1. Dilute the rIgG experimental, reference standard and fetuin from 5 to 20 mg/mL in Milli-Q water.
2. Appropriately label one 1.6-mL polypropylene freeze vial for each blank, sample, and fetuin.
3. Place 5 µL of the diluted samples and fetuin into the corresponding labeled vials.
4. Prepare a blank by placing 5 µL of Milli-Q water into the corresponding labeled vial.
5. Add 30 µL of the ammonium hydroxide-SDS-2-mercaptoethanol solution to each vial. Cap the vials and mix them on a Vortex mixer (*see* **Note 1**).
6. Heat the vials in a heating block set at 100°C for 2–3 min to denature the proteins. Remove the vials and allow them to cool to room temperature.
7. Add 5 µL of 5% Nonidet P-40 to each vial and mix the vials using a Vortex mixer.
8. Add 2 µL of the PNGase F to each vial and gently mix by aspiration in and out of the pipet tip.
9. Place the vials in a heating block that is set at 37°C for about 18 h (overnight digestion is sufficient for most glycoproteins).

3.2 Derivatization and Purification

1. Allow the blank, samples, and fetuin to cool to room temperature. Briefly centrifuge the vials at the maximum setting in a microcentrifuge to collect the solutions at the bottom of the tubes (*see* **Note 2**).
2. Add 2 µL of glacial acetic acid to each vial and mix.

3. Add 100 μL of the anthranilic acid reagent to each vial. Cap the vials tightly and mix.
4. Heat the vials for 1 h in a heating block that is set at 80°C. Allow the vials to come to room temperature before proceeding.
5. Cut one 3- or 5-mL plastic syringe with luer lock for each sample at 2-mL mark to use as a funnel for Nylon syringe filters.
6. Plug the nozzle lightly with glass wool in order to prevent the formation of air bubbles during filtration.
7. Prime the Nylon syringe filters with about 2 mL of 95% acetonitrile-water using an uncut syringe.
8. Attach cut syringe to the filter and rinse the filters with additional 2x 1 mL of 95% acetonitrile using gravity flow.
9. Dilute the oligosaccharide reaction mixture with 1.0 mL of 95% acetonitrile and vortex.
10. Transfer the diluted reaction mixture into the precut syringe and allow it to flow through the Nylon filter. The oligosaccharides will bind to the Nylon filter membrane.
11. Wash the bound oligosaccharides only two times with 1.0 mL of 95% acetonitrile. Discard the flow-through and washes (*see* **Notes 3** and **4**).
12. Appropriately label one autosampler vial for each sample.
13. Elute the bound oligosaccharides with 2 × 0.5 mL of 20% acetonitrile into the appropriately labeled autosampler vial and mix. Inject an aliquot of 50–200 μL for analysis.
14. Chromatographic runs are serially arranged as follows: blank, fetuin, reference standard, and experimental samples.

3.3 Chromatography

1. For mapping, freshly prepare eluents A and B.
2. For mapping use the gradient program as follows (Astec or Asahipak column at 50°C, flow rate of 1.0 mL/min and detection at λex = 360 nm and λem = 425 nm):

0 min, B = 30% (A = 70%)
2 min, B = 30%
82 min, B = 95% (linear increase)
97 min, B = 95% (Wash)
97.1 min, B = 30%
112.1 min, B = 30% (Equilibration)

3. Prepare eluents C and D for neutral oligosaccharide chromatography.
4. For neutral oligosaccharide chromatography (*see* **Note 5**) run the gradient program (Amide column at 25°C, flow rate of 1.0 mL/min and detection at λex = 360 nm and λem = 425 nm) (*see* **Note 6**):

0 min D = 28% (C = 72%)
20 min, D = 28%
80 min D = 45% (linear increase)
80.1 min D = 95%
90.1 min D = 95% (Wash)
90.2 min D = 28%
105.2 min D = 28% (Equilibration)

Agilent 1100 fluorescent detector settings: Peak Width Response time: >0.4 min (8s, slow); PMT Gain: 12; Baseline: Append; Reference: On; Polarity: Positive.

3.4 Acceptance Criteria for Results

Suitability of the HPLC system for the analysis is deemed acceptable when the oligosaccharide map of bovine serum fetuin compares favorably to the representative chromatogram provided in Figure 22.1. Fetuin glycan maps are routinely used as system check for good manufacturing practice (GMP) purposes. In the event of abnormal fetuin glycan map, the chromatographic runs are aborted for further investigation prior to lot analysis.

Test results are deemed acceptable when the oligosaccharide map of the experimental sample compares favorably to the reference standard. The retention times of the main peaks should have a standard deviation (SD) of less than 0.25 min and RSD of less than 15% for relative percent area in comparison to corresponding peaks in the reference standard. Main peaks are defined as peaks with relative areas more than 10% of total.

4 Notes

1. Dilute ammonium hydroxide can be used for PNGase F release of oligosaccharides from glycoproteins, and the yields from this procedure are similar to buffer systems provided by the suppliers precluding the need to prepare buffers. The profile of fetuin glycans released by PNGase F using dilute ammonium hydroxide is comparable to the hydrazine-released glycans (Fig. 22.1). However, the hydrazine released oligosaccharide mixture contains both N- and O-linked glycans, whereas the PNGase released oligosaccharides contain only the N-linked chains.

2. PNGase F reaction mixture and/or exo-glycosidase reaction mixture does not require additional purification prior to labeling. The Nylon purification step used for the removal of excess AA is sufficient to cleanup the samples (i.e., removal of SDS, Nonidet P-40, buffers, etc.) prior to chromatography.

3. Nylon purification is a crucial step that may cause loss of glycans due to excessive washing. Generally washing the filter with bound oligosaccharides with 95%

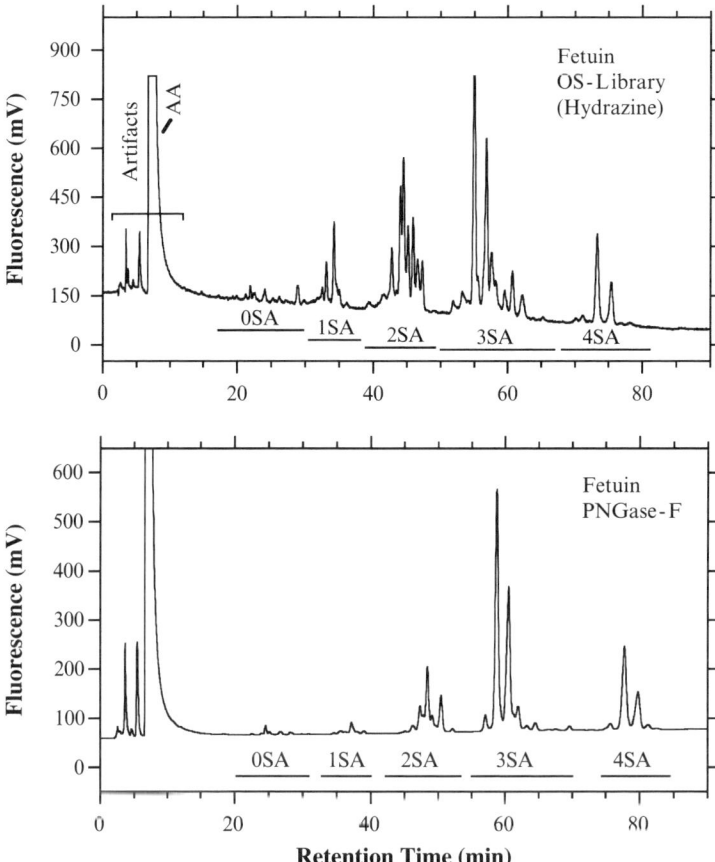

Fig. 22.1 Comparison of fetuin oligosaccharide maps: oligosaccharides released using hydrazine (top) and PNGase F (bottom). Bars indicate the regions where oligosaccharides with 0–4 sialic acids (SA) elute

acetonitrile should be limited to two as suggested in the method, and elution with 0.5 mL × 2 is sufficient for most of the glycoproteins tested.

4. AA labeling has been successfully applied to *O*-linked oligosaccharides released from bovine submaxillary mucin (BSM) using hydrazine (Fig. 22.2). The O-linked glycan profile with and without purification is similar in the case of BSM sugar chains, albeit the losses (>30%) encountered upon purification. See reference *(15)* for structures present in BSM.

5. Figure 22.3 shows the separation of neutral oligosaccharides on an amide column. The top two chromatograms are the maps of two different rIgG molecules produced in CHO cells. Although these two rIgGs have the same constant region of heavy chains, they differ in their oligosaccharide profile. Compared

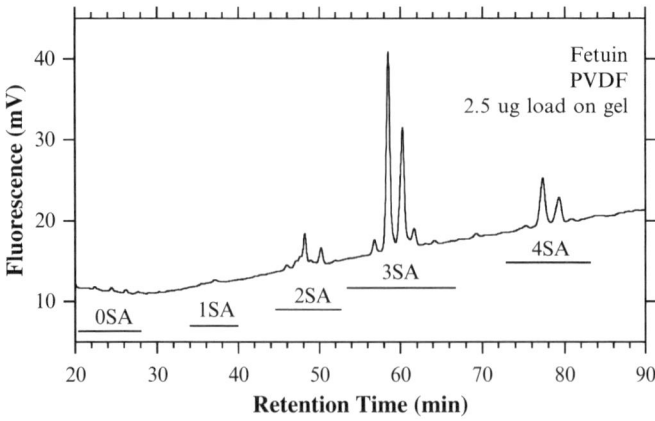

Fig. 22.2 Oligosaccharide map obtained with 2.5 µg of fetuin run on SDS-PAGE and electroblotted onto PVDF membrane. One of the three bands was used for PNGase F digestion

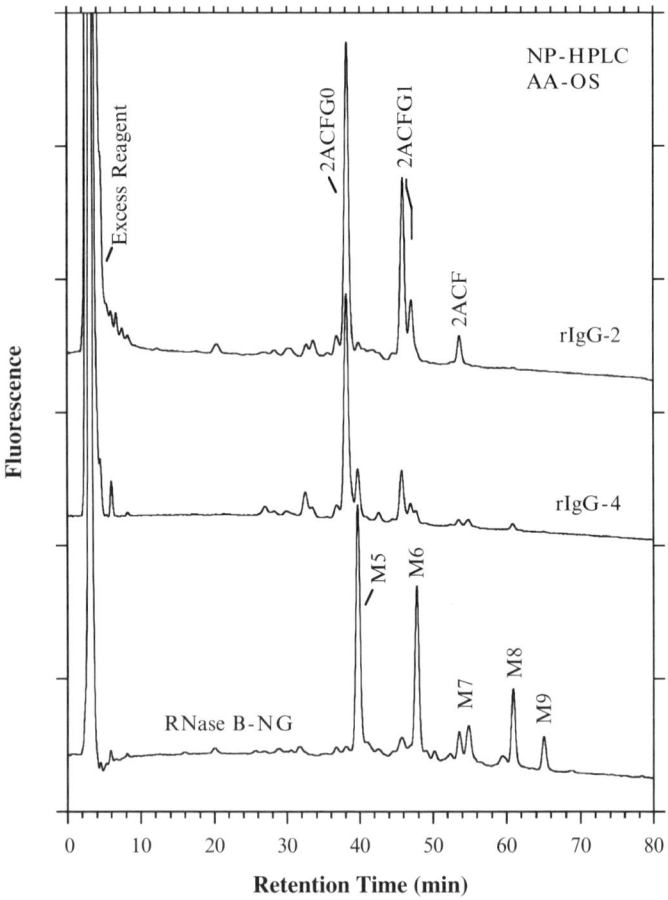

Fig. 22.3 Separation of neutral oligosaccharides on a TSK amide-80 column. Neutral oligosaccharides from two different rIgGs (top two chromatograms), in comparison with RNase B high mannose oligosaccharides

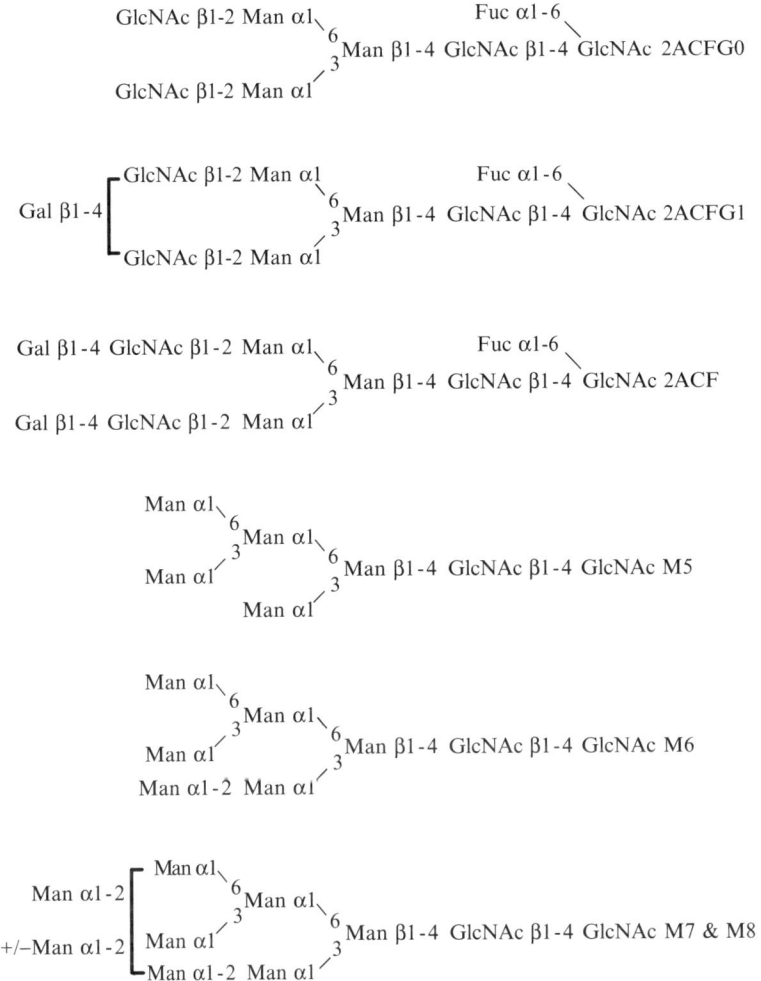

Fig. 22.4 Oligosaccharide structures for the peaks in Figures 22.3 and 22.5

to rIgG-2, rIgG-4 has diminished amount of 2ACFG 1 (biantennary complex with fucose and one galactose residue) and practically no 2ACF (biantennary complex with fucose) (see Fig. 22.4 for structures). However, it has measurable amounts of high-mannose oligosaccharide structures, Man5-Man8, similar to the glycan chains from RNase B released by PNGase F (RNase B-NG). In addition, the two isomers of 2ACFG1 are easily separated in this chromatography, which in turn are easily differentiated from the Man6 glycan chain. Such resolution of oligosaccharide structures could not be obtained by HPAEC-PAD analysis.

6. For less sensitive detectors, the excitation wavelength can be changed to 230/245 nm.

7. The oligosaccharide profile obtained by the AA fluorescent-labeling procedure is similar to the ones obtained using HPAEC-PAD except for higher resolution, which has been discussed in detail earlier (6). The fetuin oligosaccharide profiles obtained with oligosaccharide released by hydrazine and PNGase F are shown in Fig. 22.1.

8. The oligosaccharide mapping technology can be applied to glycoproteins blotted onto PVDF membrane following SDS-PAGE or directly from gels after electroelution. Commercial fetuin typically resolves into three bands of equal intensity upon SDS-PAGE. One of the bands from a 2.5 μg load was excised, minced into small pieces, and digested with PNGase F as in the case of a sample

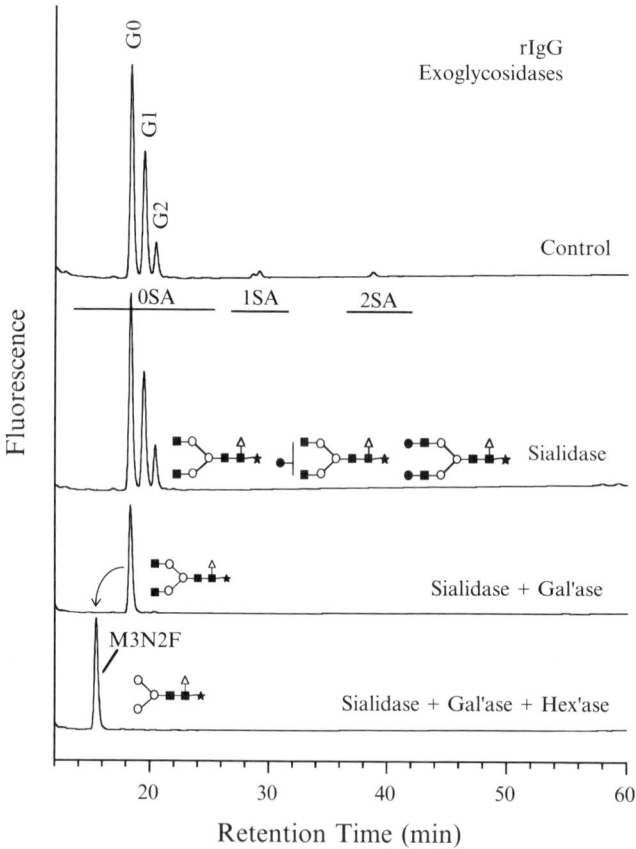

Fig. 22.5 Effect of exoglycosidases on an oligosaccharide pool from a Chinese hamster ovary (CHO) cell derived rIgG. Structures in rIgG can be easily elucidated starting with SA2 and following the action of exoglycosidases until the fucosylated 'core structure' (M3N2F, standard) is obtained. The fucosylated-core structure can be easily confirmed using Endo F3 (16) in addition to using commercial M3N2F standard. Key to symbolic representation of monosaccharides: , ● galactose; ■, N-Acetylglucosamine; ○, mannose; Δ, fucose; and *, AA

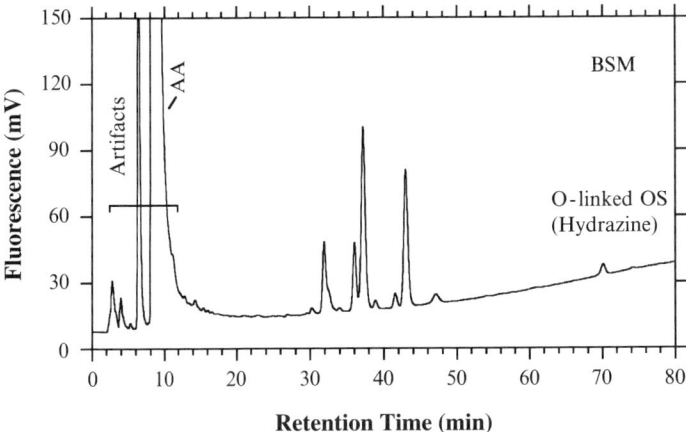

Fig. 22.6 Profile of O-linked oligosaccharides from bovine submaxillary mucin released by hydrazine. The oligosaccharides were analyzed using the mapping method

Table 22.1 Analyst to Analyst Reproducibility with Oligosaccharide Mapping Method

Peaks	Analyst 1	Analyst 2	Average
Gal0	49.93[a] (0.08)[b]	51.39 (1.38)	50.66 (2.04)
Gal1	35.32 (0.10)	34.88 (0.69)	35.10 (0.88)
Gal2	10.18 (0.35)	9.24 (3.29)	9.71 (6.85)
SA1	3.09 (0.92)	3.02 (3.99)	3.05 (1.74)
SA2	1.50 (0.47)	1.49 (2.38)	1.49 (0.47)

[a]Average of relative percent peak area from two replicates.
[b]Percent relative standard deviation (*see* **Note 10**).

in solution. The oligosaccharide map/profile obtained from a fetuin band is shown in Figure 22.5. Characteristic profile of the fetuin oligosaccharides is clearly seen even with this level ($< 1.0\,\mu g$) of glycoprotein.

9. Figure 22.6 shows the oligosaccharide maps of rIgG treated with and without exoglycosidases demonstrating the sequential loss of monosaccharide units from the nonreducing end and their comparison with the standards. Co-elution of the rIgG peaks with commercial standards confirms the structures identified in the map. Maps were obtained without any additional purification to remove the digestion buffers.

10. The oligosaccharide mapping method has been validated for several IgG proteins. It has been shown to be accurate via NMR, mass spectrometry, coelution (with glycan standards), and exo-glycosidase experiments. Precision of the

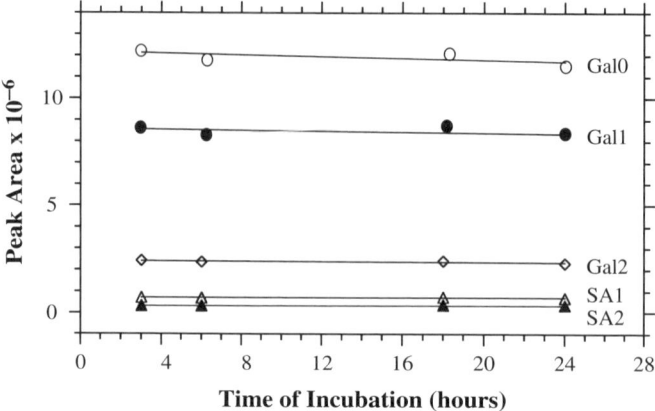

Fig. 22.7 Time course for release of oligosaccharides from a rIgG with PNGase F enzyme. Plots are obtained by averaging peak area values from two replicates at each time point. Error bars represent standard deviation of replicates

method has also been established. Ruggedness and robustness with respect to variations in HPLC systems, analysts, wavelengths, gradient slope, solvent A and B composition, temperature, and so on, has been demonstrated. Results of reproducibility experiments are summarized in Table 22.1. Relative area % RSDs are less than 15% for all peaks and pass acceptance criteria.

11. Several experiments were also carried out to test the "shelf-life" of reagents and solvents used in the oligosaccharide-mapping method. All reagents and solvents were more than 5 month old, a testimony to the ruggedness of the method. Except for the AA reagent, which was stored at −20°C, the other solvents and reagents were at room temperature. Oligosaccharide profiles obtained with these reagents/solvents were comparable to ones obtained with freshly made solvents/reagents.

12. For most glycoproteins, 3 h of incubation with PNGase F suffices for complete oligosaccharide release (Fig. 22.7). However, a time course experiment may need to be done to confirm this in individual cases. Additionally, no degradation (diminution of peak area) of any oligosaccharide species is observed at 24 h of incubation (Fig. 22.7) or even at 48 h (not shown). The protocol described here is applicable to any glycoprotein.

Acknowledgement We thank Ms. Ping Du and Ms. Mary Beth Ebert for their contributions to validation of the methods.

References

1. Varki, A. (1993) Biological roles of oligosaccharides: all of the theories are correct. *Glycobiology*, **3**, 97–130.

2. Dwek, R. A. (1996) Toward understanding the function of sugars. *Chem. Rev.*, **96**, 683–720.
3. Jefferis, R. (2005) Glycosylation of recombinant antibody therapeutics. *Biotechnol. Prog.* 21, 11–16.
4. Townsend, R. R. (1995) Analysis of glycoconjugates using high-pH anion-exchange chromatography. *Carbohydrate Analysis: High-Performance Liquid Chromatography and Capillary Electrophoresis* (El Rassi, Z., ed.), Elsevier, New York, pp. 181–209.
5. Davies, M. J. and Hounsell, E. F. (1998) HPLC and HPAEC of Oligosaccharides and Glycopeptides. *Methods in Molecular Biology, Glycoanalysis Protocols* (Hounsell, E. F., ed.), Humana Press, Totowa, NJ.
6. Anumula, K. R. (2006) Advances in fluorescence derivatization methods for high-performance liquid chromatographic analysis of glycoprotein carbohydrates. *Anal. Biochem.* 350:1–23.
7. Anumula, K. R. and Dhume, S. T. (1998) High resolution and high sensitivity methods for oligosaccharide mapping and characterization by normal phase high performance liquid chromatography following derivatization with highly fluorescent anthranilic acid. *Glycobiology*, **8**, 685–694.
8. Naka R, Kamoda S, Ishizuka A, Kinoshita M, and Kakehi K. (2006) Analysis of total N-glycans in cell membrane fractions of cancer cells using a combination of serotonin affinity chromatography and normal phase chromatography. *J Proteome Res.* 5, 88–97.
9. Patel, T., Bruce, J., Merry, A., Bigge, C., Wormald, M., Jacques, A., and Parekh, R. (1993) Use of hydrazine to release in intact and unreduced form both N- and O-linked oligosaccharides from glycoproteins. *Biochemistry*, **32**, 679–693.
10. Anumula, K. R. and Du, P. (1999) Characterization of carbohydrates using highly fluorescent 2-aminobenzoic acid tag following gel electrophoresis of glycoproteins. *Anal. Biochem.*, **275**, 236–242.
11. Dhume, S. T. and Anumula, K. R. (1998) Evaluation of various fluorescent tags for oligosaccharide characterization: a comparative study with anthranilic acid. *Glycobiology*, **8**, Abstract no. 34.
12. Anumula, K. R. (1994) Quantitative determination of monosaccharides in glycoproteins by high performance liquid chromatography with highly sensitive fluorescence detection. *Anal. Biochem.*, **220**, 275–283.
13. Anumula, K. R. (1995) Novel fluorescent methods for quantitative determination of monosaccharides and sialic acids in glycoproteins by reversed phase high performance liquid chromatography. *Methods in Protein Structure Analysis* (Atassi, M. Z. and Apella, E., ed.), Plenum Press, New York, pp. 195–206.
14. Anumula, K. R. (1997) Highly sensitive pre-column derivatization procedures for quantitative determination of monosaccharides, sialic acids and amino sugars alcohols of glycoproteins by reversed phase HPLC. Techniques in Glycobiology (Townsend R. R. and Hotchkiss, A. T., ed.), Marcel Dekkar, New York, pp. 349–357.
15. Gutierrez Gallego, R., Haseley S. R., van Miegem V. F., Vliegenthart, J. F., Kamerling, J. P. (2004) Identification of carbohydrates binding to lectins by using surface plasmon resonance in combination with HPLC profiling. *Glycobiology.* 14, 373–386.
16. Anumula, K. R. (1993) Endo beta-N-acetylglucosaminidase F cleavage specificity with peptide free oligosaccharides, J. Mol. Recog. 6, 139–145.

23

Comparability and Monitoring Immunogenic *N*-linked Oligosaccharides from Recombinant Monoclonal Antibodies from Two Different Cell Lines using HPLC with Fluorescence Detection and Mass Spectrometry

Bruce R. Kilgore, Adam W. Lucka, Rekha Patel, Bruce A. Andrien, Jr., and Shirish T. Dhume

Summary One of the most important structural features of recombinant monoclonal antibodies produced in mammalian cells is the *N*-linked oligosaccharide profile. These profiles impact recombinant therapeutics in a multitude of ways affecting distribution, efficacy, and immunogenicity. High mannose, α-gal and other oligosaccharide species are highly immunogenic and in most cases should be minimized during manufacturing. A recombinant monoclonal antibody, h5G1.1, was produced in NS0 and CHO cell lines and tested to identify changes in the *N*-linked oligosaccharide profiles caused from a change in cell line. Traditional peak analysis using HPLC with fluorescence detection was augmented by mass spectrometric analysis. Nano LC-MS following tryptic digestion corroborated HPLC findings of the presence of several α-gal oligosaccharide species in the recombinant IgG (rIgG) from NS0 cell line. Both cell lines possessed rIgGs with complex and small amounts of high mannose glycans.

Key Words Mass Spectrometry; ESI-ToF, nano LC-MS; oligosaccharide, glycans; saccharides; carbohydrates, *N*-linked, glycoprotein, glycopeptides, HPLC, fluorescent, anthranilic acid.

1 Introduction

As biopharmaceuticals are moved from target validation toward clinical entities, protein characterization or structural proteomics becomes significantly more important. A majority of the therapeutic proteins currently in the market are glycoproteins *(1)*. In turn, many of the glycoproteins are antibodies, and most of these are whole antibodies *(2)*. As several glycoproteins are introduced into the pipcline, product *N*-linked oligosaccharide characterization, fingerprinting or profiling assumes more significance *(3)*.

From: *Post-translational Modifications of Proteins.*
Methods in Molecular Biology, Vol. 446.
Edited by: C. Kannicht © Humana Press, Totowa, NJ

Oligosaccharides not only influence the conformation of proteins, but also directly or indirectly influence glycoprotein function. For example, the immune system exerts its diversity by several means, including glycosylation. The minimal core glycosylation essential for the Fcγ recognition and for C1 activation has been reported by Jefferis (4). Amino acid side chains contacting core oligosaccharides modulate Fc-effector functions. There is constant reciprocity in the structure and function of glycoproteins and their resident oligosaccharides. Even the effector ligand activation and glycosylation profile is influenced by quaternary IgG structure (5).

Role of terminal sialic acid in increasing half-life of glycoproteins is well established (6). A change from terminal sialo- to galacto-oligosaccharide structure produces rapid clearance through recognition by the asialoglycoprotein receptor (ASGP-R), a calcium-dependent C-type lectin on hepatocytes. Liver parenchymal cells mediate the receptor-mediated endocytosis of asialoglycoproteins. The receptors bind galactose-terminating glycoprotein oligosaccharides and internalize them through coated pits, removing them from circulation. The receptor-ligand complex goes through several steps of internalization, intracellular transfers and trafficking to the compartment of uncoupling of receptors and ligand (CURL). Acidification of the complex allows receptor to recycle to the cell surface with the ligand culminating into lysosomes for degradation.

Proteins containing glycans with high mannose structures will experience lower half-life because of clearance mediated through the high mannose receptor on macrophages. Calcium-dependent C-type carbohydrate recognition domains (CRD) are involved in the recognition of extracellular glycoprotein mannosides, components found on pathogens (7). Mannose-binding lectin (MBL) on macrophages are involved in the antibody-independent phagocytosis of serum glycoproteins that are not recognized as "self." These receptors possess multiple C-type CRDs allowing it to bind large, particulate ligands of microbial pathogens and other sources of bound and/or free high mannose glycoproteins.

Though more than 40 different sialic acids have been reported, mainly 2 forms, N-acetyl neuraminic acid (Neu5Ac or NANA) and N-glycolyl neuraminic acid (Neu5Gc or NGNA) predominate, especially in biopharmaceutical manufacturing using mammalian cells. Human and chicken IgG contain oligosaccharides with Neu5Ac, whereas rhesus, cow, sheep, goat, horse, and mouse IgGs contain oligosaccharides with Neu5Gc sialic acids (8). The immunological implications in humans are not fully understood but commercial recombinant human erythropoietin (EPO) produced by Chinese hamster ovary cells (CHO) is reported to contain small amounts (1% of total sialic acids) of Neu5Gc. Because chickens do not possess Neu5Gc, they were used to determine possible immunogenicity of Neu5Gc (9). Though chickens immunized with EPO did not produce significant titer of antibody, significant titer was obtained from chickens immunized with fetuin, which had 7% Neu5Gc and GM3 containing 100% Neu5Gc. Thus, an increase in Neu5Gc content correlated with enhanced antigenicity requiring close monitoring of levels of Neu5Gc in biopharmaceutical production. Also, in CHO cell-lines, sialic acid linkages are found in the α2–3 position whereas in human glycoproteins, sialic acids are linked in the α2–6 position.

N-linked glycans on plants are mostly high mannose or complex glycans containing fucose, xylose, *N*-acetylglucosamine and galactose. The initial biosynthesis of glycans is similar to the ones in yeast and mammals. Succeeding biosynthesis includes α1–3 fucose linkage to the proximal *N*-acetylglucosamine of the chitobiose core and α1–2 xylose linkage to the proximal core mannose. This gives rise to glycans that are encountered in lower animals such as molluscs, insects, and spiders, but not in mammalian glycoproteins. Hence, plant glycoproteins are immunogenic and are thought to induce allergic responses in mammals through these residues. Fucosyl moities are found in an α1–6 linkages in higher animals *(10)*. The only way transgenic plants can be used as a source of therapeutic proteins would be through the production of mutant plants that produce proteins with "humanized" glycosylation *(11)*.

One percent of circulating antibodies in humans constitute the anti-α-gal (or anti-gal) antibody *(12)*. This antibody reacts specifically against the galα1-3galβ1-4glcNAc-R epitope, also termed the α-gal or galactosyl epitope. Though most genes in humans have their animal homologues, α-galactosyltransferase is an exception. The α-galactosyltransferase enzyme is inactive in humans. Evolutionary pressures on old world monkeys, including humans possibly caused the suppression and inactivation of the α1-3galactosyltransferase enzyme. Galili et al. reported the linkage specificity of the α-gal antibody *(12)*. Rabbit red cells, which possess 2 major glycolipids: ceramide trihexoside (CTH or galα1-4galβ1-4glc-R) and ceramide pentahexoside (CPH or galα1-3galβ1-4glcNAc-R), bind human blood group AB sera. Blood groups AB sera do not possess anti-blood group A or B antibodies. Experiments with synthetic oligosaccharides in solution indicated a ten-fold higher affinity for the trisaccharide galα1-3galβ1-4glcNAc over that of the disaccharide galα1 3gal giving an insight into the combining site of the antibody for antigen recognition *(13)*. Hence, the carbohydrate specificity of α-gal antibody is well established.

The affinity of α-gal antibody between individuals ranges from 2×10^5 to 6×10^6 M^{-1} *(14)*. There are other differences within individuals attributed to blood groups. Anti-gal from individuals with A and O blood groups interacts with group B antigen, however that from blood group B individual does not because it is a self-antigen. The α-gal antibody of such individuals interacts solely with α-gal epitopes. This is because of the fact that blood group B is similar to the α-gal epitope except for an additional fucose.

Though a low affinity binding because of lack of electrostatic interactions, the affinity increases several folds during immune response against xeno-transplantation. Hence, the study of α-gal epitopes assumes importance, because of their involvement in hyperacute rejection (HAR). Organ transplantations from pigs to humans result in HAR. The IgM isotype of anti-gal binds to α-gal epitopes. This causes activation of the complement cascade and an ensuing massive cell lysis.

Many of these issues need to be dealt with during protein production using different cell-lines. Antibodies produced using CHO cells and NS0 cells predominantly possess fucosylated biantennary complex saccharides with none, one and two galactose: G0F, G1F and G2F, respectively. Bisected N-acetylglucosamine glycans are absent, which are implicated in antibody-dependent cellular cyto-toxicity (ADCC) *(4)*. Varying amounts of high mannose oligosaccharides are also present, their amounts depending on culture conditions. Both cell lines add sialic acids to galactose through an α2–3 linkage. However, the predominant sialic acid in CHO cells is Neu5Ac whereas NS0 cells possess both Neu5Gc and Neu5Ac, predominantly the former. Another important difference between the 2 cell lines include the presence of α-gal oligosaccharides in NS0 cells, which are generally absent in CHO cell lines, except in one CHO cell line reported thus far *(4)*.

Mostly, HPLC and/or mass spectrometric (MS) techniques are used to delineate protein and glycan structures. Mass spectrometry *(15)* may serve as an orthogonal analysis to traditional HPLC methods *(16–20)* for the oligosaccharide characteri-zation of recombinant antibodies. The glycoform analyses achieved via mass spectrometric analysis range from intact native glycoprotein molecular weight, glycopeptide analysis using LC-MS, MALDI-ToF (matrix assisted laser desorption ionization time-of-flight) of PNGase F released free glycans, to finally fragmenta-tion and confirmation of individual oligosaccharide sugar and linkage assignments.

We present here complementary techniques of mass spectrometry that are used to characterize recombinant monoclonal antibody glycans as glycopeptides and of HPLC with fluorescent anthranilic acid (AA) detection of free glycans. Nano liquid chromatography-electrospray ionization-time of flight mass spectrometry (nano LC-ESI-ToF MS or nano LC-MS) was carried out on tryptic digests of the recom-binant monoclonal antibody, h5G1.1. The goal was to determine and monitor the presence of several glycans, as glycopeptides, and corroborate results from PNGase-F released N-linked oligosaccharide HPLC profiles of h5g1.1-mAb observed and expected from NS0 and CHO cell lines.

2 Materials

2.1 Sample Preparation, Derivatization and Purification of Oligosaccharides

1. Peptide *N*-glycosidase F (Prozyme).
2. α(1–3, 4, 6)-galactosidase (Prozyme).
3. 100 m*M* sodium citrate/phosphate (pH 6.0) (Prozyme) (*see* **Note 1**).
4. 1 µg/mL commercial standards, e.g., NA2F (also called G2F or Gal2), NGA2F (also called G0F or Gal0) (Prozyme).
5. 5 mg/mL each (in water or buffer) commercial standards, e.g. RNAse B, Fetuin (Sigma-Aldrich); test samples: rIgGs (*see* **Note 2**).

6. 1:100 ammonium hydroxide-0.5% (w/v) SDS-1% (v/v) 2-mercaptoethanol solution (prepare fresh from respective stock solution, e.g., 10% (w/v) SDS).

7. Heating block.

8. Vacuum centrifuge.

9. 3- or 5-mL Plastic syringes with luer lock.

10. 5% Igepal (w/v) (Sigma-Aldrich).

11. 30 mg/mL anthranilic acid (Sigma-Aldrich) and 20 mg/mL sodium cyanoboro-hydride in 4% (w/v) sodium acetate trihydrate-2% (w/v) sodium borate-methanol solution (stable for several weeks at −20°C). Note: Sodium cyanoborohydride is hazardous and hygroscopic. Handle with care.

12. Nylon Acrodisc syringe filter, 0.45 μm (Gelman, cat. no. 4438).

13. Polypropylene vials (1.6 mL) with O-ring seals (Fisher, cat. no. 118448).

14. 95% acetonitrile in water for oligosaccharide purification.

15. 20% acetonitrile in water for oligosaccharide purification.

2.2 HPLC of Oligosaccharides

1. HPLC with fluorescence detector (Waters 2695 separation module with Waters 2475 fluorescence detector and thermostated column compartment).

2. Polymeric-amine bonded HPLC column, 0.46×25 cm (Asahipak NH2P-50 4E, Phenomenex, cat. no. CHO-2628).

3. Mobile phase A = 2% (v/v) acetic acid and 1% (v/v) tetrahydrofuran (inhibited) in acetonitrile (all HPLC grade). Stable for at least 3 mo.

4. Mobile phase B = 5% (v/v) acetic acid, 1% (v/v) tetrahydrofuran (inhibited) and 3% (v/v) triethylamine (all HPLC grade) in water. Stable for at least 3 mo.

5. Column prefilter with 0.2 μm prefilter insert.

6. Waters autosampler vials (amber).

7. 40% methanol.

2.3 Trypsin Digestion for Nano LC-MS

1. 50 mM ammonium bicarbonate; the solution will be close to pH 8.0 (±0.2).

2. 25 mg/mL DTT in 50 mM ammonium bicarbonate solution.

3. 200 mg/mL iodoacetic acid in 50 mM ammonium bicarbonate solution.

4. Add 800 μL of 50 mM ammonium bicarbonate to a 1 mg vial of RapiGest SF (Waters). Vortex to mix solution. Transfer the solution to a 1.5 mL siliconized microcentrifuge tube with a pipetor.

5. Test samples: rIgGs.

6. 0.5 mg/mL Trypsin (Sequencing grade modified, Promega).

7. 0.5M HCl solution.

8. 50 mL Falcon tubes.

2.4 Glycopeptide Analysis using Nano LC-MS

1. Agilent LC/MSD TOF mass spectrometer system connected to Agilent nanobore
   1100 HPLC system with thermostated column compartments.
2. Zorbax C-18 reverse phase $75\,\mu m \times 15\,cm$., $3\,\mu m$ particles, 300-Å pore size analytical column.
3. Zorbax C-18 reverse phase $300\,\mu m \times 0.5\,cm$., $5\,\mu m$ particles, 300-Å pore size
   trapping column.
4. Mobile phase A = 0.1%. formic acid in water. Stable for 3 wk.
5. Mobile phase B = 0.1% formic acid in acetonitrile. Stable for 3 wk.
6. Agilent amber glass deactivated 2-mL injection vials and PTFE-silicone caps.
7. Nitrogen gas from 230-L liquid nitrogen Dewar or equivalent.
8. ES-TOF Tuning Mix (Agilent).

3 Methods

3.1 Sample Preparation, Derivatization and Purification of Oligosaccharides

1. 2×25-µg aliquots of the test protein (only 1 aliquot of each commercial standard) are denatured for 2 min at 100°C in 30 µL of ammonium-hydroxide-SDS-2-
   mercaptoethanol solution.
2. Add 5 µL of 5% Igepal and incubate overnight with 2 µL PNGase F at 37°C.
3. Centrifuge and stop PNGAse F reaction of commercial standard(s) and one of the
   test sample aliquots using 2 µL glacial acetic acid, mix. The other aliquot of the
   test sample is treated with 5 µL α-galactosidase for 5 h at 37°C using 12 µL of
   manufacturer's buffer and buffer conditions (*see* **Note 3**). Reaction is stopped
   with 2 µL of glacial acetic acid.
4. Tagging reaction is carried out in situ with 100 µL of anthranilic acid solution at
   80°C for 1 h. 5 µL solution of each commercial standard oligosaccharides such
   as NA2F, NGA2F maybe added at this step for tagging. Cool contents and mix
   in 95% acetonitrile. Load onto 3- or 5-mL plastic syringe with luer lock on
   Nylon filtration membranes primed with 95% acetonitrile.
5. Excess anthranilic acid and other impurities are removed by washing, using
   $2 \times 1\,mL$ 95% acetonitrile.
6. Elute the bound oligosaccharides with $2 \times 0.5\,mL$ of 20% acetonitrile into appropriately labeled autosampler vials (*see* **Note 4**).

3.2 HPLC of Oligosaccharides

1. The gradient program for mapping is as follows (polymeric-amine column at
   50°C, flow rate of 1.0 mL/min and detection at $\lambda_{ex} = 360\,nm$ and $\lambda_{em} = 425\,nm$):

30% B for 2 min.; linear increase to 95% B at 82 min.; wash with 95% B for 15 min.; equilibration with 30% B for 15 min). Injection volume ranges from 50-200 µL for optimal signal, typically 100 µL (*see* **Notes 5–8**).

2. For processing results, peak width was set at 150, minimum height and area were set at 0, and the method threshold was 50, in the Millenium version 4 software. If peaks of interest are not integrated, manual integration was performed.

3. After analyzing results, if further structural studies are desired by HPLC (or MS), isolate peaks of interest. Individual glycans in HPLC eluents are dried overnight in a vacuum centrifuge, redissolved in buffer of choice for enzymatic treatment, such as α-galactosidase, and re-analyzed on HPLC as described in **section 3.1** and **3.2** (*see* **Note 9**).

4. After all samples runs are completed, use 40% methanol to wash column using several column volumes and to store column in.

3.3   Trypsin digestion for nano LC-MS

1. Add 200 µL of protein solution to a microcentrifuge tube containing the 800 µL of RapiGest solution. Mix by vortexing briefly. Final protein concentrations of 0.1–2.0 mg/mL are recommended for digests.

2. Add 50 µL of DTT solution (25 mg/mL), vortex briefly and incubate at 37°C for 1 h.

3. Alkylate the protein by adding 52 µL of iodoacetic acid solution (200 mg/mL) to the digest, vortex briefly and reincubate at 37°C for 1 h in the dark.

4. Add 4% trypsin solution to the digest (w/w = enzyme/protein). Volume of trypsin = (amount of protein × 4%) / concentration of trypsin. Incubate the digest at 37°C for 1 h.

5. Acidify a 100 µL aliquot of digest mixture to break down the Rapigest denaturant so as to not interfere with the LC-MS analysis. Add approximately 20–30 µL of 0.5*M* HCL (*see* **Note 10**).

6. Incubate the acidified digest at 37°C for 45 min. Analyze sample by mass spectrometry immediately and/or freeze aliquots at −80°C.

3.4   Glycopeptide Analysis using Nano LC-MS

1. The gradient program for intact glycoprotein analysis is as follows (Zorbax C-18 reverse phase 75 µm ID × 15 cm. analytical column with Zorbax C-18 300 µm × 0.5 cm trapping column):
0% B for 5 min.; linear increase to 65% B at 12.5 min.; linear increase to 100% B at 125.1 min; 100% B at 140 min (wash); 0% B at 140.1 min and equilibrate for 9.9 min. (end time 150 min.).

2. With the HPLC-MS system thermally stabilized, select an appropriate operating method. Example: Gradient flow rate 300 nL/min, load pump 20 μL/min, back-flushing from trap column begin at 5 min and end at 140 min and m/z range 250–3,000.
3. Check calibration of m/z axis using ES-ToF mix. Calibrate MS system within 100 ppm.
4. Set injection volume so that ~2.5 pmol is loaded on the column based on known concentration of sample.
5. Upon completion of data collection, check and record calibration.
6. Sum the spectra above full width at half maximum (FWHM) from the total ion chromatogram (TIC) elution peak.
7. Each summed electrospray spectra and other chromatograms may be included if necessary. (*see* **Note 11**).

4   Notes

1. 100 mM sodium citrate/phosphate is provided by the manufacturer as a 5× concentration.
2. Water used in these experiments was double distilled Milli-Q, Millipore, filtered, ≥18 MΩ. Suitable substitutions such as HPLC grade water can be made.
3. G0F is the most abundant glycan in IgGs. Using G0F in the calculation as the source of main substrate, the concentration of the sample is about the concentration recommended by the manufacturer for the α-galactosidase reaction. Also, using the provided buffer will bring the pH slightly above the optimum (tested by pH paper), because the reaction mixture already contains dilute ammonium hydroxide from the PNGase F reaction. The 5-h incubation time is theoretically sufficient to complete the reaction. However a complete hydrolysis of the starting substrate, glycan peak(s) in this case, is not observed. This does not cause a problem and is in fact an advantage because a diminution in substrate peak size and the corresponding hydrolyzed product peak(s) is/are easily seen, allowing one to "trace" where the product came from.
4. Oligosaccharides containing 1–4 saccharide units are not recovered when employing nylon Acrodisc membranes for the removal of excess anthranilic acid (*16,17,20*). In general this is not an issue when dealing with *N*-linked oligosaccharides from rIgGs, because low molecular weight glycans are not encountered. However, a modified method using DPA 6S columns allows the recovery and analysis of mono- to tetrasaccharides (*21*).
5. Switching production of recombinant IgG in CHO cell line from NS0 causes several changes. The presence of α-gal oligosaccharides in the latter is the most obvious change (*see* Fig. 23.1). Treatment with α-galactosidase shows diminution or elimination of several peaks identified as αGal1 to 4 (aGal1 to 4 in the Fig.) based on their sequence of elution with the fourth one in the monosialyl region, designated as SA1/aGal4. Recombinant IgG from NS0 possessed approximately 5% α-gal, which generally ranged from 2 to 10% for various lots examined (not shown).

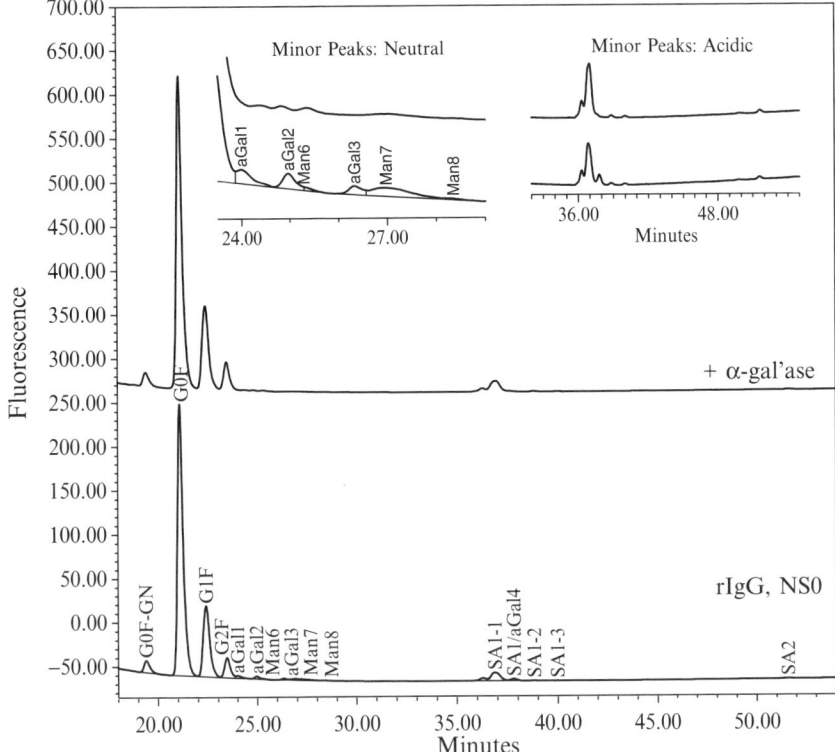

Fig. 23.1 Oligosaccharides of rIgG from NS0 cell line (bottom) are treated with α galactosidase (top). Diminution or elimination of peaks in the control is indication of α-gal containing oligosaccharides. Inset shows minor peaks in neutral and acidic portions of the chromatogram

6. As expected from published reports, no α-gal was observed in the CHO cell line rIgG. Similar treatment as above with α-galactosidase had no effect on the oligosaccharide peaks (*see* Fig. 23.2). Recombinant antibodies from both lots, CHO and NS0, showed presence of complex (G0F, G1F, G2F), high mannose, mono-and bisialyl glycans. Comparison of the oligosaccharide profiles of rIgGs from both cell lines is shown in Figure 23.3.

7. Oligosaccharides from NS0 cells are terminated by Neu5Gc and those from CHO cells by Neu5Ac sialic acid. This is easily seen in the oligosaccharide map elution profile without having to perform a separate sialic acid analysis. Monosialyl Neu5Ac glycans from CHO cell lines elute much earlier at about 32 min (*see* arrows in Fig. 23.3) whereas monosialyl Neu5Gc glycans from NS0 cells elute much later at about 36 min. NS0 monosialyl glycan heterogeneity is obvious. One of the causes of heterogeneity is the presence of an α-galactosyl moiety on the sialyl oligosaccharide. This was proven from the elimination of one sialyl peak after α-galactosidase treatment (*see* Fig. 23.1 and **Note 5**). There was low level of sialylation in the CHO cell lines, hence the sialyl peaks

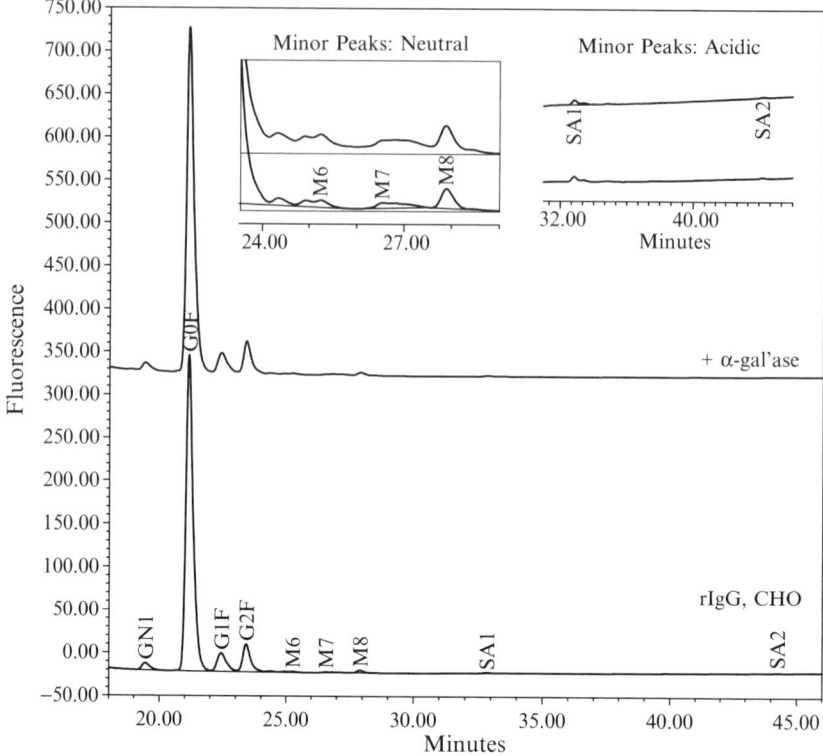

Fig. 23.2 Oligosaccharides of rIgG from CHO cell line (bottom) are treated with α-galactosidase (top) showing no effect of enzyme, confirming absence of α-gal-containing oligosaccharides in CHO cells. Inset shows minor peaks in neutral and acidic portions of the chromatogram

are barely observed. In fact, bisialyl glycan from CHO is also present, albeit at 0.04% of total glycan amount, eluting at 44 min. and not seen unless magnified several fold. Interestingly, the AA technology allows detection and quantitation at such low levels. Bisialyl glycan from NS0 elutes much later, at 52 min.

8. The presence of Neu5Gc in NS0 and Neu5Ac in CHO was confirmed independently using sialic acid HPLC assay (*see* Fig. 23.4) (*18,19*), the description of which is beyond the scope of this chapter. This corroborates the HPLC results discussed in **Note 7**. Sialic acid analysis of NS0 rIgG (second chromatogram from top) shows presence of Neu5Gc (NGNA in Fig. 23.4) and of CHO rIgG (top) shows presence of Neu5Ac (NANA). Ortho-phenylene-diamine (OPD)-derived standard sialic acid chromatogram is at the bottom. Placebo (third from top) showed artifact peak eluting near Neu5Gc elution time. Correcting for Neu5Gc amount shows negligible amounts of Neu5Gc in CHO cells. Also, note small amounts of Neu5Ac in CHO, corroborating amounts predicted from mapping.

9. Treatment of individually isolated αGal1, αGal2 and αGal3, with α-galac-tosidase enzyme sheds more light on the structure of each (*see* Fig. 23.5).

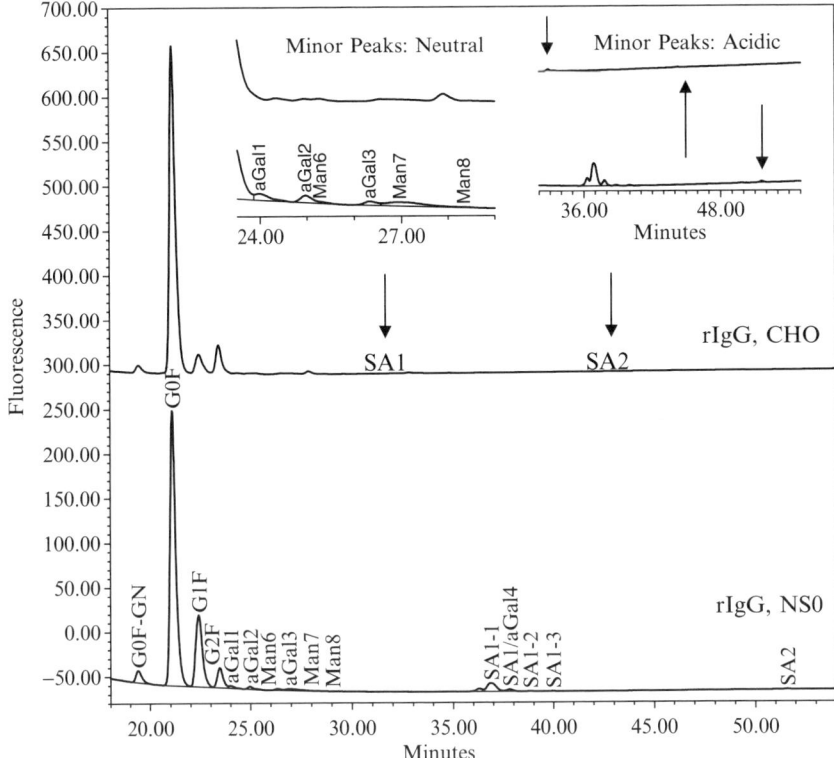

Fig. 23.3 Comparison of oligosaccharide profiles of rIgGs from NS0 (bottom) and CHO (top) cell lines. Sialylation in rIgG from CHO cell line was much lower compared to that in rIgG from NS0 cell line (shown by arrows). Also, late eluting mono- and bisialyl glycans indicate Neu5Gc sialylation (*see* NS0 rIgG glycan map) compared to early eluting Neu5Ac sialyl glycans in CHO rIgG glycan map

For example, enzyme-hydrolyzed αGal1 shows final product peak position at Gal1 (arrows point from substrate to product) indicating αGal1 is mono-α-galactosyl glycan linked to the βGal; similarly, αGal2 and αGal3 are mono- and/or bi-α-galactosyl because their enzyme-treated products co-elute at Gal2 position (there is a minor high mannose glycan contamination in the latter). Mass spectrometry results show mono- and bi-α-galactosyl glycans with doubly- and triply-charged ions on each in the rIgG tryptic LC-MS spectra (*see* **Note 11**).The fourth α-galactosyl glycan (SA/αGal4 peak) is most likely mono-α-galactosyl because the peak is completely eliminated on enzyme treatment, in Figure 23.1. One of the biantennary arm would carry the sialic acid and the other, α-gal.

10. Add 20 μL 0.5 M HCl initially and then add 5 μL at a time so the pH is not overshot. Measure pH with microelectrode making sure the pH ≤ 2.
11. When the LC is coupled to an appropriate ESI-MS instrument, the resulting LC-MS data provides multiple dimensions to the data beyond the retention time and relative intensity of an LC chromatogram. LC-MS data provides data cor-

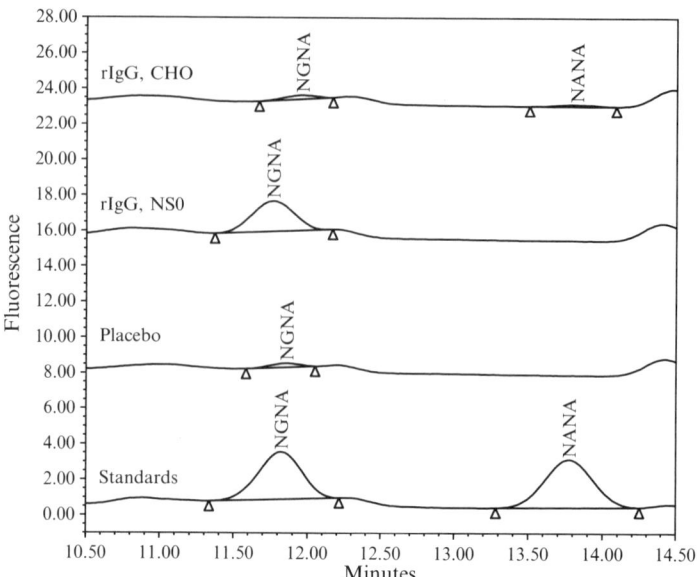

Fig. 23.4 Sialic acid analysis of oligosaccharides of rIgGs from NS0 cells (second chromatogram from top) shows presence of Neu5Gc (NGNA in Fig.) and from CHO cells (top) shows very small amounts of Neu5Ac (NANA). Standard chromatogram is at the bottom. Placebo showed artifact peak eluting near Neu5Gc; the amount is subtracted from that in CHO cells profile, giving negligible value for Neu5Gc in CHO

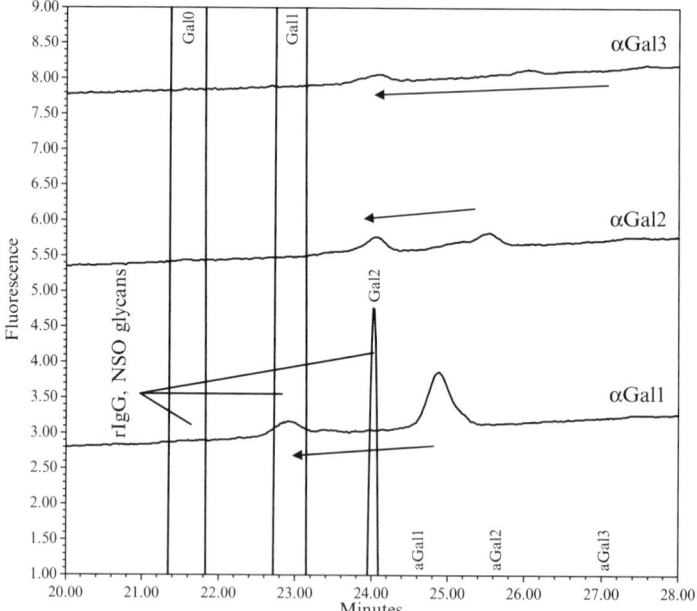

Fig. 23.5 α-Galactosidase treatment of isolated, individual glycans: αGal1, αGal2 and αGal3, were separately treated with α-galactosidase enzyme. αGal1 shows product peak position at Gal1 (arrows point from substrate to product) retention time. Similarly, αGal2 and αGal3 each generate product that elutes at the Gal2 position

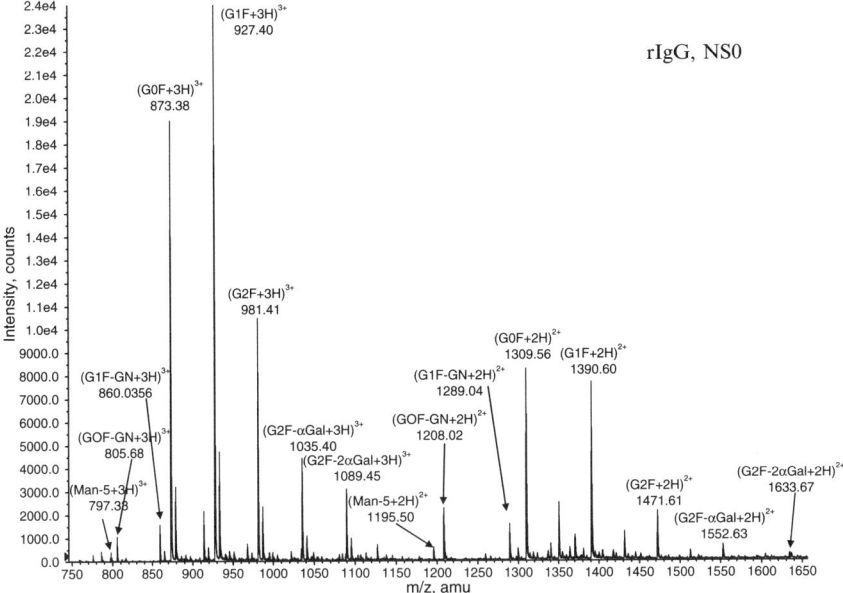

Fig. 23.6 Nano LC-MS of tryptic digest of rIgG from NS0 cell line confirms presence of predominantly complex glycans and small amounts of α-gal-containing glycans, detected as glycopeptides

responding to the masses of the species under an LC peak as well as the relative intensity of the masses. RP-HPLC retention of glycopeptides appears to be driven by the relative hydrophobic/hydrophilic nature of the peptide as opposed to the attached N-linked oligosaccharide. Because of this trait all of the glycopeptides elute from the nanobore LC-MS under a single peak. A summed spectra for the glycopeptide LC chromatogram peak from a trypsin digest of a rIgG from NS0 cell line is shown in Figure 23.6. The presence of G0F, G1F and G2F, predominant glycans of rIgG, is easily seen using nano LC-MS, as doubly- and triply-charged species confirming HPLC results. Also, 2 types of α-galactose-containing glycans are easily discerned: one with a single and the other with two α-gal, each type on a G2F oligosaccharide. The α-galactosyl glycans are also present as doubly- and triply-charged ions.

References

1. Humphreys, A. and Boersig, C. (2003) Cholesterol drugs dominate. *MedAdNews*. **22**, 42–57.
2. Glennie, M. J. and Johnson, P.W.M. (2000) Clinical Trials of Antibody Therapy. *Immunol. Today*. **21**, 403–410.
3. Gottschalk U. (2003) Biotech Manufacturing Is Coming of Age. *BioProcess International*. **1**, 54–61.
4. Jefferis, R. (2005) Glycosylation of Recombinant Antibody Therapeutics. *Biotechnol. Prog.* **21**, 11–16.

5. Lund J., Takahashi, N., Popplewell, A., Goodall, M., Pound, J. D., Tyler, R., King, D. J., and Jefferis, R. (2000) Expression and characterization of truncated glycoforms of humanized L243 IgG1: architectural features can influence synthesis of its oligosaccharide chains and affect superoxide production triggered through human FcγR1. *Eur. J. Biochem.* **267**, 7246–7257.

6. Stockert, R. J. (1995) The asialoglycoprotein receptor: relationships between structure, function, and expression. *Physiol. Rev.* **75**, 591–609.

7. Taylor, P. R., Martinez-Pomares, L., Stacey, M., Lin, H-H., Brown, G.D., and Gordon S. (2005) Macrophage receptors and immune recognition. *Ann Rev Immunol.* **23**, 901–944.

8. Raju, T. S., Briggs, J. B., Borge, S. M., and Jones A. J. S. (2000) Species-specific variation in glycosylation of IgG: evidence for the species-specific sialylation and branch-specific galactosylation and importance for engineering recombinant glycoprotein therapeutics. *Glycobiology.* **10**, 477–486.

9. Noguchi, A., Mukuria C. J., Suzuki, E., and Naiki, M. (1995) Immunogenicity of *N*-glycolylneuraminic acid-containing carbohydrate chains of recombinant human erythropoietin expressed in chinese hamster ovary cells. *Biochem. Japanese Biochem. Soc.* **117**, 59–62.

10. Wilson, I. B. H., Zeleny, R., Kolarich, D., Staudacher, E., Stroop, C. J. M., Kamerling, J. P., and Altmann, F. (2001) Analysis of Asn-linked glycans from vegetable foodstuffs: widespread occurrence of Lewis a, core α1,3-linked fucose and xylose substitutions. *Glycobiology.* **11**, 261–274.

11. Garcia-Casado, G., Sanchez-Monge, R., Chrispeels, M. J., Armentia, A., Salcedo, G., and Gomez, L. (1996) Role of complex asparagine-linked glycans in the allergenicity of plant glycoproteins. *Glycobiology.* **6**, 471–477.

12. Galili, U., Rachmilewitz, E.A., Peleg, A., and Flechner, I. (1984) A unique natural human IgG antibody with anti-α-galactosyl specificity. *J. Exp. Med.* **160**, 1519–1531.

13. Galili U. and Matta, K. L. (1996) Inhibition of anti-Gal IgG binding to porcine endothelial cells by synthetic oligosaccharides. *Transplantation.* **62**, 356–362.

14. Wang, L., Anaraki, F., Henion, T. R., and Galili, U. (1995) Variations in activity of the human natural anti-Gal antibody in young and elderly populations. *J. Gerontol. (Med. Sci.)* 50A, M227–M233.

15. Harvey, D. J. (2001) Identification of protein-bound carbohydrates by mass spectrometry. *Proteomics.* **1**, 311–328.

16. Anumula, K. R. and Dhume, S. T. (1998) High resolution and high sensitivity methods for oligosaccharide mapping and characterization by normal phase high performance liquid chromatography following derivatization with highly fluorescent anthranilic acid. *Glycobiology.* **8**, 685–694.

17. Dhume, S. T., Ebert, M. B., Saddic, G. N., and Anumula, K. R. (2002) Monitoring glycosylation of therapeutic glycoproteins for consistency using highly fluorescent anthranilic acid in *Methods in Molecular Biology, Posttranslational Modifications of Proteins.* (Kannicht, C., ed), Humana, Totowa, NJ, pp. 127–142.

18. Anumula, K. R. (1995) Novel fluorescent methods for quantitative determination of monosaccharides and sialic acids in glycoproteins by reversed phase high performance liquid chromatography in *Methods in Protein Structure Analysis* (Atassi, M. Z. and Apella, E., eds.), Plenum, New York, NY, pp. 195–206.

19. Anumula, K. R. (1997) Highly sensitive pre-column derivatization procedures for quantitative determination of monosaccharides, sialic acids and amino sugar alcohols of glycoproteins by reversed phase HPLC in *Techniques in Glycobiology* (Townsend, R. R. and Hotchkiss, A. T., eds.), Marcel Dekker, New York, NY, pp. 349–357.

20. Anumula, K.R. (2006) Advances in fluorescence derivatization methods for high-performance liquid chromatographic analysis of glycoprotein carbohydrates. *Anal. Biochem.* **350**, 1–23.

21. Neville, D. C. A., Coquard, V., Priestman, D. A., te Vruchte, D. J. M., Sillence, D. J., Dwek, R. A., Platt, F. M., and Butters, T. D. (2004) Analysis of fluorescently labeled glycosphingolipid-derived oligosaccharides following ceramide glycanase digestion and anthranilic acid labeling. *Anal. Biochem.* **331**, 275–282.

24

Mass Spectrometry and HPLC with Fluorescent Detection-Based Orthogonal Approaches to Characterize *N*-Linked Oligosaccharides of Recombinant Monoclonal Antibodies

Adam W. Lucka, Bruce R. Kilgore, Rekha Patel, Bruce A. Andrien, Jr., and Shirish T. Dhume

Summary A number of HPLC and mass spectrometric techniques are used to characterize post-translational modification in recombinant monoclonal antibodies (MAbs) using the intact glycoprotein and free glycans. LC separation utilizing fluorescent detection technique allows tentative structural assignment of MAb oligosaccharides. Intact molecular weight analysis via electrospray allows for an accurate mass determination and observation of the native glycoform mass envelope. N-linked oligosaccharides are then analyzed by MALDI-ToF. Their structures are further confirmed by analyzing the fragmentation patterns formed by MS/MS. All these techniques provide useful information when performed in isolation. However, the combined information allows for definitive and robust characterization of the N-linked glycans from recombinant MAbs.

Key Words Monoclonal; antibodies, MAbs, recombinant; IgG, *N*-linked, oligosaccharides; glycans, saccharides, HPLC, fluorescence; anthranilic acid, mass spectrometry, ESI-ToF; MALDI-ToF.

1 Introduction

Approximately 140 therapeutic proteins are approved for marketing in the US and Europe with more than 500 in clinical trials *(1)*. Among these, therapeutic monoclonal antibodies (MAbs) are among the fastest growing segment. Sixty percent of the proteins marketed are glycosylated *(2)*, and because glycosylation affects drug metabolism and in turn efficacy, complete characterization of MAbs is essential not only to controlling their function but is a regulatory requirement. Routine assays involving chromatographic profile or fingerprint comparisons to those of predetermined reference standards are used to assure similarities before clinical lot release. Post-translational modifications such as glycosylation are apt to be sensitive to cell culture conditions, so manufacturing changes and multiple

From: *Post-translational Modifications of Proteins.*
Methods in Molecular Biology, Vol. 446.
Edited by: C. Kannicht © Humana Press, Totowa, NJ

manufacturing sites present issues requiring detailed comparability analysis. Structural elucidation of glycosylation and its correlation with function, pharmacokinetics and pharmacodynamics is an additional benefit. The biological significance of oligosaccharides in proteins is detailed in a review by Varki *(3)*.

Unlike proteins or DNA where the linkage of monomers is uniformly straightforward, oligosaccharides of similar size can possess many more isomeric structures. For example 1.05×10^{12} structures are possible for a hexasaccharide alone *(4)*. Fortunately, despite the increasing number of oligosaccharides being reported, relatively few of these structures are present in nature and there are some common underlying similarities within a given species. For example, human and chicken IgG contain oligosaccharides with *N*-acetylneuraminic acid (Neu5Ac or NANA), whereas rhesus, cow, sheep, goat, horse, and mouse IgGs possess N-glycolylneuraminic acid (Neu5Gc or NGNA) *(5)*. Also, fucosyl moities are found in α1-6 linkages in animals whereas in plants the fucosyl linkage is α1-3 *(6)*.

The analysis of carbohydrates can begin with the determination of monosaccharide composition. Monosaccharide composition may give insight into the presence or absence of certain types of glycans. For example, presence of *N*-acetylgalactosamine may indicate the presence of *O*-linked sugars on glycoproteins. If acidic sugar chains are present, their determination may need to be performed along with oligosaccharide analysis, to adequately characterize the carbohydrates. Monosaccharide determination can be performed using GC *(7,8)*, HPAEC-PAD *(9)* or HPLC with chromogenic or fluorometric detection *(10)*. Recently, simultaneous HPLC determination of monosaccharides and sialic acids exploiting the reversible enzyme, sialic acid aldolase, has been described *(11)*. A modification of that method allows distinguishing between Neu5Ac and Neu5Gc *(12)*.

In most cases, especially in industrial settings, oligosaccharide mapping is the method of choice and is routinely performed. If all the peaks in the profile have been identified, mapping allows the theoretical determination of monosaccharide (including sialic acid) ratios and composition. In such instances, oligosaccharide analysis may preclude the need to perform monosaccharide composition analysis while in others the latter confirms the former.

For analysis of oligosaccharides, quantitative release of the glycan chains is usually required, though glycopeptides and intact proteins can be analyzed by mass spectroscopy as discussed later in this chapter. Chemical methods such as hydrazine *(13)* and enzymatic release such as PNGAse F *(14)* are widely used for the release of most *N*-linked oligosaccharides. As for *O*-linked glycans, alkaline sodium borohydride reduction using radioactivity gives pmol level detection *(15)* but it gives rise to alditols that cannot further react with conventional tags.

Post-release characterization of derivatized or native oligosaccharide can be done several ways. Techniques for detection include PAD *(16)*, refractive index *(17)*, intrinsic absorbance *(18)*, UV chromophore *(19)*, or fluorescent tags *(20)*. Fluorescent detection provides high pmol level detection, a few magnitudes higher sensitivity over HPAEC-PAD and UV-detection. Dansyl and other hydrazide derivatives have also been used but are not preferred because of stability issues *(21)*.

Wide choices of fluorescent tags are available for the chromatographer. Some tags are more popular than others depending on the method of separation, though there is no reason why these can't be interchangeable, at least in principle. Most of these tags have been compared head-to-head for sensitivity, sample preparation, and post-column ease of recovery after HPLC separation *(22)*. Not only was 2-aminobenzoic acid (2AA, AA or anthranilic acid), the most sensitive tag for labeling acidic and neutral oligosaccharides, but also tagging can be performed under much simpler and aqueous conditions *(23)*. A simple volatile HPLC solvent system allows ease of oligosaccharide isolation. Further characterization of carbohydrates can be done by NMR *(24)* or mass spectrometry *(25)* either subsequent to or independent of HPLC analysis.

Bioanalysis protocols of carbohydrates from glycoproteins and glycosaminoglycans using various CE methods were reviewed by Thibault and Honda *(26)*. However, HPLC provides a better means for separating complex mixtures of glycans.

One of the challenges of analytical methods is the agreement of values between methods. Though accuracy maybe difficult to establish, comparability between methods endorses results obtained from either method. Various types of mass spectrometry are used routinely to identify and possibly quantify post-translational modifications in proteins. MALDI-ToF and LC-ESI-MS using intact glycoproteins, glycopeptides or released glycans are common. Agreement between values obtained by mass spectrometry and HPLC, especially for major peaks, is one of the goals of this report.

Oligosaccharides with similar molecular weights or similar structures provide challenges for baseline HPLC separation. Recombinant IgG antibodies generally contain both high mannose and complex oligosaccharides that are difficult to resolve. Easy access to orthogonal methods such as mass spectrometry allows one to identify presence of coeluting glycans, but in the absence of this, additional chromatography runs with modified gradients or columns need to be employed.

2   Materials

2.1   Sample Preparation, Derivatization and Purification of Oligosaccharides

1. Peptide *N*-glycosidase F (Prozyme).
2. 1 µg/mL oligosaccharide commercial standards, e.g., NA2F (also called G2F or Gal2), NGA2F (also called G0F or Gal0) (Prozyme).

3. 5mg/mL each (in water or buffer) commercial standards, e.g., RNAse B, Fetuin (Sigma-Aldrich); test samples: rIgGs (*see* **Note 1**).
4. 1:100 ammonium hydroxide-0.5% (w/v) SDS-1% (v/v) 2-mercaptoethanol solution (prepare fresh from respective stock solution, e.g., 10% (w/v) SDS).
5. Heating block.
6. 3- or 5-mL Plastic syringes with luer lock.
7. 5% Igepal (w/v) (Sigma-Aldrich).
8. 30 mg/mL anthranilic acid (Sigma-Adrich) and 20 mg/mL sodium cyanoborohydride in 4% (w/v) sodium acetate trihydrate-2% (w/v) sodium borate-methanol solution (stable for several weeks at −20°C). Note: sodium cyanoborohydride is hazardous and hygroscopic. Handle with care.
9. Nylon Acrodisc syringe filter, 0.45 μm (Gelman, cat. no. 4438).
10. Polypropylene vials (1.6 mL) with O-ring seals (Fisher, cat. no. 118448).
11. 95% acetonitrile in water for oligosaccharide purification.
12. 20% acetonitrile in water for oligosaccharide purification.

2.2   HPLC of Oligosaccharides

1. HPLC with fluorescence detector (Waters 2695 separation module with Waters 2475 fluorescence detector and thermostated column compartment).
2. Polymeric-amine bonded HPLC column, 0.46 × 25 cm (Asahipak NH2P-50 4E, Phenomenex, cat. no. CHO-2628).
3. Polymeric-amine bonded HPLC column, 0.2 × 15 cm, (Asahipak NH2P-50 2D, Phenomenex, cat. no. CHO-5582) for rapid oligosaccharide mapping.
4. Mobile phase A: 2% (v/v) acetic acid and 1% (v/v) tetrahydrofuran (inhibited) in acetonitrile (all HPLC grade). Stable for at least 3 mo.
5. Mobile phase B: 5% (v/v) acetic acid, 1% (v/v) tetrahydrofuran (inhibited) and 3% (v/v) triethylamine (all HPLC grade) in water. Stable for at least 3 m.
6. Column prefilter with 0.2-μm prefilter insert.
7. Waters autosampler vials (amber).
8. 40% methanol (to wash and store polymeric-amine columns).

2.3   Simultaneous Desalting and Chromatography using LC-ESI-ToF Mass Spectrometry

1. An Agilent μbore 1100 HPLC system connected to Agilent LC/MSD ToF mass spectrometer system with thermostated column compartment.

2. C-4 reverse phase HPLC column, 0.1 × 25 cm (Vydac, 5 μm particles, 300-Å pore size).
3. Mobile phase A: 5% acetonitrile, 0.1% TFA in water.
4. Mobile Phase B: 20% water, 0.1% TFA in acetonitrile. Prepare fresh mobile phases; stable for 3 wk.
5. Amber glass deactivated 2-mL injection vials, Agilent and caps PTFE-Silicone.
6. Test samples: rIgGs.

2.4   LC-ESI-ToF Mass Spectrometry of Intact, Native rIgG Glycoprotein

1. An Agilent μbore 1100 HPLC system connected to Agilent LC/MSD ToF mass spectrometer system (same as **Section 2.3**).
2. ES-TOF tuning mix (Agilent).
3. Nitrogen gas from 230 L liquid nitrogen Dewar or equivalent.

2.5   N-linked Glycan Sample Preparation and Purification for Analysis by MALDI-ToF Mass Spectrometry

1. 2M DTT (GE Healthcare) in water.
2. PNGase F (Proteomics grade, Sigma), 1 unit/μL in water; store at 2–8°C until use.
3. Test samples: rIgGs.
4. 0.1% TFA.
5. 0.1% TFA in 25% acetonitrile.
6. 0.1% TFA in 50% acetonitrile.
7. Hypercarb TopTip (Glygen Corp) (available in 3 binding-capacity levels).
8. Dihydroxybenzoic acid (Sigma).
9. Vacuum desiccator.

2.6   MALDI-ToF Mass Spectrometry of PNGase F-Released N-Linked Glycans

1. Applied Biosystems Voyager DE-Pro MALDI-ToF-MS System.

2. ES-TOF tuning Mix (Agilent).
3. Microcentrifuge tubes 1.5 mL, sterilized and siliconized.

3 Methods

3.1 Sample Preparation, Derivatization and Purification of Oligosaccharides

1. N-linked carbohydrate from test article (rIgGs) and standard commercial glyc-oproteins such as RNAse B were released, derivatized and purified *(27)*. 25 µg of the protein is denatured for 2 min at 100°C in 30 µL of ammonium-hydrox-ide-SDS-2-mercaptoethanol solution. Blank samples can be added here (*see* **Note 1**).
2. Incubate overnight with 2 µL PNGase F at 37°C after adding 5 µL of 5% Igepal.
3. Contents are centrifuged and reaction is stopped by 2 µL glacial acetic acid.
4. Tagging reaction is carried out in situ with 100 µL of anthranilic acid solution at 80°C for 1 h. 5 µL solution of each commercially available oligosaccharide standards such as NA2F, NGA2F maybe added at this step for tagging. Cool contents and mix in 95% acetonitrile. Load onto 3- or 5-mL plastic syringe with luer lock on Nylon filtration membranes primed with 95% acetonitrile.
5. Excess anthranilic acid (AA) and other impurities are removed by using 2 × 1 mL 95% acetonitrile under gravity flow or vacuum manifold (*see* **Note 2**).
6. Elute the bound oligosaccharides with 2 × 0.5 mL of 20% acetonitrile into the appropriately labeled autosampler vial. For recovery of low molecular weight glycans (*see* **Note 3**). Inject an aliquot of 50–200 µL for analysis.

3.2 HPLC of Oligosaccharides

1. The gradient program for mapping is as follows (25-cm. column at 50°C, flow rate of 1.0 mL/min. and detection at λex = 360nm and λem = 425 nm):
   30% B for 2 min.; linear increase to 95% B at 82 min.; wash with 95% B for 15 min.; equilibration with 30% B for 15 min.

2. The gradient program for rapid mapping is as follows (15-cm. column, flow rate of 0.4 mL/min., detection, column temperature and mobile phases same as **Section 3.2**, step 1, above):

36% B for 2 min.; linear increase to 95% B at 20 min.; maintain at 95% B for 25 min.; wash for 10 min. with 95% B; equilibrate with 36% B for 10 min. (*see* **Note 4**).

3.3 Simultaneous Desalting and Chromatography using LC-ESI-ToF Mass Spectrometry

1. With the HPLC-MS system thermally stabilized, the gradient program for intact glycoprotein analysis is as follows (Agilent μbore 1100 HPLC system using a C-4 reversed phase column, 0.1 × 25 cm. column, flow rate 50 μL/min., column temperature 60°C, detection at UV 215 nm and *m/z* range 1,000–7,500):

   Linear increase over 10 min. of B from 20 to 60%; linear increase to 100% B in 1 min. followed by a 5 min. wash; decrease percent B to 20% in 1 min.; 13 min. equilibration; end time: 30 min.

3.4 LC-ESI-ToF Mass Spectrometry of Intact, Native rIgG Glycoprotein

1. Check calibration of m/z axis using ES-ToF tuning mix. Calibrate MS within 100 ppm.
2. Set injection volume so that ~10 μg or ~65 pmol is loaded on the column based on known concentration of sample.
3. Sum the spectra above full width at half maximum (FWHM) from the total ion chromatogram (TIC) elution peak.
4. Use an appropriate deconvolution software package so data would include a chromatogram and deconvoluted spectra for each summed peak. Each summed electrospray spectra and other chromatograms may be included if necessary (*see* **Note 5**).

3.5 N-linked Glycan Sample Preparation and Purification for Analysis by MALDI-ToF Mass Spectrometry

1. Dilute protein with water to a concentration of ~2 mg/mL
2. Add 2*M* DTT to a final DTT concentration of ~10 m*M*. Incubate at 37°C for 1 h.
3. Dilute the reduced protein sample with water to ~1 mg/mL.

4. To a 100 μL aliquot of above, add 7.5 μL of PNGase F (7.5 units) (see Note 6) and incubate at 37°C for 2 h. Analyze immediately and/or freeze aliquots at −80°C.

5. 20-μL aliquots of sample above are subjected to Hypercarb release (*see* **Note 7**).

6. Wet the hypercarb material 2× with 0.1% TFA in 50% acetonitrile.

7. Equilibrate material 2× with 0.1% TFA.

8. Load the entire sample 5 times for maximum binding.

9. Wash the Hypercarb TopTip 2× with 0.1% TFA.

10. Elution from the Hypercarb TopTip should be performed with a premeasured 2.5 μL 0.1% TFA in 25% acetonitrile.

11. Following elution, 0.75 μL of the samples are mixed on target with 0.75 μL of DHB matrix. Keep calibration standards and samples close together for accurate external calibration.

12. The sample plate is placed into a vacuum desiccator for sample drying.

3.6 MALDI-ToF Mass Spectrometry of PNGase F-Released N-Linked Glycans

1. An appropriate operating method for the ~1,000–3,000 *m/z* range is selected from the list of operating methods.

2. An initial *m/z* axis calibration is performed with calibration standards as close external calibrants (*see* **Note 8**).

Table 24.1 Comparison of Relative Glycan Amounts from Lot X rIgG Utilizing Values Obtained from HPLC with Fluorescence Detection and MALDI-ToF

Oligosaccharide	Relative amount by HPLC	Normalized HPLC values, G0F = 100	Relative amount by MS
GN1	1.99	2.41	2.00
G0	ND	ND	1.99
G0F	82.49	100.00	100.00
G1F	6.12	7.42	7.00
Man5	7.63	9.25	8.00
G2F	0.00	0.00	ND
Unknown	0.25	0.30	ND
Man6	0.40	0.49	ND
Man7	0.20	0.24	ND
Man8	0.57	0.69	ND
Monosialiyl	0.30	0.36	ND
Bisialyl	0.04	0.05	ND

ND = Not detected

3. Data is acquired from the calibration standard and samples as 200-shot composite spectra (4 spectra × 50 shots/spectrum). Acquire 1 set of data for each experimental sample (*see* **Note 9**).
4. Relative glycan amounts obtained (G0F = 100) agree closely with values obtained from HPLC (normalized to G0F = 100) as shown in Table 24.1 (also, *see* **Note 9**).

3.7 Fragmentation for Structural Confirmation using Post-Source Decay (PSD) MALDI-ToF Mass Spectrometry

1. Once the target ion source (TIS) is calibrated then acquisition of the post source decay (PSD) spectrum can begin, if the current spectrum is acceptable on visual inspection.
2. Collect all segments, 4 × 50 shots = 200 shot spectra.
3. Once the spectrum of a known standard is collected, the m/z axis should be internally calibrated and this internal calibration should be used as an external calibration of the m/z axis for the acquisition of all experimental samples.
4. Once the instrument is calibrated, acquisition of PSD fragmentation spectra of experimental precursor ions may take place using the same procedure to acquire the standard spectrum (*see* **Notes 10 and 11**).

4   Notes

1. For oligosaccharide mapping, a placebo or formulation blank is added to discern artifacts. Their peaks generally do not interfere with peaks of interest and need not be run each time unless there is a method, process or formulation change. Water can be used as method blank. The source of water used in all experiments was double distilled Milli-Q, Millipore, filtered, ≥ 18 MΩ. Suitable substitutions such as HPLC grade water can be made.
2. A vacuum manifold instead of gravity flow has been used for ease in the Nylon filtration step of oligosaccharide recovery. However, yields were low and most likely related to the high flow rates. Optimization of yield is possible by adjusting rate of filtration. Glycan profiles and final results were same with either filtration methods.
3. Oligosaccharides containing 1–4 saccharide units are not recovered from nylon acrodisc membranes *(27)*. This is not an issue because low molecular weight glycans are not encountered in rIgGs. However, a modified method using DPA 6S columns allow the recovery and analysis of mono- to tetrasaccharides *(28)*.
4. Recombinant IgGs from mammalian cells possess mostly complex glycans: G0F, G1F and G2F. However, a rIgG from lot X showed presence of higher

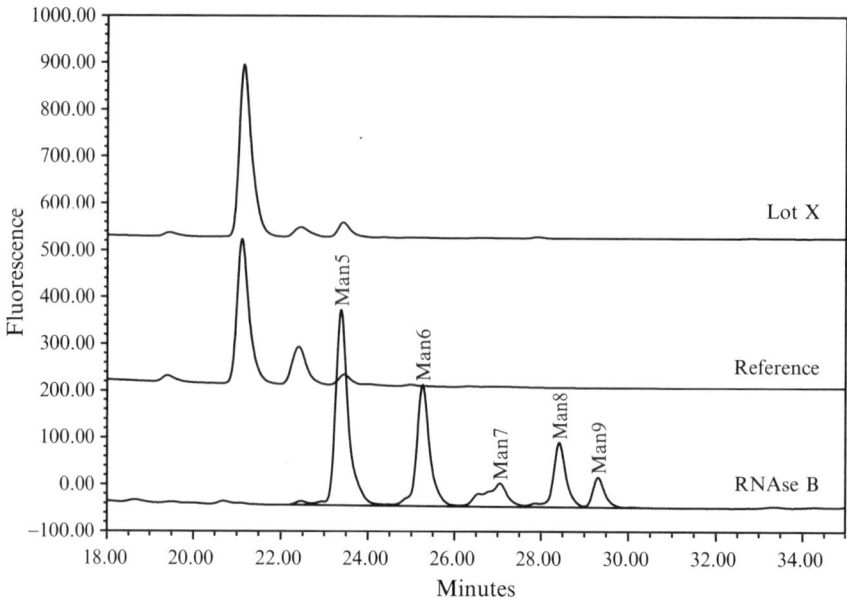

Fig. 24.1 Oligosaccharides from rIgG released using PNGase F are identified on the long 25 cm. polyamine column using Waters HPLC. RNAse B oligosaccharides (bottom) allow identification of high mannose glycans in reference rIgG (middle) and lot X rIgG (top). G2F in reference and Man5 in lot X coelute in this method

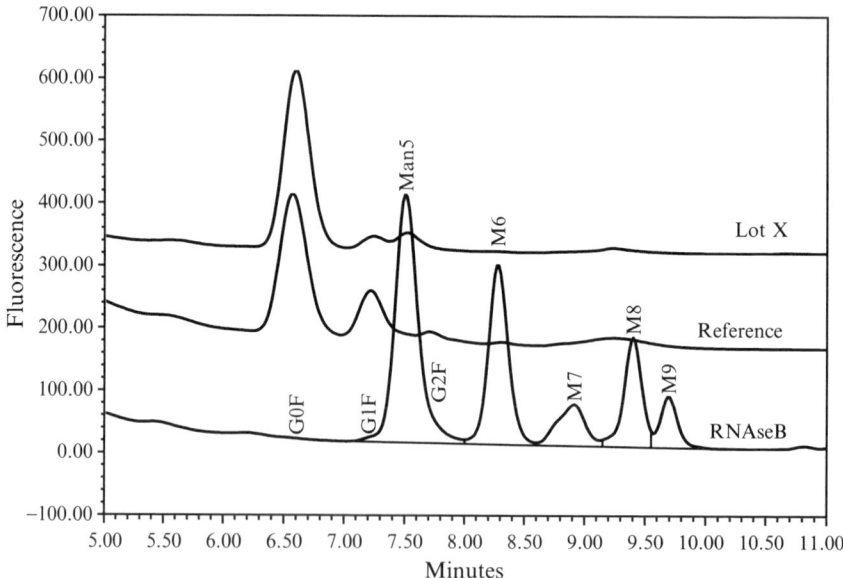

Fig. 24.2 Oligosaccharides from rIgG released using PNGase F are identified on the short 15 cm polyamine column using Waters HPLC. RNAse B oligosaccharides (bottom) allow identification of high mannose glycans in reference IgG (middle) and lot X IgG (top). G2F in reference and Man5 in lot X are easily distinguished in this method

than usual high mannose Man5 (or M5) and absence of G2F (*see* top chromatogram, Fig. 24.1) compared to reference lot with G2F (middle chromatogram). G2F and Man5 glycans co-eluted in the Waters HPLC system described here using the long 25 cm polyamine column. RNAse B was used

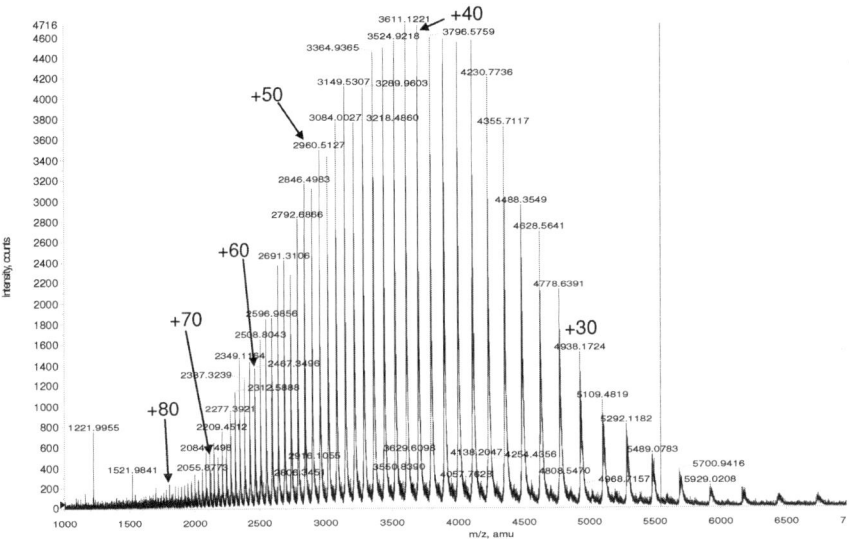

Fig. 24.3 Charge state envelope of monoclonal antibody (mAb)

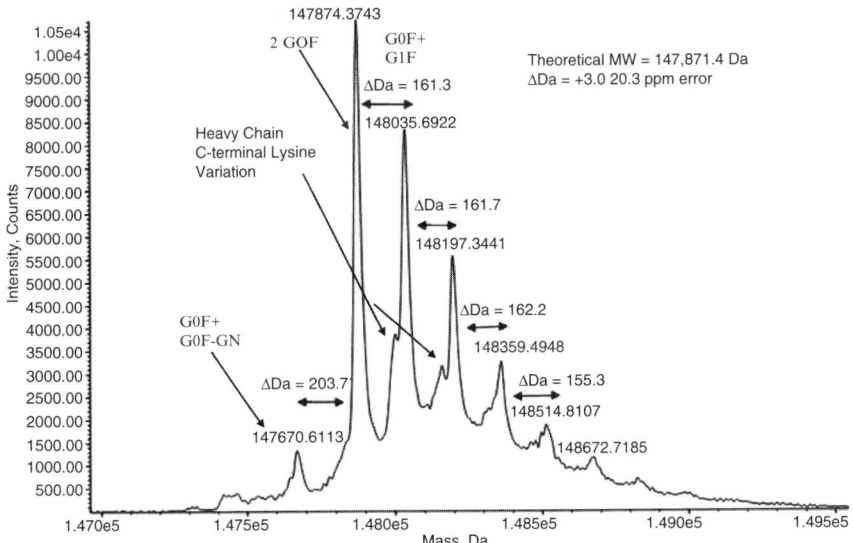

Fig. 24.4 Charge deconvoluted spectrum of mAb in Figure 24.3 allows visualization of microheterogeneity and C-terminal lysine variants of native glycoprotein

as a source of high mannose glycans (bottom chromatogram). Short 15-cm polyamine column and gradient described above easily differentiates between the 2 glycans (see Fig. 24.2).

5. The charge state envelope of the monoclonal antibody is shown in Figure 24.3 and Figure 24.4 is a native "snapshot" of the rIgG obtained by the deconvolution routine. The latter shows not only the glycosylation state of each heavy chain of the antibody, but also captures the heterogeneity caused by the presence of one or more C-terminal lysines. For example, the presence of G0F, G1F, etc., on either chain of the MAb is easily evident. The protein population spread with sequential addition of a monosaccharide unit on each heavy chain can only be obtained by ESI-ToF of native glycoprotein and not from the free glycan experiments.

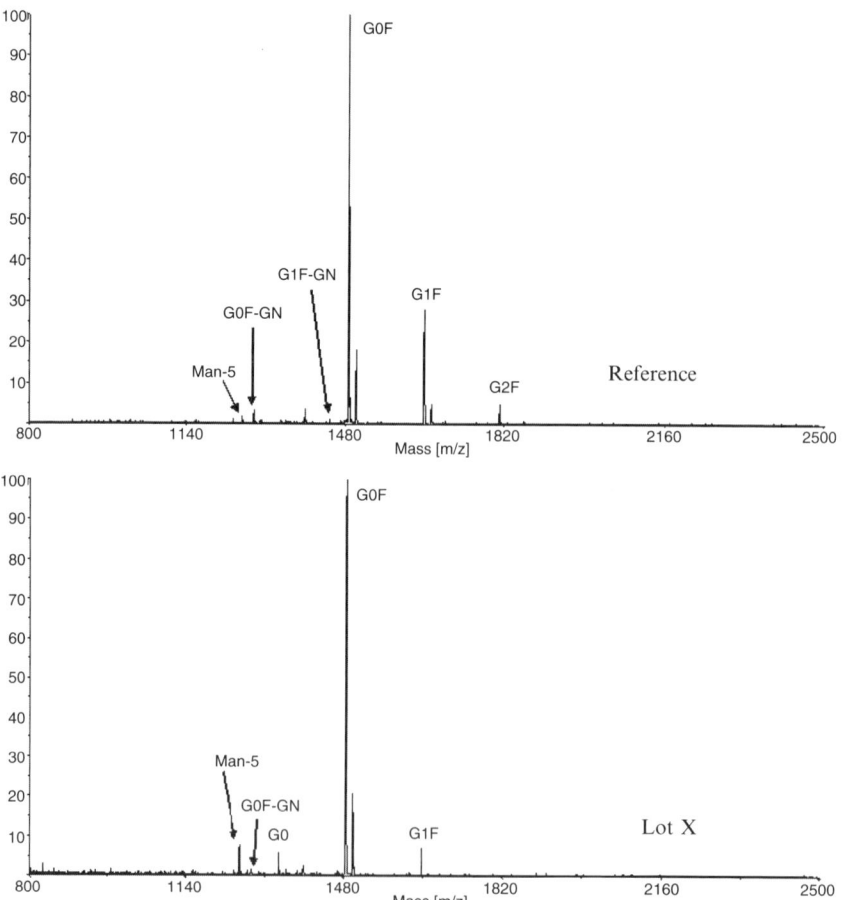

Fig. 24.5 PNGase F released N-linked oligosaccharides from reference (top) and lot X (bottom) recombinant mAb analyzed by MALDI-ToF. High mannose Man5 is easily detected by MALDI-ToF, even at low amounts in reference. Absence of G2F in lot X corroborates HPLC results

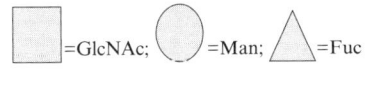

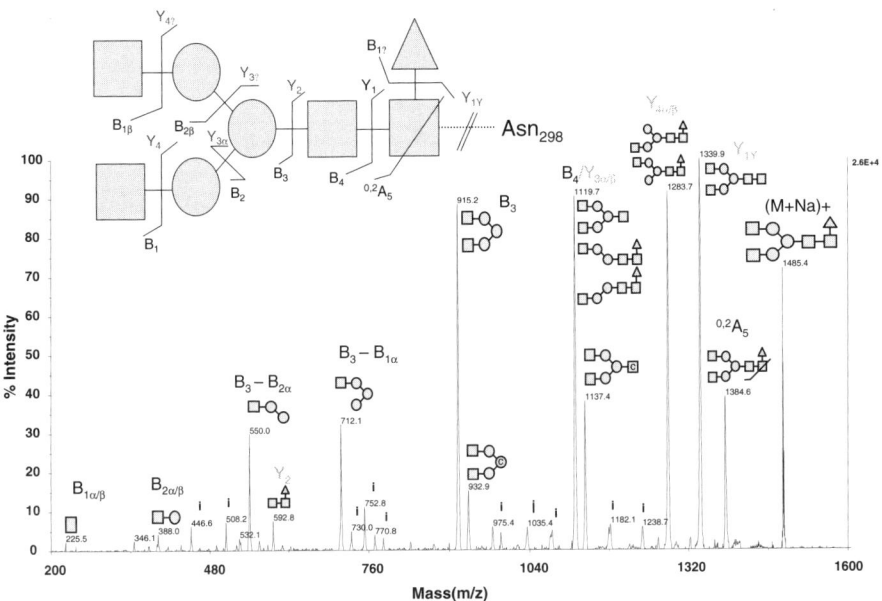

Fig. 24.6 Post-Source decay spectrum of G0F glycan selected for further structural elucidation made in MALDI-ToF experiments

6. Enzyme to protein ratio is 0.075 units enzyme/μg protein. If protein concentration varies from above, adjust amount of enzyme accordingly.

7. All solvents should be delivered through the back of the tip. Liquid should be not be aspirated up as the media is not fixed. Liquid should be forced out with air pressure using a 1-mL syringe placed in the back of the tip.

8. The mass accuracy of each of the calibrant is noted during a preacquisition calibration check and a postacquisition calibration check. Calibration should fall within 200 ppm with close external calibration.

9. A qualitative comparison of the spectra is made to ensure that the glycan peaks observed fall within the range of the proposed isoforms observed in the LC-MS intact molecular weight analysis. Presence of Man5 and absence of G2F in lot X (bottom) compared to the "normal" presence of G2F and minor amounts of Man5 in reference (top) (*see* Fig. 24.5) corroborated HPLC results. Some low abundance glycans detected by MALDI-ToF co-elute in the HPLC. Likewise, low levels of high mannose and acidic glycans easily quantitated in HPLC are not detected by MALDI-ToF. However, amounts of major glycans: G0F, G1F, and Man5 are similar by either method as shown for lot X in Table 24.1.

10. When the PSD experiment is opened, the necessary parameters for acquiring the differently focused segments of the *m/z* range are checked. The experiment keeps the individual segments that are individual acquisition files grouped into a single experiment for post-acquisition stitching. Each segment acquisition occurs just as normal acquisition does with the current spectrum considered. The number of segments (individual acquisitions) that will be stitched together to make the composite PSD spectrum ranges between 10 and 15 depending on the precursor mass. When evaluating the accumulated spectra to save for each segment it is important to maximize the signal in this stitch region and ignoring peaks that fall outside. If care is not taken to maintain a consistent signal level of the segments to be stitched, wild fluctuations in the signal level may be present in the final PSD spectrum. Once the final segment has been collected, the experiment gathers the appropriate segments from each of the acquisitions and stitches the PSD fragmentation spectrum together.

11. Post-Source Decay Spectrum of G0F glycan selected for further structural elucidation confirms rIgG glycan structural assignments made in MALDI-ToF as well as those made in HPLC experiments (*see* Fig. 24.6).

References

1. Walsh, G. (2003) Biopharmaceutical Benchmarks. *Nat. Biotechnol.* **21**, 865–870.
2. Humphreys, A. and Boersig, C. (2003) Cholesterol drugs dominate. *MedAdNews.* **22**, 42–57.
3. Varki, A. (1993) Biological roles of oligosaccharides: all of the theories are correct. *Glycobiology.* **3**, 97–130.
4. Roger A. Laine. (1994) Invited Commentary: A calculation of all possible oligosaccharide isomers both branched and linear yields 1.05×10^{12} structures for a reducing hexasaccharide: the Isomer Barrier to development of single-method saccharide sequencing or synthesis systems. *Glycobiology.* **4**, 759–767.
5. Raju, T. S., Briggs, J. B., Borge, S. M., and Jones A. J. S. (2000) Species-specific variation in glycosylation of IgG: evidence for the species-specific sialylation and branch-specific galactosylation and importance for engineering recombinant glycoprotein therapeutics. *Glycobiology.* **10**, 477–486.
6. Wilson, I. B. H., Zeleny, R., Kolarich, D., Staudacher, E., Stroop, C. J. M., Kamerling, J. P., and Altmann, F. (2001) Analysis of Asn-linked glycans from vegetable foodstuffs: widespread occurrence of Lewis a, core α1,3-linked fucose and xylose substitutions. *Glycobiology.* **11**, 261–274.
7. Sawardeker, J. S., Sloneker, J. H.and Jeanes, A. (1965) Quantitative determination of monosaccharides as their alditol acetates by gas liquid chromatography. *Anal. Chem.* **37**, 1602–1604.
8. Sweeley, C. C., Bentley, R., Makita, M., and Wells, W. W. (1963) Gas-liquid chromatography of trimethylsilyl derivatives of sugars and related substances. *J. Am. Chem. Soc.* **85**, 2497–2507.
9. Townsend, R. R. (1995) Analysis of glycoconjugates using high-pH anion-exchange chromatography, in *Carbohydrate Analysis: High-Performance Liquid Chromatography and Capillary Electrophoresis.* (El Rassi, Z., ed.), Elsevier, New York, NY, pp. 181–209.
10. Horvath, C. and Ettre, L. S. (eds.) (1993) *Chromatography in Biotechnology.* American Chemical Society, Washington, D.C.

11. Kolisis, F.N. (1986) An immobilized bienzyme system for assay of sialic acid. *Biotechnol Appl Biochem.* **8**, 148–152.

12. Yasuno, S., Kokubo, K., and Kamei M. (1999) New method for determining the sugar composition of glycoproteins, glycolipids, and oligosaccharides by high-performance liquid chromatography. *Biosci. Biotechnol. Biochem.* **63**, 1353m1359.

13. Patel T.P. and Parekh, R. B. (1994) Release of oligosaccharides from glycoproteins by hydrazinolysis. *Methods Enzymol.* **230**, 57–66.

14. Maley, F., Trimble, R.B., Tarentino, A.L., and Plummer T.H. (1989) Characterization of glycoproteins and their associated oligosaccharides through the use of endoglycosidases. *Anal. Biochem.* **180**, 195–204.

15. McLean C., Werner, D.A., and Aminoff, D. (1973) Quantitative determination of reducing sugars, oligosaccharides, and glycoproteins with (3H)borohydride. *Anal. Biochem.* **55**, 72–84.

16. Koles, K., van Berkel, P. H. C., Pieper, F. R., Nuijens, J. H., Mannesse, M. L. M., Vliegenthart, J. F. G., and Kamerling, J. P. (2004) N- and O-glycans of recombinant human C1 inhibitor expressed in milk of transgenic rabbits. *Glycobiology.* **14**, 51–64.

17. Honda, S. (1984) High-performance liquid chromatography of mono- and oligosaccharides. *Anal. Biochem.* **140**, 1–47.

18. Bergh, M. L. E., Koppen, P., and Van den Eijnden, D. H (1981) High pressure liquid chromatography of sialic-acid-containing oligosaccharides. *Carbohydr. Res.* **94**, 225–229.

19. Rosenfelder G., Mörgelin, M., Chang, J.-Y., Schönenberger, C.-A., Braun, D.G., and Towbin, H. (1985) Chromogenic labeling of monosaccharides using 4'-*N,N*-dimethylamino-4-aminoazobenzene. *Anal. Biochem.* **147**, 156–165.

20. Anumula, K.R. (2006) Advances in fluorescence derivatization methods for high-performance liquid chromatographic analysis of glycoprotein carbohydrates. *Anal. Biochem.* **350**, 1–23.

21. Alpenfels, W.F. (1981) A rapid and sensitive method for the determination of monosaccharides as their dansyl hydrazones by high-performance liquid chromatography. *Anal. Biochem.* **114**, 153–157.

22. Dhume, S. T. and Anumula, K. R. (1998) Evaluation of various fluorescent tags for oligosaccharide characterization: a comparative study with anthranilic acid. *Glycobiology.* **8**, abstract 34.

23. Anumula, K. R. and Dhume, S. T. (1998) High resolution and high sensitivity methods for oligosaccharide mapping and characterization by normal phase high performance liquid chromatography following derivatization with highly fluorescent anthranilic acid. *Glycobiology*, **8**, 685–694.

24. Leeflang B.R. and Vliegenthart, J.F.G. (2000) Glycoprotein analysis using nuclear magnetic resonance, in *Encyclopedia of Analytical Chemistry.* (Meyers, R. A. ed.), John Wiley, New York, NY, pp. 821–834.

25. Harvey, D.J. (2001) Identification of protein-bound carbohydrates by mass spectrometry. *Proteomics*, **1**, 311–328.

26. Thibault, P. and Honda, S. (eds.) (2002) *Capillary electrophoresis of oligosaccharide and complex carbohydrates.* Humana, Totowa, NJ.

27. Dhume, S. T., Ebert, M. B., Saddic, G. N., and Anumula, K. R. (2002) Monitoring glycosylation of therapeutic glycoproteins for consistency using highly fluorescent anthranilic acid, in *Methods in Molecular Biology, Posttranslational Modifications of Proteins.* (Kannicht, C. ed) Humana Press, New Jersey, 127–142.

28. Neville, D. C. A., Coquard, V., Priestman, D. A., te Vruchte, D. J. M., Sillence, D. J., Dwek, R. A., Platt, F. M., and Butters, T. D. (2004) Analysis of fluorescently labeled glycosphingolipid-derived oligosaccharides following ceramide glycanase digestion and anthranilic acid labeling. *Anal. Biochem.* **331**, 275–282.

25

Web-based Computational Tools for the Prediction and Analysis of Post-translational Modifications of Proteins

Vladimir A. Ivanisenko, Dmitry A. Afonnikov, and Nikolay A. Kolchanov

Summary The increase in the number of Web-based resources on post-translational modification sites (PTMSs) in proteins is accelerating. The paper presents a set of computational protocols describing how to work with the Internet resources when dealing with PTMSs. The protocols are intended for querying in PTMSs related databases, search of the PTMSs in the protein sequences and structures, calculating the pI and molecular mass of the PTM isoforms. Thus, the modern bioinformatics prediction tools make feasible to express protein modification in broader quantitative terms.

Key Words Amino acid sequence; protein 3D structure, protein post-translational modification prediction; Web-based resources; database.

1 Introduction

Prediction of post-translational modifications (PTMs) of a protein is an important task in modern computer proteomics. A large proportion of proteins are subject to PTMs after synthesis and the PTMs may control the state of activity of proteins *(1)*, their localization *(2)*, turnover *(3)* and the partners of protein or nucleic-acid binding *(4)*. Phosphorylation on serine, threonine, and tyrosine residues by enzymes is the most abundant and consequential modulator of protein function *(5)*. The N-terminal ends of the molecules or their C-terminal end (glycosylation and phosphorylation), or their side chains (amidation and prenylation) can be modified post-translationally. Some PTMs are dependent on the local context of amino acid sequences. These dependencies can be observed as amino acid patterns (motifs) that characterize the post-translational modification sites (PTMSs) applicable to their recognition in the protein primary structures. These context signals have been identified and described as motifs for the phosphorylation *(6)*, acetylation *(7)* and for many other sites (for a review, see, for example, *(8)*). PROSITE is a database of numerous annotated motifs of these sites *(9)*. The more complex dependencies can be identified by

From: *Post-translational Modifications of Proteins.*
Methods in Molecular Biology, Vol. 446.
Edited by: C. Kannicht © Humana Press, Totowa, NJ

using machine learning algorithms, and they have proven to be useful for PTMSs recognition in the protein sequences *(10)*.

Post-translational modification sites can also be described in the 3-dimensional (3D) structures of proteins *(11)*. The description is advantageous, because it enables the localization of site forming residues, even when they are so far apart in primary structure that their 1-dimensional (1D) patterns elude description.

There is another facet of proteomics related to PTMs. Protein modifications can cause changes in the protein pI because of addition, removal or alteration of titrable groups, in their molecular mass as well; for these reasons, the shifts can affect the 2D gel electrophoresis patterns of the protein isoforms. Theoretical estimates of the pI and molecular mass of the PTM isoforms, therefore, would be helpful in protein modification studies.

We present here a set of computational protocols describing how to work with the Web-based resources when dealing with PTMSs. There is a variety of numerous resources. Several types are distinguishable among those listed in Table 25.1: (i) the databases contain information about the different PTMSs in both the primary and tertiary structures; (ii) the programs for PTMS search allow to localize PTMSs in protein sequences and structures; (iii) programs for estimating the pI shift and molecular mass for different protein isoforms. The resources are provided with interfaces that, as a rule, have much in common. For this reason, it appeared appropriate to describe in some detail how some, not all, the proposed protocols are implemented. The protocols are intended for (i) querying in PROSITE *(9)* and PDBSite *(11)* databases; (ii) search of the PTMSs in the protein sequences using the PROSITE signatures *(13)* and in the 3D structures using the PDBSiteScan program *(26)*; (iii) to calculate the pI and molecular mass of the PTM isoforms using the ProMoST program *(27)*. A comprehensive review of the algorithms currently used for machine learning and of the programs developed for the recognition of the phosphorylation and glycosylation sites can be found in *(10)*. A detailed description of the algorithms, program capabilities, Web-based interfaces can be found in the literature references compiled in Table 25.1.

2 Materials

2.1 *Hardware*

A computer connected to the Internet.

2.2 *Software*

An up-to-date Web browser such as Microsoft Internet Explorer, Mozilla, Opera or FireFox.

Table 25.1 Internet Resources Related to Protein Post-Translational Modification Sites

Resource Name	Resource topic	WWW reference	Paper reference
Databases			
PROSITE	Active sites in protein sequences	http://us.expasy.org/prosite/	(9)
Phospho.ELM	Phosphorylation	http://phospho.elm.eu.org	(12)
DSDBASE	Disulphide bonding	http://www.ncbs.res.in/~faculty/mini/dsdbase/dsdbase.html	(13)
O-GLYCBASE	Glycosylation	http://www.cbs.dtu.dk/databases/OGLYCBASE/	(14)
RESID	Protein modifications annotations	http://www.ebi.ac.uk/RESID http://www.ncifcrf.gov/RESID	(15)
PDBSite	Various sites in 3D structures of protein	http://wwwmgs.bionet.nsc.ru/mgs/gnw/pdbsite/	(11)
Programs			
PROSITE	General motif search in protein sequence	http://us.expasy.org/prosite/	(16)
NetPhos	Phosphorylation	http://www.cbs.dtu.dk/services/NetPhos/	(17)
NetPhosK	Phosphorylation	http://www.cbs.dtu.dk/services/NetPhos/	(10)
ScanSite	Phosphorylation	http://scansite.mit.edu	(18)
PredPhospho	Phosphorylation	http://www.ngri.re.kr/proteo/PredPhospho.htm	(19)
NetAcet	Acetylation	http://www.cbs.dtu.dk/services/NetAcet/	(20)
MeMo	Methylation	http://www.bioinfo.tsinghua.edu.cn/~tigerchen/memo.html	–
Methylator	Methylation	http://bio.dfci.harvard.edu/Methylator/	–
NetOGlyc	Glycosylation	http://www.cbs.dtu.dk/services/NetOGlyc/	(21)
NetNGlyc	Glycosylation	http://www.cbs.dtu.dk/services/NetNGlyc/	–
DictyOGlyc	Glycosylation in *Dictyostelium discoideum* proteins	http://www.cbs.dtu.dk/services/DictyOGlyc/	(14)
YinOYang	Glycosylation	http://www.cbs.dtu.dk/services/YinOYang/	–
big-PI	GPI-anchor	http://mendel.imp.univie.ac.at/sat/gpi/gpi_server.html	(22)
The Sulfinator	Sulfation	http://ca.expasy.org/tools/sulfinator/	(23)
GDAP	Disulfide bond formation	http://www.doe-mbi.ucla.edu/~boconnor/GDAP/	(24)

(continued)

Table 25.1 (continued)

Resource Name	Resource topic	WWW reference	Paper reference
DIpro	Disulfide bond formation	http://contact.ics.uci.edu/dipro2.html	(25)
PDBSiteScan	Sites in 3D structures of proteins	http://wwwmgs.bionet.nsc.ru/mgs/systems/fastprot/pdbsitescan.html	(26)
ProMost		http://proteomics.mcw.edu/promost	(27)
3D visualization of macro-molecules			
RasMol		http://www.umass.edu/microbio/rasmol/index2.htm	(28)
Protein Explorer		http://www.umass.edu/microbio/chime/pe/protexpl/frntdoor.htm	–
Other resources			
SRS	Database integration and search	http://srs.ebi.ac.uk	(29)
PDB	Database of 3D structures of macromol-ecules	http://www.pdb.org	(30)

2.3   The PROSITE Database

The PROSITE database *(9)* is a collection of signatures that describe the different functional sites in the protein sequences. The signatures are of two types: (i) the PROSITE pattern (regular expression), a simplified representation of amino acid occurrence at functional region positions; (ii) position-specific amino acid profile, a probabilistic representation of amino acid occurrence. Each PROSITE entry contains the name of the functional site, a text description of biologically meaningful information, site signature. By October 2005, PROSITE contained 1,377 documentation entries that described 1,329 patterns and 552 profiles. The database is provided with the ScanProsite system for the search of sites in the user's sequences and for interactive representation of the results for the ScanProsite search *(16)*. The PROSITE database is available at *http://us.expasy.org/prosite/*.

2.4   The PDBSite Database and PDBSiteScan Program

The PDBSite database *(11)* provides versatile structural and functional information about various protein sites (post-translational modification, catalytic active, organic and inorganic ligand binding in protein-protein, protein-DNA, protein-RNA binding) in the Protein Databank *(30)*. The database is composed of annotated sites from records of the protein spatial structures and also of calculated spatial, physicochemical, and mutational characteristics of sites and their spatial environment. Furthermore, it contains the structural template of the site, i.e. spatial coordinates of the N, CA, C atoms of the main chain of the residues that form a site. Each entry in the database contains information about a particular site that was described in the PDB structure. The latest PDBSite version (by October 2005) contained information about 14,460 such sites. The database stores information for PTMSs of the following types: myristylation, phosphorylation, acetylation, cleavage, glycosylation, lipoylation. The information retrieval interface of PDBSite database is based on the SRS system *(29)*.

The PDBSiteScan program *(26)* makes it feasible to search for the functional sites in a query protein structure using the method of spatial superimposition of atoms of a functional site structural template from the PDBSite over the N, CA, C atoms of residues in a query protein. A site is deemed to be detected, if its amino acid types match with the corresponding amino acids in template and that the deviation of the template-query coordinates upon optimum superimposition is not above the user-defined threshold. The program outputs search results as a query protein spatial structure with superimposed templates in a file in PDB format.

2.5   The ProMoST Program

The ProMoST program *(27)* allows the estimation of changes in the pI and protein molecular mass caused by PTMs. ProMoST draws the estimation results as a graph that bears similarity to the 2D gel image output.

2.6 Software for the Protein 3D Structure Visualization

To visualize the query protein superimposed over the PDBSite templates, the program for 3D protein structure visualization that accepts PDB file format can be used. These programs include Rasmol *(28)* and Protein Explorer (*http://www. umass.edu/microbio/chime/pe/protexpl/frntdoor.htm*), among others.

3   Methods

3.1 Protocols for the PTMSs Information Search in the Databases

Information about the PTMSs deposited in the databases is accessible through the Internet (Table 25.1). The databases are collections of documents that contain information about sites of post-translational modifications. To make the information in the database machine-readable and convenient for the user, the data are structured and contain fields, including, among others, the name and the type of a site, literature reference from which the information was taken, its primary, secondary and tertiary structural features, links to other relevant biological databases. As a rule, the information stored in the databases is text-based.

The user's main requirement is search for particular information in the database. To meet the search criterion, an interface for the text-based search is provided. The interface, as a rule, allows choosing records, containing query words, their free combinations in defined fields from the database. Boolean operators (AND, OR, NOT) can be used to combine query words. For example, the "phosphorylation AND kinase" query allows to choose documents that contain both "phosphorylation" and "kinase" words. The "phosphorylation OR kinase" query allows to choose the documents that contain either "phosphorylation" or "kinase," the "phosphorylation NOT kinase" query enables to choose the documents that do contain the word "phosphorylation," but do not contain "kinase." Exemplary protocols of search in the databases containing information about PTMSs are given below.

3.1.1 Search of the Information about PTMSs in the PROSITE Database

1. Open the browser and go to the PROSITE home page (Fig. 25.1).
2. Enter query word or Boolean combination of words in the text box [1] (Fig. 25.1A).
3. You may want to search query text as a part of a word in the database. If so, place a check in the 'Prefix and append wildcard to words' check box [2]. In this case, any word containing query text will be considered as match. For example, in the "phosph" search, "phosphorylation" and "cyclodiphosphate" match.

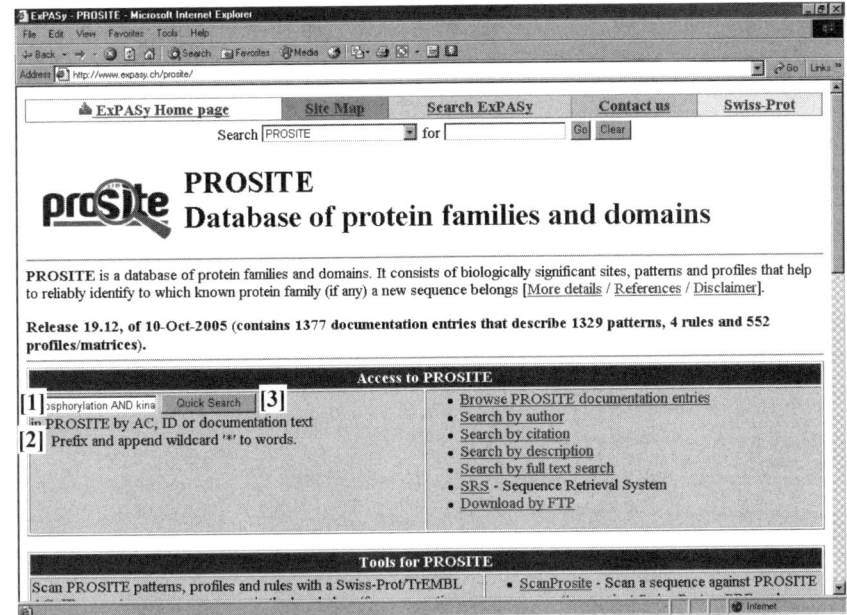

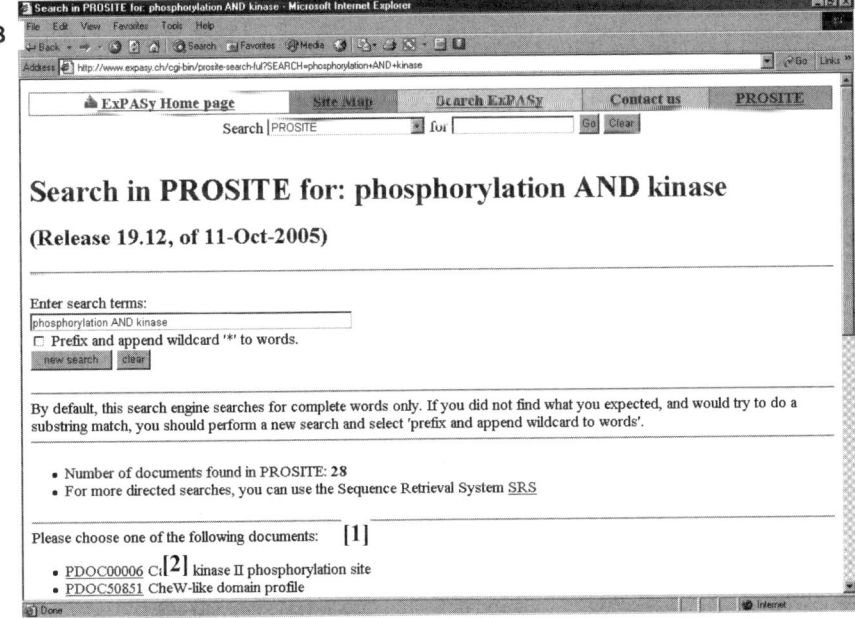

Fig. 25.1 Search of the information about PTMSs in PROSITE database. (**A**) input page; (**B**) search results page; (**C**) document page

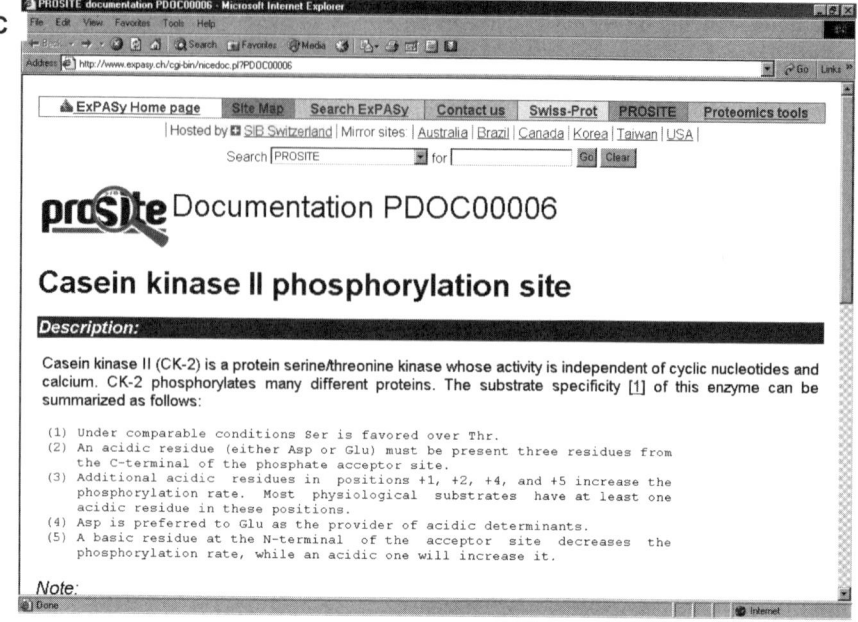

Fig. 25.1 (continued)

4. To perform a query, press the 'Quick Search' button [3].
5. Search results page (Fig. 25.1B) will list identifiers of documents [1] from the PROSITE database that meet the query criterion. Every identifier is a hyperlink by which the corresponding record can be obtained.
6. For document retrieval, click on the hyperlink [2] (Fig. 25.1B). Results page is shown in Figure 25.1C.

3.1.2   Search of the Information about PTMSs in the PDBSite Database. Standard Query Form

1. Begin with the PDBSite home page (Fig. 25.2A).
2. Go to the SRS page of the access to the PDBSite database by the hyperlink [1] (Fig. 25.2A). As a result, you will obtain information regarding its current state (Fig. 25.2B). Table [1] presents the current list of fields and their brief description.
3. To work further, press the "Search" button [2] (Fig. 25.2B). You will go to the standard query form (Fig. 25.2C).
4. At the standard query form page, select the type of the field in the database you intend to search from the drop-down list [1]. Enter the query term in the text box [2].

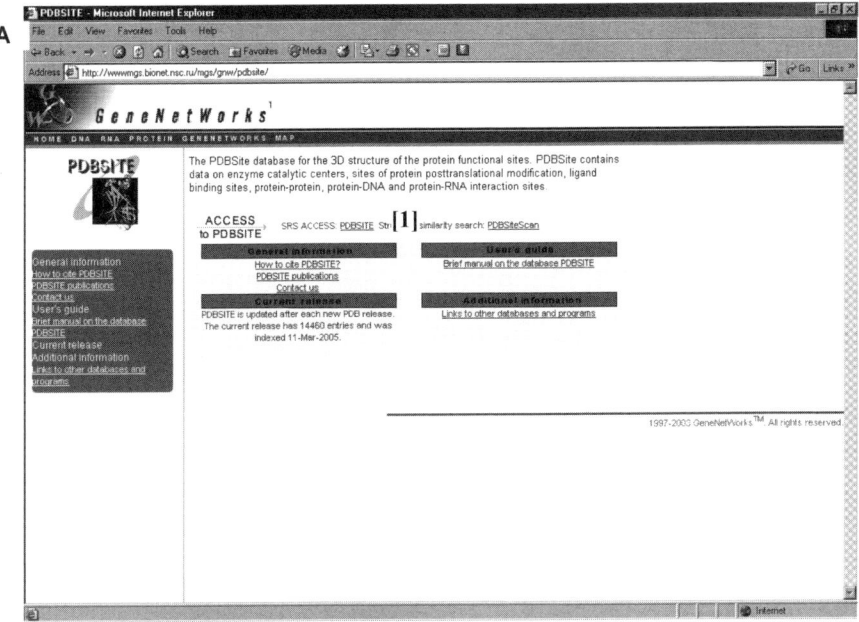

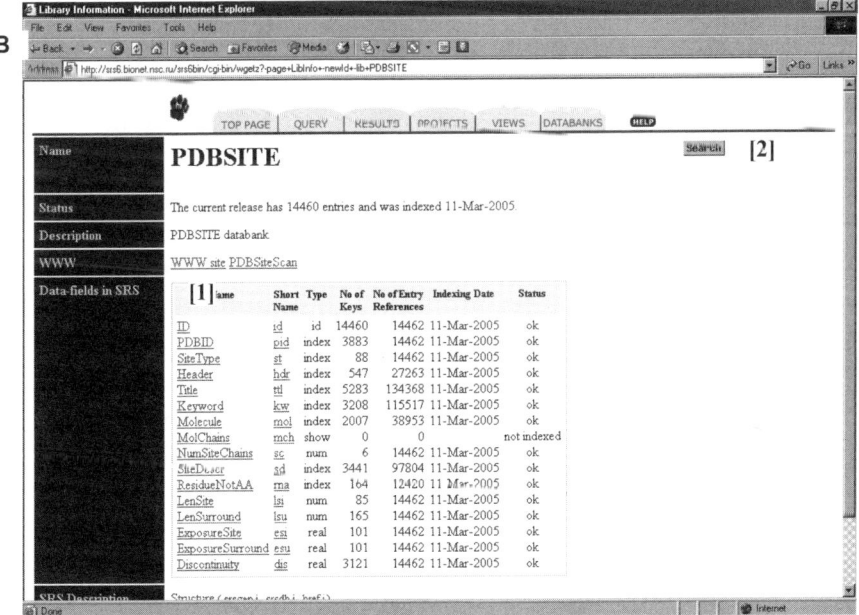

Fig. 25.2 Search of the information about PTMSs in the PDBSite database: (**A**) the PDBSite home page; (**B**) database status page; (**C**) standard query form; (**D**) query results page; (**E**) entry page; (**F**) download page; (**G**) extended query form

C

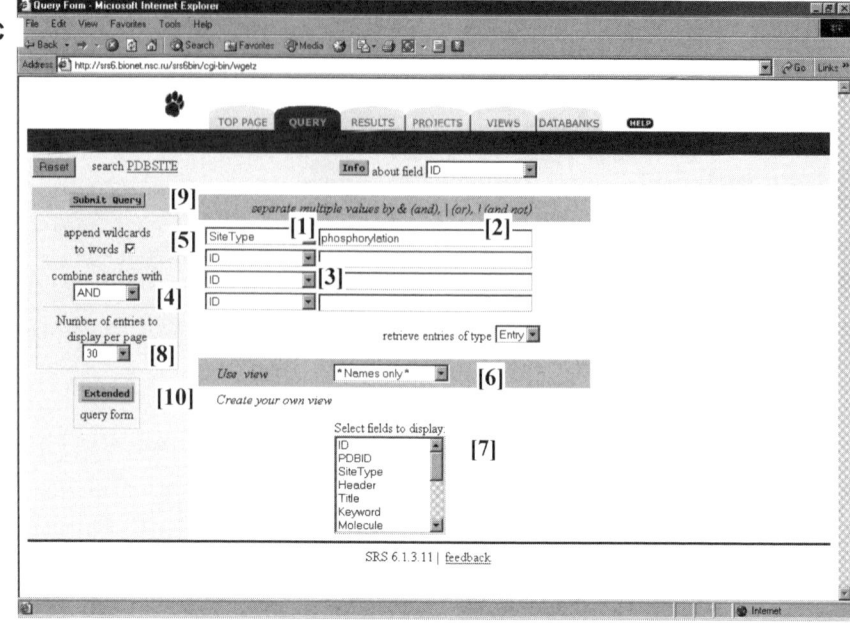

D

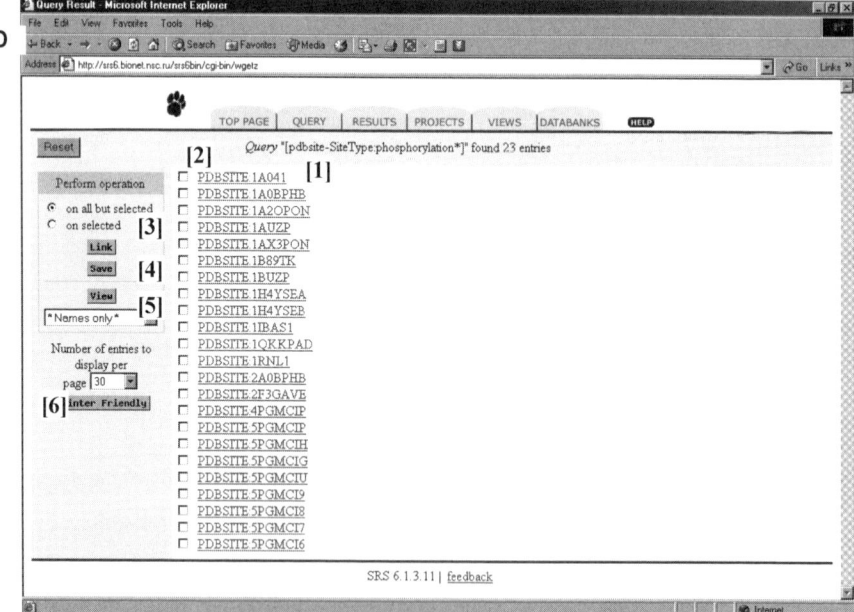

Fig. 25.2 (continued)

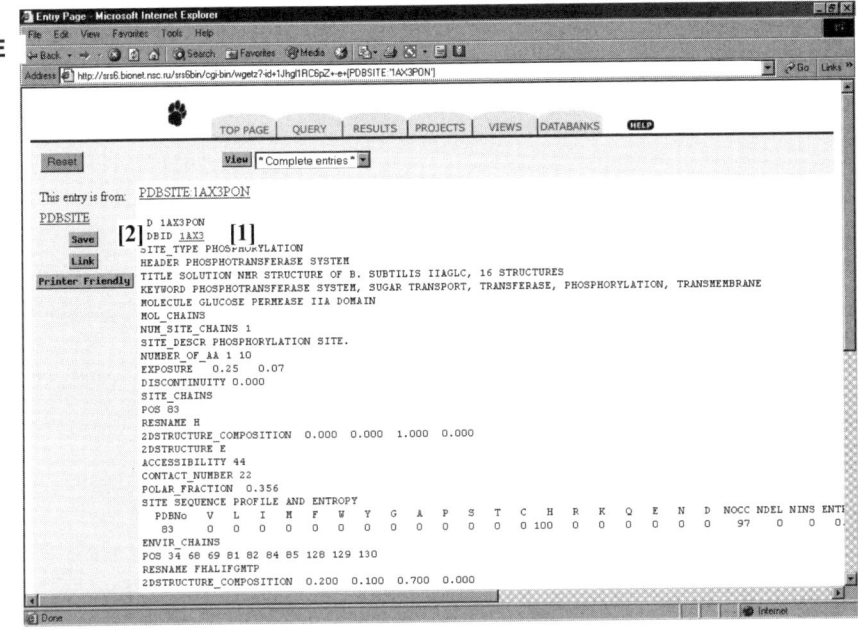

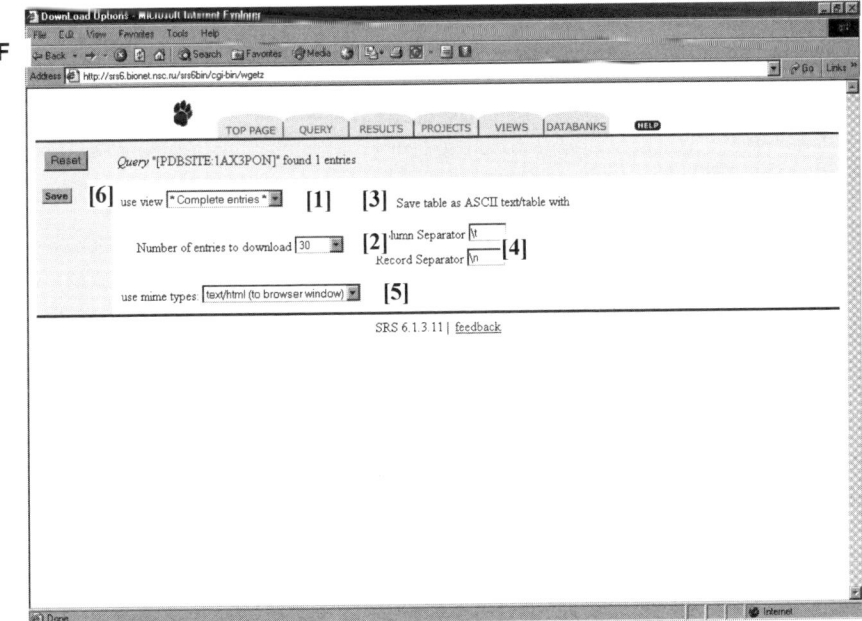

Fig. 25.2 (continued)

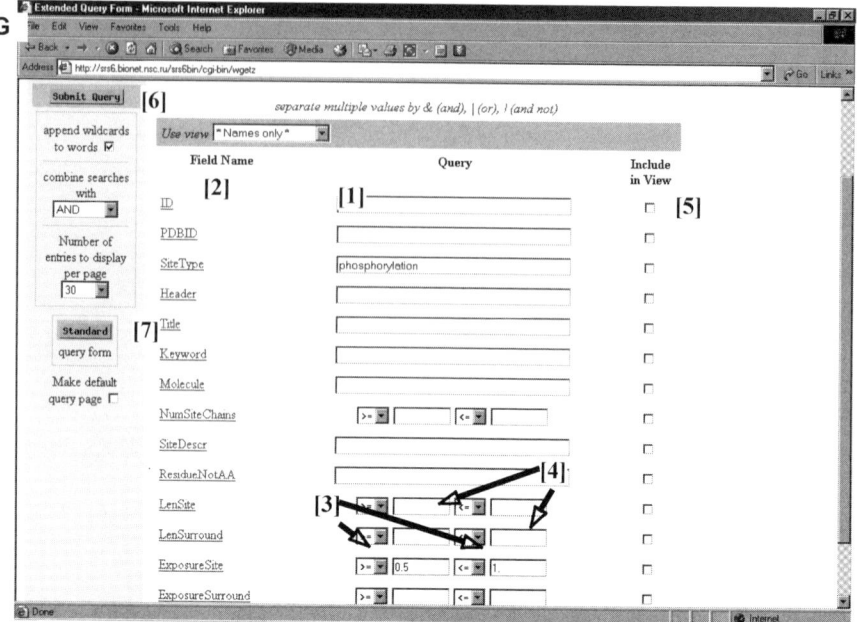

Fig. 25.2 (continued)

5. The SRS interface allows you to make a query that contains the different words in different fields at the same time. To do so, the query can be extended by adding the corresponding "field name"-"query word" pairs [3]. Choose logical operation to combine query from the drop-down list [4].

6. If you want use the wildcard extension, place a check in the 'append wildcards to words' check box [5].

7. The output format of query results can be chosen from the menu at the bottom of the page. To choose the standard formats, use the drop-down list [6]; to visualize the other fields, as you wish, use the menu below [7].

8. To limit the number of records downloaded per results page, use the drop-down list [8].

9. To perform a query, press the "Submit query" button [9].

10. The results are displayed as a list of record indices meeting the query criterion. Each line is the hyperlink by which the full text can be obtained [1] (Fig. 25.2D). The SRS interface allows you to obtain data from several records. To do so, place checks in the check boxes that correspond to several records [2]. You can make several operations with checked or unchecked entries depending on the radio-button settings [3]: download data to your local computer as a text file (the 'Save' button [4]); change the display format (the "View" button and the drop-down list [5]). The left panel of the page can be removed by clicking on the 'Printer friendly' button [6].

11. The entry page (Fig. 25.2E) contains text information about the name, structural properties of a site, hyperlink to the protein 3D structure where it was described [1], also, it has buttons to save information on the local computer hard discs [2].

12. To save information on the local disc (after steps 10 or 11 of this protocol), press the "Save" button on the left panel of the page [2] (Fig. 25.2F). You will go to the download page (Fig. 25.2E).

13. At the download page (Fig. 25.2E), the "use the view" drop-down list [1] allows to choose the data format, the drop-down list [2] limits the number of entries to be saved. To save data as a text file, place a check in the "Save table as ASCII text/table" check box [3] and choose the column and record separators [4]. The drop-down list [5] allows you to download data to the browser window or directly to the local disk. Finally, press the "Save" button [6].

3.1.3 Search of Information about PTMSs in the PDBSite Database. Extended Query form

Perform steps 1-3 as described in **Section 3.1.2.**

4. At the standard query form page (Fig. 25.2C), click on the 'Extended' button [10] to go to the extended query form page (Fig. 25.2G).

5. Input query words in the text boxes [1]. Each text box on this page corresponds to a PSBSite database field [2].

6. You can set limits for values of the numerical fields by the drop-down menus [3] and the text boxes for limiting values [4].

7. If you want the database field value to be shown on the results page, place a check in the corresponding check box [5].

8. To perform query, press the 'Submit query' button [6]. To return to the standard query form, press the "Standard" button [7].

9. You can work with query results as described in **Section 3.1.2** (steps 10–13).

3.2 Protocols for PTMSs Prediction in Proteins

3.2.1 Prediction of the Sites Described in the PROSITE Database in Protein Sequence

1. Begin at the PROSITE home page (Fig. 25.1).

2. Enter the query sequence in FASTA format in the text area [1] (Fig. 25.3A). An exemplary sequence in FASTA format is shown in the callout [2].

3. To exclude patterns whose occurrence is unacceptably frequent, place a check in the check box [3].

4. To perform query, press the "Quick Scan" button [4]. To clear the query form before making a new query, press the "Clean" button [5].

5. After the search has been done, the server outputs the results page (Fig. 25.3B). The page gives the number of detected signatures [1], the query sequence [2], also its

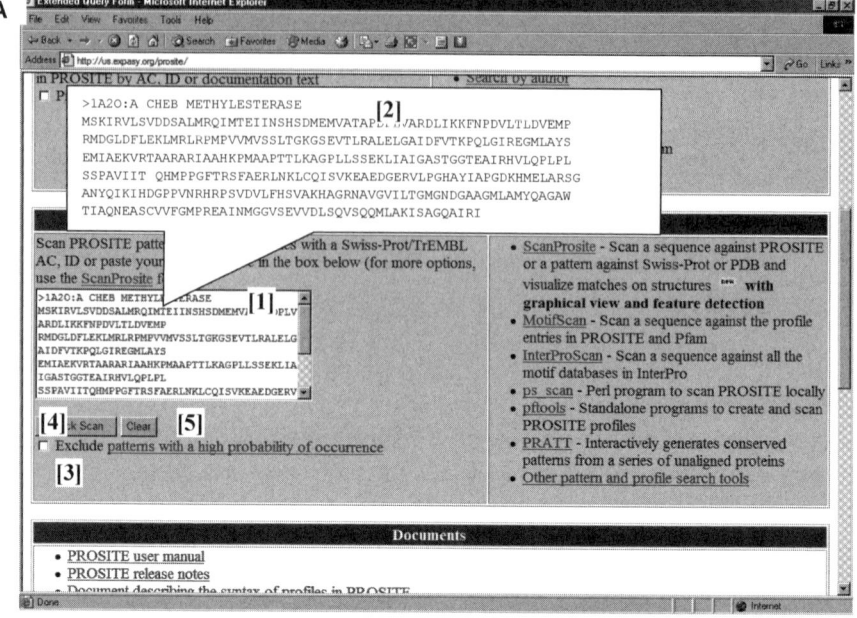

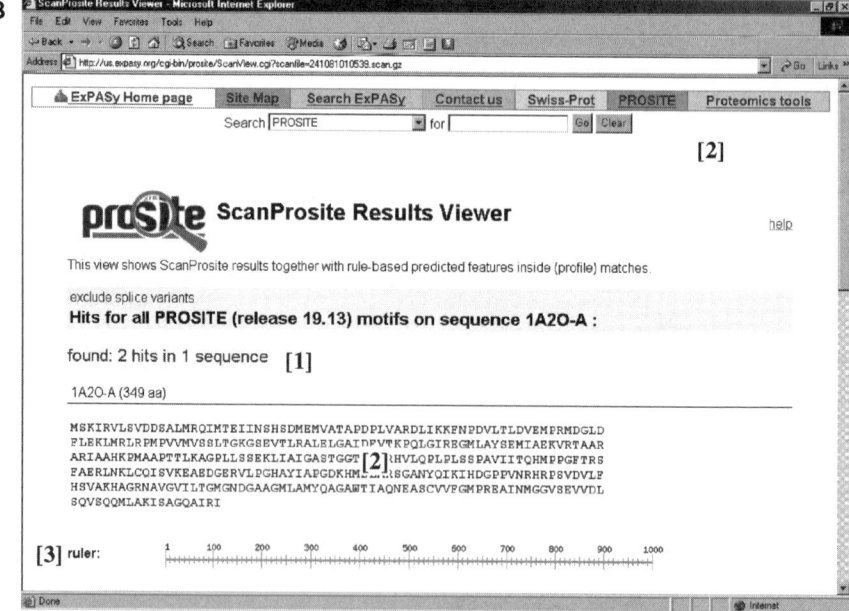

Fig. 25.3 Search of the sites described in the PROSITE database in protein sequence: (**A**) input query page; (**B**) query results page (sequence); (**C**) query results page (sequence information and detected sites)

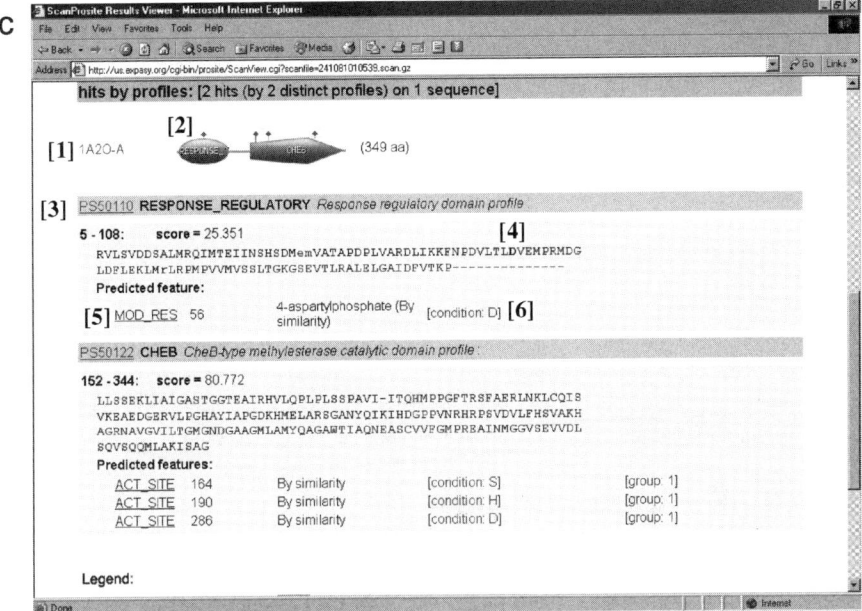

Fig. 25.3 (continued)

annotation in interactive form. For convenient display on the screen, the positions of the sequence are presented as a ruler [3]. If the searched sequence is detected as one of the already known, its identifier is indicated. The identifier is a hyperlink to the database that contains sequence annotation (Fig. 25.3C, [1]). In the case the protein functional domains are detected by homology in a query sequence, their localization in the sequence is displayed in graphical form (Fig. 25.3C, [2]). Each element of the graph is a hyperlink to the information about the corresponding domain. The location of the determined sites is shown by small diamonds (the display parameters of the graphical elements can be changed by pressing the control buttons at the bottom of the HTML page). There follow: a description of the functional sites; reference to the PROSITE database [3]; the name of the functional domain where the site was identified; its position in the sequence, the similarity score to the profile of the given domain; the location of the site in sequence in text form (highlighted in color or bold letters) [4], the name of the predicted motif and its location in the sequence [5], also the additional conditions for its identification [6].

3.2.2 Prediction of the PTMSs in Protein 3D Structures using the PDBSiteScan Program

1. Begin with the PDBSiteScan home page (Fig. 25.4A).
2. Enter the name of the file with the 3D protein structure in PDB format in the text box [1]. To search the file on a local computer, press the 'Browse' button [2].

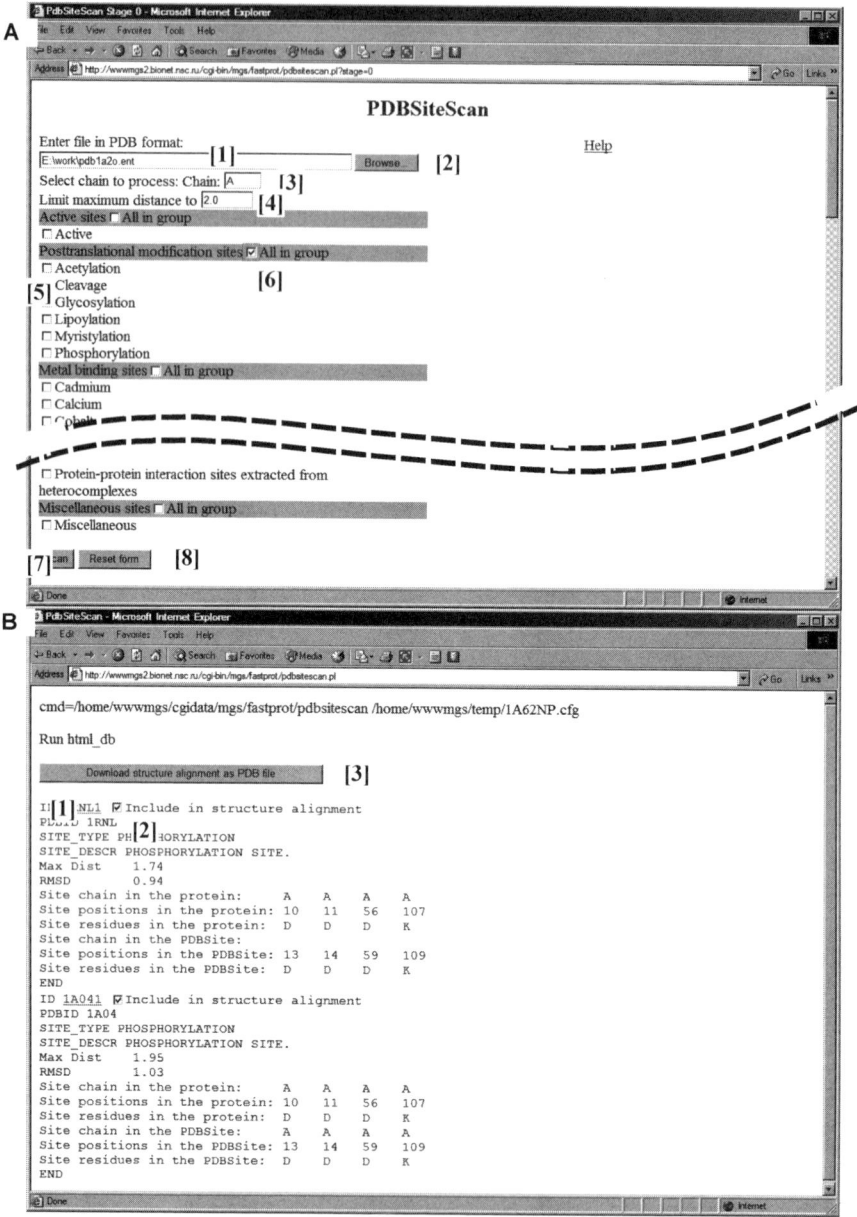

Fig. 25.4 Search of the PTMSs in protein 3D structures using the PDBSiteScan program: (**A**) query input form; (**B**) query results page.

3. In the "Chain" text box [3], enter identifier of the query protein chain. If there is no chain identifier, enter the underline "_".

4. Set the threshold that gives the similarity value of the site in the query structure and structural template in the PDBSite database in the 'Limit maximum distance to' text box [4]; the value of 2.0 Å is advised.

5. Choose the types of PTMSs for the search by placing checks in the check boxes [5]. Search through all the possible sites can be done by placing a check in the 'All in group' check box [6].

6. To perform search, press the "Scan" button [7]. To clear query form before interrogating a new query, press the "Reset form" button [8].

7. After the search has been done, the search results appear (Fig. 25.4B). The page gives a list of site templates from the PDBSite database that were detected in the query structure. The following is indicated for each detected site:

- the site identifier with the hyperlink to the PDBSite entry [1];
- the identifier of protein 3D structure where the site template is located;
- the site type and its brief description (SITE_TYPE and SITE_DESCR database fields);
- the maximum distance mismatch (MDM) and root mean square deviation (RMSD) between site in query structure and site template atomic coordinates after their optimum superimposition;
- the chain identifiers, site residue numbers and types for query structure, residue types for site template.

8. The user can download the 3D structure of query protein with superimposed template atoms. To select templates for downloading, place checks in corresponding check boxes [2]. Press the "Download structure alignment as PDB file" button [3].

9. After the file in the local disk is saved, it can be viewed by any program for macromolecular spatial structure visualization for standard PDB format. The file contains several chains; of these, the first marked with the "A" index is the query protein structure, the others arc the site template structures chosen by the user. Figure 25.5 provides an example of the file for the visualization of superimposed structures by the Rasmol program (28).

4 Protocol for Estimation of Changes in the Protein pI and Molecular Mass because of PTMs

4.1 Estimation of Changes in the Protein pI and Molecular Mass because of PTMs using the ProMoST Program

1. Begin with the ProMoST home page (Fig. 25.6A).

2. Enter the protein sequence or its identification index in the text area [1] and, depending on this, set the radio buttons [2]. You can upload the file from the local disk by indicating its name in the text box [3] (you can choose the file by pressing the "Browse" button [4]). The "Advanced" hyperlink [5] provides the change-over to the advanced input form.

3. The protein modification types and parameters for the results output are set at the bottom of the page (Fig. 25.6B): terminal ends blocking [1], deamidation [2], phosphorylation [3] or additional modifications defined by the user [4]. Set the

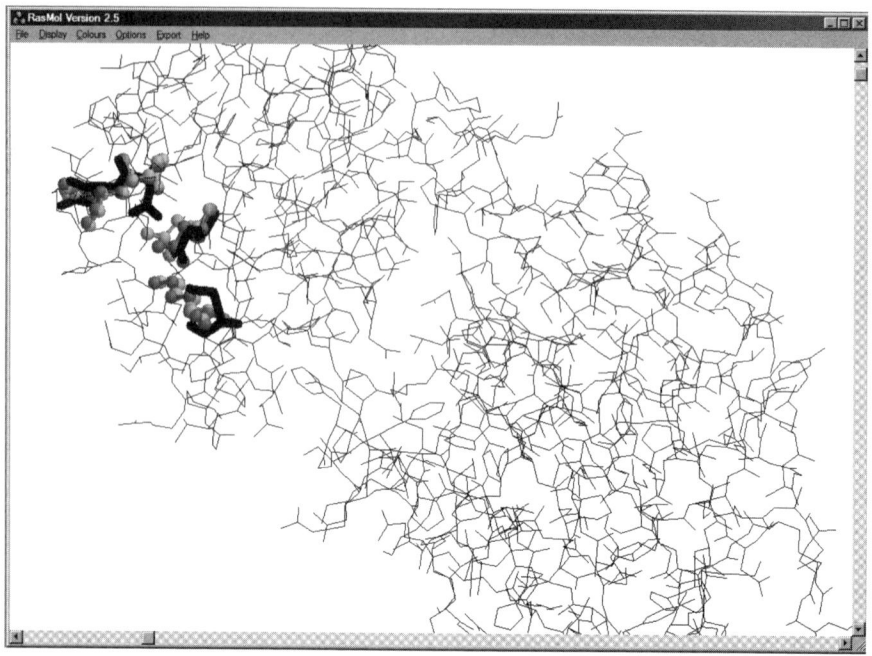

Fig. 25.5 Visualization of the PSBSiteScan results by the Rasmol program. The query protein structure (CHEB methylesterase, PDB ID 1A2O, chain A) is shown as wireframe in black color. Two
site templates 1RNL1 and 1A041 were identified by PDBSiteScan program (see Fig. 25.4a–b).
Both template atoms superimposed over D10, D11, D56 and K107 residues of the query protein
(shown as sticks in black color). Template atoms are shown as balls and sticks in grayscale
colors.

parameters of the sequence output by placing checks in the "Display options"
check boxes [5], the parameters of the graphical result output by placing checks
in the "Output locations" check boxes [6]. To display pI and molecular mass
changes in graphical form similar to 2D gel image, place a check in the "Plot gel
image" check box [7], also, set the limitations on the displayed pI values and
molecular mass in the text boxes [8]. Additional options are set by placing
checks in the "Output options" check boxes [9].

4. Click on the 'Send' button' [10]. Click on the "Clear" button [11] to reset the
parameters.

5. Once implemented, the program will display the results page (Fig. 25.6C) that
presents the estimates for the pI and molecular mass of unmodified protein (the first
line in the table) and its isoforms (modifier number in the last column of the table).

6. In the case the output graphical version was set, the virtual 2D gel image (Fig.
25.6D) appears on the results page. The position of the peak of unmodified protein is displayed as white spots, while those of the peaks of the protein isoforms
come up as colored spots. The color depends on the type of PTM (please see the
legend to the left of the figure).

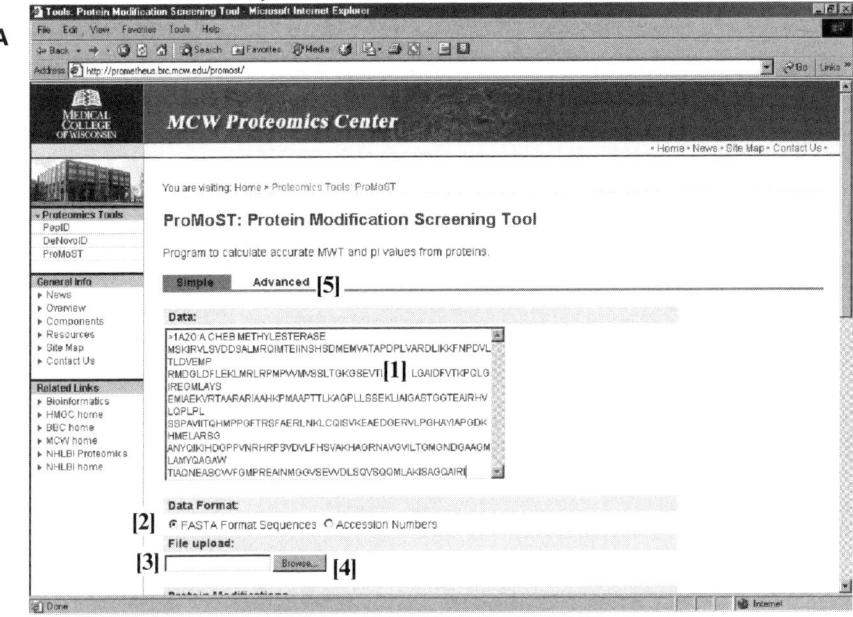

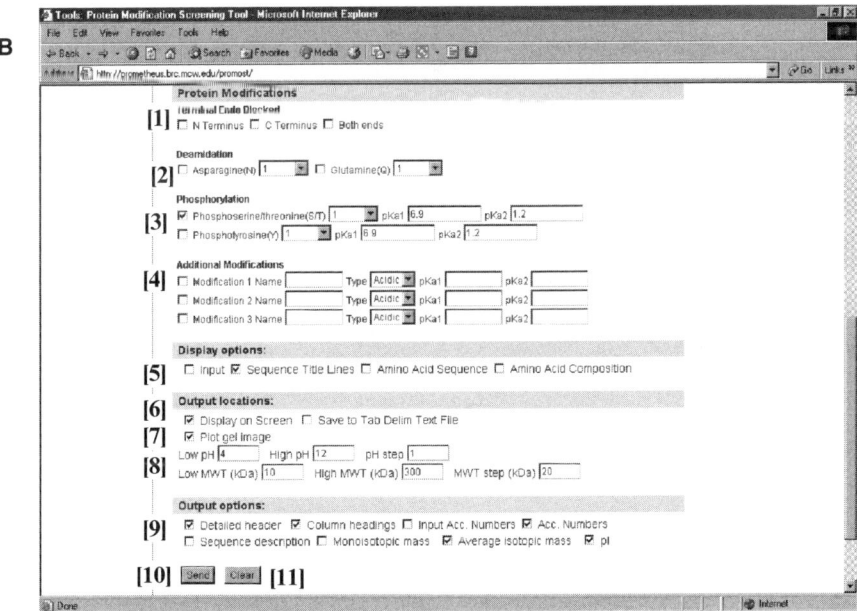

Fig. 25.6 Estimation of changes in the protein pI and molecular mass because of PTMs using the ProMoST program: (A,B) program input form; (C,D) program output page

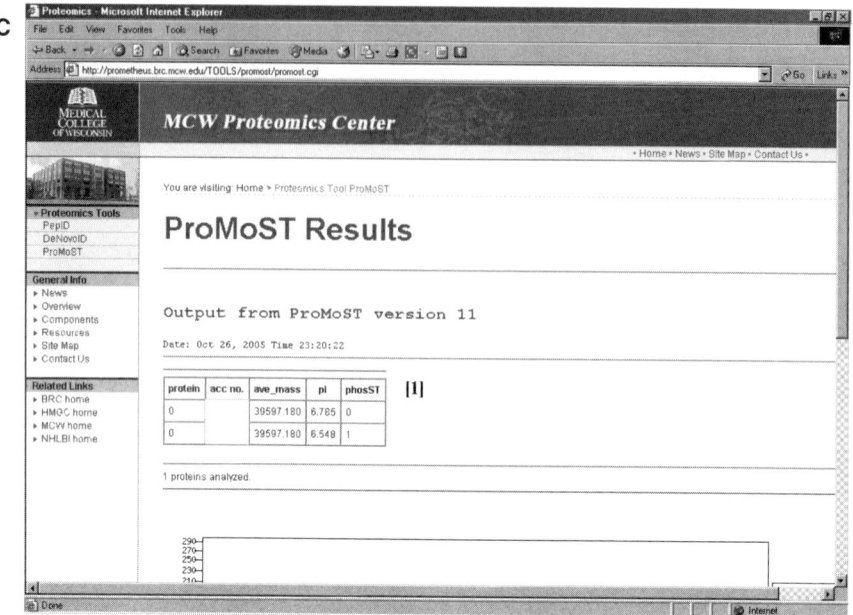

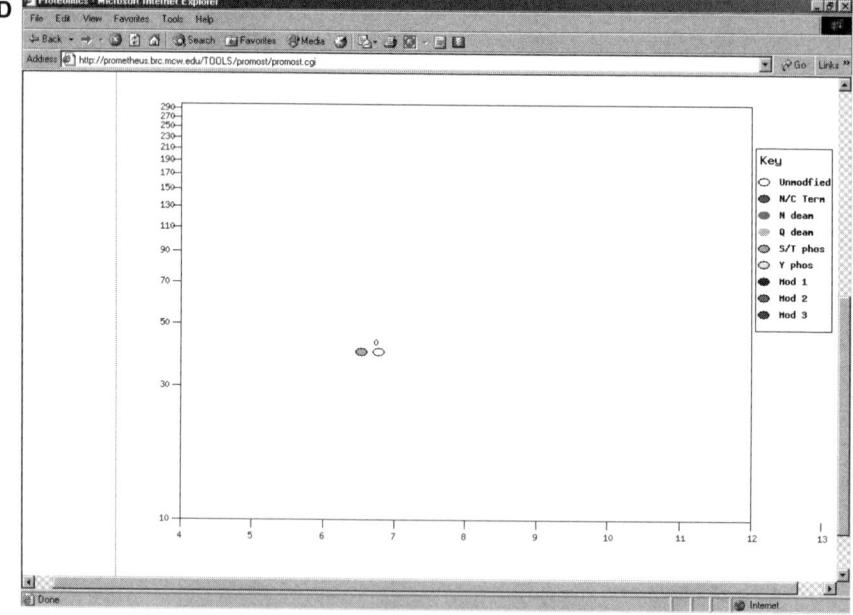

Fig. 25.6 (continued)

Acknowledgements The author is grateful to Fadeeva A.N. for English translation. The work is supported by the U.S. Civilian Research & Development Foundation for the Independent States of the Former Soviet Union (CRDF) within the Basic Research and Higher Education Program (REC-008, Y1-B-08-20) and Rup2-2629-NO-04, the program of the Ministry of Education of the Russian Federation "The Development of the Higher School Scientific Potential" (N8274, 4224), Russian Foundation of the Basic Research (05-04-49141-a, 05-04-49283-a) and by Governmental contract with Federal Agency for Science and Technology "Identification of potential targets for novel medicinal drugs based on reconstructed gene networks", priority direction "Living system", by Governmental contract project 02.434.11.3004.

References

1. Kaiser, W. M. and Huber, S. C. (2001) Post-translational regulation of nitrate reductase: mechanism, physiological relevance and environmental triggers. *J Exp Bot.* **52**, 1981–1989.
2. Rocks, O., Peyker, A., Kahms, M., Verveer, P. J., Koerner, C., Lumbierres, M., Kuhlmann, J., Waldmann, H., Wittinghofer, A., and Bastiaens, P. I. (2005) An acylation cycle regulates localization and activity of palmitoylated Ras isoforms. *Science* **307**, 1746–1752.
3. Goldberg, A. L. (2003) Protein degradation and protection against misfolded or damaged proteins. *Nature* **426**, 895–899.
4. Tootle, T. L. and Rebay, I. (2005) Post-translational modifications influence transcription factor activity: a view from the ETS superfamily. *Bioessays* **27**, 285–298.
5. McLachlin, D. T. and Chait, B. T. (2001) Analysis of phosphorylated proteins and peptides by mass spectrometry. *Curr Opin Chem Biol.* **5**, 591–602.
6. Kemp, B. E. and Pearson, R. B. (1990) Protein kinase recognition sequence motifs. *Trends Biochem Sci.* **15**, 342–346.
7. Persoon, B , Flinta, C., von Heijne, G., and Jornvall, H. (1985) Structures of N-terminally acetylated proteins. *Eur J Biochem.* **152**, 523–527.
8. Han, K. K. and Martinage, A. (1992) Post-translational chemical modification(s) of proteins. *Int J Biochem.* **24**, 19–28.
9. Hulo, N., Sigrist, C. J., Le Saux, V., Langendijk-Genevaux, P. S., Bordoli, L., Gattiker, A., De Castro, E., Bucher, P., and Bairoch, A. (2004) Recent improvements to the PROSITE database. *Nucleic Acids Res.* **32**, D134–D137.
10. Blom, N., Sicheritz-Ponten, T., Gupta, R., Gammeltoft, S., and Brunak, S. (2004) Prediction of post-translational glycosylation and phosphorylation of proteins from the amino acid sequence. *Proteomics* 4, 1633–1649.
11. Ivanisenko, V. A., Pintus, S. S., Grigorovich, D. A., and Kolchanov, N. A. (2005) PDBSite: a database of the 3D structure of protein functional sites. *Nucleic Acids Res.* **33**, D183–D187.
12. Diella, F., Cameron, S., Gemund, C., Linding, R., Via, A., Kuster, B., Sicheritz-Ponten, T., Blom, N., and Gibson, T. J. (2004) Phospho.ELM: a database of experimentally verified phosphorylation sites in eukaryotic proteins. *BMC Bioinformatics* **5**, 79.
13. Vinayagam, A., Pugalenthi, G., Rajesh, R., and Sowdhamini, R. (2004) DSDBASE: a consortium of native and modelled disulphide bonds in proteins. *Nucleic Acids Res.* **32**, D200–D202.
14. Gupta, R., Jung, E., Gooley, A. A., Williams, K. L., Brunak, S., and Hansen, J. (1999) Scanning the available Dictyostelium discoideum proteome for O-linked GlcNAc glycosylation sites using neural networks, *Glycobiology* **9**, 1009–1022.
15. Garavelli, J. S. (2004) The RESID Database of Protein Modifications as a resource and annotation tool. *Proteomics* 4, 1527–1533.
16. Gattiker, A., Gasteiger, E., and Bairoch, A. (2002) ScanProsite: a reference implementation of a PROSITE scanning tool. *Appl Bioinformatics* **1**, 107–108.
17. Blom, N., Gammeltoft, S., and Brunak, S. (1999) Sequence and structure-based prediction of eukaryotic protein phosphorylation sites. *J Mol Biol.* **294**, 1351–1362.

18. Obenauer, J. C., Cantley, L. C., and Yaffe, M. B. (2003) Scansite 2.0: Proteome-wide prediction of cell signaling interactions using short sequence motifs. *Nucleic Acids Res.* **31**, 3635–3641.

19. Kim, J. H., Lee, J., Oh, B., Kimm, K., and Koh, I. (2004) Prediction of phosphorylation sites using SVMs. *Bioinformatics* **20**, 3179–3184.

20. Liu, Y. and Lin, Y. (2004) A novel method for N-terminal acetylation prediction. *Genomics Proteomics Bioinformatics* **2**, 253–255.

21. Julenius, K., Molgaard, A., Gupta, R., and Brunak, S. (2005) Prediction, conservation analysis, and structural characterization of mammalian mucin-type O-glycosylation sites. *Glycobiology* **15**, 153–164.

22. Eisenhaber, B., Bork, P., and Eisenhaber, F. (1999) Prediction of potential GPI-modification sites in proprotein sequences. *J Mol Biol.* **292**, 741–758.

23. Monigatti, F., Gasteiger, E., Bairoch, A., and Jung, E. (2002) The Sulfinator: predicting tyrosine sulfation sites in protein sequences. *Bioinformatics* **18**, 769–770.

24. O'Connor, B. D. and Yeates, T. O. (2004) GDAP: a web tool for genome-wide protein disulfide bond prediction. *Nucleic Acids Res.* **32**, W360–W364.

25. Baldi, P., Cheng, J., and Vullo A. (2005) Large-Scale Prediction of Disulphide Bond Connectivity, in *Advances in Neural Information Processing Systems (NIPS 2004)* vol. 17 (Saul, L., Weiss, Y., and Bottou, L., eds.), MIT press, Cambridge, MA, pp. 97–104.

26. Ivanisenko, V. A., Pintus, S. S., Grigorovich, D. A., and Kolchanov, N. A. (2004) PDBSiteScan: a program for searching for active, binding and post-translational modification sites in the 3D structures of proteins. *Nucleic Acids Res.* **32**, W549–W554.

27. Halligan, B. D., Ruotti, V., Jin, W., Laffoon, S., Twigger, S. N., and Dratz, E. A. (2004) ProMoST (Protein Modification Screening Tool): a web-based tool for mapping protein modifications on two-dimensional gels. *Nucleic Acids Res.* **32**, W638–W644.

28. Sayle, R. A. and Milner-White, E. J. (1995) RasMol: Biomolecular graphics for all. *Trends in Biochemical Sciences* **20**, 374–376.

29. Zdobnov, E. M., Lopez, R., Apweiler, R., and Etzold, T. (2002) The EBI SRS server – recent developments. *Bioinformatics* **18**, 368–373.

30. Berman, H. M., Westbrook, J., Feng, Z., Gilliland, G., Bhat, T. N., Weissig, H., Shindyalov, I. N., Bourne, P. E. (2000) The Protein Data Bank. *Nucleic Acids Research* **28**, 235–242.

Index

A

Acetylation, 151ff, 200, 262, 231, 363, 365
α1-Acid glycoprotein, 200, 204, 243ff, 251,
    255ff, 311
Acylation, 163ff
Adenosine 3'-phosphate 5'-phosphosulfate
    (PAPS), 48
Affi-Gel Blue, 237
Alkylation, 35, 38, 295, 297, 309
α-Amidation, 67ff, 363
Amino acid
    acidic hydrolysis, 50, 100ff
    analsis, 50, 88ff, 95ff
    hydrophobic, 18
    phospho, 34
    sequence, 35, 36, 37, 363
2-Aminobenzamide (2-AB), 231ff, 246,
    260, 319
ε-Amino group, 96, 131, 312
2-Aminopyridine (2-AP), 241
AMINOSep, 101
Anion-exchange chromatography, 54, 89, 186,
    241, 248ff, 285ff
Anthranilic acid (AA, 2-aminobenzoic acid),
    215ff, 317ff, 333ff, 347ff
α1-Antitrypsin, 200, 204ff
Arabinose, 299
Arylsulfatase, 49, 55, 58, 61
Aspartate, *see* isoAspartate
2,2'-Azino-di-3-ethylbenzthiazolinesulfonate
    (ABTS), 269

B

Bio-Gel
    P4, 185, 187ff
    P6, 233
    P10, 284, 288

Blott, *see* Electrophoretic transfer
Bolton hunter reagent, 53, 57
Bradykinin, 3

C

Carbodiimide, 51, 55, 63, 73
Carbohydrate
    composition analysis, 215ff, 246, 252,
    320, 348
Carbopak column, 193, 244, 247
γ-Carboxyglutamic acid (Gla), 85
γ-Carboxylation, 85
Ceruloplasmin, 200, 204, 208
Chinese hamster ovary cells (CHO), 222, 319,
    320, 325, 328, 333ff
Chloramine T, 63
Chymotrypsin, 285, 290
Coagulation factor, 85
Collagen, 95ff, 282
    crosslinking, 95
    type I, 95
    type III, 106
Complex type oligosaccharide,
    251, 256
Coomassie Blue (CB), 21ff, 38, 100, 136, 178,
    275, 277
Creatine phosphokinase, 3, 5
CTD110.6 antibody, 277, 278
1-Cyano-4-dimethylamino-pyridinium
    tetrafluoroborate (CDAP), 2ff
α-Cyano-4-hydroxy-*trans*-cinnamic acid
    (ACCA), 4, 52, 54, 64, 73, 81, 83, 110,
    159, 285, 287, 289
Cyanylation, 1ff
Cycloheximide, 166, 172
Cysteine, 1ff, 62, 131, 163ff
    alkylation, 35, 38

D

Dansyl hydrazone, 284, 286ff
Datura stramonium lectin (DSA), 200, 203
Deamidation, 379
Deamination, 96, 185ff
Deglycosylation, 200, 205ff
  chemical, 231
  enzymatical, 231ff
Dephosphorylation, 185ff, 191
Desialylation, 200ff, 257ff, 319
2,5-Dihydroxybenzoic acid (DHB), 55, 83,
  252, 258, 289, 319, 351
Dihydroxylysinonorleucine (DHLNL),
  98, 99
Disulfide bond, 1ff, 106, 178, 365, 366
Dopamine beta monooxygenase
  (DBM), 69
Dot-blot, *see* Immunoblotting
Dowex 187, 188, 192, 233, 235, 257,
  284, 287

E

Electrophoretic transfer, 145, 273
β-Elimination, 2, 3, 5, 8, 9, 13, 15, 17, 34,
  288, 289
Endocrine
  peptides, 67
Endoproteinase, 49
  Asp-N, 98, 285, 290
  Lys-C, 98, 285, 290
Enzymatic sequencing, 255ff *see also*
  Sequencing
Enzyme-linked immunosorbent assay
  (ELISA), 267ff, 277
Epidermal growth factor (EGF), 85, 86
Epimerization, 240
Escherichia coli, 42, 135, 151ff, 177, 233,
  237, 257
Ethanethiol (ET), 87
1-Ethyl-3-(3-dimethylaminopropyl)carbodiimi
  de (EDC), 70, 73, 74
Exoglycosidase, 186, 188, 192, 239, 240, 248,
  255ff, 288, 294, 299, 308, 328, 329
  substrate specificity, 261, 301

F

Factor VIII, recombinant (rhFVIII), 296ff, 311
Fast protein liquid chromatography (FPLC),
  54, 58, 59, 137, 285, 289 *see also*
  Liquid chromatography
Fatty acid, 163ff, 169ff, 176ff, 168,
  189, 193ff

Fetuin, 200, 223, 237, 308, 309, 312, 320,
  322ff, 334, 336, 350
Fibrinogen, 47, 200, 204, 205
Fibronectin, 200, 204, 205
Fluorescence, 101, 104, 107, 216, 218, 256
  detection, 223ff, 228, 317, 320ff, 333, 337,
  347, 350
  immunofluorescence, 276ff
  labeling, 236ff, 239ff, 246
Fluorophore-assisted carbohydrate
  electrophoresis (FACE), 265, 294,
  305, 307
Fucose, 218, 224, 232, 237, 247, 256, 261ff,
  282, 284, 286, 293, 294, 299, 318, 320,
  327, 335
α-Fucosidase, 257, 259ff, 288
  substrate specificity, 261, 301

G

β-Galactosidase, 257, 259ff, 308, 309, 336,
  338ff, 344
  substrate specificity, 261, 301
Gastrin, 51, 52, 58, 59, 63
Gel electrophoresis, *see* Polyacrylamide gel
  electrophoresis
Glutaraldehyde, 51, 56, 63, 271, 276, 277
Glycopeptide, 232, 234, 237, 258, 290, 292,
  293ff, 302ff, 311ff, 333, 336, 338, 339,
  345, 349
  identification, 298ff
Glycoprotein, 48, 177, 183, 200, 204ff, 231ff,
  239, 240, 243ff, 255, 256, 258ff, 282ff,
  286ff, 305, 311, 313, 333ff, 347ff, 357,
  358
  carbohydrate composition analysis, 215ff
  deglycosylation, 231ff
  mass determination, 284
  mass spectrometry, 351ff
  reduction and alkylation, 295, 297
  therapeutic, 317ff
GlycoFree, 201
GlycoSep C, 232, 242ff, 252, 253
Glycosylation, *see* N- and O-glycosylation
Glycosylphosphatidylinositol (GPI), 183ff

H

Heat-inactivated horse serum (HIHS), 76
Hexosyl-N-acetylhexosamine, 295
HGAC 85 antibody, 268, 277, 278
High mannose oligosaccharide, 232, 250ff,
  256, 263, 264, 318ff, 326, 327, 333ff,
  341, 343, 349, 356ff

High performance anion exchange
  chromatography (HPAEC),
  185, 188, 192, 193, 215, 216, 232,
  240ff
  with pulsed amperometric detection
    (PAD), 232, 240ff, 244ff, 252, 318,
    319, 327, 328, 348
High performance liquid chromatography
  (HPLC), 4, 5, 8ff, 70, 87ff, 101ff,
  215ff, 232, 258, 290ff, 293, 295ff, 312,
  317ff, 333ff, 347ff
  microbore, 296ff
  nanoflow, 36, 40
  reversed phase (RP), 49, 52, 54, 55, 57,
    59ff, 72ff, 234, 284ff
  separation of N-glycans, 239ff
α2-HS-glycoprotein, 200, 205
Hybrid type oligosaccharide, 251, 252, 256,
  263, 264, 300
Hydrazine, 231, 284, 287, 291, 324, 325, 328,
  329, 348
Hydrolysis, 86ff, 196, 340
  acidic, 50, 100ff, 171, 216, 220ff,
    227, 286
  alkaline, 47, 48
  enzymatical, 231ff
Hydrophobic interaction, 232
β-Hydroxyaspartate (Hya), 85ff
Hydroxylamine, 166, 167, 170, 177ff
Hydroxylation, 85ff, 95ff
Hydroxylysine (Hyl), 96, 104
Hydroxylysinonorleucine (HLNL), 98,
  99, 105ff
Hydroxyproline (Hyp), 96, 104, 282

I

Iminothiazolidine, 3, 5, 6
Immunoblot, 132, 135, 212, 277
Immunochemical screening, 267ff
Immunoglobulin G
  recombinant (rIgG), 222, 226, 227, 319,
    320, 322, 325ff, 333ff, 350ff
Immunohistochemistry, 270, 274
Immunoprecipitation, 132, 140, 141, 143, 145,
  147, 171ff, 180
In-gel digestion, 21, 31, 32, 34, 35, 38,
  113, 213
Insulin, 3, 4, 268
Interferon, 139, 140
Iodination, 53, 57, 63, 67, 75, 82
ISGylation, 139ff
Isoelectric foccusing, 23, 24, 26, 31, 202, 208,
  209, 213, 214, 270, 271, 273

K

K562-cells, 249, 251

L

Lectin, 199, 200, 203ff, 211, 214, 295, 334
Lewis X, 255ff, 311
Ligase
  E3, 114, 131, 132, 139
Liquid chromatography-mass spectrometry
  (LC-MS), 34, 36, 37, 39ff, 153, 268,
  318, 333, 336ff, 343, 345, 359, see also
  Mass spectrometry
  Nuclear magnetic resonance (LC-NMR),
    318
Lysine, 41, 95ff, 109, 114ff, 122, 124, 125,
  128, 131, 132, 137, 357, 358
  hydroxylation, 95, 103, 104, 282
Lysobacter enzymogenes protease (Lys-C), 49,
  114, 232, 285, 290, 312

M

Maleimidobenzoyl-succinimide ester,
  52, 56
α-Mannosidase, 196, 301
  Aspergillus saitoi, 188
  Jack bean, 188
  substrate specificity, 301
β-Mannosidase
  Helix pomatia, 301
  substrate specificity, 301
Mass spectrometry (MS), 1ff, 33ff, 109ff,
  255ff, 293ff, 333ff, 347ff
  collision-induced dissociation (CID), 34,
    158, 291
  delayed extraction, 51
  diagnostic ion, 109, 116ff, 293, 295, 299ff,
    306, 311ff
  ion trap, 36, 39ff, 44, 45, 158, 312
  liquid chromatography/electrospray
    ionization (LC/ESI), 33, 36ff, 118,
    157, 289, 343, 349
  matrix assisted laser desorption time-of-
    flight (MALDI-TOF), 2, 4ff, 17, 18,
    42, 47, 50, 54, 55, 60, 62, 64, 73, 80ff,
    110, 111, 113, 124, 126, 127, 151ff,
    237, 252, 255, 256, 258ff, 265, 284,
    285, 287, 289, 319, 336, 347, 349, 351,
    353ff, 358ff
  post-source decay (PSD), 355, 360
  selective Ion Monitoring (SIM), 102
  total Ion Current (TIC), 298
  tandem, 291, 295

Matrix assisted laser desorption (MALDI), *see* Mass spectrometry
Membrane protein, 61, 86, 158, 163ff, 177, 178, 183, 184, 267
Metabolic labeling, 163, 164, 168, 171, 183, 185ff, 194
Methylation, 118, 139, 151, 152, 155ff, 356
β-Methylthiolation, 155
Monoclonal antibody, 49, 64, 73, 137, 180, 215, 267ff, 272, 277, 317, 318, 333ff, 347ff, 357, 358
Monosaccharide, 96, 215, 235, 237, 239, 246, 255, 267, 281, 329, 358
  composition analysis, 215ff, 246, 252, 282, 318, 248
  linkage, 256
  sequence, 255ff, 290
Mucin, 226, 281, 325, 329
MUD 50 antibody, 268, 277
Myoglobin, 3, 4, 153, 156
Myristoylation, 164

N
β-N-acetylhexosaminidase
  Jack bean, 257, 259, 260
  substrate specificity, 261
N-acetylgalactosamine, 282, 288, 291, 293, 348
N-acetylglucosamine, 232, 261, 267ff, 291, 293, 335
  immunological detection, 267ff
N-acetylhexosamine, 231, 295, 298
  mass difference, 299
β-N-acetylhexosaminidase, 257, 259
  substrate specificity, 261
N-acetyl neuraminic acid (NeuAc), 217, 224, 256, 261, 299, 334, 348
Neuraminic acid, 231, 303, 304
Neuraminidase, 252, 308, 309
  substrate specificity, 301
N-glycan *see* N-linked glycan
N-glycolyl neuraminic acid (NeuGc), 217, 224, 256, 299, 334, 348
N-glycosylation, *see* N-linked glycan
N-hydroxysuccinimide ester, 52, 62
2-Nitro-5-thiocyanobenzoic acid (NTCB), 3
N-linked glycan, 239ff, 256, 293ff, 289, 318, 347ff
  charge profiling, 251
  composition analysis, 215ff
  enzymatic sequence analysis, 255ff
  fluorescent labeling, 231ff
  mass, 300

  monitoring, 333ff
  separation, 239ff
  site-specific, 293ff
N-trismethyl-2-aminoethanesulfonic acid (TES), 101

O
O-GlcNAc (*see* O-linked N-acetyglucosamine)
O-glycosylation, 267, 281ff, 293, 318, 325, 329, 348 *see also* O-linked
O-linked
  core sequences, 283
  fucose, 282, 293
  glucose, 282, 293
  N-acetylglucosamine (O-GlcNAc), 267ff, 282, 293
  oligosaccharides, 281
O-phenylenediamine (OPD), 215, 216, 342
O-phthalaldehyde (OPA), 87
O-Sulfation, 47ff
Ovalbumin, 134, 137
Oxidation
  glycopeptides, 295
  methionine, 62, 309
  oligosaccharide, 241
  sulfhydryl group, 1

P
Palmitoylation, 163ff
  detection, 167
Peptide
  extraction, 29ff, 38, 44
  mapping, 292
Peptide-N-glycanase A (PNGase A), 232
Peptide-N-glycanase F (PNGase F), 231ff, 293, 307, 318, 324ff, 336, 338, 348ff
Peptidyl-α-hydroxyglycine α-amidating lyase (PAL), 69
Peptidylglycine α-amidating monooxygenase (PAM), 69
Peptidylglycine α-hydroxylating monooxygenase (PHM), 69
Phosphatase, 33, 267
Phospholipase A_2 (PLA_2), 185, 186
Phospholipase C, 185, 187
Phospholipase D, 185, 187
Phosphopeptide, 33ff, 268
Phosphoprotein, 21ff, 33ff *see also* Phosphorylation
Phosphoproteome, 21ff, 33, 35
Phosphorylation, 21ff, 33ff, 50, 62, 111, 139, 363, 365

Plasma protein, 85, 87, 200, 205
Plasminogen, 200, 204, 205
Polyacrylamide gel electrophoresis (PAGE),
    38, 74, 97, 102, 112, 135, 141, 271, 326
    2D, 21ff, 199ff
Polyethylene glycol, 72
    precipitation, 76
Polymerase chain reaction (PCR), 164, 167,
    173ff
Polyvinylidene fluoride (PVDF) membrane,
    100, 102, 219, 223, 319, 326
Proline (Pro), 293
    hydroxylation, 96, 104, 282
ProQ Diamond, 21ff
PROSITE database, 367ff
Protein
    C, 85, 87, 91
    database, 263ff
    3D structure, 363ff
Protozoa, 183
P-Selectin glycoprotein ligand-1
    (PSGL-1), 48
Pseudomonas fragi protease (Asp-N), 49, 232,
    285, 290, 312
Pull down
    Ni²⁺, 133ff
Pulse-chase experiment, 172
Pulsed amperometric detection (PAD), 215,
    240ff, 318, 327, 348

R
Radioimmunoassay (RIA), 49, 51, 55ff, 67,
    70ff
Reduction, 27, 103, 185, 187, 190, 247, 295,
    297
    disulfide, 5, 6ff
    in-gel, 35, 38
Ribonuclease A, 3
Ribosomal Proteins, 151

S
S-Acylation, 163ff
Sambucus nigra lectin (SNA), 200, 203ff
Sephadex
    A25, 188, 192
    G10, 52, 56
    G25, 52, 56, 233, 252, 285, 289
    G75, 137
SepPak C₁₈, 75
Sequencing, 34 *see also* N-linked glycan
    sequencing
    glycans, 188, 192, 255ff, 301

peptides, 40, 45, 62, 114, 125, 232, 290,
    295
Sialic acid, 196, 203ff, 215ff, 232, 237, 250,
    258, 264, 295, 299, 304, 306, 308,
    312, 318, 320, 325, 334, 341ff, 348
    see also Neuraminic acid and N-acetyl
    neuraminic acid
Sialidase, 288, 328
    Arthrobacter ureafaciens, 201, 204ff, 257ff
    newcastle disease virus, 265
    substrate specificity, 261
Site-specific
    glycosylation, 293ff
    mutagenesis, 164, 174
Size-exclusion chromatography, 186, 191, 196
Sodium cyanoborohydride, 217, 226, 236,
    321, 337, 350
Staphylococcus aureus
    V8 protease (Glu-C), 49, 232
    sphingomyelinase,188, 193
Sulfation, *see* Tyrosine O-sulfation
Sulfhydryl group, 1ff, 297
Sulfotransferase, 47
SUMO (small ubiquitin related modifier),
    131ff
Sumoylation, 131ff
Superdex HR10/30, 284, 288

T
Transfection, 139ff
Transferrin, 200, 204, 205
Trypanosoma brucei, 183
Trypsin, 25, 30, 36, 38, 49, 111, 114, 115ff,
    141, 143, 232, 285, 290, 296, 309, 312
    digest, 40, 233ff, 297, 337, 339, 345
    miscleavage, 41, 42
2D-Electrophoresis (2DE), 21ff, 152, 199ff,
    270, 364
2D-Polyacrylamide gel electrophoresis (2D-
    PAGE), *see* 2D-Electrophoresis (2DE)
Tyrosine, 22, 34, 75, 82, 363
    O-Sulfation, 47ff
    phosphorylation, 44, 125, 268, 363
Tyrosylprotein sulfotransferase, 47

U
Ubiquitin, 109ff, 131, 139, 140, 153, 159
    ubiquitination, 109ff

V
Vitamin K, 85

W
Web-based database, 363ff
Western-blot, *see* Immunoblotting

X
Xylose, 256, 284, 286, 299, 335